Chemistry of
Nucleosides
and Nucleotides

Volume 1

Chemistry of Nucleosides and Nucleotides

Volume 1

Edited by

Leroy B. Townsend
University of Michigan
Ann Arbor, Michigan

Plenum Press • *New York and London*

Library of Congress Cataloging in Publication Data

Chemistry of nucleosides and nucleotides / edited by Leroy B. Townsend.
    p.   cm.
  Includes bibliographical references and indexes.
  ISBN 0-306-42871-7
  1. Nucleotides. 2. Nucleosides. I. Townsend, Leroy B.
QD436.N85C47 1988   88-22359
547.7′9—dc19   CIP

© 1988 Plenum Press, New York
A Division of Plenum Publishing Corporation
233 Spring Street, New York, N.Y. 10013

Printed in the United States of America

Contributors

Morio Ikehara, Faculty of Pharmaceutical Sciences, Osaka University, Osaka, Japan 565

Rich B. Meyer, Jr., MicroProbe Corporation, Bothell, Washington 98021

Eiko Ohtsuka, Faculty of Pharmaceutical Sciences, Osaka University, Osaka, Japan 565

Roland K. Robins, Nucleic Acid Research Institute, Costa Mesa, California 92626

Prem C. Srivastava, Nuclear Medicine Group, Health and Safety Research Division, Oak Ridge National Laboratory, Oak Ridge, Tennessee 37831

Toshiki Tanaka, Faculty of Pharmaceutical Sciences, Osaka University, Osaka, Japan 565

Tohru Ueda, Faculty of Pharmaceutical Sciences, Hokkaido University, Sapporo 060, Japan

Seiichi Uesugi, Faculty of Pharmaceutical Sciences, Osaka University, Osaka, Japan 565

Preface

The present volume is the first of a projected four-volume treatise. This volume contains the following chapters: "Synthesis and Reaction of Pyrimidine Nucleosides," "Synthesis and Properties of Purine Nucleosides and Nucleotides," and "Synthesis and Properties of Oligonucleotides."

These three chapters were selected for inclusion in Volume 1 because the areas have provided the basis and impetus for the initiation and development of the other areas of research, which will be described in subsequent volumes. Each chapter is rather comprehensive in nature and should provide a ready reference source for not only the novice but also the experienced investigator or researcher. The chapters have been prepared by authors with considerable experience in each particular area of research, and this has resulted in a lucid presentation of each well-defined area.

These volumes were designed with medicinal chemists, medicinal organic chemists, organic chemists, carbohydrate chemists, physical chemists, and biological chemists in mind. However, because of the tremendous recent interest in this research area owing to the biological and chemotherapeutic evaluation of nucleosides and nucleotides as anticancer, antiviral, and antiparasitic agents, these volumes should also be valuable additions to the libraries of virologists, biochemical pharmacologists, oncologists, and pharmacologists.

We would like to thank the authors for their enthusiasm and help in making these volumes available to the scientific community.

Leroy B. Townsend

Ann Arbor

Contents

Chapter 2

Synthesis and Properties of Purine Nucleosides and Nucleotides

Prem C. Srivastava, Roland K. Robins, and Rich B. Meyer, Jr.

Chapter 3

Synthesis and Properties of Oligonucleotides

Morio Ikehara, Eiko Ohtsuka, Seiichi Uesugi, and Toshiki Tanaka

Chapter 1

Synthesis and Reaction of Pyrimidine Nucleosides

Tohru Ueda

1. Introduction

Nucleic acids play a fundamental role in the transformation of genetic information. DNAs carry the genetic information and these are replicated during cell division. The genetic information, the sequence of four nucleobases, adenine (A), guanine (G), cytosine (C), and thymine (T) of the DNA strands, are transcribed to RNAs (messenger RNA, mRNA). These are then translated to furnish specific proteins coded by the mRNA, on the system called polysome, which consists of various ribosomal RNAs, proteins and transfer RNAs (tRNA) carrying specific amino acids to be connected in a specific sequence.

Throughout these transformations, the complementary hydrogen bond formation between A-T (U) and G-C is the essential chemical event. Therefore, for the study of genetic control, the chemistry of nucleosides becomes very important. In addition, nucleosides in the form of phosphate esters, such as ATP and other nucleoside polyphosphates, play an essential role in the energy transfer and regulation of metabolic systems.

A large number of new pyrimidine and purine nucleosides have been synthesized and tested for their biological activities. It is reasonable to expect that the nucleoside analogues will exhibit various activities through the interactions at many enzymatic pathways of synthesis and metabolism of nucleosides, nucleotides, DNA, and RNA. Thus, the nucleosides are major candidates for chemotherapeutic agents, especially for virus and cancer.

This chapter deals with the synthesis of pyrimidine nucleosides (Section 2) and reactions of preformed nucleosides (Sections 3–5). Excellent books and reviews related to the present subjects are available.[1-10]

Tohru Ueda • Faculty of Pharmaceutical Sciences, Hokkaido University, Sapporo 060, Japan.

2. Synthesis of Pyrimidine Nucleosides

A nucleoside was originally designated as the N-ribosyl (or 2-deoxyribosyl) derivative of pyrimidines and purines and exists in nature as the constituent of nucleic acids and as nucleoside antibiotics. Some nucleosides are found as naturally occurring substances. The C-nucleosides are the subject of another chapter.

The usual ribonucleosides and 2′-deoxyribonucleosides are now commercially available and there is no need to elaborate on their synthesis. However, for the synthesis of pyrimidine nucleosides modified at either the sugar and base moieties—as is often found in the nucleoside antibiotics—the N-glycosylation procedure is efficient.

Recent progress on the synthetic studies of pyrimidine nucleosides has provided a general method, the Vorbrüggen procedure. Therefore, this section is devoted primarily to this method with very brief discussions of the previous procedures of pyrimidine nucleoside synthesis.

2.1. The Vorbrüggen Method

The usual pyrimidine bases, cytosine (1) and uracil [or thymine, (2), are more than ambident nucleophiles to be glycosylated. Therefore, the first practical synthetic procedure was the Hilbert–Johnson reaction.[11] This procedure used a 2,4-dialkoxypyrimidine (3) for the base, which was then condensed with a protected sugar halide (4) to afford the nucleoside (5). In structure (3), nitrogen-1 is the most nucleophilic and usually the first to be alkylated. From compound (5), the cytosine nucleoside (6) and uracil nucleoside (7) were prepared by ammonolysis and hydrolysis, respectively. While the 2,4-dialkoxypyrimidine cannot be prepared directly from the parent base, for example, from uracil, the trimethylsilylation procedure does afford the 2,4-bis-*O*-trimethylsilylated derivative.

Thus, treatment of (1) or (2) with excess of hexamethyldisilazane (HMDS) and chlorotrimethylsilane gives 2,4-bistrimethylsilylcytosine (8) and 2,4-bistrimethylsilyluracil (9), respectively. These derivatives are soluble in most organic solvents. A condensation of these silylated bases with the protected ribosyl halides gives the sugar-protected nucleosides and constitutes a modified silyl–Hilbert–Johnson reaction.[12] The major problem in these condensations is the use of generally unstable protected sugar halides. Vorbrüggen and co-workers have overcome this problem by using a stable acyl sugar and certain Lewis acids for an activation of the sugar portion in the condensation with the silyl base. For example, the treatment of bistrimethylsilylated 5-substituted uracils (10) with 1-*O*-acetyl-2,3,5-tri-*O*-benzoyl-D-ribose (11) and stannic chloride in an inert solvent such as 1,2-dichloroethane or acetonitrile at room temperature afforded the benzoylated nucleosides (12) in high yields.[13] This method is adaptable for the synthesis of 6-azauridine, 2-thiocytidine, and 2-thio-5-butyluridine. 6-Methyluridine,[14] 5-azacytidine, [15] 2- and 4-pyridone, and 2-pyrimidinone ribosides[16] were prepared in a similar manner. The disaccharide nucleosides were also prepared by this method.[17]

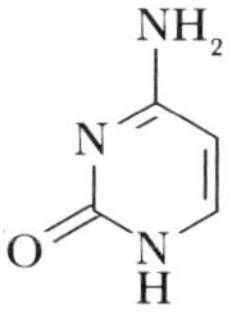

1

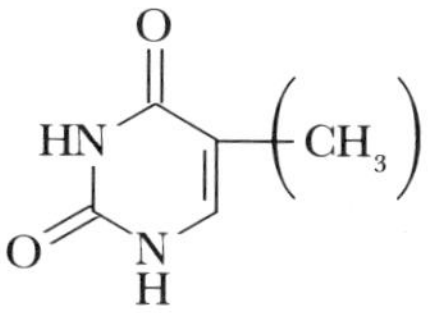

2

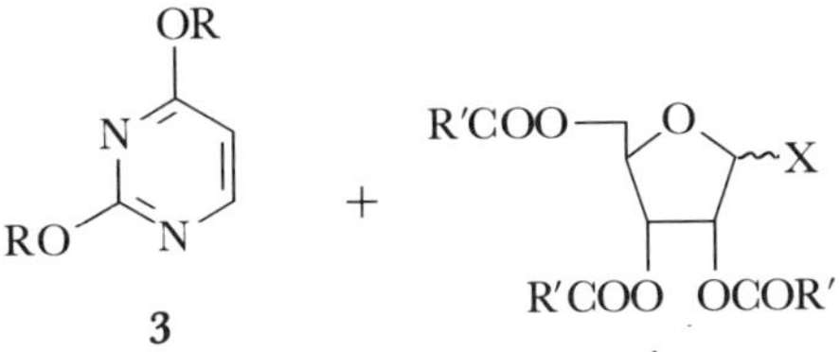

3 **4**

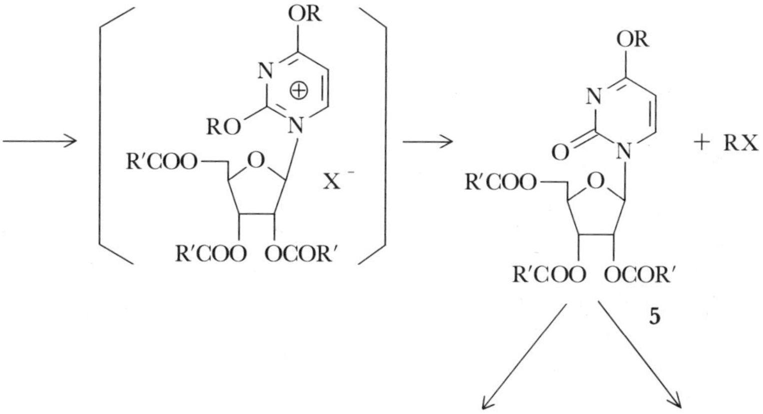

5

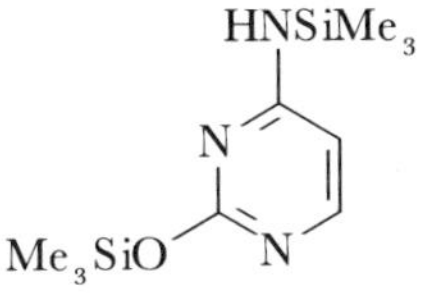

8

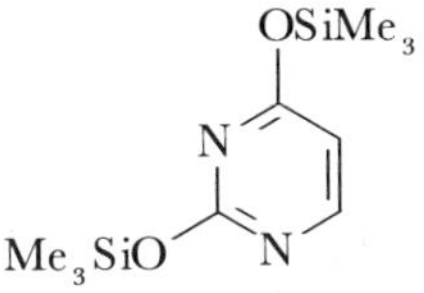

9

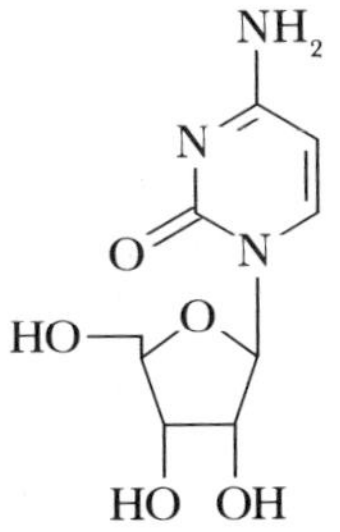

6

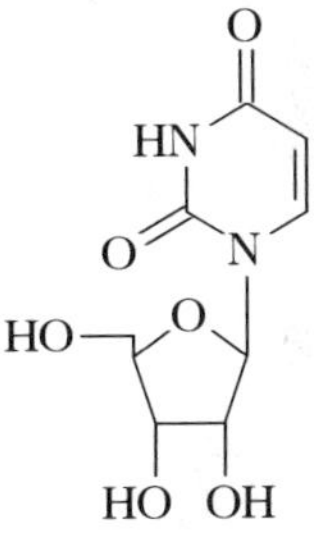

7

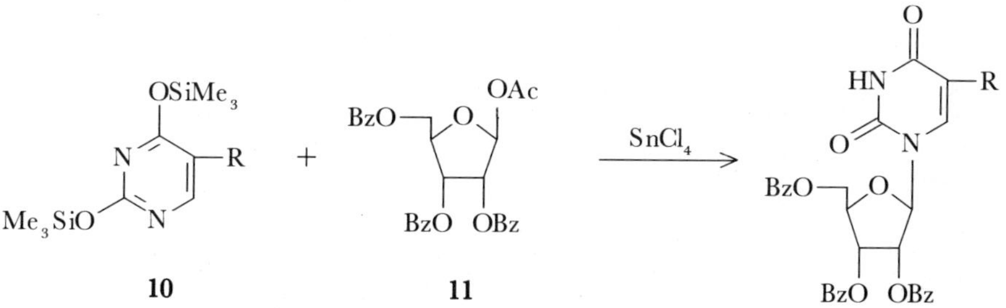

10 **11** **12**

In the case of 2′-deoxynucleosides, the use of per-O-acyl-2-deoxyribose or methyl di-O-acyl-2-deoxyriboside and a Lewis acid was not so successful. The very widely used 3,5-di-O-tolyl-2-deoxyribosyl chloride (**13**) was condensed with the silyl base in the presence of a catalyst to afford, for example, 5-ethyl-2′-deoxyuridine (**14**).[13] In this case, both α- and β-nucleosides were formed in comparable ratio. This is because there is no functional group controlling the stereoselectivity at the 2′ position such as the 2′-O-acyl group in the ribonucleoside synthesis. The influence of solvents and catalysts for the regioselectivity of the ribosylation of various pyrimidines has been studied.[18] More recently, Vorbrüggen *et al.* have found that instead of stannic chloride, trimethylsilyl trifluoromethanesulfonate (TMS-triflate) or trimethylsilyl perchlorate can be a very effective catalyst for the condensation.[19] This reagent also catalyzed the equilibration of anomers in some of the deoxynucleosides.[19b]

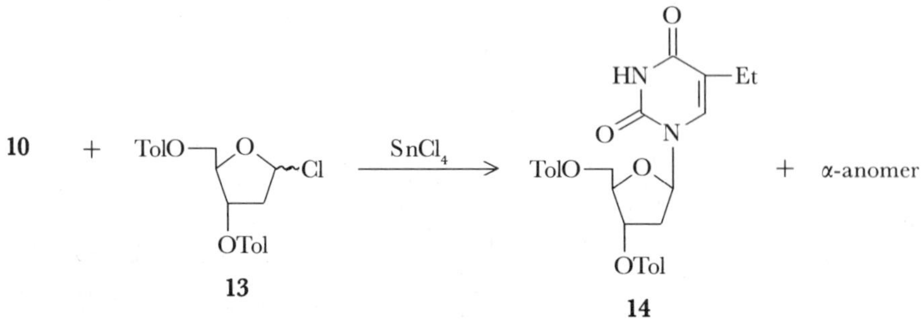

The mechanisms of the catalysis for TMS-triflate or SnCl$_4$ are generally the same. The formation of a 1,2-acyloxonium ion (**15**) is the key step, which is followed by an attack by the N-1 of the silyl base to afford the β-anomer (**16**) together with the regeneration of the active catalyst.[20] The structure of (**15**), which was formed by the reaction of tri-O-benzoylribosyl bromide and silver trifluoromethanesulfonate in nitromethane at −18°C, was confirmed by NMR.[21]

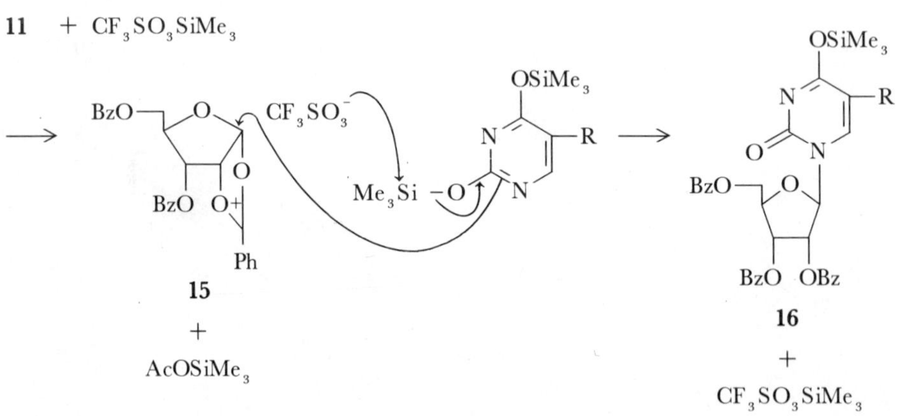

More recently, a simplified method was developed in which both per-fluoroalkanesulfonic acid and the pyrimidine base were simultaneously trimethylsilylated and condensed with 1-*O*-acetyl-2,3,5-tri-*O*-benzoyl-D-ribose in a one pot reaction.[22]

Another catalyst, iodotrimethylsilane, was reported to be useful in the preparation of uridine and cytidine. The benzoylated ribosyl iodide is suggested to be the active sugar species in this specific reaction.[23] The combination of chlorotrimethylsilane and sodium iodide was reported to provide an *in situ* formation of the iodosilane in this type of reaction.[24]

The synthesis of pyrimidine nucleosides with modified sugar and base moieties have been obtained by adaptations or slight modifications of the Vorbrüggen method. The fundamental methodology is essentially the same and is not described here except for a few cases that deal with condensations of sugars having nonparticipating substituents at the 2 positions. Treatment of the anomeric methyl 5-deoxy-2,3-*O*-isopropylidene-ribofuranosides with trimethylsilylated 5-fluorouracil and TMS-triflate as a catalyst afforded the α-anomer of 5'-deoxy-2',3'-*O*-isopropylidene-5-fluorouridine.[25] On the other hand, a condensation of 1,3,5-tri-*O*-benzoyl-2-*O*-methyl-α-D-ribose with persilylated uracil and SnCl$_4$ afforded only the β-nucleoside in high yield.[26] The use of 2-*O*-*t*-butyldimethylsilyl-1,3,5-tri-*O*-benzoylribose in this reaction gave a similar result. These anomeric preferences seem to be a consequence of the thermodynamic stability of the products in each case. However, a number of additional experiments will be necessary to clarify the mechanism fully. For example, a recent investigation on the factors determining the α : β ratio of the product in 2'-deoxyribonucleoside synthesis revealed that the reaction proceeds by an S$_N$2 mechanism with an inversion of configuration at the anomeric carbon of the chlorosugar. Therefore, the β-anomer is obtained almost exclusively if the chlorosugar is the α form (as is usually the case) and the condensation is carried out in chloroform. Addition of the catalyst tends to anomerize the chlorosugar, resulting in the formation of an anomeric mixture of nucleosides.[27]

2.2. The Mercuri Procedure

Fox and co-workers adapted the mercuri procedure, which had previously been used for purine nucleoside synthesis, for the preparation of ribofuranosyl-thymine in 1956.[28] Treatment of thymine with mercuric chloride, in the presence of alkali, afforded the mercury salt "dithyminylmercury." This salt was condensed with 2,3,5-tri-*O*-benzoyl-D-ribosyl chloride in toluene at reflux temperature to afford the nucleoside. In a similar manner, mercuri-N^4-acetylcytosine was condensed with the appropriate sugar to afford cytidine, after deprotection.[29] This procedure was the general method until the silyl–Hilbert–Johnson method or Vorbrüggen method was introduced. A modification of the mercuri procedure was developed in which the free pyrimidine base and a halosugar were condensed in the presence of mercuric cyanide in nitromethane.[30] Thus, this procedure eliminates the prior preparation of mercuri pyrimidines. In addition, this method has demonstrated some advantage for the synthesis of 2-thiopyrimidine nucleosides such as 2-thiouridine[31,32] and 2-thiocytidine[32] as well as 6-methyl derivatives of uridine[33] and cytidine.[34] The mechanism of the mercuri reaction has been reviewed by Watanabe *et al.*[35]

It must be emphasized that the use of mercuri compounds in organic synthesis should be limited, because of the high toxicity of the mercuri compounds, except when there is no alternative synthesis for these compounds. The primary reason for this is because the nucleosides prepared by this method are sometimes contaminated by trace amounts of mercuri compounds.

2.3. Ring Closure of N-Substituted Ribosylamines

The above-mentioned methods involve the condensation of a sugar component and a pyrimidine base as the final step of nucleoside synthesis (followed by the deprotection). There have been several alternate routes starting from the ribosylamine or ribosylurea derivatives with the pyrimidine ring formation occurring as the last step. Shaw and co-workers reported the synthesis of uridine by a treatment of 2,3,5-tri-*O*-benzoyl-D-ribosylamine (**17**) with β-ethoxy-*N*-ethoxycarbonylacrylamide followed by debenzoylation.[36] By using α-methyl-β-methoxyacryloyl isothiocyanate, 2-thio-5-methyluridine (**18**) was obtained. Various 5-substituted uridines were likewise prepared in a similar manner. In these cyclizations, the more stable β-anomers were obtained even though the starting ribosylamine was an anomeric mixture. An improvement involved the treatment of ribopyranosylamine, readily obtained by treatment of ribose with methanolic ammonia, with 2,2-dimethoxypropane and *p*-toluenesulfonic acid in acetone which gave 2,3-*O*-isopropylidene-β-D-ribofuranosylamine as the tosylate.[37] Various anomeric pyrimidine nucleosides, including 5-cyano-, 5-acetyl-, 5-carboxyuridines, and 5-methyl-2-thiouridines, were prepared from this ribosylamine. Xylosylamine also gives the 3,5-*O*-isopropylidene-β-D-xylofuranosylammonium tosylate.

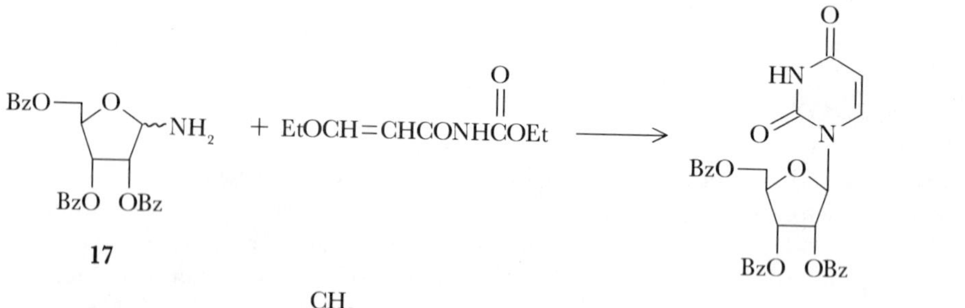

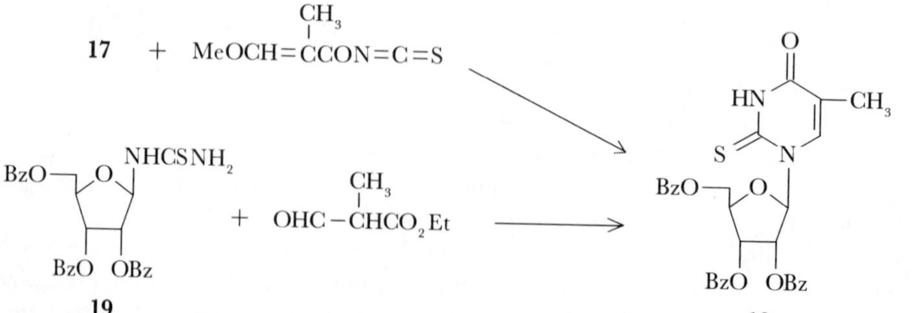

The ring closure of 2,3,4,6-tetra-*O*-acetyl-D-glucosylurea with α-methyl-β-methoxyacryloyl chloride gave the glucosylthymine.[38] Similarly, 2,3,5-tri-*O*-benzoyl-D-ribosylthiourea (**19**) with ethyl α-formylpropionate afforded ribosyl-2-thiothymine.[39] Ethyl cyanoacetate[40] or diethyl malonate[41] was condensed with 2,3,5-tri-*O*-benzoyl-D-ribosylurea to give the 6-aminouridine or ribosyl-barbituric acid, respectively. The reaction of D-ribofuranosyl isocyanate with *O*-methoxypseudourea gave a 4-methoxytriazine, which on amination furnished 5-azacytidine.[42] 2,3,5-Tri-*O*-benzoyl-β-D-ribosyl isothiocyanate was converted to 5-aza-2-thiocytidine by a similar condensation with guanidine followed by a ring closure with ethyl orthoformate and deprotection.[43]

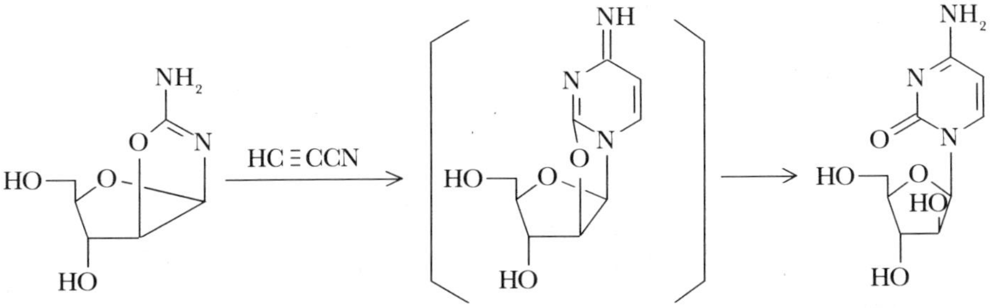

Perhaps the most interesting reaction in the synthesis of pyrimidine nucleosides is the one introduced by Sanchez and Orgel in 1970.[44] Treatment of D-arabinose with cyanamide in a basic aqueous solution gave 2-amino-β-D-arabinofurano[1′,2′ : 4,5]-2-oxazoline (**20**). Condensation of (**20**) with cyanoacetylene afforded arabinofuranosylcytosine (**22**), a potent antileukemic nucleoside. Subsequently, a condensation in dimethylacetamide followed by the treatment of the intermediate with aqueous acetic acid gave O^2, 2′-cyclocytidine (**21**) acetate.[45] Condensation of (**20**) with methyl propiolate gave O^2, 2′-cyclouridine (**23**). The α-anomers of cytidine and uridine have been prepared in a similar sequence starting from D-ribose.[45] The L-enantiomers of pyrimidine nucleosides are also prepared from the appropriate L-sugars.[46] D-Fructofuranose was converted to its oxazoline derivative, which was then condensed with methyl propiolate to give, after cleavage of the cyclo linkage with chloride ion and successive reduction, 3′-deoxypsicofuranosyluracil (**24**).[47] Various other applications of this "oxazoline method" have been reported but are not discussed further here.

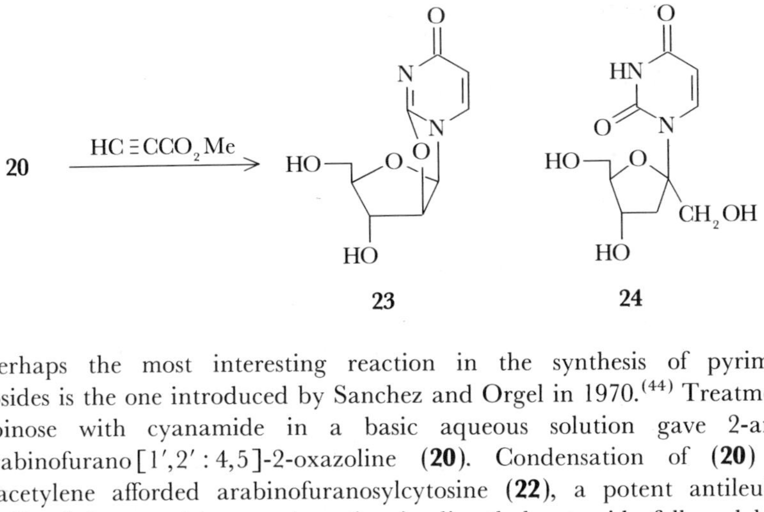

3. Reaction of Pyrimidine Nucleosides in the Heterocyclic Moiety

Since a nucleoside is composed of a heterocyclic moiety and a carbohydrate moiety, many reactions reported in the heterocyclic area and also in the carbohydrate area are directly applicable to the nucleoside area, or vice versa. However, a nucleoside, *per se*, has its own characteristic reactivities that arise from an interaction of both the sugar and base moieties in one reaction, for example, cyclonucleoside formation.

The reactivity of pyrimidine (1,3-diazine system) in terms of aromatic substitution reactions is generally summarized as follows:

1. The lone pair of electrons of the ring nitrogens is usually the most susceptible toward electrophiles such as proton and alkylating agents.

2. Positions 2, 4, and 6 are susceptible to nucleophilic attack and if a leaving group is present on these positions the reaction proceeds as a nucleophilic substitution.

3. In terms of electrophilic substitution, the only possible position is C-5. Although this reactivity is suppressed by the 1,3-ring nitrogens, substituents with a $+M$ effect such as amino and hydroxyl groups at position 2 and/or position 4 facilitate the electrophilic reaction at C-5, which is the case for uridine and cytidine.

4. As found recently, the lithiation at position 6 of uridine and cytidine is possible and enhances electrophilic reactions at this position.

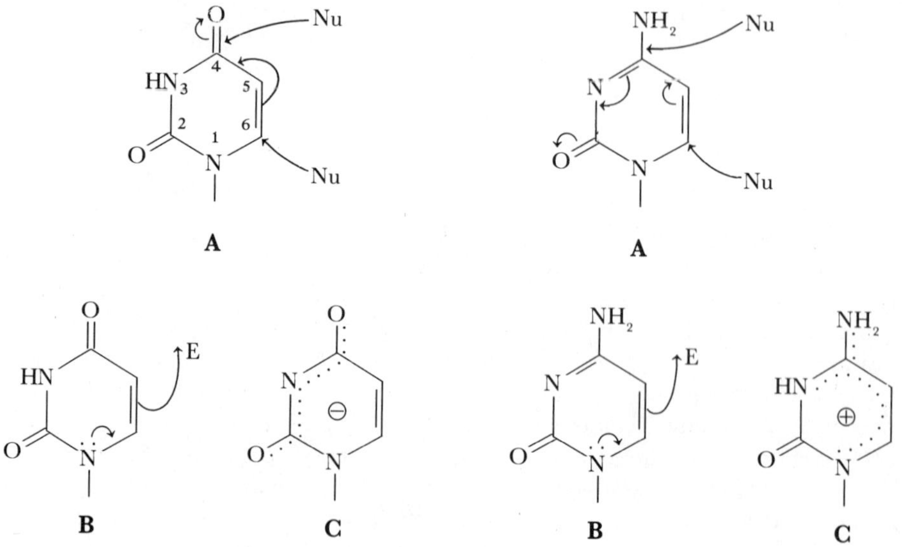

The most common pyrimidine nucleosides are derivatives of the 1-substituted pyrimidin-2-one system, (25) and (26), and are therefore less stabilized than the pyrimidine system *per se*. Also the double bond at the 5,6 position cannot be regarded as simply as part of a benzenoid system but rather that of a conjugated lactam, **(25A)** and **(26A)**, which makes the C-6 position a definite target for nucleophilic attack. It is also apparent that the C-5 position is activated by the electron flow from N-1, as in enamine systems—**(25B)** and **(26B)**. These mobilities of electrons in the base moieties of uridine and cytidine are illustrated as **(25)** and **(26)**. An additional characteristic feature of most pyrimidine nucleosides is that a functional group in the sugar moiety is usually close enough to be involved in a reaction at either position 2 or 6 of the aglycon, for example, cyclonucleoside (or anhydronucleoside) formation.

A vast number of reactions are known to occur in pyrimidines and their nucleosides. These are roughly classified by the nature of the reactions and described as either electrophilic or nucleophilic reactions. The radical and photoreactions are also known in many instances and these are described in relation to the ionic reactions involved.

3.1. Protonation

Protonation of cytidine occurs at N-3 ($pK_a = 4.5$), **(26C)**. The positive charge can then be delocalized throughout the N-1, N-3, and exocyclic nitrogens.[48,49] This makes position 5 of cytidine less reactive, as compared with the C-5 in uridine, to the electrophiles that may be generated in acidic medium. Uridine has the least basic character among the naturally occurring nucleosides. Protonation seems to occur only in strong acid and then primarily at O-4 rather than O-2, since better delocalization of the charge can be established in this manner.[50] Dissociation of the N-3 hydrogen of uridine occurs readily ($pK_a = 9.5$) **(25C)** and thus makes nucleophilic attack less favorable than that observed for cytidine, if the nucleophiles are of a basic character.

3.2. Alkylation and Related Reactions

Since a number of methylated nucleosides have been found as minor constituents in nucleic acids, especially in transfer RNA, a simple and selective method for obtaining such alkylated nucleosides became essential for clarifying their functions in tRNAs. In addition, studies on the alkylation of nucleosides and nucleotides are also devoted to the investigation of the mutagenic action of various alkylating agents. A recent review of alkylation by Singer is now available.[51]

Both the heterocyclic moiety and the sugar hydroxyls are susceptible to alkylation in pyrimidine nucleosides. As a result of recent methodological developments, one can select a suitable procedure for specific alkylation of either moiety.

In the case of cytidine (and 2′-deoxycytidine), treatment with dimethyl sulfate in dimethylformamide gives N^3-methylcytidine methosulfate **(27)**.[52] Benzylation of cytidine with benzyl bromide in dimethylformamide also occurs to give N^3-benzylcytidine.[53] These are quarternization reactions and are consistent with the protonation studies in terms of the site of reaction. However, ethylation of cytidine or 2′-deoxycytidine with ethyl iodile in dimethylsulfoxide gave a small amount of N^4-ethylcytidine in addition to the expected N^3-ethyl derivative.[54] Direct N^4-ethylation must have occurred since it was established that the former compound was not formed by the rearrangement of the latter under the reaction conditions. The bulkiness of the alkylating agent seems to affect the rate and site of alkylation of cytidines. Alkylation of cytidine with 7-bromomethyl-benz[a]anthracene in aqueous solution also gave the N^3- and N^4-alkyl derivatives.[55] Trimethyl phosphate was also reported to alkylate cytidines in dimethylformamide to give the N^3 derivative along with the N^4 derivative as a minor product.[56] Treatment of 2'-deoxycytidine with an excess of sodium hydride and methyl iodide in dimethylformamide gave the N^4, N^4-dimethyl derivative due to the dissociation of the 4-amino protons by the strong base.[57]

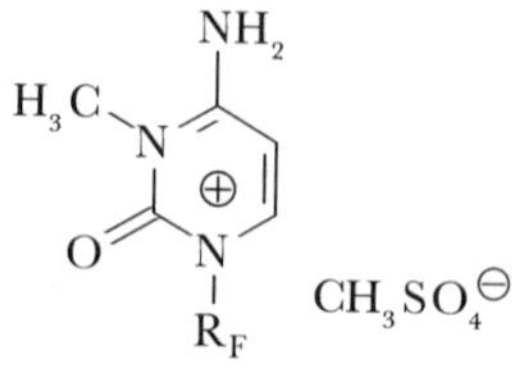

27 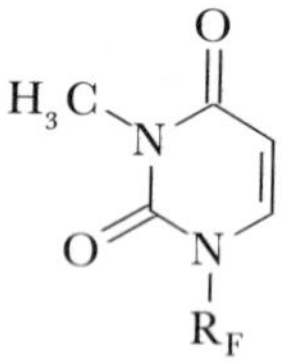   28

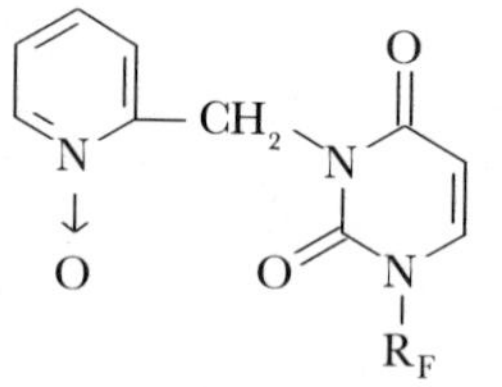

29 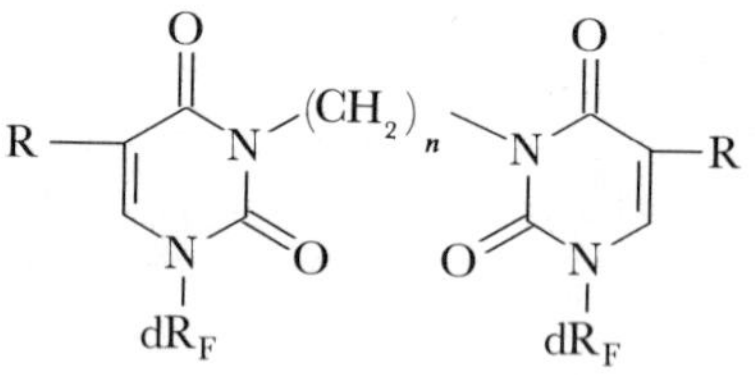   30

Uridine undergoes alkylation at N^3 by a variety of reagents. Treatment of uridine,[58,59] thymidine,[59,60] or 2',3'-O-isopropylideneuridine[61] with diazomethane gave the respective N^3-methyl derivatives (**28**). A recent study on the reaction showed some formation of O^4-alkyl derivatives such as in thymidine[62] and 1-(β-D-arabinofuranosyl)-5-fluorouracil.[63] The product distributions obtained by diazomethane on uridine were $16:2:1$ for the N^3-, O^4-, and O^2-methyl derivatives, respectively. The ratio changes in the diazoethane reaction to $2:1:1$.[64] Action of N-methyl (and ethyl)-N-nitrosourea with cytidine in neutral aqueous solution gave the O^2-alkyl derivative in addition to the N^3 and N^4 derivatives.[65] The formation of N^4-carbamoylcytidine was also noted.[66] The use of 1-oxido-2-pyridyldiazomethane for N^3-alkylation is unique in that the 3-(1-oxido-2-pyridylmethyl) group in (**29**) can be removed by treatment with acetic anhydride followed by alkaline treatment.[67] Meerwein reagents (trialkyl-oxonium tetrafluoroborate), which usually alkylate the oxygen in an amide function, also gave the N^3-methyl (or ethyl)-uridine.[68] Alkylation of uridines in the presence of alkali, designed to alkylate specifically the sugar 2'-hydroxyl group, also tend to alkylate the base portion. Treatment of uridine with benzyl bromide and sodium hydride in dimethylformamide gave a mixture of 3-benzyluridine and its 2'-O-benzyl derivative.[69] Protection of the 2',3'-hydroxyl groups gave the N^3-benzyl derivative as shown in the case of 2',3'-O-(2,4-dimethoxybenzylidene) uridine.[70] The use of methyl iodide and silver oxide usually gives a complex mixture of products that are methylated in the base and sugar portions of the uridine derivatives.[71–73] Alkylation of 2',3'-O-isopropylideneuridine with 3,3-diethoxypropyl bromide in the presence of potassium carbonate afforded the 3-alkyl derivative, which was then converted to 3-(3-amino-3-carboxypropyl) uridine, a minor constituent of tRNA.[74]

Studies on the alkylation of both sites of pyrimidine nucleosides with trimethyloxosulfonium hydroxide[75] and trimethyl phosphate[76,77] under various conditions have been presented. Methylation of uridine with trimethyl phosphate or dimethylsulfate in the presence of tetrabutylammonium fluoride in tetrahydrofurane gave N^3-methyluridine without any alkylation of the sugar hydroxyls. In the case of cytidine, the N^3, N^4-dimethyl derivative was obtained[78,79] when an excess of the fluoride was added. This method was adapted to the synthesis of dimers (**30**) of uridine and thymidine connected through the N-3 positions by methylene groups.[80] Studies of alkylation of the sugar hydroxyl groups are discussed in detail in Section 5.1.

The combination of a primary alcohol and dicyclohexylcarbodiimide at an elevated temperature affords another approach for the N^3-alkylation of uridine and thymidine. The yields of the N^3-alkyl derivatives decrease with the increase in chain length of the alcohol.[81] A much better yield of N^3-ethylation and methylation of uridine, 6-azauridine, and thymidine was obtained by the use of dimethylformamide diethyl (or dimethyl) acetal under milder conditions.[82] This reagent was originally used for the protection of the exocyclic amino group of nucleosides by forming dimethylaminomethylene derivatives. In the case of ribonucleosides, the 2',3'-cis-diol system is also blocked as dimethylaminomethylidene derivatives.[83] By a similar approach, uridine was hydroxyethylated at N^3 with 2-dimethylamino-1,3-dioxolane.[84] Ethylene oxide also reacts with uridine, uridine 5'-phosphate,[85] and 1-methylcytosine[86] to yield the corresponding N^3-(2-hydroxyethyl) derivatives. Acrylonitrile reacts with all common nucleosides at pH 11.5 to give the 2-cyanoethyl derivatives. Pseudouridine is preferentially cyanoethylated at N^1 then at N^3 at pH 8.5–8.8,[87] which made a selective modification of this moiety in certain tRNAs possible.[88] The reaction of propynal with 5-substituted 2'-deoxyuridines and other nucleobases gave the Michael adducts, such as 3-(3-oxopropen-1-yl) derivatives of 2'-deoxy-5-iodouridine and thymidine. Their cytotoxicities have been reported.[89] These propenal groups can be eliminated with 2-mercaptoethanol by an addition–elimination mechanism. Other conjugated acetylenes, such as cyanoacetylene and 1-butyn-3-one, likewise react to give the respective Michael adducts.

Alkylation of cytidine with chloroacetaldehyde occurred at N-3, which then cyclized, *in situ*, by a Schiff base formation between the 4-amino group and the aldehyde to give the N^3, N^4-etheno derivative (**32**).[90,91] Recent kinetic and spectroscopic studies have confirmed the rapid formation of the hydroxyethano intermediate (**31**).[92] Bromoacetaldehyde reacts much faster in the etheno derivative formation.[93] This cyclization also takes place with adenosine derivatives to give the highly fluorescent 1,6-etheno-adenosines. A similar cyclization to furnish the N^3, N^4-(2-buteno)cytosine derivative was observed when acetylvinyl triphenylphosphonium bromide was reacted with cytosine and 2′, 3′, 5′-tri-O-acetylcytidine.[94] Treatment of cytidine with methoxymethylene-cyanamide in hexamethylphosphorotriamide afforded a pyrimido [1,6,a]-1,3,5-triazinylidene-guandine derivative (33), another potential fluorophore probe of nucleosides.[95] A recent report involves the use of bromomalonaldehyde to prepare a fluorescent ethenocytidine carboxaldehyde (34) by treatment with cytidine.[96] Structures for the reaction products of malondialdehyde with adenosine and cytidine have also been reported.[97]

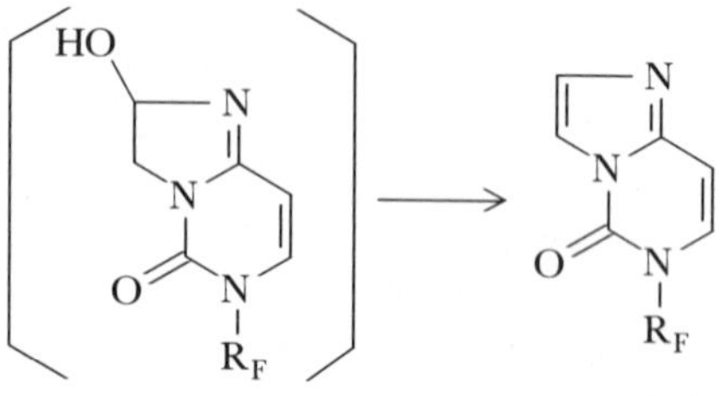

31   32

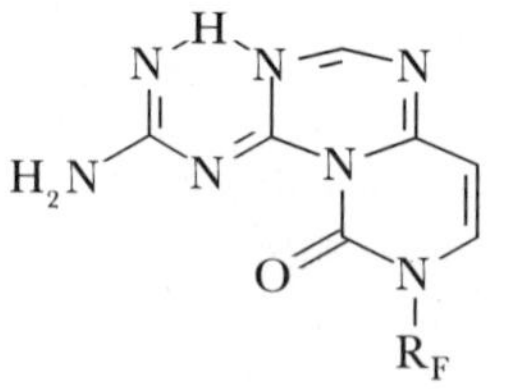

33 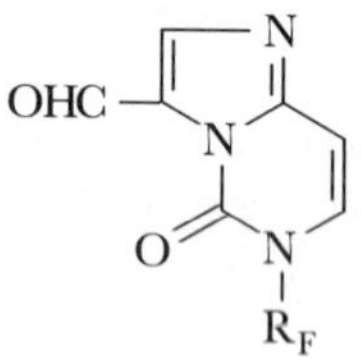   34

A somewhat peculiar but interesting reaction of perfluoroalkyl (butyl to decyl) iodide with uracil, uridine, and 2′-deoxyuridine was reported. Treatment of a uracil derivative with perfluorobutyl iodide in the presence of copper–bronze in dimethylsulfoxide gave the 5-perfluorobutyluracil derivative.[98] In this reaction 5-iodouracil did not react, which means that the reaction is different from the usual Ullmann type reaction described later.

Methylation of 5,6-dihydrouridine with methyl iodide and dimsyl carbanion afforded 3,5,5-trimethyl derivative with some methylation on the sugar hydroxyl groups.[99] Methylation of 3-(β-D-ribofuranosyl)uracil by methyl iodide and methoxide gave the N^1-methyl derivative.[100]

3.3. N-Oxidation and N-Amination

Aromatic nitrogen heterocycles are usually susceptible to N-oxidation by peroxyacids. The characteristic features of such N-oxides are that the electron-attracting nature of the ring nitrogen is reversed by N-oxidation.[101] Brown and co-workers observed the formation of adenosine N^1-oxide by treatment of adenosine with peroxyacetic acid[102] and subsequently the N^3-oxidation of cytidine by *m*-chloroperbenzoic acid in acetic acid to give (35).[103] This compound was also obtained by the treatment of cytidine with perphthalic acid.[104] 1-β-D-Arabinofuranosylcytosine gave the N^3-oxide under similar conditions.[105] Action of diacyl peroxide on cytidine under heterolytic conditions afforded the N^3-benzoyloxy and N^4-benzoyl derivatives along with a trace amount of the N^3-oxide.[106] Although various transformation reactions starting from adenosine N^1-oxide have been accomplished,[107] very little research has been done using cytidine oxide. Treatment of cytosine N^3-oxide with acetic anhydride afforded N^4-acetoxycytosine by a rearrangement similar to the Dimroth rearrangement.[103] A study on the rate and mode of the reaction of *m*-chloroperbenzoic acid with nucleic acid components as a function of pH has been presented.[108]

| 35 | 36 | 37 | 38 |

Hydroxylamine-*O*-esters or -*O*-aryl ethers are regarded as amino cation sources for aromatic electrophilic substitution, and with the appropriate purine nucleosides, N^1-aminoguanosine[109] and N^1-aminoadenosine[110] have been prepared. Treatment of the sodium salt of uridine, 2′,3′-*O*-isopropylideneuridine, or 5-bromouridine with 2,4-dinitrophenoxyamine in dimethylformamide at 37°C for 4 days gave the corresponding N^3-aminouridines (36) in moderate to high yield.[111] Hydroxylamine-*O*-sulfonic acid gave the *N*-amino derivative of uridine in low yield. Cytidine was converted to 3-aminocytidine (37) with dinitrophenoxyamine, which was isolated in the form of a hydrochloride.[110] Cyclization of this nucleoside with triethyl orthoformate afforded the triazolopyrimidine (38).[111] It is of interest that amination of 1,3-dimethyluracil with hydroxylamine-*O*-sulfonic acid in aqueous solution at pH 2 gave the 5-aminouracil derivative.[112]

3.4. Nitration

As reported in the historical work on the structure determination of uridine by Levene and co-workers, nitric acid treatment of uridine afforded "5-nitrouridine-5′-carboxylic acid" (39).[113] Nitration of 2′, 3′, 5′-tri-*O*-(3,5-dinitrobenzoyl)uridine with nitric acid gave, after deblocking, 5-nitrouridine (40).[114] 5-Nitrouridines are reported to be unstable in alkaline medium, and this instability is presumably initiated by the addition of a hydroxide ion to the

6 position. In accord with this property, treatment of the ethyl ester of (39) with sodium borohydride gave the 5,6-dihydro derivative.[115] Likewise, nitration of arabinosyluracil also gives the 5-nitro derivative.[116] The nature of nitric acid as a powerful oxidant limits its use. Cytidine is resistant to nitration due to a facile protonation. Therefore, 5-nitrocytidine (41) was prepared by the ribosylation of preformed 5-nitrocytosine.[117] Torrence and co-workers have recently used nitronium tetrafluoroborate in the nitration of pyrimidine nucleosides and nucleotides. Treatment of 2'-deoxyuridine 5'-phosphate with nitronium tetrafluoroborate in sulfolane afforded the 5-nitro derivative, which was then dephosphorylated to afford the nucleoside. 2'-Deoxyuridine itself is very labile with the glycosyl bond cleavage usually occurring during the nitration.[118,119] dCMP was also nitrated to afford the 5-nitro derivative (in the form of a 3'-O-nitrate).[120] The 5-nitro group of nitrouridines can be converted to the 5-amino group by a catalytic hydrogenation.[114]

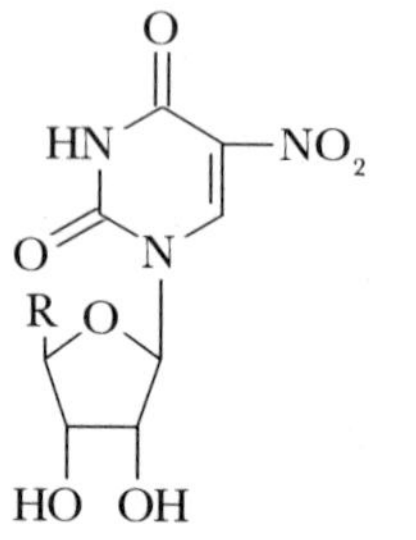

39, R = COOH
40, R = CH$_2$OH

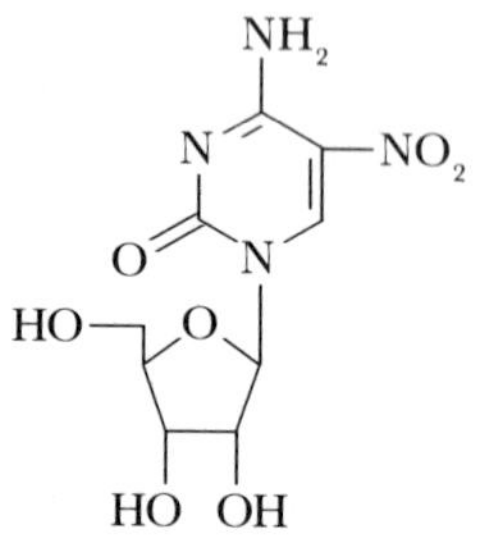

41

3.5. Hydroxymethylation and Related Reactions

Hydroxymethylation of uracil with formaldehyde in potassium hydroxide solution gave 5-hydroxymethyluracil.[121] This compound was readily converted to 5-chloromethyluracil by treatment with concentrated hydrochloric acid.[122] While hydroxymethylation of uridine and 2'-deoxyuridine with formaldehyde in acidic conditions afforded the 5-hydroxymethyl derivatives in relatively low yield,[121] treatment of 2',3'-O-isopropylideneuridine (42) with formaldehyde in alkali gave the 5-hydroxymethyl derivative (44) in high yield.[123] 2'-Deoxyuridine gave the 5-ethoxymethyl derivative by a similar hydroxymethylation followed by ethanolic hydrochloric acid treatment.[124] 5'-Deoxy-5'-fluorouridine gave various 5-alkoxymethyl derivatives by a similar treatment.[125] Treatment of uridine with formaldehyde and diethylamine gave 5-diethylaminomethyluridine, and subsequent hydrogenation with Pt catalyst gave ribosylthymine.[126] As shown by Santi and Brewer,[127] this hydroxymethylation is accelerated by an initial nucleophilic attack of the 5'-hydroxymethyl group of (42) at C-6 to generate a C-5 carbanion (43), which then attacked the formaldehyde carbon to yield (44). This was also demonstrated in the synthesis of the 5-dimethylaminomethyl derivatives of 2′,3′-*O*-isopropylideneuridine or 2′-deoxyuridine-photohydrate with formaldehyde and dimethylamine.[128] The type of reaction that is initiated by the C-6 nucleophilic addition is discussed later. Hydroxymethylation of arabino-furanosyluracil goes smoothly, by 2′-hydroxyl group participation, to give the 5-hydroxymethyl derivatives, which was further derivatized to the 5-aminomethyl derivatives of arabinofuranosyluracil and arabinofuranosylcytosine.[129]

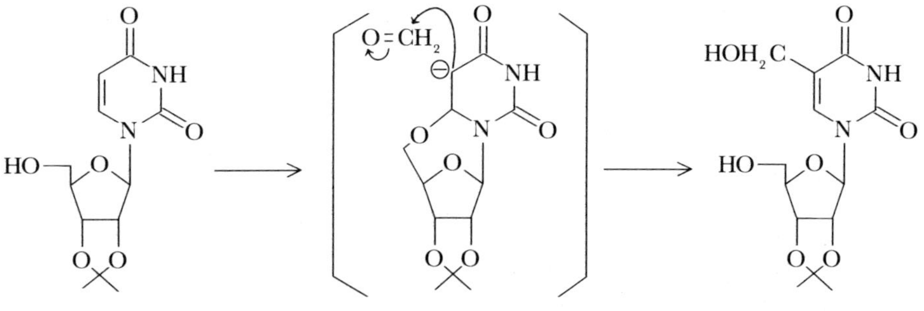

42   **43**   **44**

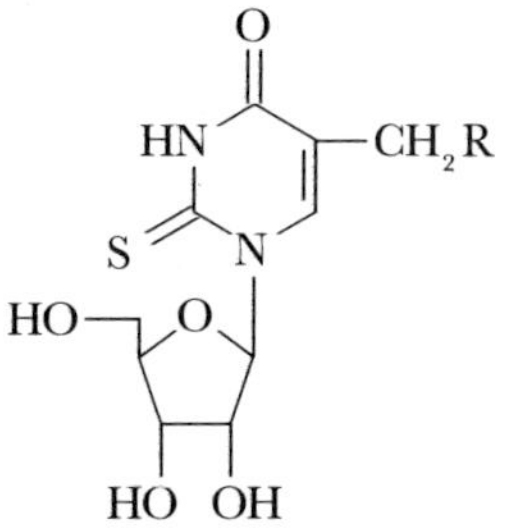

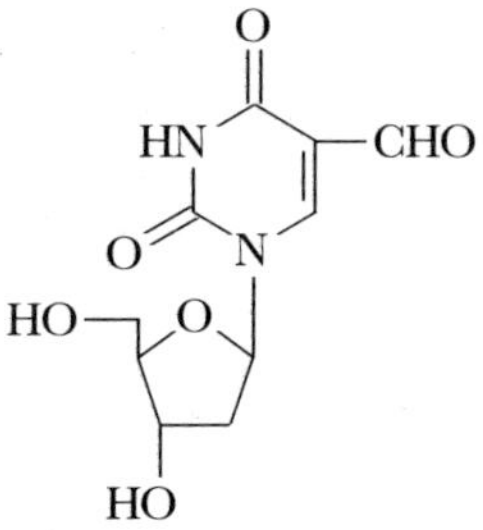

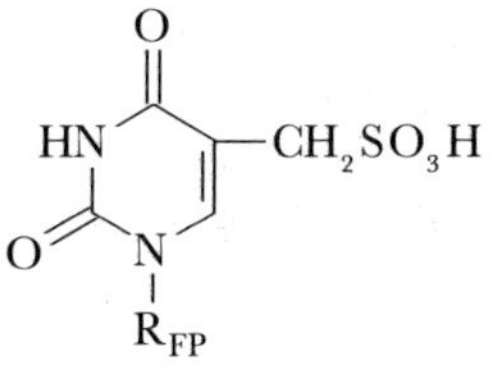

45, R=NHCH$_3$
46, R=COOCH$_3$
47, R=NHCH$_2$COOCH$_3$

48   **49**

By the same principle, 2′,3′-*O*-isopropylidene (or ethoxymethylidene)-2-thiouridine was converted to the 5-hydroxymethyl derivative with formaldehyde and triethylamine [130] or potassium hydroxide.[131] This nucleoside was further converted to the minor nucleosides of tRNA, 5-methylaminomethyl-2-thiouridine (**45**), 5-methoxycarbonylmethyl-2-thiouridine[130] (**46**), and 5-(*N*-methoxycarbonylmethyl)aminomethyl derivative[132] (**47**) via the 5-chloromethyl derivative. As reported recently, the pyrrolidinylmethylation of 2′,3′-*O*-isopropylideneuridine and -2-thiouridine with formaldehyde and pyrrolidine was performed. Successive methylation (at the pyrrolidine nitrogen) and further treatment with a glycine ester then furnished, after deprotection, 5-(*N*-methoxycarbonylmethyl)amino-methyluridine and -2-thiouridine (**47**).[132a] The action of toluenethiol with 5-pyrrolidinylmethyluridine gave the 5-tolylthiomethyluridine,[132a] which was then desulfurized with Raney Ni to give, after deprotection, 5-methyluridine.[133] The hydroxymethyl group in the 5 position of uridine has been hydrogenated to a

5-methyl group[121,123] or oxidized by air and Pt catalyst to the 5-formyl or
5-carboxyl group with some concomitant oxidation of the 5' primary alcohol.[134]
Oxidation of (44) with manganese dioxide gave the 5-formyl derivative in high
yield, which was then converted to 5-formyluridine 5'-phosphate.[135] 5-Hydroxy-
methyl-2'-deoxyuridine was also oxidized to the 5-formyl derivative (48) by
manganese dioxide.[136] 5-Formyl-2'-deoxyuridine was further converted to the
5-oxime, 5-dithiolane, and 5-cyano derivatives.[137] The Schiff-base formation of
5-formyluridines with various amines followed by reduction of the products with
sodium borohydride afforded the appropriate 5-alkylaminomethyl derivatives.[138]
Various 5-(substituted)methyl uridine derivatives have been prepared *via* the
5-chloromethyl intermediates.[122,124,130,136b,139,140–144]

　　　Although the direct hydroxymethylation of cytosine nucleosides has not
been reported and 5-hydroxymethylcytidine is prepared by the Vorbrüggen
method,[145] treatment of 2'-deoxycytidine 5'-phosphate (and CMP) with formaldehyde in strong alkaline conditions afforded the 5-hydroxymethyl-cytidine
derivative, though the yields were very low.[146] Deamination of these nucleotides
by nitrous acid gave the 5-hydroxymethyl-UMPs.

　　　Treatment of 5-hydroxymethyl derivatives of CMP and UMP with sodium
hydrogen sulfite gave the respective 5-methanesulfonic acid derivatives (49).[147]
Nucleophilic substitution at the carbon of the 5 position is closely related to the
thymidylate synthetase action.[142] In this relation it is interesting to note that the
reaction of cyanide to 5-acetoxymethyluridine (50) gives 6-cyano-5-methyl-
uridine[148] (51) instead of the 5-cyanomethyl derivative.[130]

50   51   52

　　　A reaction that is mechanistically close to the hydroxymethylation, but is
actually a different type of reaction, involves the treatment of cytidine with
ninhydrin. Shapiro and Agarwal observed that treatment of cytidine with an
excess of ninhydrin gave the stable product (52), in which C-5 and N-4 of cytidine
were involved in the reaction. Cytosine and cytidine phosphates gave the
polynuclear products, whereas uridine was unreactive toward ninhydrin under
similar reaction conditions.[149]

3.6. Halogenation

　　　Halogenation of nucleosides is perhaps the most extensively studied electrophilic reaction. This is primarily because of the potent activity of 5-halogenated
pyrimidine nucleosides as anticancer and antiviral chemotherapeutic agents. The
5-halogenopyrimidines are also useful intermediates for further conversion of
nucleosides.

Uridine and cytidine undergo a facile halogenation at position 5. Treatment of uridine with bromine water gives the intermediate, 5-bromo-6-hydroxy-5,6-dihydrouridine (**53**), which on heating in acidic ethanol, yields 5-bromouridine (**54**).[113] 2'-Deoxyuridine can be brominated by a similar treatment[150,151] or by bromine in acetic anhydride (of the 3',5'-di-*O*-acetate).[152] *N*-Bromosuccinimide is also reported to afford 5-bromouridine, 5-bromocytidine, and 8-bromo-guanosine from the respective nucleosides[153] or nucleotides.[154] The reaction of chlorine with uridine or 2'-deoxyuridine in acetic acid gives 5-chlorouridine[155] and 5-chloro-2'-deoxyuridine,[156] respectively. *N*-Chlorosuccinimide with uridine or cytidine gives the corresponding 5-chloro derivatives.[157] Iodination of uridine has been accomplished with iodine and nitric acid in chloroform to give the 5-iodo derivative.[158] 2'-Deoxyuridine or 2', 5'-dideoxyuridine under similar conditions gave the appropriate iodo derivatives.[159,160] *N*-Iodosuccinimide[161] and iodine monochloride[162] were also used in the iodination of uracil and cytosine nucleosides. Chlorination of 3',5'-di-*O*-tolyl-2'-deoxyuridine by iodobenenedichloride to give the 5-chloro derivative was recently reported.[163]

Uridine ⟶ [**53** structure] ⟶ **54** structure, **55** structure

53   54   55

Bromination of cytidine proceeds smoothly to give the 5-bromo derivative[164,165] (**55**) by treatment with bromine in acetic acid–pyridine. In the classical work, photoactivation was employed in this system[166,167] but was later found to be unnecessary. Iodine monochloride with cytidine (or uridine) was reported to give the intermediate, 5-iodo-6-hydroxy-5,6-dihydrocytidine (or uridine).[168] The reaction of sodium hypochlorite with the various nucleosides from RNA and DNA was studied. In the case of cytidine, a dichlorocytidine compound was detected with the second chlorine atom presumably being attached to N^3 or N^4.[169] In the study of *N*-oxidation of nucleosides, Ryu and MacCoss found that the action of *m*-chloroperbenzoic acid and hydrogen chloride with cytidine, 2'-deoxycytidine, and arabinofuranosylcytosine in dimethylacetamide gave the corresponding 5-chloro derivatives in good yields. Uridine and adenosine also gave the expected chloro derivatives.[170]

Treatment of 2'-deoxycytidine with iodine and iodic acid gave 5-iodo-2'-deoxycytidine (**57**) and 5,5-diiodo-O^6, 5'-cyclo-5,6-dihydro-2'-deoxyuridine[171] (**58**), a compound that has previously been reported as 6-iododeoxycytidine.[172] Alkaline treatment of (**58**) gave O^6, 5'-cyclo-5-iodo-2'-deoxyuridine (**59**).[171] A similar reaction was later found, in the iodination of arabinofuranosylcytosine with the same reagent, to yield the O^6, 2'-cyclo-5-iodo derivative.[173] Formation of the 5,5-diiodo-5,6-dihydro intermediate (**58**) may be the result of over-iodination and hydrolytic deamination of (**57**) or the initially formed cyclo intermediate (56). The reaction of bromine or chlorine in methanol with 5-fluoro-2'-deoxyuridine gave the respective 5-fluoro-5-halogeno-5,6-dihydro derivatives.[174] A detailed mechanism for the bromination of cytosine derivatives is presented.[175]

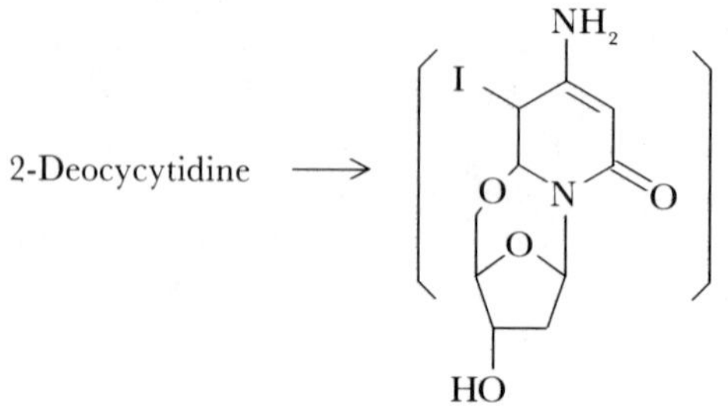

2-Deocycytidine $\longrightarrow$

56

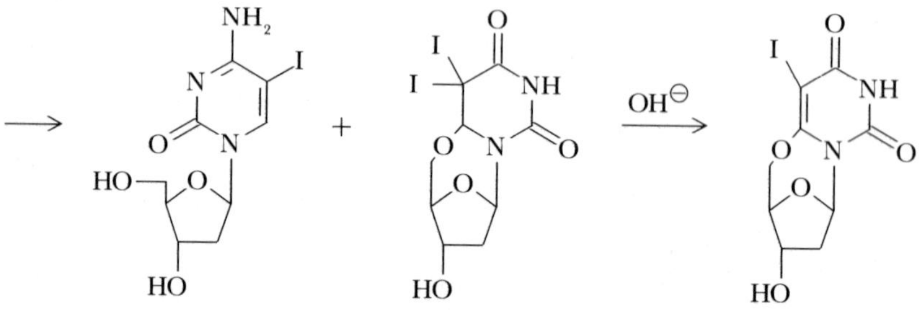

57 + 58 $\xrightarrow{OH^{\ominus}}$ 59

Treatment of thymidine (60) with *N*-iodosuccinimide in dimethylsulfoxide in the presence of trifluoroacetic acid as an accelerator gave a diastereomeric mixture of 5-iodo-O^6,5'-cyclo-5,6-dihydrothymidine—(61A) and (61B). This must be the result of a trans addition of iodine and the 5'-hydroxyl group to the 5,6 double bond. One of the diastereomers [endo form, (61A)], when illuminated, isomerized to *cis*-iodohydrine (61C), which on treatment with base or silver nitrate then gave O^6,5'-cyclothymidine (62). [176] This sequence of reactions made it clear that the electrophilic addition reaction and successive elimination at the 5,6 bond of pyrimidine nucleosides were both a trans process. Therefore, in the formation of 5-halogenopyrimidines, the epimerization at C-5 in the dihydro intermediate must have occurred so as to promote trans elimination of water, whereas in the formation of O^6,5'-cyclothymidine, photoactivation was neccessary for the C-5 epimerization. The addition of bromine to thymidine in acetic acid gave a mixture of diastereomers of the trans addition products, which on hydrolysis with aqueous pyridine gave the C-5 inverted *cis*-5,6-dihydroxy-5,6-dihydrothymidine.[177] 5-Iodo-6-hydroxy-5,6-dihydrouridine, prepared by the action of *N*-iodosuccinimide with uridine, was also converted to the (+)- and (−)-6-hydroxy-5,6-dihydrouridine.[178] Reduction of the 5-bromo-6-hydroxy-5,6-dihydrothymidine with zinc and acetic acid gave 5,6-dihydro-6-hydroxy-thymidine,[179] the photohydrate of thymidine.

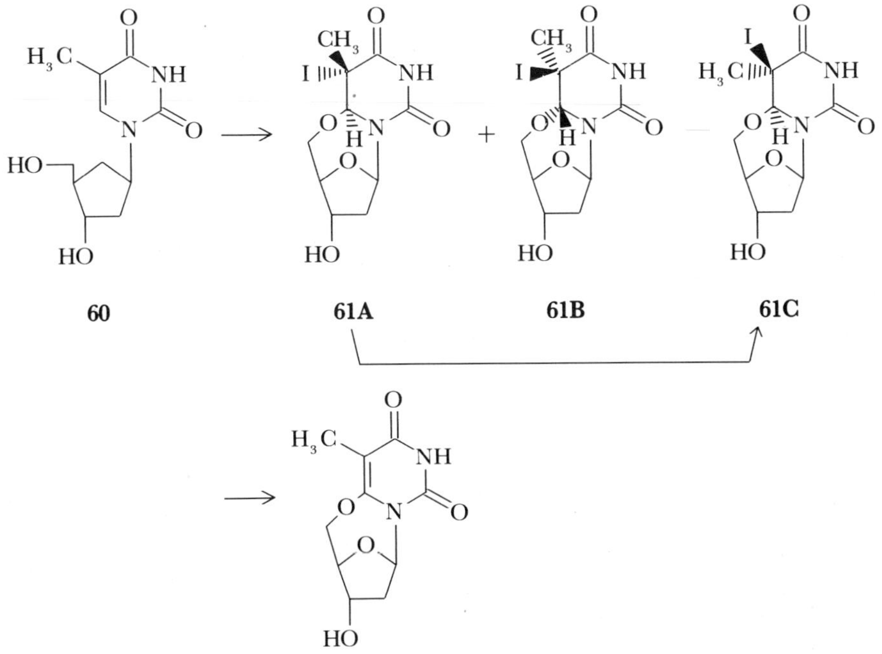

60   61A   61B   61C

62

5-Fluorouracil and certain 5-fluorouracil derivatives are potent anticancer agents.[180] The direct fluorination of uracil to afford 5-fluorouracil by trifluoromethyl hypofluorite has been accomplished.[181,182] Treatment of acetylated uridine or 2′-deoxyuridine with the same reagent in methanol and fluorotrichloromethane at −78°C gave the addition product (63), which by elimination with triethylamine treatment and deblocking afforded the respective 5-fluorouridines (64).[182] Cytosine, fully acetylated cytidine, and arabino-

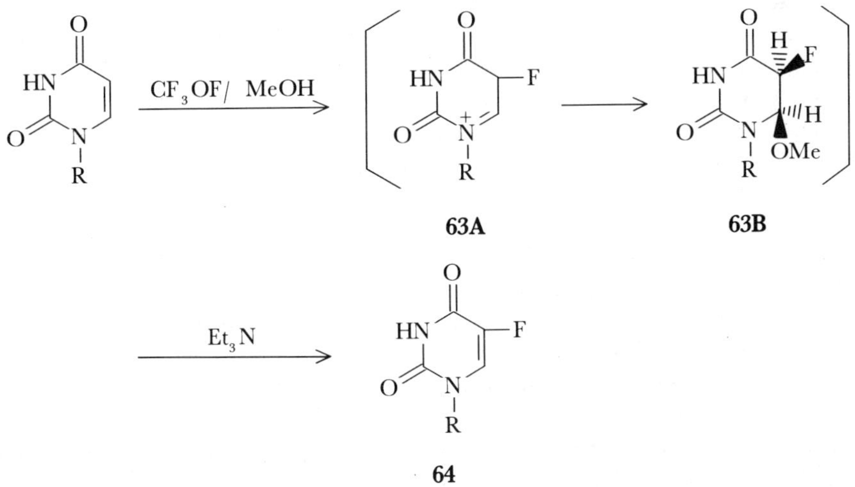

63A   63B

64

furanosylcytosine also gave the corresponding 5-fluoro derivatives[183,184] under similar conditions. This method has been adapted for the fluorination of the uracil moieties in mononucleotides and dinucleoside phosphates.[185,186] Intensive studies on the mechanism of fluorination in this system[187] showed that the configuration of the 5-fluoro-6-methoxy-5,6-dihydrouracil intermediate was the cis relation, which is in direct contrast to the previously mentioned general statement. Robins *et al.* explained this fact by assuming the presence of an initial intermediate (**63A**), which will give the stable second intermediate (**63B**) as the result of an attack of methanol. This second form would exist as a boat form in which the fluoro and methoxy functions occupied equatorial and axial positions (the gauche orientation), respectively. By treatment of (**63B**) with triethylamine, a facile trans elimination of methanol affords the 5-fluorouracil derivative (**64**).[187]

2′-Amino-2′-deoxyuridine was fluorinated in a similar manner by this system after prior trifluoroacetylation of the 2′-amino group.[188] For other direct fluorinations, fluorine in pyridine-Freon-11 at −78°C was reported to be effective for the preparation of 5-fluorouridine.[189] Fluorine in acetic acid followed by treatment with triethylamine was also efficient as demonstrated in the preparation of 5-fluoro-6-methyluridine and its 2′-deoxy derivative,[190] as well as various 2′-substituted 2′-deoxyurdines.[191] Formation of 5-fluoro-6-fluoromethyluridine was observed in the fluorination of 6-methyluridine, which was stable in the deprotection step.[190]

A recent report using cesium fluoroxysulfate ($CsSO_4F$) in methanol is noteworthy since the fluorination can be carried out at room temperature and converts uridine to 5-fluorouridine in high yield.[192] The stereochemistry of the 5-fluoro-6-methoxy-5,6-dihydrouracil intermediate was also a cis form as discussed above.

The bromination of 2′,3′-*O*-isopropylideneisocytidine with *N*-bromosuccinimide in methylene chloride gave the 5-bromo derivative.[193] The pseudohalogen thiocyanogen chloride also reacted with acetylated derivatives of uridine and 2′-deoxyuridine in acetic acid to give the respective 5-thiocyanatouridines (**65**).[194] Treatment of the 5-thiocyanato derivative with dithiothreitol afforded the corresponding 5-mercaptouridines. Later work[195,196] showed that no prior protection of the sugar hydroxyls was necessary and that the 5-thiocyanato group was converted to the 5-alkylthio group by *in situ* alkylation in the presence of mercaptoethanol. Oxidation of the 5-methylthio group to the 5-methylsulfone was also reported.[196] 5-Mercapto-2′-deoxyuridine, prepared by the glycosylation method, was *S*-alkylated by various alkyl halides containing unsaturations in the alkyl chains.[197]

As described in the beginning of this section, the action of bromine with uridine gives the 5-bromo-6-hydroxy-5,6-dihydro intermediate (**53**). Treatment of (**53**) with lead oxide,[198] or better with aqueous pyridine,[199] afforded 5-hydroxyuridine (**67**). In the latter conditions, the replacement of the 5-bromo group of (**53**) by a hydroxyl group proceeded before dehydration occurs to give the 5,6-dihydroxy intermediate (**66**). This type of intermediate can be isolated in uridine and 2′-deoxyuridine, although its conversion to the product (**67**) was not attempted.[200] The 5-hydroxylation *via* the bromination and aqueous pyridine treatment was applied to 2′-deoxyuridine,[201] arabinofuranosyluracil,[116] and

65 **66** **67**

cytidine.[202] By a similar principle, treatment of 5-bromo-6-methoxy-5,6-dihydro-2′-deoxyuridine with sodium disulfide[203] or sodium diselenide[204] gave the 5-mercapto or 5-hydroseleno uridine derivatives (after a reduction of the corresponding disulfide or diselenide). The 5-mercapto derivatives of cytidine and 2′-deoxycytidine were likewise prepared.[205]

68

67 **67A** **69**

71 **72A** **72** **70**

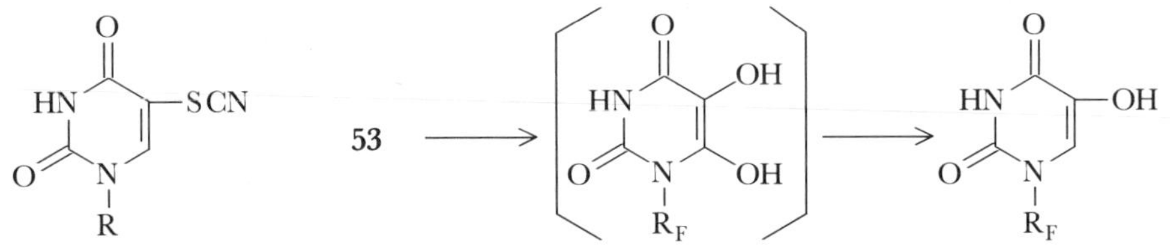

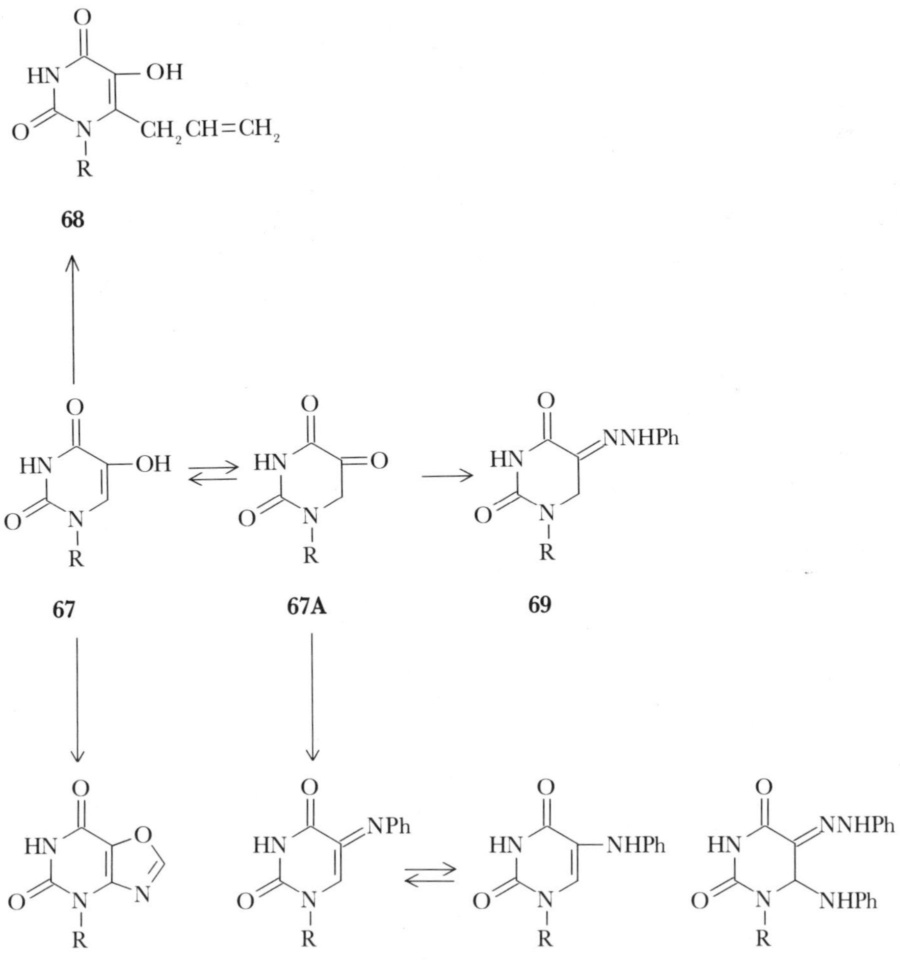

In 5-hydroxyuridine, the keto-enol tautomerization is possible as shown, **(67)** ⇌ **(67A)**. The 5-*O*-alkylation by allyl bromide gave the 5-allyloxyuridine, which underwent a Claisen rearrangement to yield 6-allyl-5-hydroxyuridine **(68)**.[206] Treatment of 5-hydroxyuridine with dimethyl sulfate in alkaline solution gave 5-methoxyuridine (along with some of the N^3-methyl derivative).[207] This is a minor nucleoside found in some tRNA. Various 5-alkoxy, 5-allyloxy, and 5-propynyloxy derivatives of 2'-deoxyuridine were prepared and tested for their antiherpes activities.[208] Treatment of 2'-deoxy-5-hydroxyuridine with iodoacetonitrile gave 5-cyanomethoxy-2'-deoxyuridine, another candidate for antiherpes activity.[209] Treatment of 5-hydroxyuridine with phenylhydrazine gave the 5-phenylhydrazone **(69)**.[210] Bromine oxidation of **(67)** followed be treatment with phenylhydrazine gave 5-phenylazouridine and with excess phenyl-hydrazine gave the adduct **(70)**.[210] This compound was formerly represented as 5,6-bisphenylhydrazinouridine by Levene.[211] The azo coupling of **(67)** with benzenediazonium salt, followed by careful hydrogenation and subsequent cyclization with formamide, gave an oxazolopyrimidine nucleoside **(71)**.[212] As found recently, 5-hydroxy-2'-deoxyuridine reacts with primary amines, including aniline, to give the 5-alkylaminouridines **(72)** due to the stability in the amine–imine tautomerism with **(72A)**.[213] The reaction was used for spin-labeling of deoxyuridines.[213]

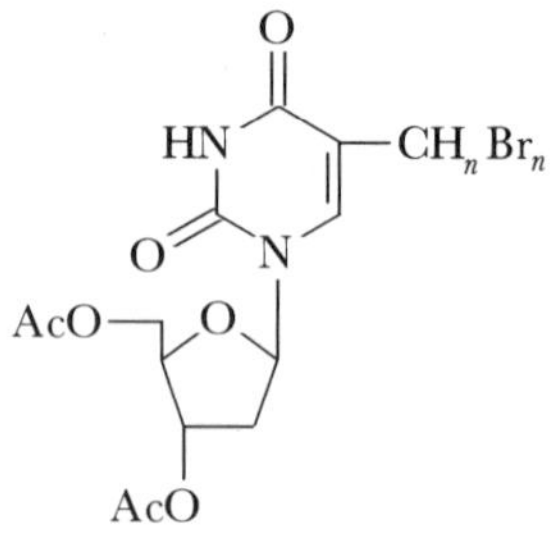

73, $n = 1,2$

A selective bromination of the 5-methyl group in 3',5'-di-*O*-acetylthymidine to afford the 5-mono- or dibromomethyl derivatives **(73)** was reported using bromine or *N*-bromosuccinimide under UV irradiation in carbon tetra-chloride.[214,215] These are unstable and easily hydrolyzed to the 5-hydroxymethyl and 5-formyl derivatives, respectively. The bromination of the 5-ethyluridine derivative by this procedure gave the 5-vinyl derivative.[214] The use of chlorine instead of bromine gave the 5,6-dichloro-5,6-dihydrouridine derivative. The 5-bromomethyl derivative was further converted to the 5-aminomethyl derivative *via* the azidomethyl intermediate.[216] As described in Section 2.4, the 5-substituted-methyluridines may readily be prepared by the hydroxymethylation of uridines.

3.7. Mercuration, Metallation, and Related Reactions

In recent years, metallation of nucleosides has been successfully applied to transformations of the base moiety, mainly the introduction of carbon units into position 5 or 6. The 5-mercuripyrimidine nucleosides are used and converted to the active palladium intermediates, which are then reacted with various alkenes or alkenyl halides. The organopalladium intermediates can also be prepared from 5-iodopyrimidine nucleosides. Another approach involves the use of a 6 (or 5)-lithio derivative, which is then reacted with alkyl halides or various carbonyl compounds. An excellent review is presented by Bergstrom.[217]

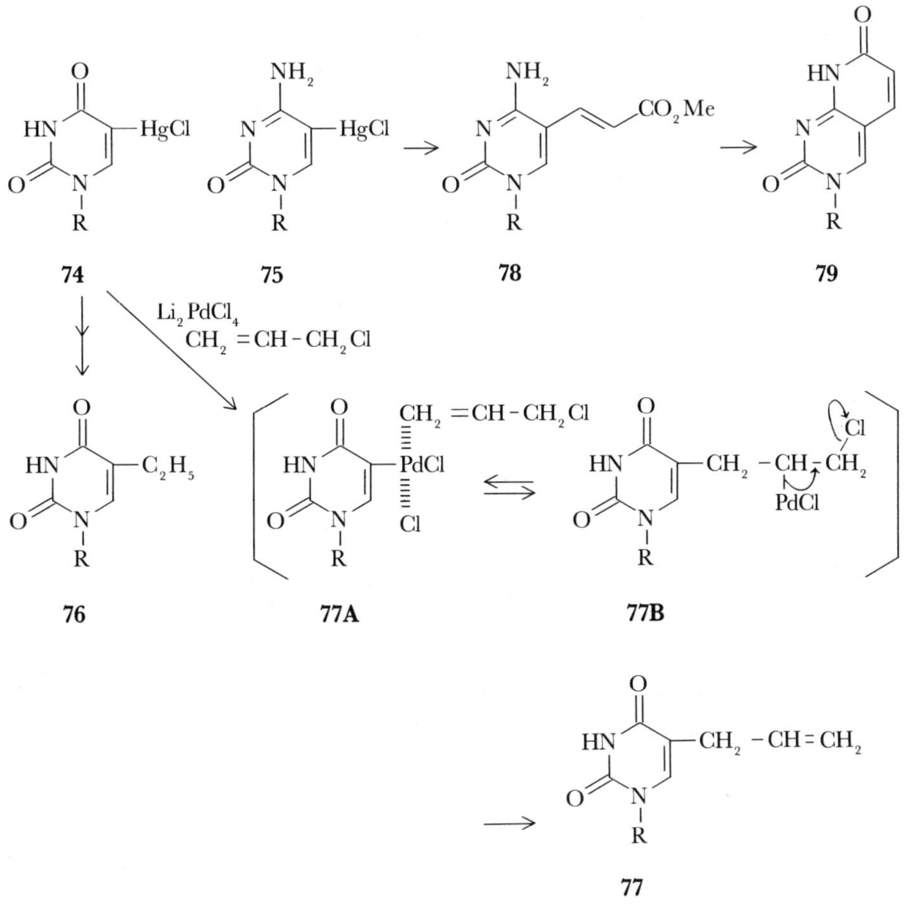

In an aqueous buffer, near neutrality, mercuric acetate reacts with UMP, CMP, and their 5'-triphosphates to give, after washing with sodium chloride solution or through column chromatographic separation, 5-chloromercuri derivatives of UMP (**74**), CMP (**75**), and their 5'-triphosphates in high yields.[218,219] This mercuration can also be carried out on polynucleotides.[220] Treatment of the mercuri compounds with iodine in ethanol gave the 5-iodo derivatives. This has furnished a versatile method for the preparation of 5-iodopyrimidine nucleotides.[221] Reduction of the 5-mercuripyrimidine nucleotides with sodium borotritiolide gave the 5-tritiated derivatives.[221] The method of preparation for the 5-chloromercuri derivatives of uridine, cytidine, and their 2'-deoxy derivatives has been described in detail.[222] Although the treatment of 2'-deoxycytidine gives the 5-acetoxymercuri derivative with mercuric acetate, uridine did not give the 5-mercuri product by using this reagent.[222]

Treatment of (74) with Li_2PdCl_4 in methanol under an atmosphere of ethylene, followed by a reduction of the product with sodium borohydride and hydrogen with palladium catalyst, afforded 5-ethyluridine (76) in 86% yield. Other alkenes and alkenols can be used instead of ethylene to yield the corresponding 5-alkyl or hydroxyalkyl uridine derivatives.[223,224] Allyl chloride[225] and allyl alcohols[226] are also condensed by this system to give 5-allyl-2'-deoxyuridine (77) and -cytidine derivatives. The pathway illustrated by (77A) and (77B) was suggested for the formation of (77). 3,3,3-Trifluoropropene reacts in this system with 5-chloromercuri-2'-deoxyuridine to give 5(E)-(3,3,3-trifluoropropenyl)-2'-deoxyuridine.[227] Styrene derivatives are also coupled with mercuri-uridine nucleosides and nucleotides by Pd(II) to give the 5-styryl derivatives.[228] The preparation of 5-hydroxymethyl-2'-deoxyuridine 5'-phosphate labeled at the 5-hydroxymethyl carbon with ^{13}C or ^{14}C was accomplished by using the labeled styrene in the above reaction, followed by oxidative cleavage of the styrene function with osmium tetroxide–sodium perchlorate and successive borohydride reduction of the 5-formyl group.[229]

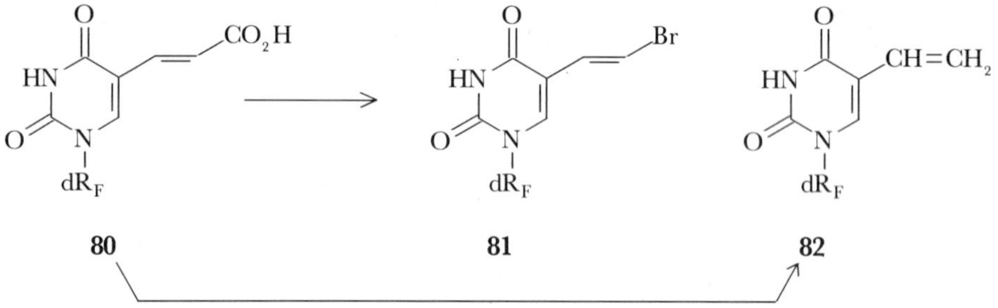

The reaction of 5-chloromercuricytidines with methyl acrylate in the presence of Li_2PdCl_4 gave the 5-(E)-acrylate (78). Compound (78) was cyclized by photoisomerization to give 3-(D-ribofuranosyl)-2,7-dioxopyrido[2,4]pyrimidine (79).[230,225] 5-(E)-Carboxyvinyl-2'-deoxyuridine (80) was converted to 5-(E-2-bromovinyl)-2'-deoxyuridine (81), a very potent antiherpes agent,[231] by treatment with N-bromosuccinimide.[232] A conversion of the geometry of (81) by photoirradiation to the (Z) form was reported and this compound was found to be less active.[233] Thermal decarboxylation of (80) gave the 5-vinyl derivative (82).[234] Treatment of 5-chloromercuri-2'-deoxyuridine with substituted iodobenzenes in the presence of Li_2PdCl_4 afforded the 5-phenyluridine derivatives in low yields.[235]

The conversion of 5-iodopyrimidine nucleosides, in place of the 5-mercuri derivatives, to the 5-alkyl derivatives has also been accomplished by a metal-mediated reaction. Treatment of 3',5'-di-O-acetyl-2'-deoxy-5-iodouridine with iodotrifluoromethane in the presence of copper powder gave 5-trifluoromethyl-2'-

deoxyuridine **(83)**,[236] an active antitumor nucleoside. In a similar manner, protected derivatives of 5-iodouridine and 5-iodocytidine were converted to the 5-trifluoromethyl derivatives by treatment with trifluoromethylcopper.[237] It is interesting to note that the 5-(perfluoroalkyl)-2′-deoxyuridines are prepared from 2′-deoxyuridine with iodo(perfluoroalkanes) in the presence of a zinc–copper couple.[98] Treatment of 2′-deoxy-5-iodouridine with cupric cyanide gave the 5-cyano derivative in low yield. This reaction was used for the ^{14}C isotope labeling.[238]

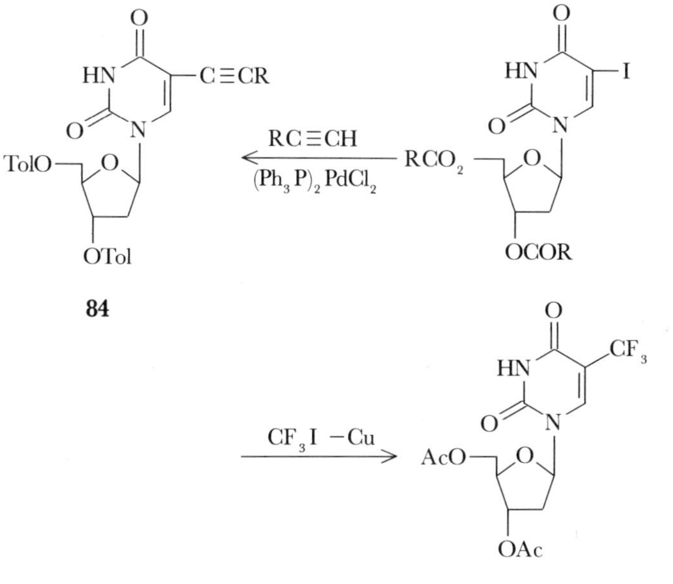

84

83

The palladium-mediated coupling of 5-iodopyrimidines is more effective. Treatment of 3′,5′-di-*O*-tolyl-2′-deoxy-5-iodouridine with various 1-alkynes and $(Ph_3P)_2PdCl_2$ in the presence of cuprous iodide and triethylamine afforded the 5-alkynyl derivative **(84)** in high yield.[239] Similarly, treatment of 2′-deoxy-5-iodouridine with vinyl acetate in the presence of palladium acetate–triphenylphosphine complex gave 5-vinyluridine.[234] The cytidine counterpart was prepared either by this procedure or by a conversion of the uracil to cytosine *via* the 4-imidazolyl procedure.[234] The 5-ethynyl-2′-deoxyuridine was hydrated by dilute sulfuric acid to give the 5-acetyl derivative in high yield.[240]

A coupling of 5-halogeno-pyrimidine nucleosides with arenes is sometimes successful by photoactivation. Irradiation of 2′,3′-*O*-isopropylidene-5-iodouridine in acetonitrile-benzene gave, after deprotection, 5-phenyluridine **(85)**.[241] Pyrene instead of benzene also gave the strongly fluorescent 5-pyrenyluridine.[241] A coupling of the 5-iodo-2′-deoxyuridine with benzene or toluene by photo-activation gave similar results.[235] Of special interest are the coupling reactions of 5-bromouridine and tryptophan derivatives by photoirradiation[242] in various conditions to give **(86)**. This may be the consequence of the cross-linking of protein and DNA containing 5-bromouracils under photoirradiation. The photocycloaddition of vinyl acetate to the 5,6 bond of pyrimidines has been reported in many cases and a conversion of 5-fluorouracil to 5-acetonyluracil with vinyl acetate is one example.[243] As recently reported, treatment of trimethylsilylated 2′-deoxy-5-iodouridine with organozinc-olefines in the presence of Pd(II) in tetrahydrofuran afforded the 5-vinyl derivatives.[244] The action of phenylzinc complex with the 5-iodouridine also gave the 5-phenyluracil derivatives, though the yield was low.[245] Photolysis of 5-iodouridine and 2′deoxy-5-iodouridine with allyltrimethylsilane in aqueous acetonitrile gave 5-allyluridine and 2′-deoxy-5-allyluridine, respectively.[246]

85   86

Another recently developed useful reaction is the "electrophilic" substitution at the position 6 of uridine or cytidine *via* lithiation. Treatment of fully trimethylsilylated derivatives of cytidine or 2′-deoxycytidine with butyllithium at −45°C, followed by treatment with methyl iodide or carbon dioxide, gave 6-methylcytidines or cytidine-6-carboxylic acid.[247] The lithiation of protected 5-bromouridine or 2′-deoxyuridine, followed by treatment with methyl iodide, gave a mixture of 5- and 6-methyluridines.[248,249] Lithiation of 3′,5′-di-*O*-trimethylsilyl-5-iodo-2′-deoxyuridine with butyllithium was followed by treatment with chlorotrifluoroethene, perfluoropropene, and perfluorocycloalkenes to give the respective 5-ethenyl, 5-propenyl, and 5-cycloalkenyl derivatives, though the yields were low.[250] Although Pichat *et al.* observed that the lithiation of fully trimethylsilylated uridine by butyllithium occurs both at the 5 and 6 positions,[251] Tanaka *et al.* found a selective alkylation *via* the 6-lithiation. Treatment of 2′,3′-*O*-isopropylideneuridine with lithium disopropylamide in terahydrofuran at −78°C and the subsequent addition of an alkyl halide, resulted in the formation of 6-alkyl-(methyl to *n*-butyl) uridines (**87**).[252] Similarly, the reaction of aromatic aldehydes with 2′,3′-*O*-isopropylidene-5′-*O*-methoxymethyl-6-lithiouridine gave the 6-hydroxybenzyl derivative, which was converted to the 6-aroyl derivatives (**88**) by oxidation with manganese oxide.[253] The reaction of diphenyl disulfide with 6-lithiouridine gave the 6-phenylthiouridine.[253] This compound was converted to various 6-alkylthio derivatives by treatment with alkanethiols.[254] In a similar manner, 6-lithio derivatives of some 5-halo- and 5-methyluridines were converted to the corresponding 6-phenylthio and 6-iodo derivatives by treatment with diphenyl disulfide and iodine, respectively.[255] A direct 6-aroylation was also accomplished by treatment of the 6-lithio derivatives with aroyl chloride.[256] The lithiation of 5,6-dihydrouridine was followed by the addition of an acyl chloride to give the 5-acyl-5,6-dihydrouridines. This compound was then converted to the 5-acyluridines (**89**) by phenylselenation and successive oxidative elimination.[257]

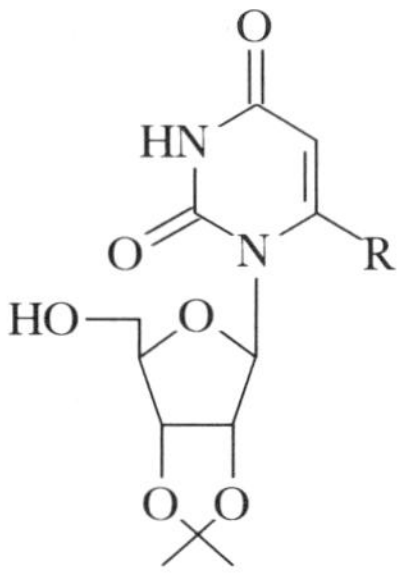

87 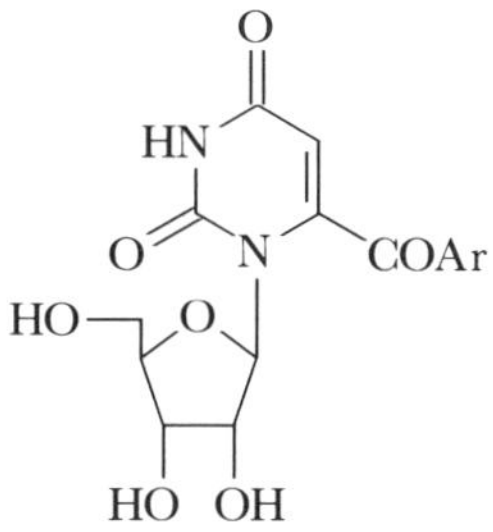   88 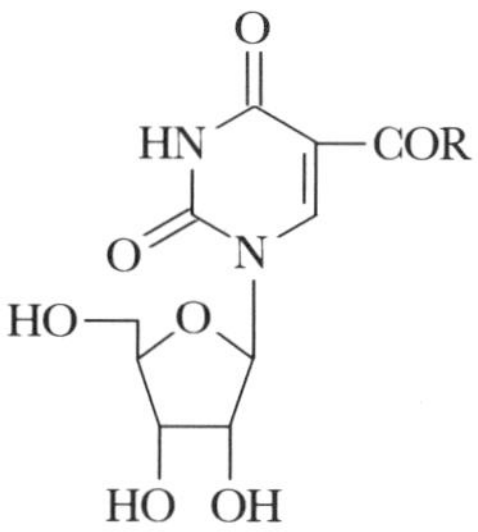   89

2', 3'-*O*-Isopropylidene-5'-*O*-methoxymethyl-6-iodouridine, prepared from the 6-lithio derivative by the above procedure, was phtoirradiated with arenes, such as *N*-phenylpyrrole, to give a highly fluorescent tetracyclic nucleoside (**90**).[258]

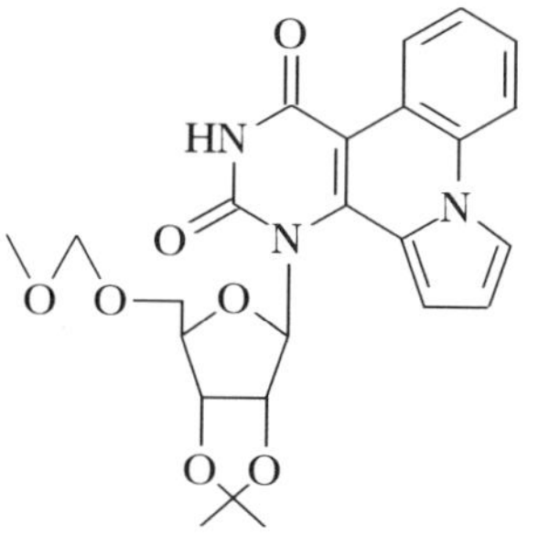

90

3.8. Nucleophilic Additions

As illustrated in structures (**25A**) and (**26A**), nucleophiles tend to attack position 6 of the aglycons. Some typical examples are given in the following discussion.

5'-Deoxy-5'-mercapto-2', 3'-*O*-isopropylideneuridine (**91**) exists as the 6,5'-epithio form (**92**).[259,260] As is evident by making a framework molecular model, the fixation of the 2', 3'-*cis*-diol by an isopropylidene group results in a favorable conformation ($O^{1'}$-exo sugar-puckering) in which the 5'-substituent is sterically accessible to add to the C-6 (as well as to C-2). The 5'-mercapto group in the 2',3'-diol form of (**91**), or of thymidine, adds very slowly at or below pH 7,[261] since these do not possess such $O^{1'}$-exo sugar-puckering. A similar intramolecular adduct formation (**93**) was observed for 5'-amino-5'-deoxy-2', 3'-*O*-isopropyl-

ideneuridine, in which the acidic conditions cleave the epimino cyclo linkage.[262]
This type of participation of the 5′-hydroxyl group was also suggested to occur in
many nucleophilic substitutions at C-4 as well as electrophilic or nucleophilic
substitutions at C-5, which are discussed in Section 3.9. 2′,3′-O-Isopropyl-
ideneuridine is also converted to the 6,5′-O-cyclodihydrouridine (**94**) *via* the
6-methoxy-5,6-dihydro derivative by acid treatment.[263] The photoirradiation of
2′-deoxyuridine also affords the O^6,5′-cyclodihydrouridine along with the normal
5,6-hydrate.[264] As is well known, the hydroxylic solvents add to the 5,6 double
bond of uridines by the photoirradiation.[265,266]

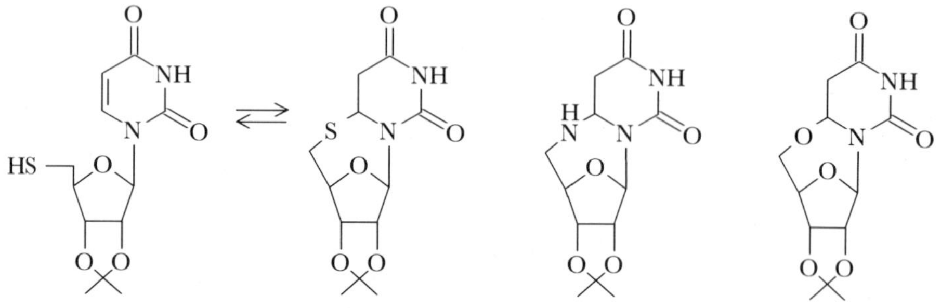

91   92   93   94

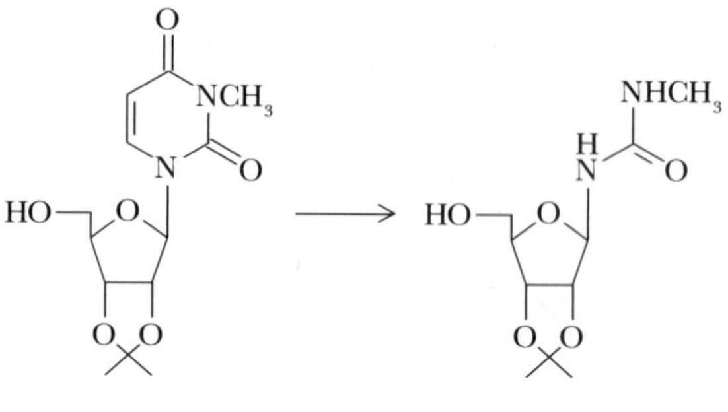

95   96

The rate of hydroxymethylation of 2′,3′-O-isopropylideneuridine at C-5 is
accelerated by the free 5′-hydroxyl group.[267] The rate of isotope exchange of the
proton at C-5 was also accelerated.[267] The extreme rapid alkaline degradation of
2′,3′-O-isopropylidene-3-methyluridine (**95**) to furnish the crystalline ribosylurea
derivative (**96**) is another example of participation of the 5′-hydroxyl to the 5,6
double bond.[268] In fact, protection of the 5′-hydroxyl group as the 5′-methoxy
group in (**95**) decreased the rate of degradation in a striking fashon. Of course, the
instability of 1,3-dialkyluracils in an alkaline medium as compared to that of
1-alkyluracil had been observed many years ago.[269]

Treatment of uracil and uridine with sodium bisulfite at pH 7 gave the
diastereomeric 5,6-dihydrouridine-6-sulfonate (**97**).[270,271] This compound was
fairly stable and could be isolated as the sodium salt. On acidification, it reverted
to the starting material. During this equilibrium the hydrogen at C-5 was

replaced with deuterium when the reaction was carried out in deuterium oxide solution. Treatment of the 5,6-dihydrouracil 6-sulfonate with sodium borohydride afforded, as generally found in the 5,6-dihydrouracils, a ureidopropanol-3-sulfonate.[272] The reaction of 5-chloro-, 5-bromo-, or 5-iodouracils[273] and 5-bromouridine[274] or 5-bromo-2′-deoxyuridine[275] with bisulfite gave uracil or its nucleosides (**99**) as a result of the reductive dehalogenation of the 5,6-dihydro-5-halo-6-sulfonate intermediate (**98**). Precise studies of this elimination by NMR measurements showed that the elimination should occur when the configuration of the 5-halo and 6-sulfonate groups is trans.[276] Decarboxylation of the 5-carboxy moiety on the uracil portion of polyoxins was also assisted by the addition of bisulfite to the 5,6 bond.[277] The equilibrium of the bisulfite addition to thymidine lies in the direction of the starting system.[272,278]

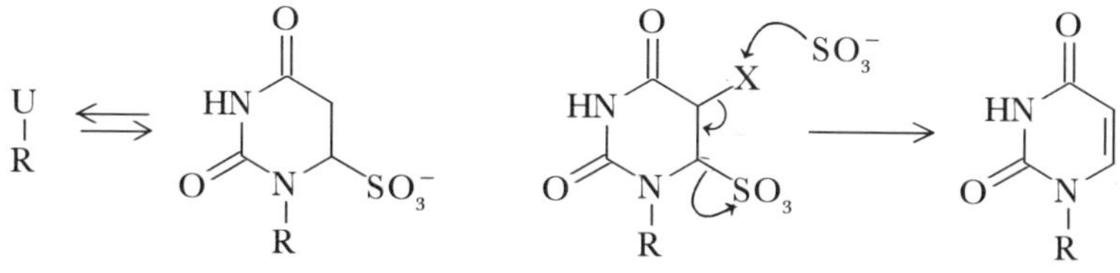

97   98   99

Pseudouridine gave a stable adduct in solution with a large excess of bisulfite, which on alkaline treatment, underwent degradations rather than reverting to the starting nucleoside.[278]

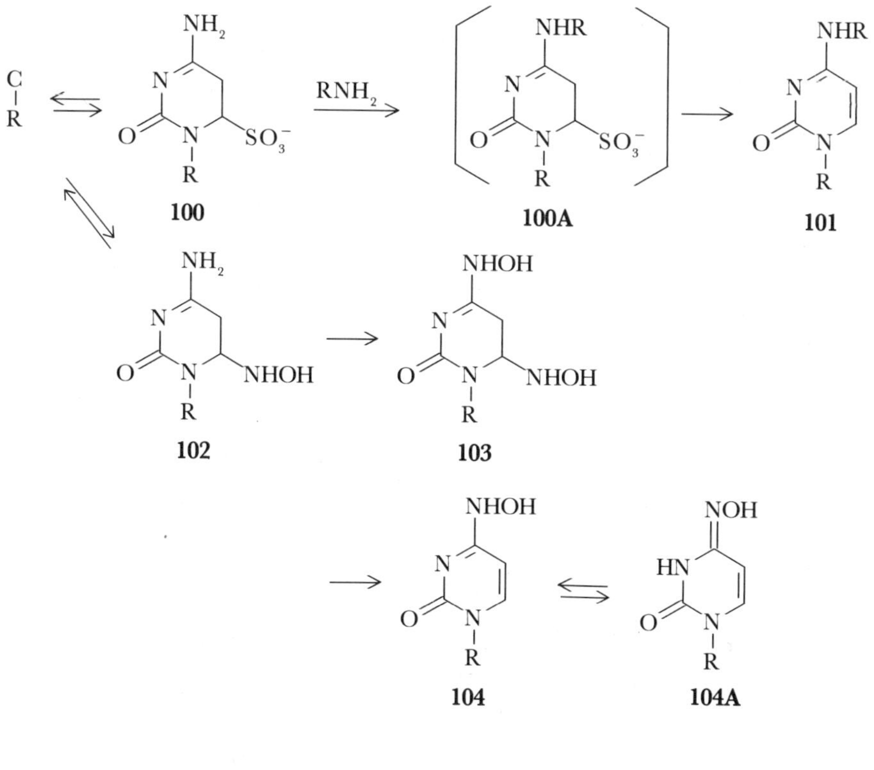

100   100A   101

102   103

104   104A

Cytosine, cytidine, and cytidylic acids react very readily with bisulfite to yield the 5,6-dihydrocytosine-6-sulfonate derivatives (**100**).[270,271] At this stage, the deamination takes place rapidly giving the dihydrouracil-6-sulfonates (**97**). The mechanistic studies of this deamination has been presented.[279,280] 5-Methyl-2'-deoxycytidine undergoes bisulfite addition much slower, as expected.[271b,279] In the dihydro intermedidate (**100**), ready exchange amination takes place to give N^4-substituted intermediates (**100A**), which are then converted to N^4-substituted cytidines (**101**) with various amines including aniline and hydrazine,[281] dimethylamine and glycine,[282] methoxyamine,[283] and semicarbazide.[284] As recently reported, N^4-amino-2'-deoxycytidine, prepared by the improved exchange amination, exhibits strikingly strong mutagenic action.[285] It is interesting to note that the 4-hydrazino group can be eliminated oxidatively to give the pyrimidin-2-one nucleosides by treatment with silver oxide or manganese oxide.[286]

This bisulfite reaction can also be applied to the modification of uracil and cytosine moieties in polynucleotides and this has been described in some detail.[287,288] It is noted that the mode of the reaction of bisulfite at low concentrations (10^{-2} M), in the presence of oxygen, is a radical reaction. Using these conditions, 4-thiouridine was converted to the 4-sulfonate derivative.[289]

Hydroxylamine is known to be a chemical mutagen.[290] Studies on the action of hydroxylamine to nucleobases have been conducted extensively. Cytidine gives N^4-hydroxycytidine (**104**) by two routes.[291] Treatment of cytidine with hydroxylamine at pH 6.5 gave the bishydroxylamino derivative (**103**) *via* the monoadduct (**102**), which on heating with acid yielded (**104**).[292] The direct substitution of hydroxylamine at C-4 of cytidine leading to (**104**) was also observed, especially in the case of 5-methylcytidine.[293] Compound (**104**) exists as the tautomer, uracil-4-oxime (**104A**), which should behave like thymine in the transcription of the genetic code, thus resulting in a mutation. *O*-Methylhydroxylamine reacts similarly with cytidine at pH 4–5 to give the N^4-methoxycytidine and bismethoxyamino derivative.[294] Other alkoxyamines react similarly.[295] Photoactivation of N^4-hydroxy- and N^4-methoxycytidines resulted in the formation of cytidine with a cleavage of the N⁴—O bond.[296] Uridine does react with hydroxylamine, at a higher pH ($\geqslant 10$), to yield an isoxazolone (**105**) and ribosylurea (**106**). The latter was further converted to ribose oxime by excess hydroxylamine.[297]

Similar pyrimidine ring contraction had been observed in the hydrazinolysis of uracil or cytosine nucleosides and nucleotides. This gives the pyrazolones or aminopyrazole, respectively, while DNAs give "apyrimidinic acid" by treatment with hydrazine.[298] RNAs are also degraded by hydrazine.[299] These reactions are now used in the sequence determination of DNA (the Maxam–Gilbert procedure[300] and RNA (the Peattie procedure[301]).

Uridine and 4-thiouridine react with propionhydroxamic acid to give the 4-O-hydroxamate. This reaction is specific to uridine and did not occur with cytidine.[302]

A number of "direct" exchange aminations of the 4-amino group of cytidine derivatives have been documented. In all cases, the nucleophiles involved in the exchange at C-4 could also attack the C-6 at the same time, and in such a 5,6-dihydro intermediate, the exchange at C-4 is enhanced. The 4-amino group in 5,6-dihydrocytidine is in fact very labile and is rapidly hydrolyzed to dihydrouridine. Hydrogenation of cytidine, in the presence of glycine at pH 9.8, gave 5,6-dihydro-N^4-carboxymethylcytidine in high yield.[303] Hydrolytic deamination of cytidine derivatives has been demonstrated,[304,305] and a much higher rate of hydrolysis for arabinofuranosylcytosine to arabinofuranosyluracil in comparison to the ribosyl derivative is presumably due to a participation of the 2'-hydroxyl group at the C-6 position.[306]

The direct exchange amination (or transamination) of cytosine,[307] cytidine, or 2'-deoxycytidine with various amino derivatives—such as butylamine,[308] semicarbazide,[309] Girard-P reagent,[310] 2,4-dinitrophenylhydrazine,[311] isonicotinic acid hydrazide,[312] acyl hydrazides,[313,314] methylamine and aniline,[315] and various aromatic amines[316]—has been reported. However, these reactions generally require more stringent conditions when compared to the bisulfite-mediated exchange aminations.

Photohydration of the double bond of uridine to give 6-hydroxy-5,6-dihydrouridine has been well known.[317] The photohydrate of cytidine has now been characterized fully.[318] The diastereomers of the uridine hydrate were prepared (by another route[178]) and separated. They epimerize at pH 5 and 30°C, probably by the cleavage and recyclization of the N^1—C^6 linkage. An

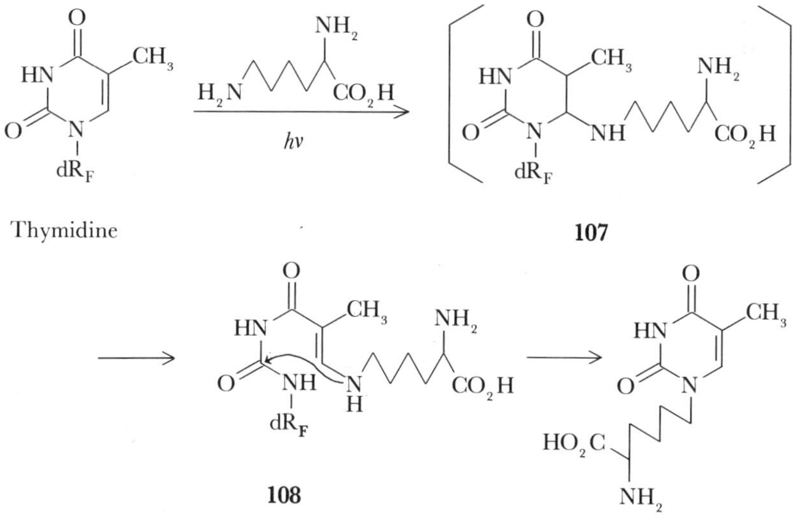

Thymidine   107

108   109

interesting nucleophilic addition of amines to the double bond of thymidine, by photoactivation, was recently reported. Irradiation of a mixture of thymidine and lysine resulted in the formation of 5-amino-5-carboxypentylthymine (109) and deoxyribose.[319] This reaction proceeds by an inital attack of the Ω-amino group of lysine to the C-6 position of thymidine. This intermediate (107) then gave a ring-opened compound (108) that recyclized to afford (109) with elimination of deoxyribosylamine. This reaction was applied to the cleavage of a DNA strand at the thymine moiety. Other amines have reacted similarly to give 1-alkylthymines. The [2 + 2] photocycloaddition of pyrimidine nucleosides with various alkenes is also well known. For studies of the chemical basis of the mutagenic action of coumarins and psoralenes, their [2 + 2] photoadditions to thymidine are reported in recent papers.[320,321]

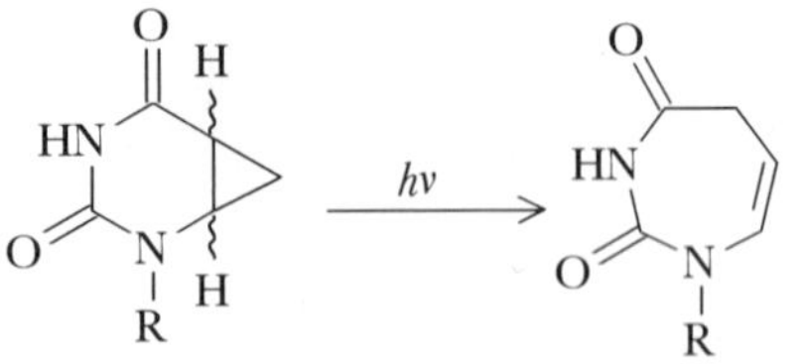

110   111

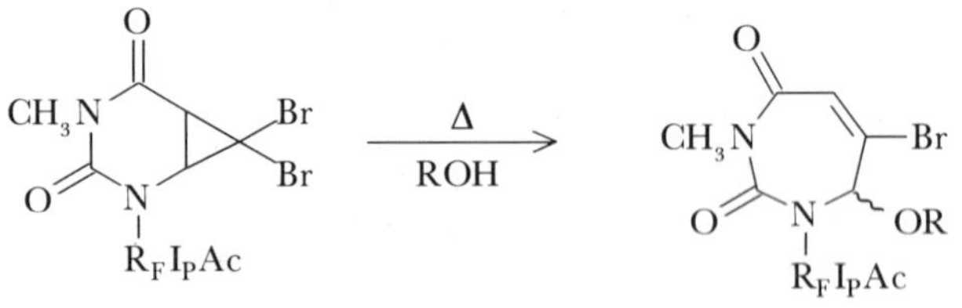

113

The reaction of dimethyloxosulfonium methylide with uridine derivatives afforded the 5,6-dihydro-5,6-methylene derivatives (110). These derivatives were photoisomerized to the corresponding diazepinone derivatives (111).[322] The 5,6-methylene compound was also prepared by the photoaddition of a suitable alcohol with 5'-O-acetyl-2', 3'-O-isopropylideneuridine to give the 6-hydroxyalkyl intermediate. This intermediate was mesylated and cyclized to furnish the expected 5,6-cyclopropane derivative along with the formation of the 6-alkyluridines.[323] The photoaddition of alcohols to the 6 position of uridine has been observed in many cases such as isopropanol.[324] The addition of dihalogenocarbenes to a protected uridine was recently reported to give the 5,6-cyclopropane derivatives (112). These derivatives were then converted to the diazepinedione nucleosides (113) along with a 5-formyluridine by heating in alcohols.[325]

3.9. Reactions of 5-Halogenopyrimidine Nucleosides with Nucleophiles and Related Reactions

The heating of 5-bromouridine (**113**) or 5-bromocytidine with ammonia in a sealed tube gave 5-aminouridine (**115**) or 5-aminocytidine, respectively.[198] Similarly, the reaction of 5-bromouridine with dimethylamine[61] and morpholine[199] gave the respective 5-aminouridines. Heating 5-bromo-2'-deoxyuridine with the appropriate amine also gave 5-amino,[152] 5-methylamino, 5-dimethylamino,[326,327] and the 5-ethylamino[327] derivatives of 2'-deoxyuridine. 5'-Deoxy-5'-fluoro-5-bromouridine under similar conditions also gave the appropriate 5-aminouridines.[328] 5-Bromo-2'-deoxycytidine gave the 5-methylamino derivative.[329] Apparently, the 5-bromo group in these nucleosides has been involved in a direct substitution by amines. However, in view of the general mechanism of aromatic nucleophilic substitution, the halogeno group at the C-5 position in uracil or cytosine moieties should be on the least-reactive position when compared to a halogeno group in positions 2, 4, or 6. If the initial attack were at C-5, the σ complex formed should be rather destabilized. Therefore, a different mechanism must operate for this facile substitution.

114   115

As described in Section 3.8, numerous examples have been recorded for the high-reactivity C-6 of pyrimidine nucleosides toward nucleophilic addition. If an electron-withdrawing group is located at C-5 (as in the case of 5-halogenopyrimidine nucleosides) such tendencies should be enhanced. Once the nucleophile attacks at C-6, the C-5 halogeno group should now become a very active leaving group, as in the α-haloamide system, for aliphatic nucleophilic substitution or elimination reactions. Such a sequence of reactions has now been established and is described next.

Otter and Fox[330] observed an interesting ring contraction of a 5-fluorouracil derivative to afford an imidazoline carboxylic acid compound. Arabinofuranosyl-5-fluorouracil (**116**), when treated with 0.1 N sodium hydroxide at 60°C for 30 min., gave the ring-opened O^6, 2'-cyclo nucleoside (**117**). This product (**117**) was recycled by further alkaline treatment to afford 1-β-D-arabinofuranosyl-2-oxo-4-imidazoline-4-carboxylic acid (**118**). A similar ring contraction to afford an imidazoline derivative has been observed using 1-β-D-arabinofuranosyl-5-bromouracil as the starting material. However, treatment of the same starting material with sodium methoxide in methanol gave O^6, 2'-cyclouridine (**119**).[330b] 5-Bromo(fluoro or iodo)-2',3'-O-isopropylideneuridine (**120**) also behaved in a similar fashion. Treatment of (**120**) in alkaline medium produced five products,

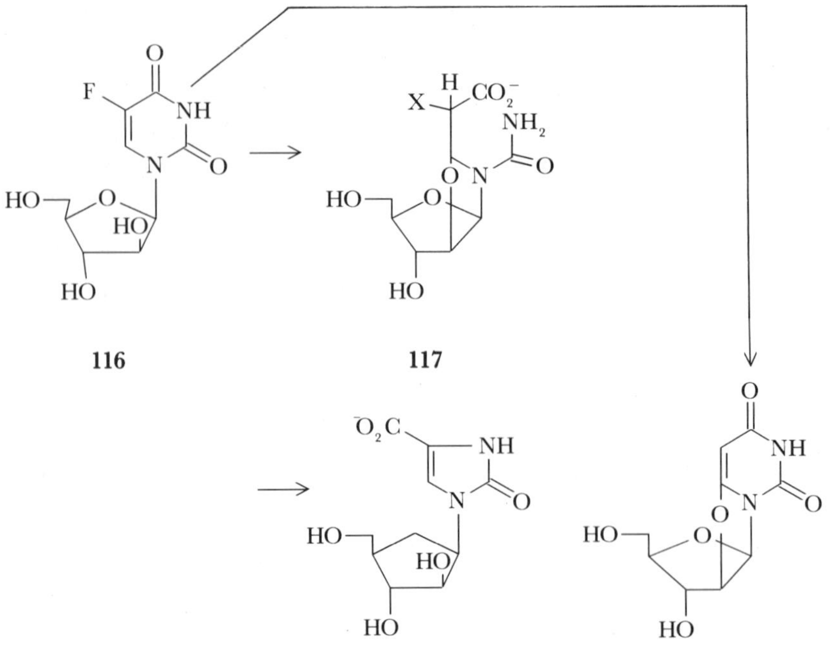

116
            117

118
            119

(42) and **(121)**–**(124)**.[331] In these transformations, a participation of the 5′-hydroxyl group may not be essential for the formation of all products. In fact, it was found in later studies[332] that 2′,3′-O-isopropylidene-5-hydroxyuridine was converted to an imidazoline derivative **(124)** without a participation of the 5′-hydroxyl group. 1-Methyl-5-hydroxyuracil was rearranged in a similar manner. The mechanism operating in this transformation seemed to be that of a benzilic acid rearrangement as shown in the sequence **(122)**–**(124)**, through the intermediate **(122A)**, **(122B)**, and **(122C)**.[332] The formation of 5-hydroxyuridine from 5-bromouridine had been substantiated by the conversion of 5-bromouracil to 5-hydroxyuracil by mild base treatment.[333]

In the formation of 5-hydroxyuridine **(122)** from **120**, 5-bromo-6-hydroxy-5,6-dihydrouridine or O^6, 5′-cyclo-5-bromo-5,6-dihydrouridine **(120A)** could be the intermediate. As described, the 5-bromo group in either intermediate becomes very reactive toward nucleophiles, in the present case a water molecule, leading to the 5,6-dihydroxy-5,6-dihydro intermediate, which on dehydration gave 5-hydroxyuridine. Related reactions involving the formation of 5-hydroxyuridine have been discussed in Section 3.6. If this is the case, the use of nucleophiles other than water should give other 5-substituted pyrimidine nucleosides. This premise was tested in a reaction of the 1-substituted 5-bromouracil derivative **(125)** with sodium hydrogen sulfide to give 5-mercaptouracil derivative **(126)**.[334] In the case of a sodium hydrogen sulfide reaction with **(125)**, the formation of some debrominated 1-substituted uracils **(127)** was observed.[334] A similar reaction of 5-bromo-2′-deoxyuridine with cysteine, to afford S-(2′-deoxyuridin-5-yl)cysteine

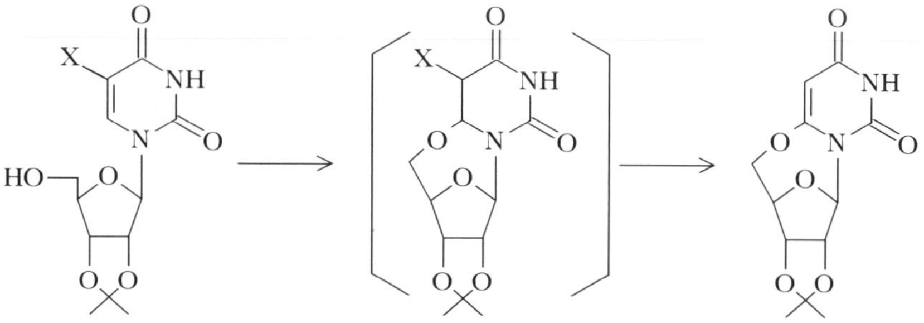

120   **120A**   **121**

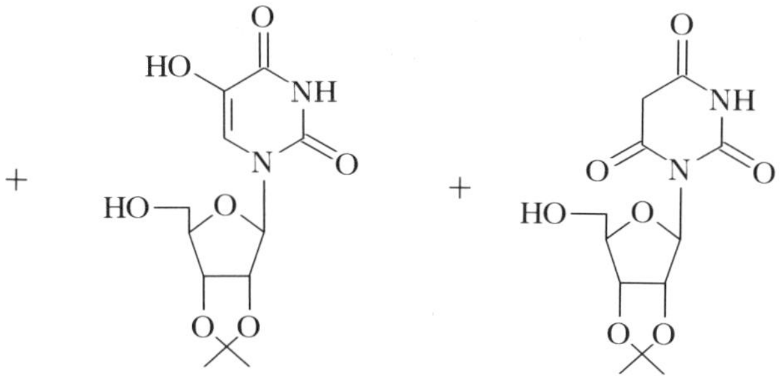

122

123

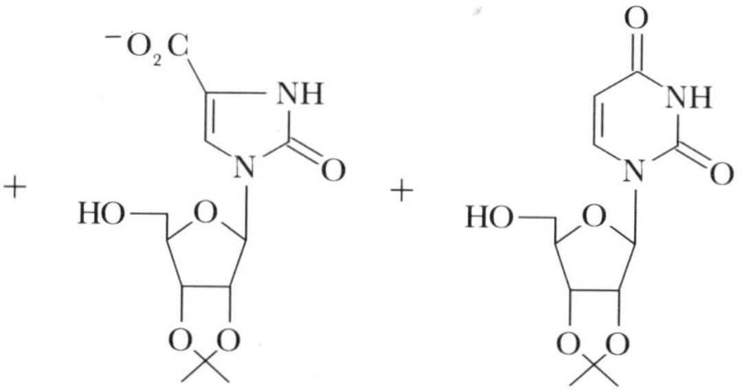

124   **42**

(**126**), was reported.[335,336] The formation of the uracil derivative (**127**) in these reactions was explained by the elimination of dihydrogen disulfide from the 5,6-dithiol intermediates (**125B**).[334,336,337]

On the other hand, it has been demonstrated[331b,338] that the treatment of 2′,3′-*O*-isopropylidene-5-bromouridine with ethoxide in ethanol at reflux gave the *O*⁶,5′-cyclouridine derivative (**121**).[331b] Treatment of 5-iodocytidine (or 2′-deoxycytidine) with strong base such as *t*-butoxide or tetrabutylammonium

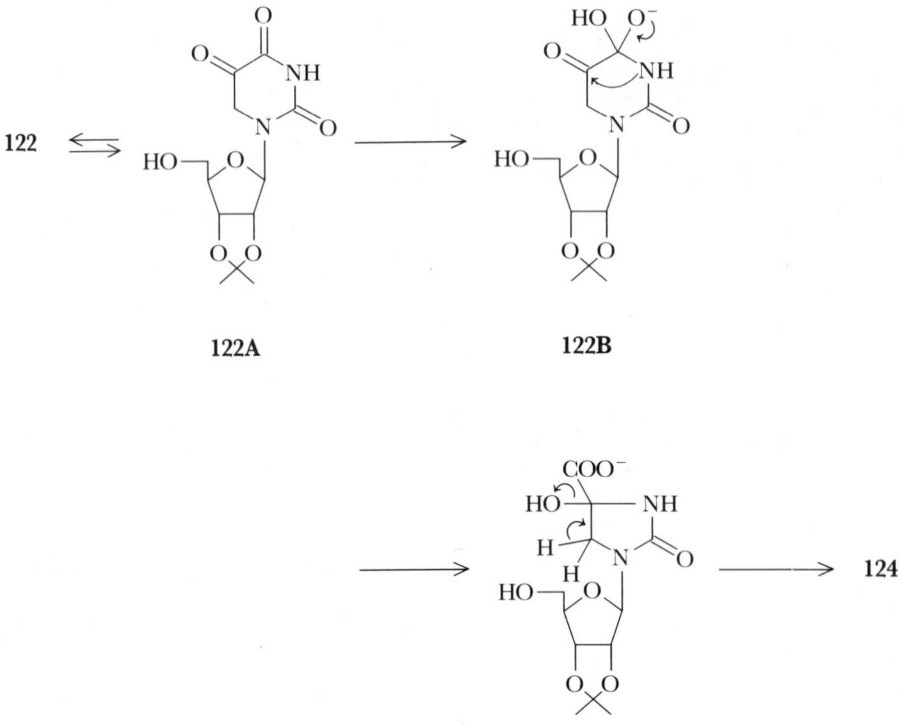

122A   122B

122C

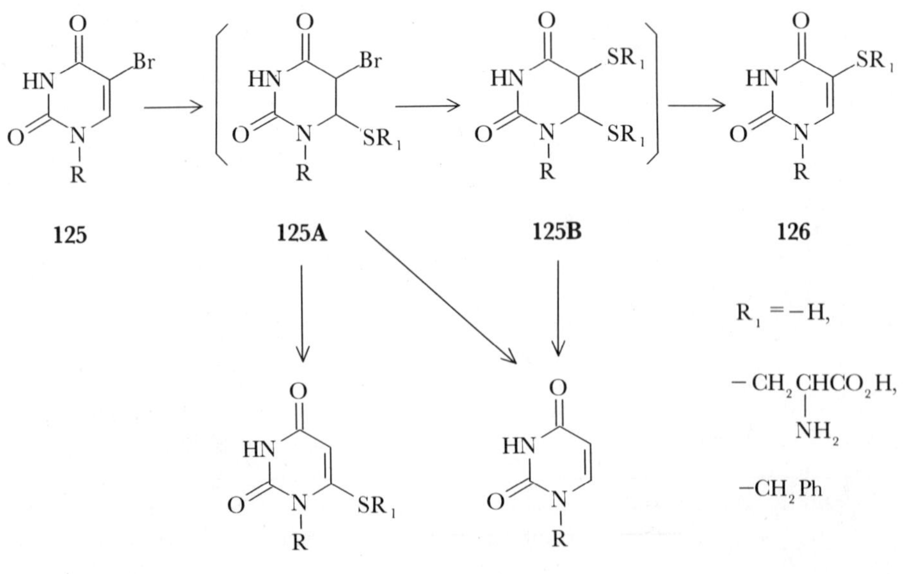

125   125A   125B   126

$R_1 = -H,$

$-CH_2CHCO_2H,$
$\quad\quad NH_2$

$-CH_2Ph$

128   127

hydroxide in dimethylsulfoxide gave $O^6,5'$-cyclocytidine.[338] These transformations show that in the intermediate, for example, (120A) and (125A), the elimination of hydrogen bromide was possible, depending on the nature of the 5-halogen, the 6-substituent, the strength of the alkali used, or the temperature of the reaction. Therefore, the preparation of 6-substituted pyrimidine nucleosides from 5-halogenopyrimidine nucleosides should be possible, not only in the above-mentioned intramolecular reaction but also in some intermolecular reactions. The reaction of 5'-O-acetyl-2',3'-O-isopropylidene-5-bromouridine (125) with ben-

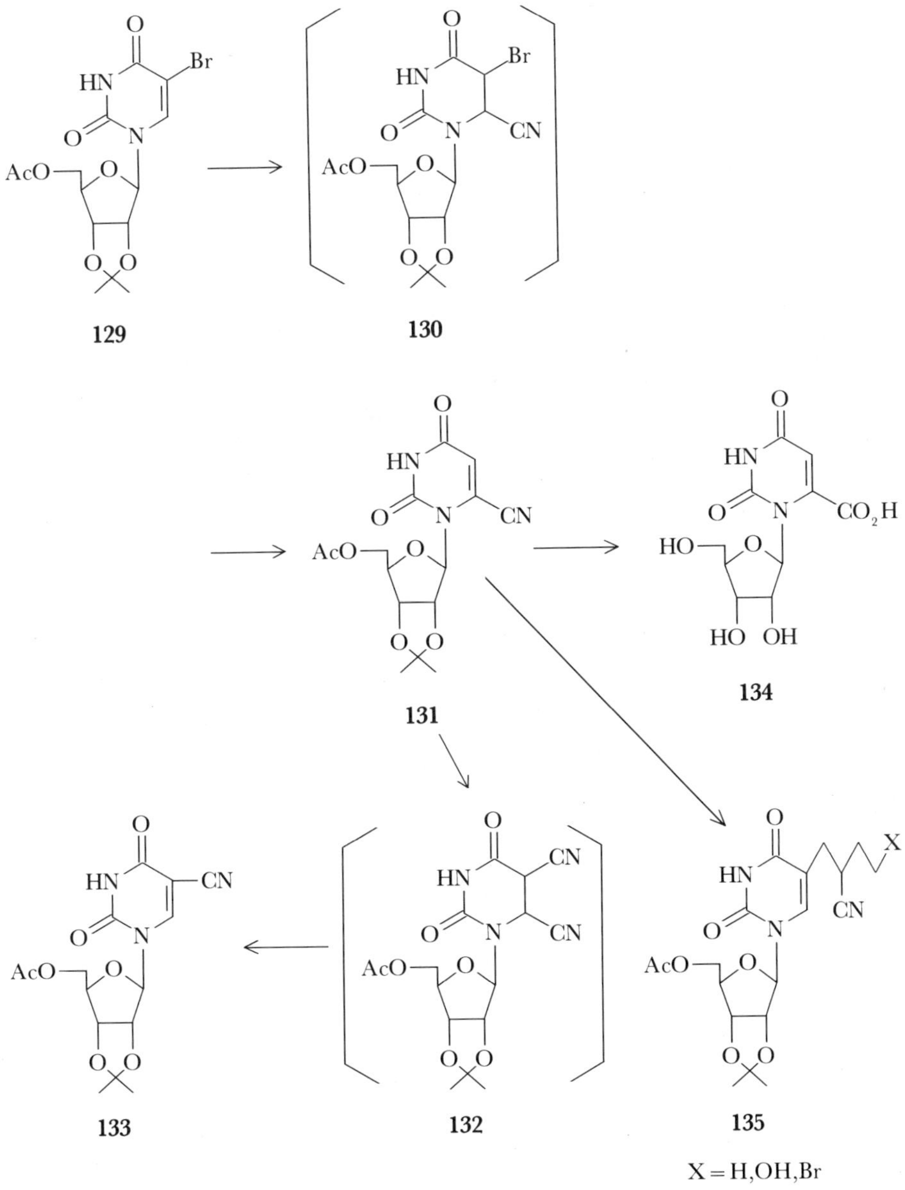

X = H,OH,Br

zylmercaptan in pyridine furnished the 6-benzylthiouridine derivative [(128), R_1 = benzyl] in addition to the expected 5-benzylthiouridine [(126), R_1 = benzyl].[339] In this reaction, formation of the dehalogenated derivative (127) was negligible. From this result, together with that already described above,[334] it is assumed that the initial adduct (125A) is a common intermediate for (126) or (128) and the second intermediate (125B) gives (126) exclusively and not (128).

A very good example of the formation of 6-substituted derivatives from 5-halogenopyrimidine nucleosides was illustrated in the reaction with cyanide ion.[340-342]

Treatment of 5'-O-acetyl-2',3'-O-isopropylidene-5-bromouridine (129) with sodium cyanide in dimethylformamide at room temperature gave almost exclusively the 6-cyanouridine derivative (131) in high yield. The compound (131), when treated with sodium cyanide at an elevated temperature, was converted to the 5-cyanouridine derivative (133).[341,342] Similarly, acylated 5-bromo or 5-iodo-2'-deoxyuridine afforded 6-cyano-2'-deoxyuridine[343] and 5-cyano-2'-deoxyuridine,[342] respectively. The sequence of the reaction was explained by the repetition of trans addition–elimination reactions. The first addition of cyanide to C-6 of (129) to afford (130) makes the hydrogen at C-6 acidic enough to be dehydrobrominated, which then furnishes the 6-cyanouridine derivative (131). Position 5 of (131) is now activated and undergoes a Michael addition by cyanide on heating to yield the 5,6-dihydro-5,6-dicyano intermediate (132), where the C-5 proton becomes more acidic and elimination of hydrogen cyanide then furnishes the 5-cyanouridine (133).[342] Hydrolysis of 6-cyanouridine afforded orotidine[340,342] and this sequence can be applied to the synthesis of orotidylic acid from uridylic acid.[344]

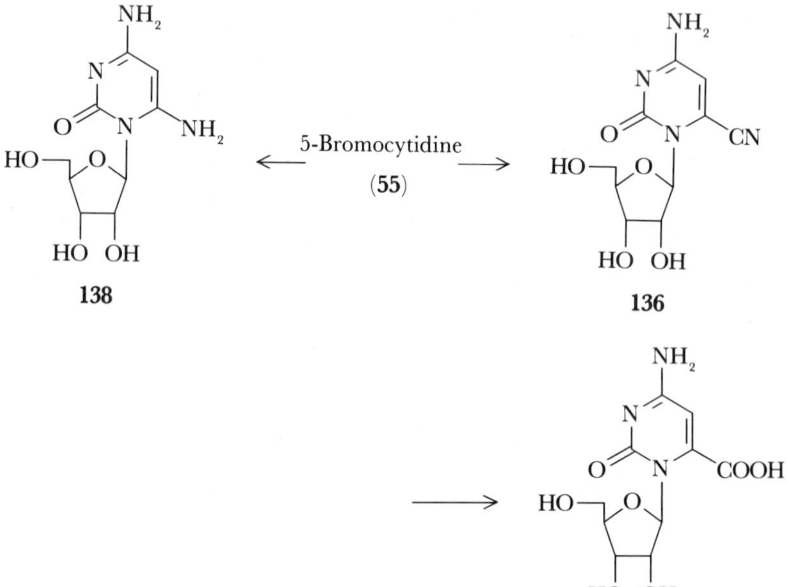

138   5-Bromocytidine   136
                           (55)

                                                          137

Photolysis of (**131**) with the 1-pentene derivatives gave the 5-(2-cyanopentyl) derivatives (**135**). A [2 + 2]-cycloaddition mechanism was presented for this reaction.[345]

The reaction of 5-bromocytidine with cyanide also gave 6-cyanocytidine (**136**),[341,346] which was converted to cytidine-6-carboxylic acid (**137**) by methanolysis followed by hydrolysis. Treatment of 5'-*O*-trityl-2',3'-*O*-isopropylidene-6-cyanocytidine with liquid ammonia or alkoxide gave the 6-aminocytidine or 6-alkoxycytidine derivative, respectively. In these cases, the 6-cyano group behaved as a leaving group.[341] Further transformation of the 6-cyano group of uridine and cytidine afforded various 6-substituted derivatives including the 6-chloromethyl, 6-methyl, 6-cyanomethyl, and 6-carboxymethyl derivatives of uridine[342] and 6-tetrazolyl derivatives of cytidine.[346]

In view of the above studies, the previous mechanism concerning the transformation of 5-bromouridine (**114**) to 5-aminouridine (**115**) should be reevaluated. The reaction is most likely initiated by an addition of ammonia to the 6 position of (**114**) followed by a substitution of the 5-bromo group with ammonia to give the 5,6-diamino-5,6-dihydro intermediate. The subsequent elimination of ammonia from this intermediate would furnish 5-aminouridine. Consequently, in the reaction sequence, it is very possible to form 6-aminouridine by selecting the appropriate reaction conditions. In a preliminary survey on the amination of 5-bromocytidine, the formation of 6-aminocytidine (**138**) has been confirmed.[347] In fact, the same result was recently reported by Goldma and Kalman.[348]

Treatment of 5-aminouridine with nitrous acid gave 5-diazouridine.[198] The structure of this nucleoside has been established as $O^6,5'$-cyclo-5-diazo-5,6-dihydrouridine (**139**), in which the 5'-hydroxyl group was again added to the 6 position because of the strong electronegative property of the 5-diazonium group.[349] An interesting ring contraction of (**139**) during a hydrolysis to afford the ribosyltriazole derivative (**140**) was reported.[350] By using a tracer [(**141**), ^{18}O at 4-C = O] and other model compounds, the mechanism of this reaction has been established as shown in schemes (**141**) to (**142**) and (**143**) in the methanolysis. The C-4 carbonyl oxygen stays on at the 4-carboxamide or 4-methoxycarbonyl group of (**142**) and (**143**), which confirmed this pathway of rearrangement.[350b]

Another interesting example is the reaction of 5-nitrouridine derivatives with azide ion. Treatment of (**144**) with sodium azide in dimethylformamide gave the 8-azapurine riboside (**145**). The reaction proceeded by an apparent attack of azide ion to give (**144A**), and this was followed by an intramolecular cyclization to (**144B**) in which the 5-nitro group was then the leaving group and gave the product (**145**).[351] 5-Nitrocytidine also afforded an 8-azapurine nucleoside by a similar treatment. The intramolecular azide addition reaction to give the triazolopyrimidine cyclonucleoside (**146**), from 5'-azido-5'-deoxy-2',3'-*O*-isopropylidene-5-bromouridine, was recently reported.[352] The other instances of the azide coupling to the pyrimidine portion are described in Section 4.3.

Treatment of common pyrimidine nucleosides with a strong base (10% NaOD in dimethylsulfoxide or NaOMe) abstracts the proton at C-6, as observed by isotope labeling at C-6.[353] In this labeling process, the proton at C-5 was also exchanged by the already described addition reaction. The introduction of a

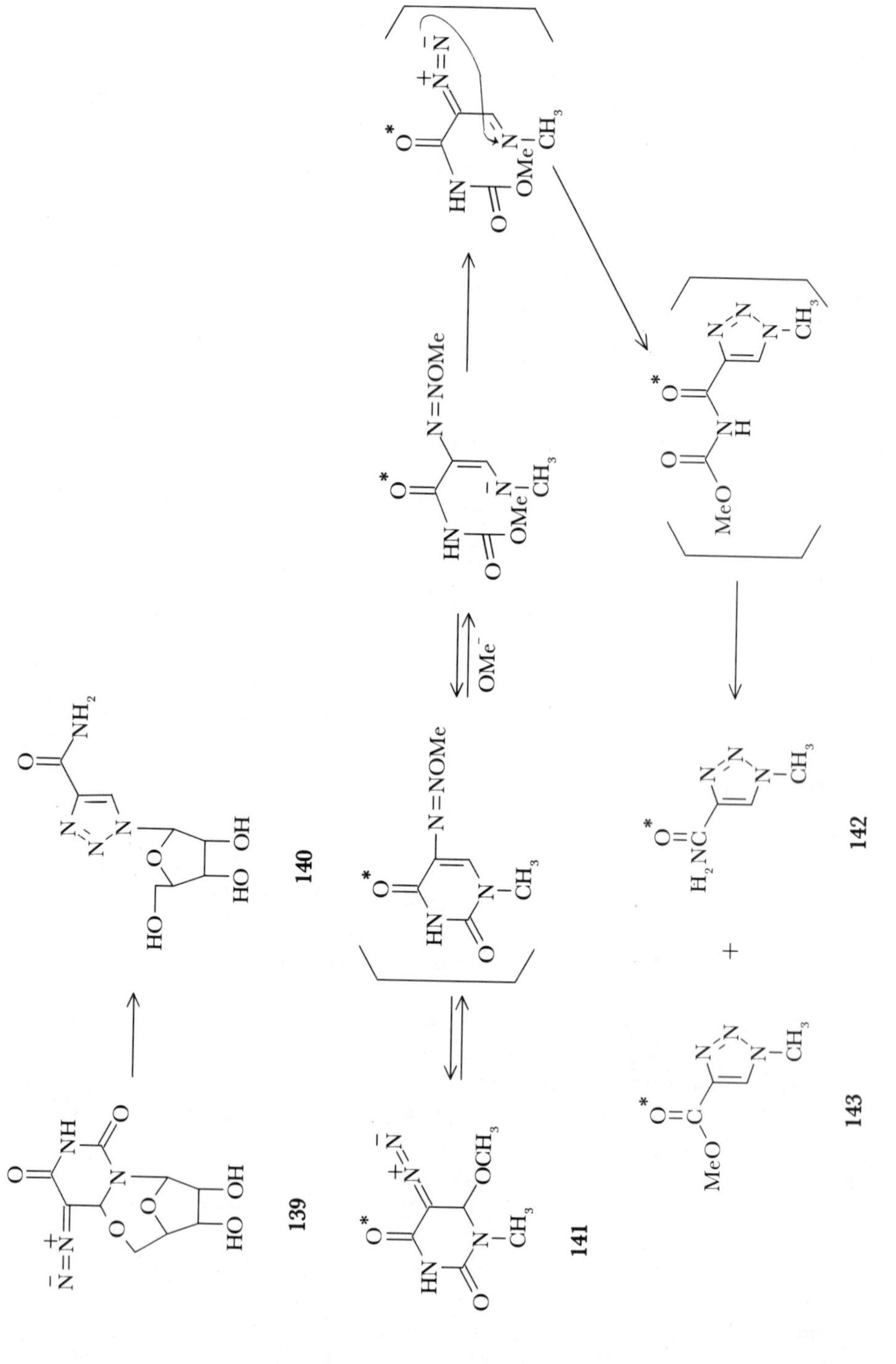
139
140
141
142
143
OMe⁻

carbon unit to the C-6 by lithiation has already been discussed in Section 3.6. The proton exchange reactions of uridine, cytidine, and thymidine have been observed by a platinum-catalyzed reaction with deuterium in deuterium oxide.[354] Uridine and cytidine gave the 5,6-dideuterio derivatives, which can then be converted to the 6-deuterio derivatives. Thymidine gave the 5-trideuteriomethyl derivative exclusively.

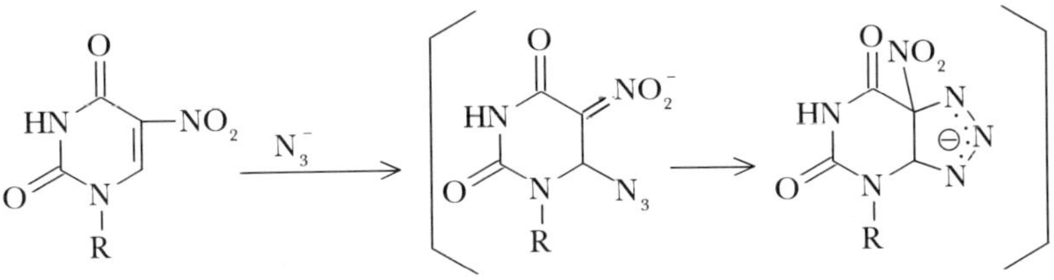

144   144A   144B

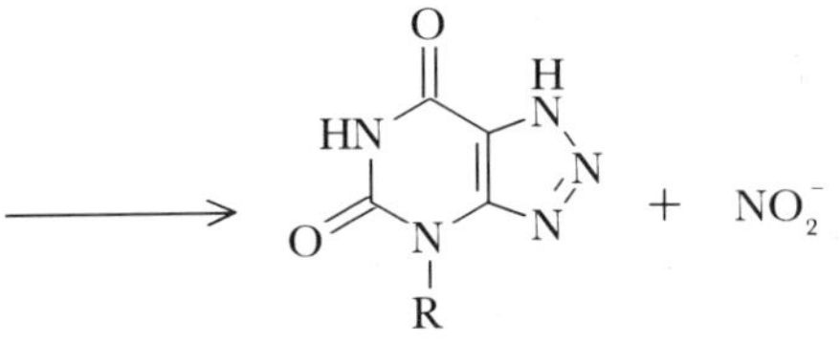

145

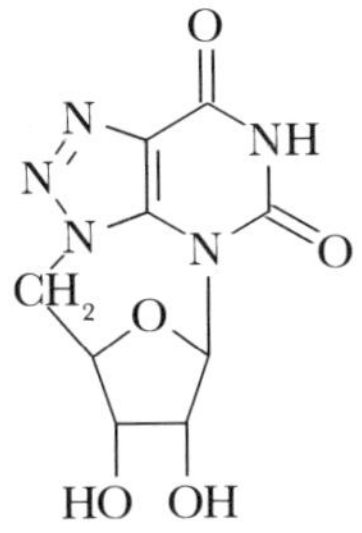

146

3.1.0. Substitution at C-4 and C-2 and Related Reactions

Various examples of nucleophilic substitution at the 4-carbonyl group of uridines and 4-amino group of cytidines have been reported. As described in Section 3.8, the substitution at C-4 of cytidine is often accelerated by a prior addition of the nucleophile at the 5,6 double bond. Deamination of cytidine by nitrous acid treatment gives uridine. However, more nitrogen was evolved than was expected in the deamination of cytosine, probably due to the decomposition of the cytosine ring.[355] Although the solvolysis of cytosine derivatives to afford uracil derivatives proceeds at almost any pH,[316,356] the bisulfite-mediated hydrolytic deamination[270,271] was superior, as demonstrated in the deamination of 6-methylcytidine to 6-methyluridine.[357]

Treatment of 2',3',5'-tri-O-benzoyluridine with phosphorus pentasulfide in pyridine at reflux affords the 4-thiouridine derivative (147) in high yield. The debenzoylation of (147) gave 4-thiouridine, which was then converted to various cytidine derivatives (148).[358] By this procedure, thymidine,[358] 5-fluorouridine and its 2'-deoxy and arabino derivatives,[359,360] 2'-amino-2'-deoxy-5-fluorouridine,[361] and arabinofuranosyluracil[362] were converted to the respective cytidine derivatives. In the substitution of 4-thiothymidine with dimethyl(or diethyl)-amine the N^4-dialkylamino derivatives were obtained in very poor yields. The monodealkylated derivatives were also obtained in this reaction, probably due to the steric hindrance of the expected products.[363]

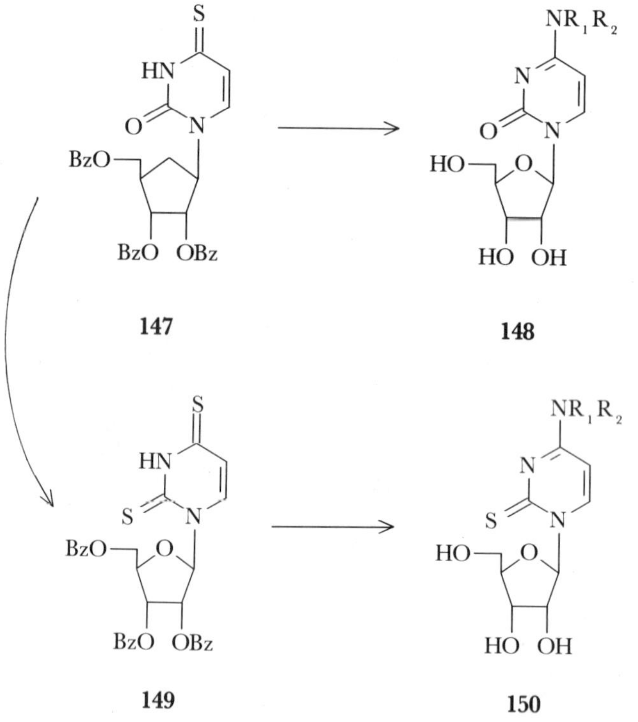

When the 4-thio group was converted to the 4-methylthiogroup a more facile replacement with amines was accomplished.[364,365] The thiation of 2-thio derivatives of arabinosyluracil[366] or thymidine[367] gave the corresponding 2,4-dithio derivatives.

The thiation procedure with solvents other than pyridine was also reported. A mixture of toluene and pyridine was used in the thiation of 5'-O-tritylated uridines.[368,369] The thiation of 3',5'-di-O-benzoyl-O^6,5'-cyclouridine and 3-(2,3,5-tri-O-benzoyl-D-ribosyl)-6-methyluracil, to furnish the respective 4-thio derivatives, was accomplished in dioxane.[370,371] This resulted in a very short reaction time as compared to the standard procedure, and the workup was much simplified. Thiation of tri-*O*-benzoyl-4-thiouridine (**147**) or its 5-methyl derivative with phosphorus pentasulfide in tetralin at 160°C afforded, after debenzoylation, 2,4-dithiouridine[372] (**149**) and 5-methyl-2,4-dithiouridine.[367] Amination of (**149**) afforded 2-thiocytidine and its N^4-methyl and N^4-dimethyl derivatives (**150**), the former being a minor nucleoside from certain tRNAs. 5,6-Dihydrouridine and its 2-thio derivative were also converted to the 4-amino derivatives by the standard procedure.[373,374] 5'-Deoxy-3'-*O*-benzoyl-O^2, 2'-cyclouridine was thiated to furnish the 4-thio-cyclo derivative.[375]

As have been described, cytidine underwent solvolysis with water (to uridine) and amines (exchange amination). The solvolysis of cytidines with liquid hydrogen sulfide was also found to be useful. Treatment of cytidine and its 5'-phosphate (**151**) with liquid hydrogen sulfide–pyridine in a sealed tube afforded 4-thiouridine and its phosphate (**152**) in high yield.[376,377] This method is readily applicable to nucleotides containing cytidine[378] and to polynucleotides.[379] N^3-Methylcytidine, 2-thiocytidine,[377] 2'-deoxycytidine,[380] and N^3-amino-cytidine[111] were also converted to the respective 4-thio derivatives in high yields. In this conversion, a participation of the 5'-hydroxyl group was again observed in the comparison of the rate of sulfhydrolysis of cytidine and its 2',3'-*O*-isopropylidene derivative.[378]

The sulfydrolytic procedure seems to be advantageous in that no prior protection of the sugar moiety is necessary for the reaction to occur, which is especially convenient for the preparation of 4-thiouridine nucleotides from cytidine nucleotides. By a similar principle, selenohydrolysis of cytidine, 2'-deoxycytidine, arabinosylcytosine, isocytidine, and arabinosylisocytosine has been reported to give the corresponding 4-seleno (**153**) or 2-seleno derivatives.[380] 4-Selenouridine was previously prepared by the treatment of 4-chloropyrimidin-2-one riboside with selenourea.[381]

As 4-thiouridine is a unique minor constituent of tRNAs, extensive studies on the chemical modification of the 4-thiouracil moiety have been accomplished. The 4-thiouridines are also versatile intermediates for the synthesis of other pyrimidine nucleosides. Treatment of 4-thiouridine with cyanogen bromide gave the 4-thio-cyanato derivative. This selective reaction was adapted for the modification of the 4-thiouracil moiety of tRNAs.[382] The course of the reaction of cyanogen bromide with 4-thiouridine has been studied,[383] which revealed the initial reaction product as the disulfide of 4-thiouridine. This was further converted to the 4-thiocyanato derivative by a reaction with excess cyanogen bromide. The action of ethylenimine with 4-thiouridine gave the 4-aminoethylthio derivative, which was rapidly replaced intramolecularly to the 2-mercaptoethylamino derivative followed by the second attack of ethylenimine to furnish (154).[384,385] Various fluorophores can be introduced by the alkylation of the 4-thio group.[386,387] Alkylation of 4-thiouracil and other thio or seleno derivatives with dialkyl disulfides or diselenides and triphenylphosphine was reported to give the S- or Se-alkyl derivatives.[388] Treatment of 4- or 2-thiouridine with chloroacetaldehyde afforded the S-alkyl derivative, which was degraded to the uracil derivative through the hydroxyethano intermediate.[389]

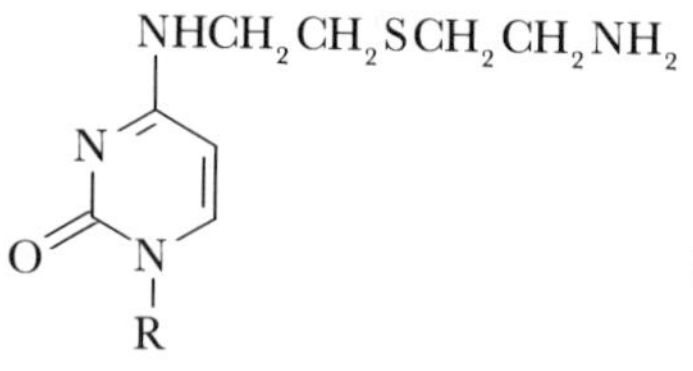

154 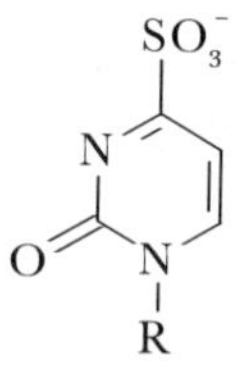   155 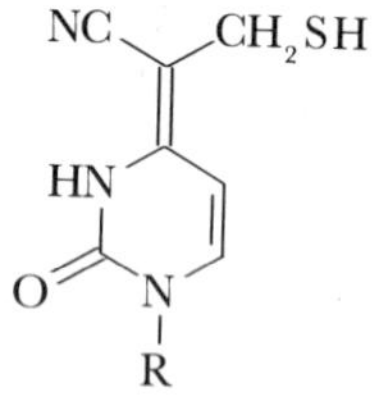   156 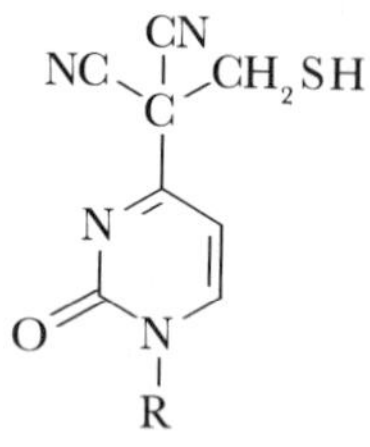   157

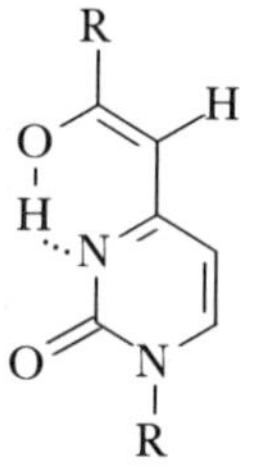

158, a R = Ph   159 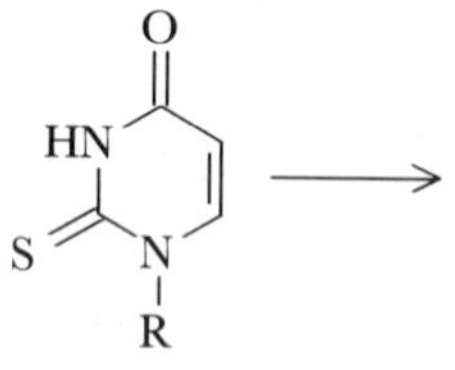   160 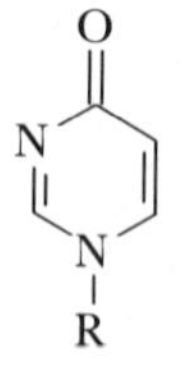   161

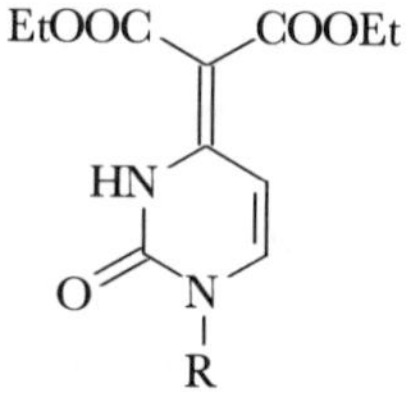

     b R = Me

Oxidation of 4-thiouridine with permanganate[390] or periodate[391] afforded the 4-sulfonate (155), which was a more active leaving group than the thio or methylthio group. Sodium hydrogen sulfite also oxidizes the 4-thio group to the corresponding 4-sulfonate. 2,4-Dithiouridine 5'-phosphates were converted to the 4-sulfonate by this method, which gave the 2-thiocytidine derivatives by ammonolysis.[289,392] By a similar procedure, 2'-chloro-2'-deoxyuridine was converted to the cytidine counterpart *via* the thiation, sulfite reaction, and successive ammonolysis.[393] Treatment of 4-thiouridine with dilute aqueous hydroxylamine at pH 7 gave N^4-hydroxycytidine.[394]

     The photoreaction of a sugar protected 4-thiouridine or 4-thiothymidine with unsaturated nitriles afforded C-4 carbon linked derivatives such as (156) and (157).[395] The sulfide contraction of sugar-protected 4-phenacylthiouridine with triphenylphosphine and *t*-butoxide at elevated temperature was reported to give (158a).[396] Similar reaction was observed by treatment of 4-acetonylthiouridine to give (**158b**).[397] Treatment of the 4-thiouridine derivative with diethyl bromomalonate gave the 4-(diethoxycarbonyl)methylene derivative (**159**).[397] Desulfurization of 2-thiouridine (**160**) to afford the ribosylpyrimidin-4-one (**161**) was observed by a treatment with dipotassium diazenedicarboxylate.[398] The same reaction with 4-thiouridine resulted in the 4-hydrazino compound (or uridine-4-hydrazone).[398] The 4-thiogroup in 4-thiouridine derivatives is generally unstable toward acidic conditions and gives the uracil derivatives. Treatment of 4-(or 2-, or 2,4-di)thiouridine with dimethyl selenoxide also gives uridine.[399]

The well-known conversion of acid amides to imidoyl chlorides by phosphoryl chloride was adaptable to the uracil nucleosides. Treatment of 2′,3′,5′-tri-*O*-benzoyluridine and the arabinosyl epimer (**162**) with phosphoryl chloride and diethylamine hydrochloride in ethyl acetate afforded the 4-chloropyrimidin-2-ones (**163**), which were then converted to the respective cytosine nucleosides (**164**).[400] Treatment of tri-*O*-acetyluridine or tri-*O*-acetylarabinosyluracil with thionyl chloride and dimethylformamide followed by ammonia also gave the ribosyl or arabinosyl cytosines.[401] This thionyl chloride conversion was an adaptation of the established method for converting 6-azauridine to 6-azacytidine.[402]

The one-pot conversion of uridine to cytidine was accomplished by simultaneous protection of the sugar hydroxyl groups and activation of the C-4 by trimethylsilylation, followed by ammonolysis with various amines such as pyrrolidine, morpholine, aniline, dimethylaminoethylamine, and hydroxylamine

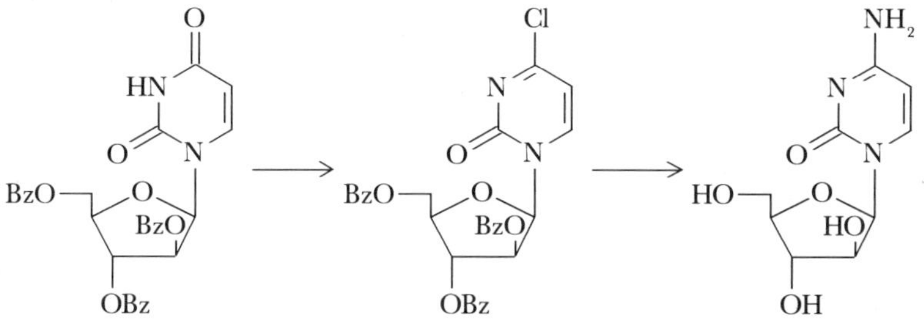

162   163   164

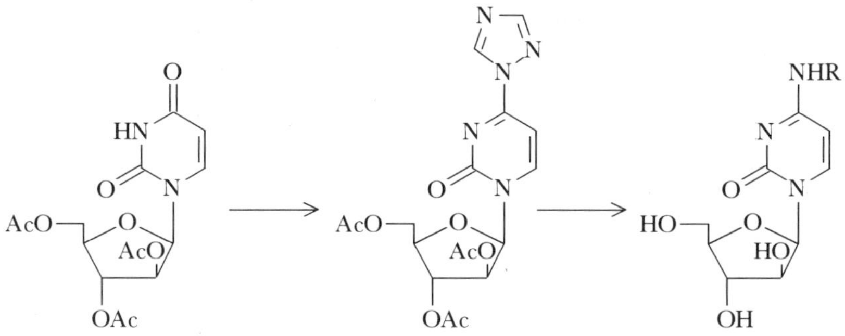

165   166   167

at high temperature.[403] Thymidine, 6-azauridine, 2-thio-6-azauridine, and UMP were likewise converted to the appropriate N^4-substituted cytidine derivatives.[403b]

A recent development in the method of conversion of uridines to cytidines is derived from the studies of oligonucleotide synthesis by Reese and co-workers.[404] Treatment of tri-*O*-acetyl-arabinofuranosyluracil (165) with tri(1*H*-1,2,4-triazol-1-yl)phosphine oxide in the presence of 1,2,4-triazole and triethylamine afforded the crystalline 4-triazolo intermediate (166), which was then converted to the cytidine derivatives (167) by ammonolysis with various amines including morpholine and aniline.[405] Treatment of (166) with *p*-toluenethiol afforded the 4-*p*-tolylthio derivative. The C-4 position was also activated by treating with diphenyl phosphorochloridate and 3-nitro-1,2,4-trazole to give the nitrotriazolo derivative of (166). Action of mesitylenesulfonyl-4-nitrotriazolide with tri-*O*-acetyluridine also gave the 4-nitrotriazolyl uridine.[406] These procedures are now most frequently used in the conversion of uridines to cytidines and 4-substituted 2-pyrimidinone nucleosides, since both the activation and substitution steps proceed under very mild conditions.

3.11. Oxidations

Oxidative reactions of the 5,6 double bonds of pyrimidine nucleosides are well established. Since the oxidation is generally of an electrophilic character, thymidine will be the most rapidly oxidized among the pyrimidine nucleosides. Hayatsu and Ukita established reaction conditions that resulted in the selective oxidative degradation of thymidine and its 5'-phosphate by permanganate.[407] At pH 3.9, permanganate oxidation of thymidine gave a quantitative yield of the 5-hydroxy-5-methylbarbituric acid (168).[408] This compound was also generated by the action of *N*-iodosuccinimide with O^6,5'-cyclothymidine. Compound (168) undergoes further degradation with an excess of permanganate. Osmium tetroxide reacts in a similar fashion[409] and the relative rates of oxidation of various pyrimidine nucleosides were compared with that of thymidine, which is one of the fastest, in 0.4 N ammonia at 0°C. The primary product of oxidation of the pyrimidines with osmium tetroxide in the presence of pyridine is a cyclic osmate ester of the *cis*-5,6-dihydroxy-5,6-dihydro derivative (169).[410,411] In this case the sugar 2',3'-*cis*-diol is also esterified with osmate if it is present.

The oxidation of uracil, thymine, and TMP with hydrogen peroxide and other hydroperoxides proceeded with either a polar or free radical mechanism.

Thymidine ⟶ (60)

168   169   170

Near neutrality, the free radical pathway is the most important, whereas at more alkaline pH the predominant reaction was promoted by the nucleophilic attack of the peroxide ion at C-6 of the pyrimidine, which then goes through the 5,6-epoxide and 5,6-diol derivative as the degradation intermediate.[412] Oxidation of 1,3-dimethylthymine with *m*-chloroperbenzoic acid gave the 5,6-*cis*-diacyloxy derivatives.[413] Oxidation of cytidine with permanganate proceeds by 5,6-hydroxylation and finally leads to the formation of a ribosylurea.[414] The *N*-oxidation of cytidine has already been discussed in Section 3.3.

The degradation of nucleic acid components with ozone has been studied.[415] Thymine and guanine are the bases most rapidly affected by ozone. In the case of 1,3-dimethylthymine, the pyruvylureide was detected among various ozonized products and exists as the hydroxyhydantion (**170**).[416] Therefore, it is certain that ozone attacks the 5,6 double bond to give the expected ozonation product, which is then degraded by further oxidation.

Treatment of thymidine with potassium persulfate in the presence of cupric sulfate and 2,6-lutidine in aqueous acetonitrile afforded 5-formyl-2'-deoxyuridine (**48**).[417] This reagent also oxidized 5-hydroxymethyluridine to afford 5-formyl-uridine.

Oxidation of 2',3'-*O*-isopropylideneuridine and 2',3'-*O*-isopropylidenecy-tidine by lead tetraacetate gave the O^6,5'-cyclo-5-acetoxy-5,6-dihydropyrimidine

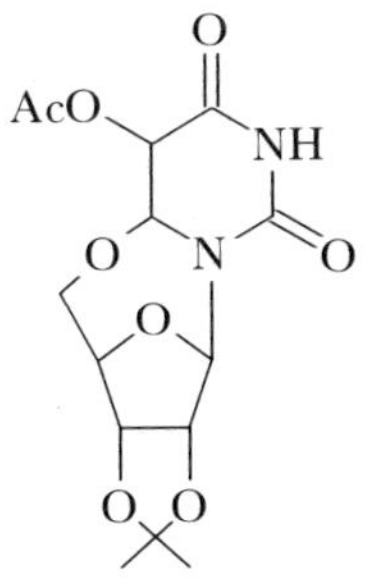

171 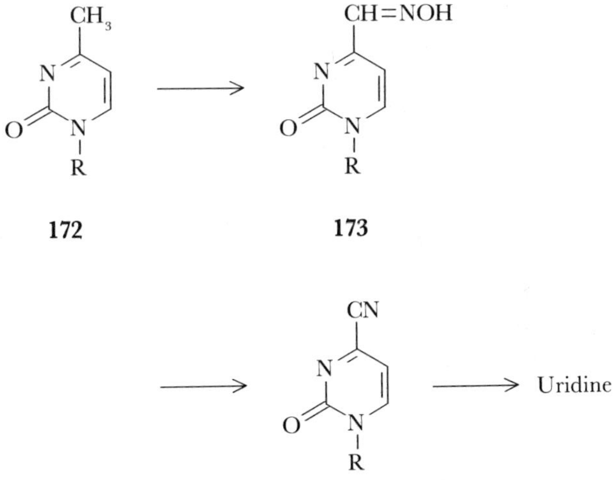   172   173

174

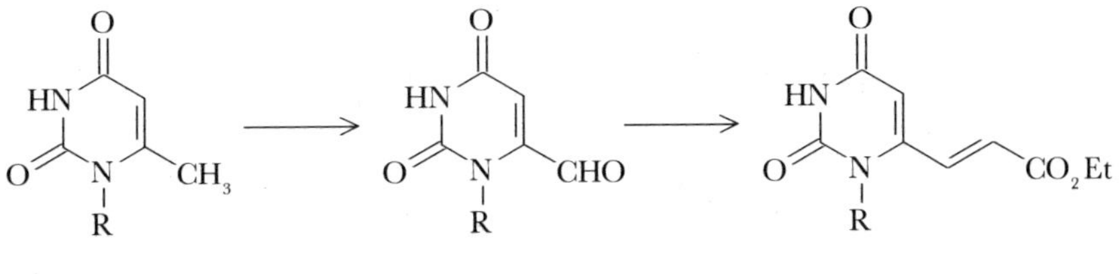

175   176   177

nucleosides (171) in high yields.[418] γ Irradiation of uridine gave the 5,6-hydroxy-hydroperoxide intermediate, which was further degraded to *N*-ribosylformamide. This intermediate was also produced by the treatment of uridine with trifluor-operacetic acid.[419]

Oxidation of methyl groups on the 4 or 6 position of pyrimidine nucleosides has been reported. 4-Methylpyrimidin-2-one riboside (172) is converted to the 4-aldoxime derivative (173) on treatment of (172) with nitrous acid.[371] Compound (173) was then converted to the 4-cyano derivative (174) and subsequently to uridine. The methyl group of 6-methyluridine (175) was oxidized to the 6-formyl derivative (176) by treatment with selenium dioxide. This aldehyde underwent the Wittig reaction with ethoxycarbonylmethylenetriphenylphosphorane to give *trans*-3-(uridinyl)acrylic acid ester (177).[357]

3.12. Reductions

Catalytic hydrogenation of uridine and its phosphates with rhodium on alumina catalyst gave the 5,6-dihydrouridine derivatives (178), which on alkaline treatment followed by acidic hydrolysis gave a ureidopropionic acid (179) and ribose or ribose phosphates. This sequence was used in a determination of the position of the phosphoryl moiety of pyrimidine nucleotides derived from RNA.[420,421] 5,6-Dihydrouridine is often found in tRNAs.[422] Although the hydrogenation of thymidine was reported to give (*S*)-(−)-5,6-dihydro-thymidine,[423] hydrogenation of 2′,3′-*O*-isopropylidene-5-methyluridine on Rh-alumina gave a mixture of the 5(*S*)- and 5(*R*)-5,6-dihydro derivatives, which were separated as the 5′-*O*-tosylate. These were further converted to the O^2,5′-cyclodihydro derivatives.[424] A facile reduction of the 5,6 double bond of uridine was also performed by photoirradiation in the presence of sodium borohydride.[425,426] Cytidine was resistant toward reduction under these conditions. Treatment of 5,6-dihydrouridine with sodium borohydride gave the ring-opened ureidopropanol derivative.[426] Thymidine underwent a similar reduction and reductive ring opening under these conditions. Photoirradiation of uridine in formic acid also yielded 5,6-dihydrouridine.[427]

Hydrogenation of the isocytidine derivatives to obtain the 5,6-dihydro compounds was also accomplished by a Rh catalyst.[428,429]

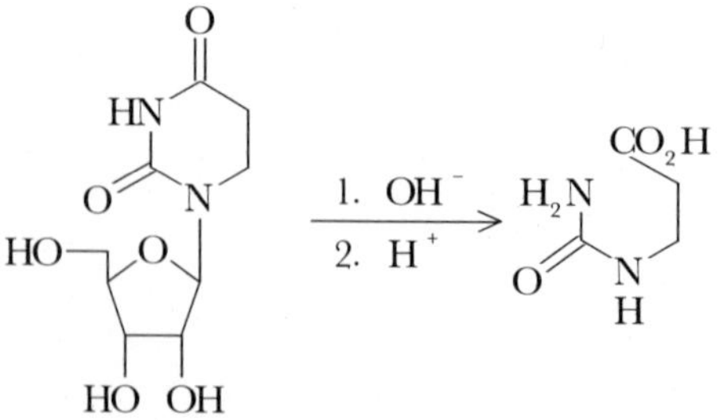

178   179

Hydrogenation of the cytosine ring with platinum oxide and hydrogen gave trimethyleneurea and ammonia.[430] A precise study of the hydrogenation of cytidine and uridine by a Rh catalyst revealed that both uridine and thymidine absorbed 1 mol of hydrogen to give their respective crystalline 5,6-dihydro derivatives. On the other hand, cytidine absorbed 2 mol of hydrogen to yield tetrahydrocytidine (**181**) rather than the dihydrocytidine (**180**). Compound (**181**) was then hydrolyzed to give tetrahydrouridine (**182**), a potent cytidine deaminase inhibitor. The tetrahydrouridine is readily autooxidized to the dihydrouridine. Further hydrogenation of tetrahydrocytidine gives the trimethyleneurea derivative (**183**).[431]

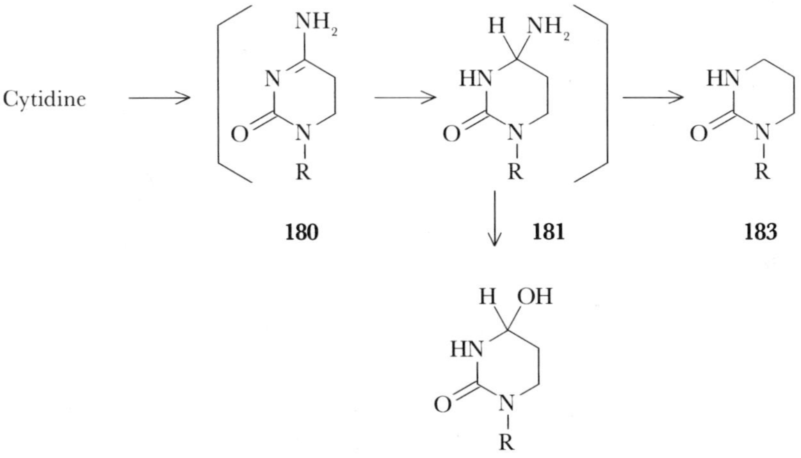

The trimethyleneurea derivative (**183**) has also been obtained by desulfurization of the 4-thiouracil and 4-thiothymine derivatives with Raney nickel catalyst.[432] Hydrogenation of ribofuranosylpyrimidin-2-one with Pd–C catalyst also gave (**183**).[433] Treatment of thymidine and deoxyuridine with sodium amalgam in dilute acetic acid gave the respective pyrimidin-2-one derivatives.[434] The same compounds were obtained in better yields from the 4-hydrazinopyrimidin-2-one derivatives as described by Cech and Holy.[286] Reduction of 5-acetyl-2'-deoxyuridine with sodium borohydride gave the 5-(1-hydroxyethyl) derivative.[435]

4. Synthesis and Reactions of Cyclonucleosides

The reaction in pyrimidine nucleosides where both the aglycon and the sugar moieties have participated in one reaction is generally referred to as "cyclonucleoside formation." In the literature the term "anhydronucleoside" is also currently used. In this chapter we use "cyclo" for the compound that has an extra linkage between the aglycon and the sugar portion. For nomenclature, an atom directly connected to the base is expressed by a capital italic with superior suffix of the number of the atom in the pyrimidine ring, and a number for the sugar portion making the cyclo linkage, such as $O^2,2'$-cyclouridine and $S^2,5'$-cyclo-2-thiouridine.

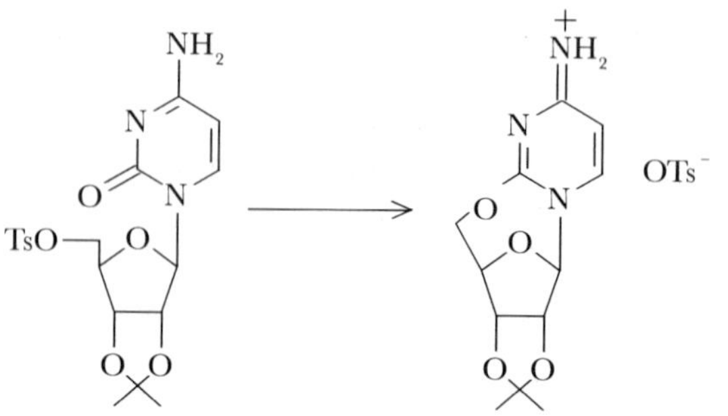

184   **185**

Since the initial discovery in 1951 by Todd and co-workers[436] of the formation of a cyclonucleoside, $2',3'$-O-isopropylidene-$O^2,5$-cyclocytidine (**185**) by heating $2',3'$-O-isopropylidene-$5'$-O-tosylcytidine (**184**) in acetone, a tremendous amount of work has been focused in this area because of the utility of the cyclonucleosides for the preparation of derivatives that are otherwise difficult to access. The finding that $O^2,2'$-cyclocytidine is an effective antileukemic agent[437] also stimulated the research in this field. Apart from the usefulness for providing various intermediates for nucleoside conversions,[438] the cyclonucleoside can be regarded as the fixed conformers of the glycosyl torsion angles of pyrimidine nucleosides and nucleotides.

4.1. Synthesis of Cyclonucleosides

4.1.1. $O^2, 2'$-Cyclonucleosides

The general principle for cyclonucleoside formation is to activate the sugar hydroxyl group as a leaving group and make use of the 2-carbonyl oxygen, which is usually a very poor nucleophile, as an attacking group. This is possible only because the 2-carbonyl group is in such close proximity to the sugar carbon(s) carrying the leaving group. From the accumulated examples in the literature, it appears that the ease of cyclonucleoside formation is in the decreasing order of $O^2,2'$-$> O^2,3'$-$> O^2,5'$-.[439] The $O^2,2'$-cyclonucleosides can now be prepared directly from uridine or cytidine by the use of various reagents. Treatment of uridine with thiocarbonyldiimidazole in toluene at reflux temperature afforded $O^2,2'$-cyclouridine (**187**).[440] 5-Fluorouridine was also converted to the 5-fluoro derivative of (**187**), which was further converted to the arabinosyl 5-fluorocytosine.[441] The 5-methyl[442] and 5'-deoxy[375] derivatives of uridine are also cyclized by this method.

The reaction involves the intermediary formation of a 2',3'-cyclic thiono-carbonate (**186**),[443] followed by an attack of the 2-carbonyl oxygen atom at the 2' position with the elimination of COS. 1,1'-Carbonyldiimidazole also gives the 2',3'-cyclic carbonate that is eventually converted to the $O^2,2'$-cyclouridine.[444,445] 1,1'-Carbonyldi(2-methylimidazole) was also used to afford (**187**).[446] Uridine 2',3'-thionocarbonate was converted to the 2',3'-dimethyl-orthocarbonate by treatment with silver oxide in methanol, which was further converted to the cyclic carbonate by mild acid treatment.[447]

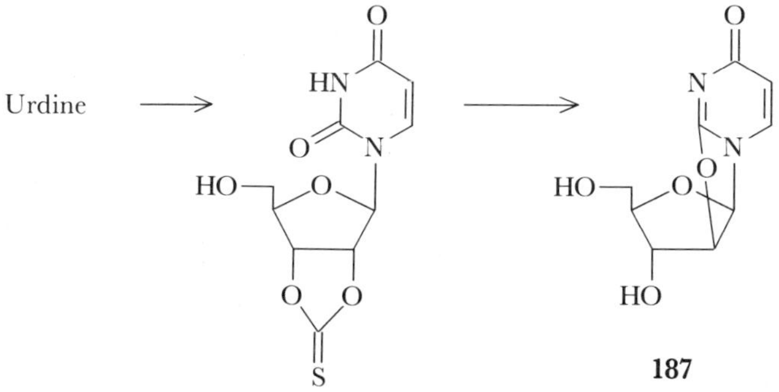

Urdine $\longrightarrow$

186

187

Diphenyl carbonate with a trace of sodium bicarbonate in hot dimethyl-formamide[448] or in hexamethylphosphorotriamide[449] likewise converts uridine to (**187**). The 2',3'-cyclic carbonate intermediate was obtained by the use of *p*-nitrophenyl chloroformate in pyridine, and this intermediate gives (**187**) when heated with a trace amount of sodium bicarbonate.[450] Thionyl chloride reacts with uridine in acetonitrile to afford (**187**), *via* the 2',3'-cyclic sulfinate.[451] Treatment of uridine with partially hydrolyzed phosphoryl chloride in boiling ethyl acetate gave 3',5'-di-*O*-acetyl-$O^2,2'$-cyclouridine (**189**) in fairly good yield.[452] In this case the active intermediate should be a 2',3'-acetoxonium ion (**188**), to which the oxygen atom of the 2-carbonyl group attacks the more available

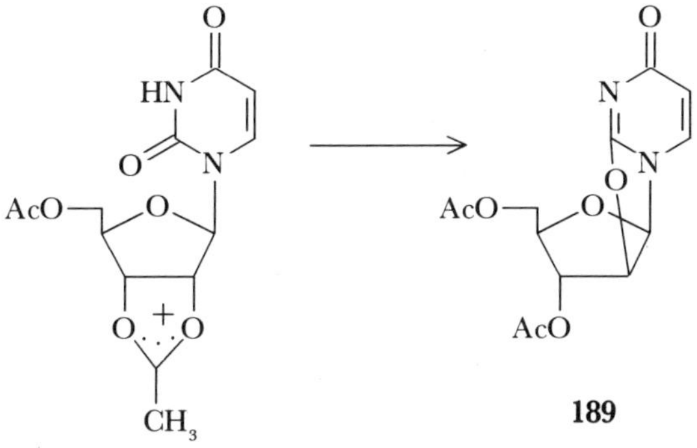

188

189

2′ position. It has already been shown that 2′,5′-di-*O*-benzoyl-O^2,3′-cyclouridine (**190**) can be rearranged on heating to afford the O^2,2′-cyclo derivative (**192**) through the 2′,3′-benzoxonium ion intermediate (**191**).[453]

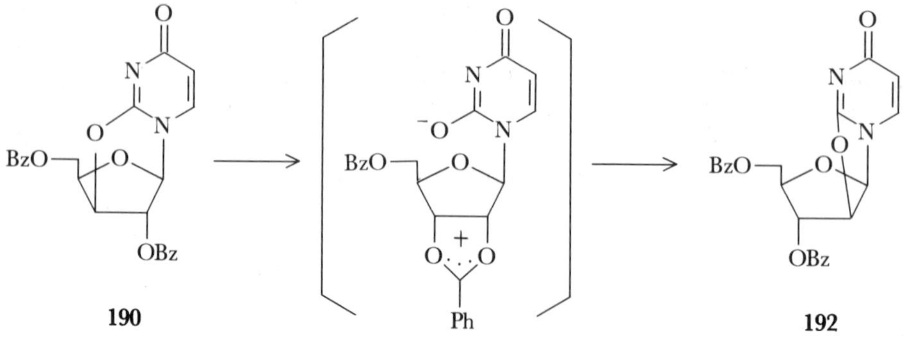

The finding that 1-(*β*-D-arabinofuranosyl)cytosine (araC) possessed anti-tumor activity prompted several efforts to improve on the synthesis of araC. Originally, araC was obtained by the treatment of cytidine with polyphosphoric acid at an elevated temperature to afford the 3′,5′-diphosphate of O^2,2′-cyclocytidine (**193**), which after successive dephosphorylation and hydrolysis gave araC.[454-456] Several procedures have now been developed for the preparation of (**193**). Treatment of cytidine with partially hydrolyzed phosphoryl chloride in ethyl acetate,[315,457,458] thionyl chloride,[451] Vilsmeier–Haak reagent ($POCl_3$- or $SOCl_2$-dimethylformamide),[157,459-461] 2-acetoxyisobutyryl chloride, and other 2-acyloxyisobutyryl halides,[164,462] 2-acetoxybenzoyl chloride (aspirin chloride),[463] tetrachlorosilane or tetraacetoxysilane and boron trifluoride,[464] and acyl chloride and boron trifluoride[465-467] gives (**193**), a monoacyl and/or a diacyl derivative. These reagents also cyclize various cytidine derivatives as expected.

Treatment of cytidine with sulfuryl chloride in acetonitrile was reported to give 5′-chloro-5′-deoxy-O^2,2′-cyclo-5-chlorocytidine as a major product.[468] However, treatment of uridine under similar conditions afforded 5′-chloro-5′-deoxy-*β*-D-lyxofuranosyl-5-chlorouracil.

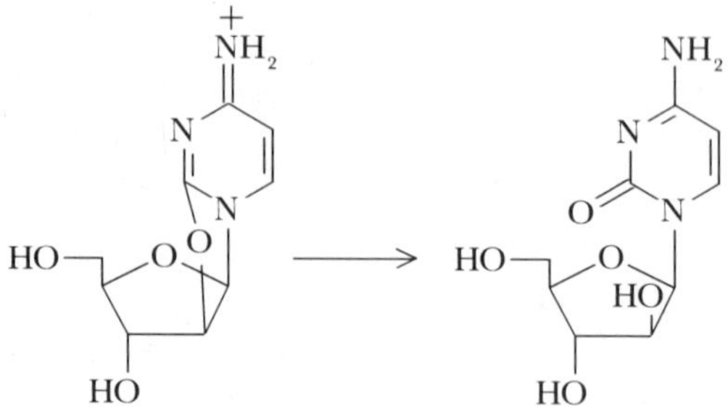

The salt form of cyclocytidine is stable in neutral or acidic medium, but attempts to isolate the neutral form resulted in a rapid cleavage of the cyclo linkage to afford araC (**194**).[469] A unique method (oxazoline method) of preparing cyclocytidines and cyclouridines has been described in Section 2.1.3.

4.1.2. $O^2,3'$-Cyclonucleosides

The first $O^2,3'$-cyclonucleoside was obtained by the treatment of 3'-deoxy-3'-iodothymidine with silver acetate in acetonitrile to give $O^2,3'$-cyclothymidine (**195**).[470] Alkaline hydrolysis of (**195**) gave 2'-deoxy-xylofuranosylthymine. Treatment of 3'-O-mesylthymidine with boiling water, while adjusting the pH to 4–5, gave (**195**).[471] $O^2,3'$-Cyclouridine was prepared by the *t*-butoxide treatment of 3'-O-tosyluridine.[472] A better procedure was the treatment of 3'-O-mesyl-2',5'-di-O-trityluridine with sodium benzoate in dimethylformamide at 130°C, followed by detritlation with hydrochloric acid in chloroform.[453] The direct formation of $O^2,3'$-cyclothymidine and other 2'-deoxynucleosides was reported by Kowollik and co-workers to occur by treatment of the appropriate nucleoside with (2-chloro-1,1,2-trifluoroethyl)diethylamine in dimethylformamide.[473] The 3',5'-di-O-nitrate of 5-fluoro-2'-deoxyuridine was converted to the $O^2,3'$-cyclo derivative by treatment with boiling ethanol containing potassium hydroxide. The 5'-O-nitrate group was then removed by hyrogenation to afford, after hydrolysis,[474] the lyxosyl derivative.

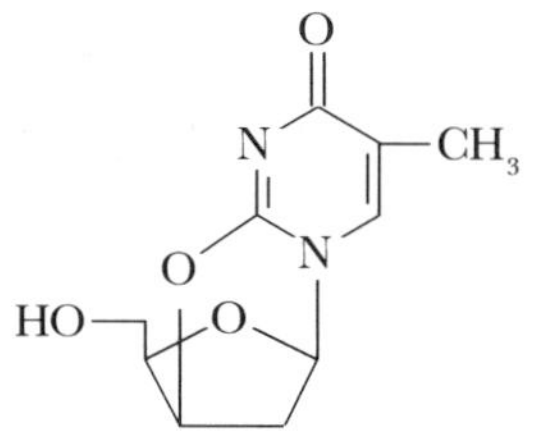

195 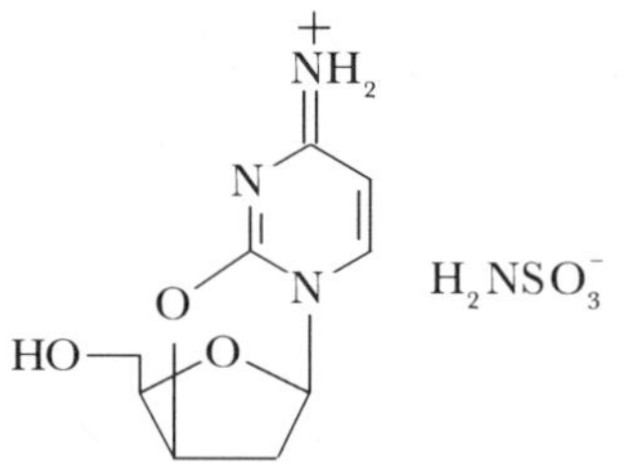   196

$O^2,3'$-Cyclo-2'-deoxycytidine (**196**) has been prepared by simply heating the 3'-O-sulfamate of 2'-deoxycytidine.[475] In the conversion of cytidine to lyxofuranosylcytosine *via* 5'-O-trityl-2',3'-di-O-mesylcytidine, by treatment with alkali, the transient $O^2,3'$-cyclo bond must be generated.[476]

4.1.3. $O^2,5'$-Cyclonucleosides

The first example of this type of nucleoside, 2',3'-O-isopropylidene-$O^2,5'$-cyclouridine (**197**), was accidentally formed in an attempt to prepare the 5'-O-acetate derivative from 5'-deoxy-5'-iodo-2',3'-O-isopropylideneuridine by treatment with silver acetate.[477] Compound (**197**) can be prepared by the treatment of 5'-tosyl-2',3'-O-isopropylideneuridine with potassium *t*-butoxide,[472] 4-morpholino-*N*,*N*'-dicyclohexylcarboxamidine,[478] or 1,5-diazabicyclo-

[5.4.0]undec-5-ene (DBU).[479] $O^2,5'$-Cyclothymidine was also prepared from the 5'-O-mesylate with t-butoxide or DBU as described. The direct formation of a $O^2,5'$-cyclization of 2'-deoxyuridine was accomplished by the treatment with naphthalenesulfonyl chloride followed by base treatment, although the yield was rather unsatisfactory.[480]

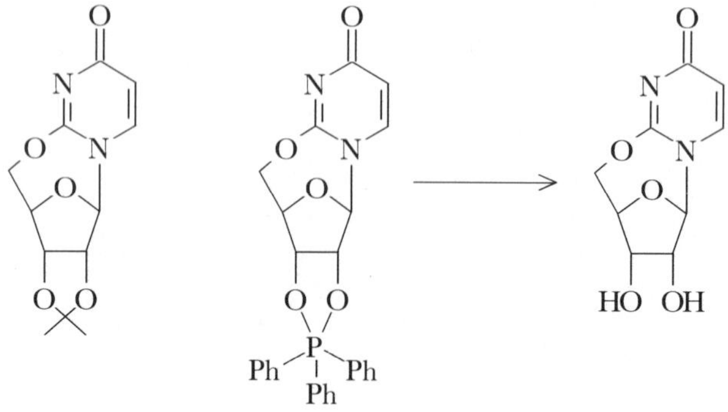

197   **198**   **199**

Treatment of 2',3'-O-isopropylideneuridine with triphenylphosphine and diethyl azodicarboxylate gave (**197**) in high yield.[481] The reaction of this combination of reagents with uridine afforded a $O^2,5'$-cyclouridine with a 2',3'-O-triphenylpholane group (**198**).[482] This group was removed by treatment with water to give $O^2,5'$-cyclouridine[483] (**199**) and thus established a direct preparation of the cyclouridine from uridine. It is assumed that the prior formation of the 2',3'-O-triphenylpholane promoted the cyclization. By a similar procedure, $N^2,5'$-cycloisocytidine and $S^2,5'$-cyclo-2-thiouridine were prepared in high yields from isocytidine and 2-thiouridine, respectively, as is described in more detail in Section 4.3.[483]

Whereas, the treatment of 5'-halogeno-2',3',-O-isopropylideneuridine with silver acetate afforded the $O^2,5'$-cyclo derivative,[477] a similar treatment of 5'-halogenouridine gave 2-O-methyluridine[484] as a result of the methanolysis of once formed $O^2,5'$-cyclouridine. It is therefore evident that the stability of the $O^2,5'$-cyclo linkage is enhanced by 2',3'-O-isopropylidenation.

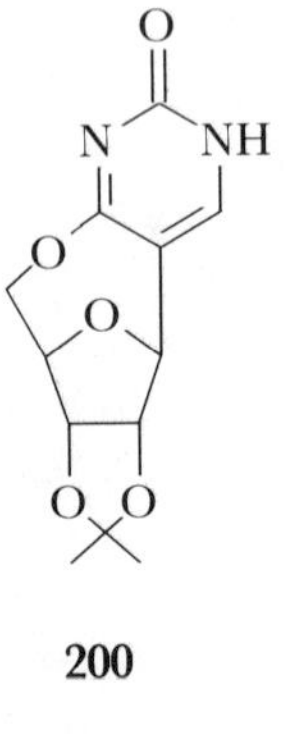

200

Formation of the $O^2,5'$-cyclo linkage provides direct proof of the anomeric configuration of the nucleosides in question as being β.[477] Formation of the $O^4,5'$-cyclo derivative (**200**) of pseudouridine provided evidence for the determination of anomeric configuration of this minor constituent of tRNA.[485]

5'-Deoxy-5'-iodo-2',3'-O-isopropylidene-5,6-dihydrouridine was also cyclized to the $O^2,5'$-cyclo derivative.[373]

The tosylation of 2',3'-O-isopropylidenecytidine gave the $N^4,5'$-O-ditosylate in addition to the expected 5'-O-tosylate (**184**).[436] To avoid the formation of the undesirable N-tosyl derivative, N^4-acetyl-2',3'-O-isopropylidenecytidine was mesylated and then cyclized by heating the compound in acetone to afford the $O^2,5'$-cyclocytidinium mesylate (**185**).[486] It was reported[476] that the mesylation of 5'-O-tritylcytidine afforded the 2',3'-di-O-mesylate with the 4-amino group intact, which would imply that no protection of the N-4 is necessary in the case of mesylation of cytidine.

Recently, the formation of $O^2,4'$-cyclouridines has been reported and this is discussed in Section 5.5.2.

4.1.4. $O^6,5'$-, $O^6,2'$- and $O^6,3'$-Cyclonucleosides

The cyclonucleosides, *vide supra*, are anhydro derivatives of the naturally occurring pyrimidine nucleosides (uridine, cytidine, and thymidine). The $O^6,2'$-, $O^6,3'$-, or $O^6,5'$-cyclonucleosides of uridine or cytidine are in fact the anhydro derivatives of the nucleoside of barbituric acid or 4-aminobarbituric acid. However, we simply describe them as cyclouridine or cyclocytidine derivatives.

The treatment of 2'-deoxycytidine with iodine and iodic acid[171] gave $O^6,5'$-cyclo-5-iodo-2'-deoxyuridine (**59**), while the treatment of 5-iodocytidine with a strong base gave $O^6,5'$-cyclocytidine (**201**).[338] The reaction of ethoxide with

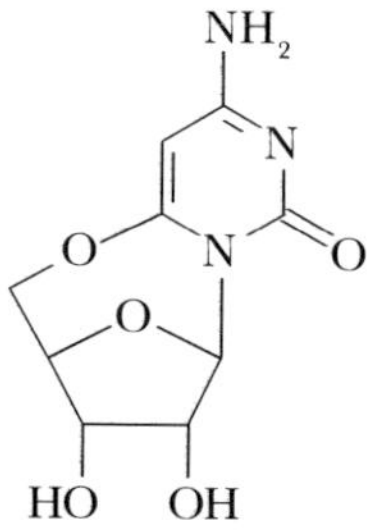

201

2',3'-O-isopropylidene-5-bromouridine gave the $O^6,5'$-cyclo derivative (**121**),[331] and acidic cleavage of (**121**) gave 1-(β-D-ribofuranosyl)barbituric acid. A similar type of conversion of 1-(β-D-arabinofuranosyl)-5-halogenopyrimidines afforded $O^6,2'$-cyclouridine (**119**).[173,330,370] The formation of $O^6,5'$-cyclothymidine (**62**), by iodination of thymidine followed by photolysis and dehydroiodination, has been described.[176]

The rearrangement of the cyclo linkage from $O^6,2'$- to $O^2,2'$- has been observed. The treatment of 4-methylthio-O^6,2′-cyclouridine or O^6,2′-cyclocytidine (**202**) with ammonia gave the O^2,2′-cyclo derivative (**204**). The intermediate of this conversion would be the 2′,3′-ribo epoxide (**203**).[370] Treatment of 3-(5-O-mesyl-2,3-O-isopropylidene-β-D-ribofuranosyl)-6-methylthiouracil (**205**) with *t*-butoxide gave both the O^2,5′- and the O^4,5′-cyclo derivatives—(**206**) and (**207**)—in a ratio of 6.6 : 1.[487] In the case of 5′-O-mesyl-2′,3′-O-isopropylidene-3-β-D-ribofuranosyluracil, cyclization gave the O^2,5′-cyclo derivative (**208**) as the sole product.[488] In the above examples, the 2-keto group seems to be the more nucleophilic under basic conditions. It is also conceivable that the pyrimidine system in the O^2,5′- (or O^2,2′-) cyclo structure is intrinsically more stable than that in the O^6,5′- (or O^6,2′-) cyclo structures.

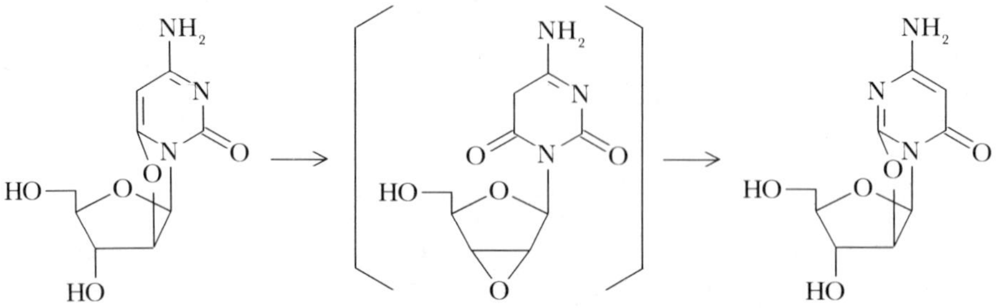

202   203   204

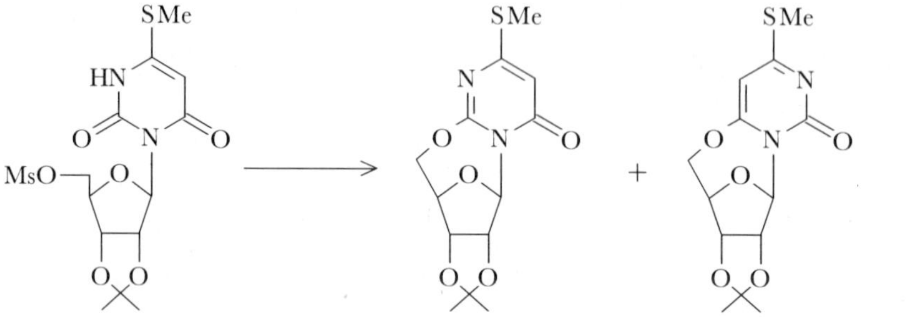

205   206   207

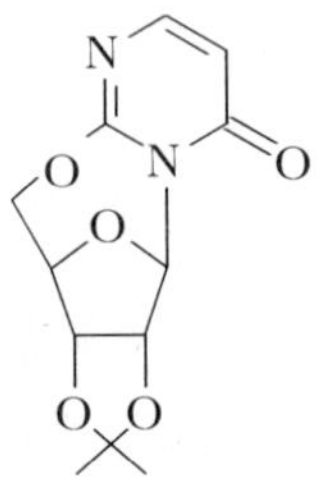

208

Acid treatment of 3-(β-D-arabinofuranosyl)-6-amino(or methylthio)uracil in 1 N hydrochloric acid gave $O^6,2'$-cyclouridine (**119**) as the sole product. The product seems to exist as a cyclo form rather than as the arabinosyl-barbituric acid.[370]

An interesting conversion of 5-iodo-araC (**209**) to $O^6,2'$-cycloisocytidine (**210**), along with the expected $O^6,2'$-cyclocytidine (**211**) and its 5-dimethylsulfonium derivative,[489] was observed by treatment of (**209**) with *t*-butoxide in dimethylsulfoxide. It seems that the ring opening and recyclization at the N^3—C^4 must have occurred on the $O^6,2'$-cyclodihydro intermediate (**209A**) *via* (**209B**).

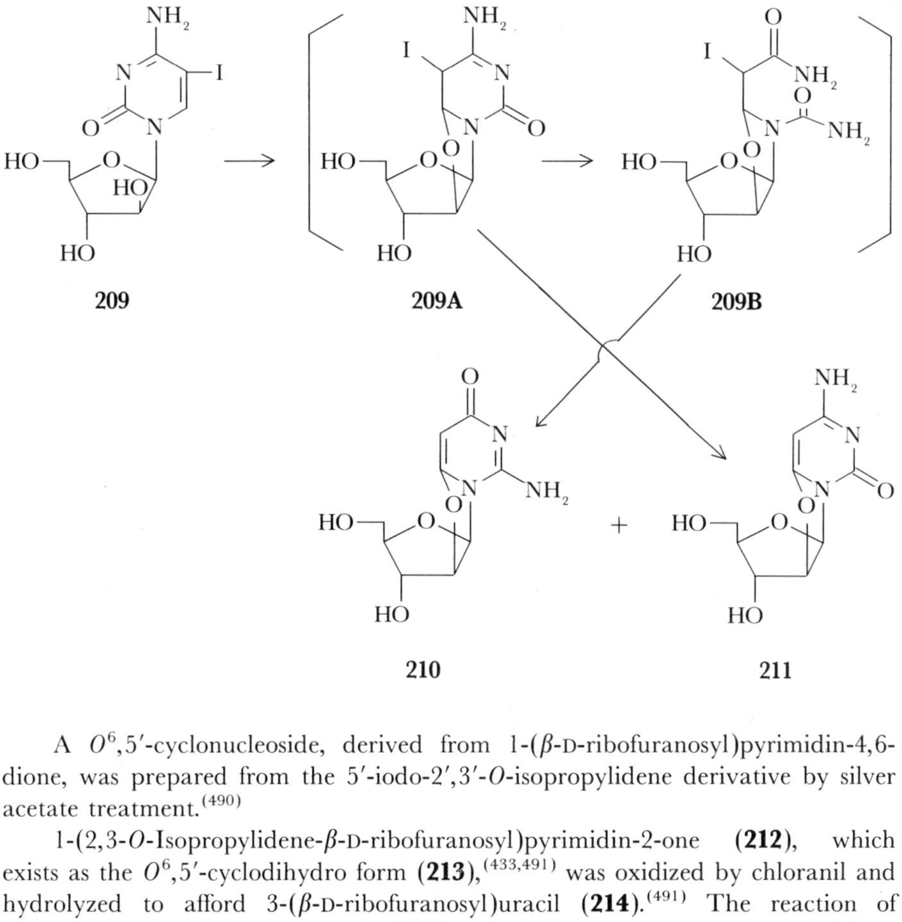

A $O^6,5'$-cyclonucleoside, derived from 1-(β-D-ribofuranosyl)pyrimidin-4,6-dione, was prepared from the 5'-iodo-2',3'-O-isopropylidene derivative by silver acetate treatment.[490]

1-(2,3-O-Isopropylidene-β-D-ribofuranosyl)pyrimidin-2-one (**212**), which exists as the $O^6,5'$-cyclodihydro form (**213**),[433,491] was oxidized by chloranil and hydrolyzed to afford 3-(β-D-ribofuranosyl)uracil (**214**).[491] The reaction of phenyltrihalomethylmercury with (**213**) furnished a 4,5-cyclopropane adduct. This adduct was ring expanded to the diazepinone derivative (**215**) by a reduction of the halogeno group with tri-*n*-butyltin hydride followed by acid treatment in methanol.[492] As previously described, the photoirradiation of 2'-deoxyuridine in water gave, in addition to the 5,6-hydrate, the $O^6,5'$-cyclodihyro derivative.[264]

Treatment of 1-(5-O-trityl-β-D-xylofuranosyl)-5-iodouracil with sodium methoxide afforded the $O^6,3'$-cyclouridine, which was then converted to the corresponding cytidine derivative (**216**) by the standard method. The same com-

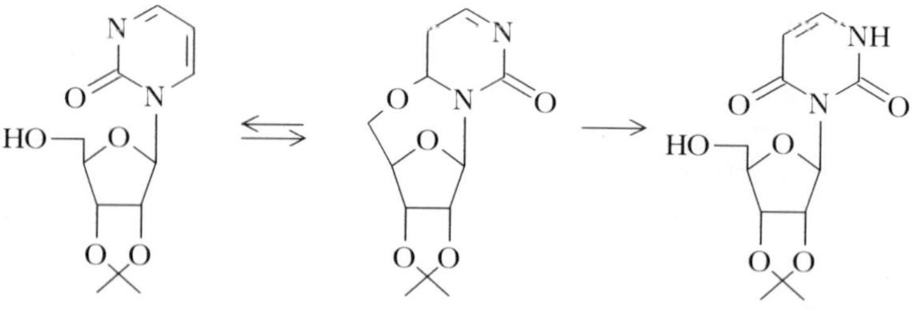

212   213   214

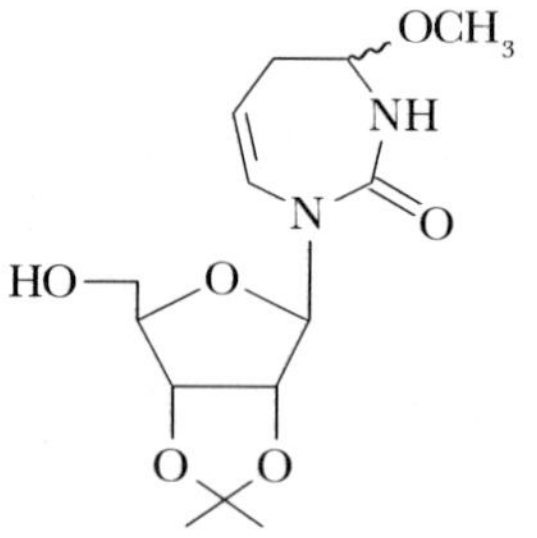

215 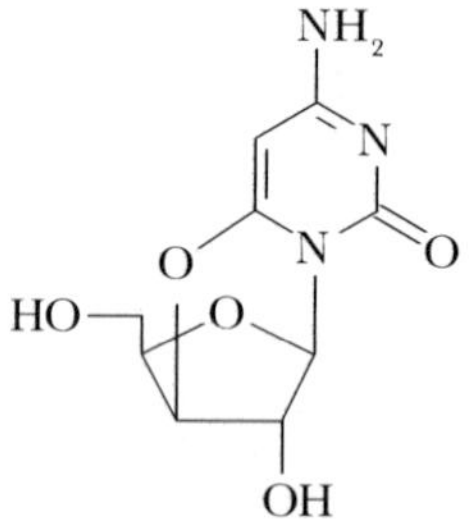   216

pound was also obtained from xylofuranosylcytosine. The stability of the O^6-cyclopyrimidine nucleosides to this type of hydrolysis was also studied.[493] Bromination of $O^6,3'$ (or $2'$)-cyclouridine with bromine (or N-bromosuccinimide) gave the 5-bromo derivative, which underwent further bromination to give the 5,5-dibromo-$O^6,3'$(or $2'$),$5'$-bicyclodihydrouridine (**217**).[494] The structure of (**217**) was confirmed by an x-ray diffraction analysis.[495]

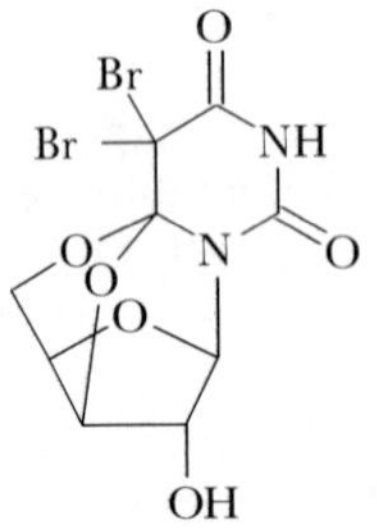

217 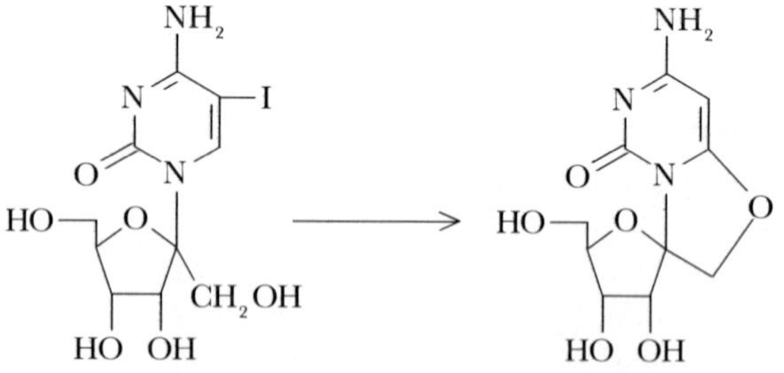   218   219

A cyclonucleoside fixed in the syn form was reported.[496] Treatment of psicofuranosyl-5-iodocytosine (**218**), prepared by the condensation method and iodination, with *t*-butoxide in dimethylsulfoxide afforded the $O^6,1'$-cyclo derivative (**219**). This compound exhibited a positive CD band. By similar approach, the syn form of the adenosine counterpart was also prepared.

4.2. Reaction of Cyclonucleosides

The cleavage of $O^2,2'$-cyclouridine (**187**), in aqueous acid,[497] gives 1-(β-D-arabinofuranosyl)uracil [spongouridine, (**220**)]. The reaction of alcoholic ammonia[498] or various amines[499] with (**187**) gave the appropriate arabinosylisocytosine (**221**). Similarly, a cleavage of (**187**) with hydrogen sulfide afforded the arabinosyl-2-thiouracil (**222**).[500,501] The ammonolysis reaction is a reversible one as observed in the treatment of arabinosyl isocytosines with acid or with heat.[502]

The cleavage products of (**187**) by alkyl-oxygen fission have provided some useful structural analogues of 2'-deoxyuridine. Treatment of (**187**) with various hydrogen halides in anhydrous dioxane gave the corresponding 2'-deoxy-2'-halogenouridines (**223**).[503,504] The 5-methyl-, 6-methyl, and 5-fluoro derivatives of (**187**) were also converted to the 2'-chloro derivatives.[505,506] Treatment of (**187**) with acyl halides also gave the 3',5'-di-O-acetate derivative of the 2'-halo-2'-deoxyuridines.[507] The direct formation of 2'-deoxy-2'-halogenouridines from uridines by treatment with various acyl halides is discussed separately in Section 5.2. The 2'-halogeno group can be removed to afford the 2'-deoxy derivatives by catalytic hydrogen_____[504b,508] ___ by tri-*n*-butyltin hydride.[505-507] The 2'-chloro derivatives of cytidine and uridine, in a solution of pH 8.9, cyclized back to the $O^2,2'$-cyclocytidine and $O^2,2'$-cyclouridine derivatives, repectively.[504,509]

Treatment of (**187**) with sodium azide in hexamethylphosphorotriamide gave 2'-azido-2'-deoxyuridine (**224**), which was then reduced to the 2'-amino derivative.[449] Various 2'-aminoacyl derivatives were prepared for biological evaluation.[510] Treatment of the di-O-acetate derivative of (**187**) with thioacetic acid in dioxane gave the 2'-deoxy-2'-mercapto derivative (**225**), after careful alkaline treatment.[511] Treatment with thiocyanic acid in dioxane gave the 2'-thiocyanato drivative.[512] Cleavage of this cyclouridine with various alkanethiolates afforded the corresponding 2'-alkylthio derivatives (**225**).[513] However, the attack of sodium ethylsulfide on (**187**) did not give the expected 2'-ethylthio derivative but instead resulted in a formation of the 3'-deoxy-3'-ethylthio-xylosyl derivative (**226**). This was confirmed by a conversion of (**226**) to 3'-deoxyuridine (**227**) by desulfurization. The 2',3'-ribo-epoxide was postulated as the intermediate.[498] However, the cleavage of (**187**) with thiophenolate ions proceeded to give the expected 2'-arylthio derivatives (**225**).[514]

Treatment of the L-enantiomer of di-O-benzoylated (**187**) with boron trifluoride in methanol gave 2'(3'),5'-di-O-benzoyluridine.[515] Treatment of the dibenzoylated derivatives of 6-methyl- and 6-ethyluridines with stannic chloride in methanol gave similar results.[516] Since various 6-alkyl-O^2, 2'-cyclouridines can be prepared by the oxazoline method, the above method may be useful for the synthesis of ribonucleosides or 2'-deoxyribonucleosides from arabinose. The neighboring group participation of the 3'-acetyl group in the acid cleavage of $O^2,2'$-cyclouridine had been observed many years ago.[508b]

Cleavage of the cyclo linkage of 5'-O-trityl-3'-O-acetyl-$O^2,2'$-cyclouridine, with 2-lithio-1,3-dithiane at low temperature, gave the 2-(1,3-dithian-2-yl)-

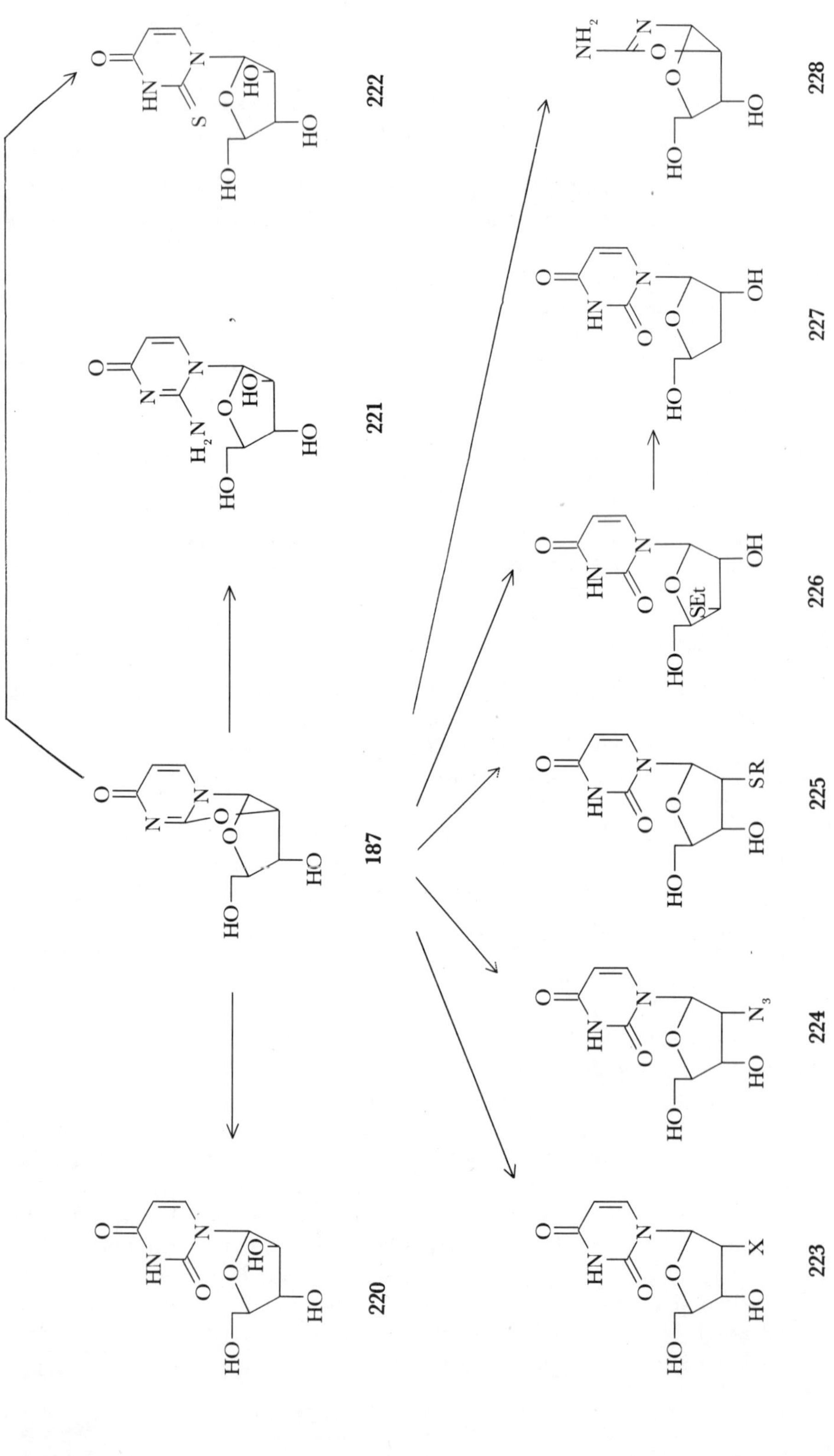
187
220
221
222
223
224
225
226
227
228
HO
HN
O
N
S
H_2N
NH_2
X
N_3
SR
SEt
OH

pyrimidine derivative along with the C-6 adduct of the cyclouridine. The yields were low in every case.[517]

Treatment of (187) with bromine water degraded the pyrimidine ring, with retention of the cyclo linkage, to give a 2-amino-oxazoline sugar (228), which was isolated as the acetate or as the oxazolin-2-one.[518] The amino-oxazoline compound was also obtained by a permanganate oxidation of the diacetate of (187). Bromine in chloroform converted (187) to 2'-bromo-2'-deoxy-5-bromouridine, which was then cyclized back to $O^2,2'$-cyclo-5-bromouridine.[518]

As described already, alkaline hydrolysis of $O^2,2'$-cyclocytidine (193) gave araC.[469] The 2'-O-nitrate of araC has been prepared from the triacetylated cyclocytidine, which was resistant to cytidine deaminase.[519] Treatment of 5-halogenated $O^2,2'$-cyclocytidines with alkali afforded other compounds in addition to the expected araC derivatives. In the case of $O^2,2'$-cyclo-5-fluorocytidine the hydrolysis at pH 10 gave an oxazoline sugar[520] (228) identical with an intermediate observed during the cyclocytidine synthesis by the oxazoline method. Whereas, the dissolution of $O^2,2'$-cyclo-5-chlorocytidine, in pH 10 solution, gave the expected arabino compound, $O^2,2'$-cyclo-5-hydroxycytidine was obtained by a treatment in 0.1 N sodium hydroxide.[157] The same compound was obtained by the oxidation of cyclocytidine with hydrogen peroxide in a neutral buffer.[521] Further treatment of 5-hydroxy-cyclocytidine with alkali effected the formation of arabinosyl-5-hydroxycytosine.[157]

Treatment of $O^2,2'$-cyclocytidine (193) with ammonia gave the arabinosyl-2,4-diaminopyrimidinium salt. Cleavage of the cyclo linkage with various amines also gave the N^2-alkyl-2,4-diaminopyrimidinium derivatives.[522] A reversible conversion to the cyclo derivatives was also observed for this type of diamino-pyrimidines.[502] Ammonolysis of the 5-bromo[458] or 5-fluoro[523] derivatives of cyclocytidine gave the respective 2,4-diaminopyrimidinium derivatives, as expected.

Heating the acetate or hydrofluoride salt of cyclocytidine (193) in dimethylformamide afforded 2'-O-acetyl- or 2'-deoxy-2'-fluorocytidine.[524,525] An attempt to cleave the cyclo linkage of (193) with hydrogen sulfide gave a low yield of the expected arabinosyl-2-thiocytosine. However, the main product was the oxazoline-2-thione derivative.[512] Treatment of (193) with phosphorus penta-

229

X = SH
NH$_2$
N$_3$

230

sulfide gave 2′-deoxy-2′-mercaptocytidine 2′,3′-cyclic (thio)phosphate (**229**), which was dephosphorylated to give the parent alkaline labile 2′-deoxy-2′-mercaptocytidine.[526]

5′-*O*-Trityl-O^2,3′-cyclothymidine was treated with potassium phthalimide or potassium thiobenzoate, in the presence of an appropriate acid in dimethylformamide at reflux, followed by deblocking to give 3′-amino-3′-deoxythymidine (**230**) and the disulfide of 3′-deoxy-3′-mercaptothymidine, respectively.[308] The attack of sodium azide with the 5′-*O*-trityl derivative of (**195**) afforded the 3′-azido derivative [(**230**), X = N_3]. Compound (**230**) was then hydrogenated to furnish the 3′-amino derivative.[527] The 3′-azidothymidine can also be obtained by a nucleophilic substitution of 5′-*O*-trityl-3′-*O*-mesyl-2′-deoxy-1-(β-D-xylofuranosyl)thymine with lithium azide followed by detritylation.[528]

An interesting rearrangement of O^2,3′-cyclouridine (**231**) to 5′-deoxy-5′-iodo-xylosyluracil (**233**) by treatment with sodium iodide has been reported.[529-531] Compound (**233**), when treated with silver acetate, gave (**231**) again. Therefore the O^2,5′-cyclo-xylosyluracil (**233**) must be the intermediate in this rearrangement.[531] Treatment of (**231**) with hydrofluoric acid and aluminum fluoride in dioxane gave the 3′-deoxy-3′-fluorouridine (**234**) and 2′-deoxy-2′-fluorouridine (**235**) in 31 and 47% yield, respectively.[532] The direct attack of fluoride ion at the 3′ position of (**231**) occurred in this case. Formation of the 2′-fluoro derivative (**235**) would be due to a sequence involving a rearrangement of (**231**) to O^2,2′-cyclouridine at first, then fluoride attack at the C-2′ position. O^2,3′-

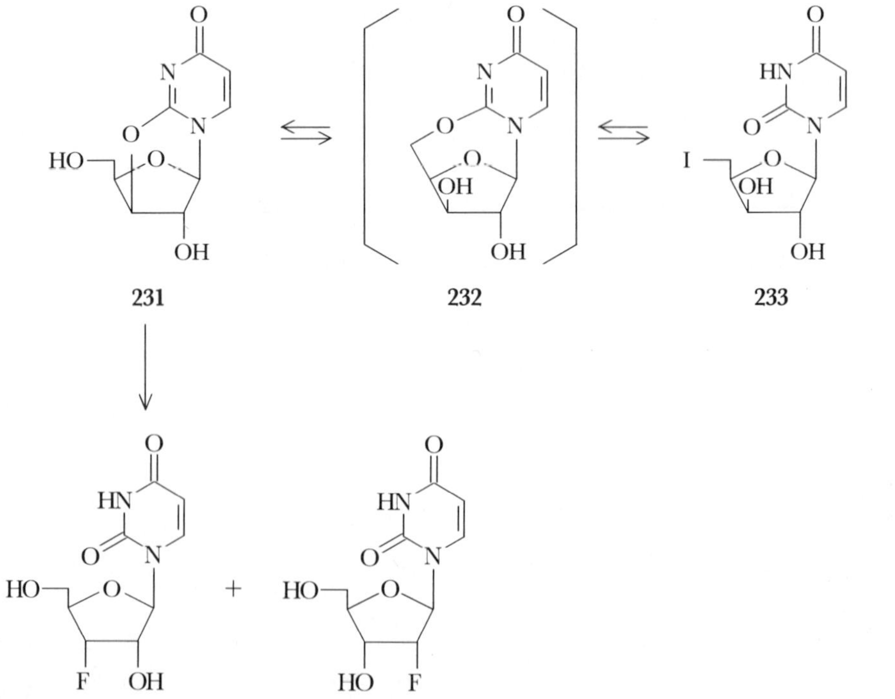

231   232   233

234   235

Cyclo-2′-deoxy-5-fluorouridine was cleaved by treatment with hydrofluoric acid and aluminum fluoride to give the 3′-fluorouridine derivative.[533] The solvolysis of 5′-O-mesyl-arabinosyluracil, at around pH 5, gave the 2′,5′-anhydro derivative and O^2,2′-cyclouridine (187). The latter compound is most likely derived from the O^2,5′-cycloarabinosyl derivative as the intermediate. The 2′ "up" hydoxyl group of the cyclo intermediate is in a suitable juxtaposition for the ready formation of the O^2,2′-cyclo derivative. Similarly, 5′-O-mesylxylosyluracil gave O^2,3′-cyclouridine by a solvolysis along with the expected 3′,5′-anhydro derivative.[534] Various instances have been reported[502,535–538] on the conversion of the sulfonyl groups of the sugar moiety involving the cyclonucleoside intermediates. Formation of nucleosides with an unsaturated sugar portion from certain cyclonucleosides are dealth with in Section 5.5.

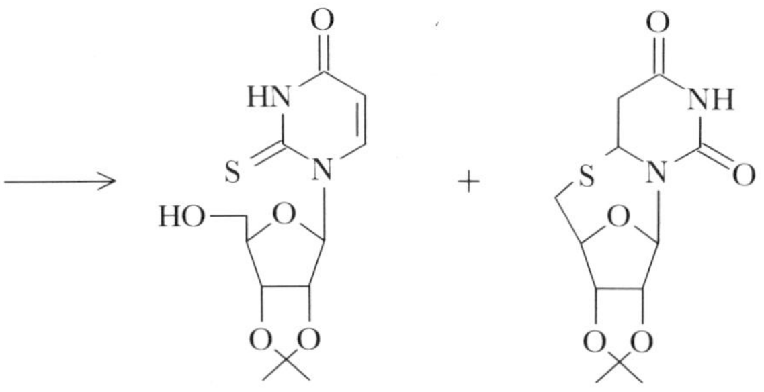

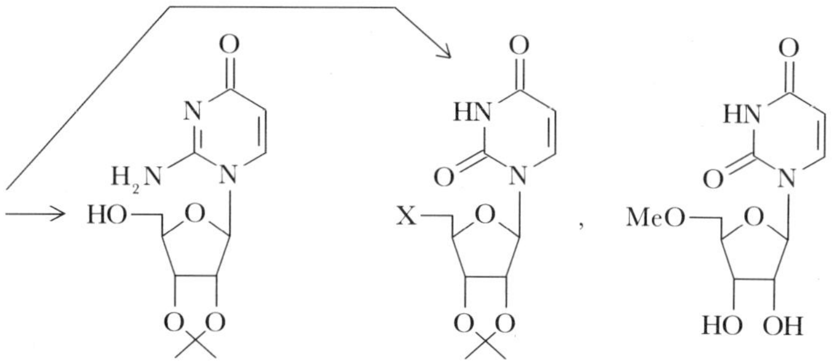

Cleavage of the O^2,5′-cyclo bond of (197) with a large excess of liquid hydrogen sulfide in pyridine gave 2′,3′-O-isopropylidene-2-thiouridine (236) in high yield,[539] while treatment of (197) with hydrogen sulfide and triethylamine in dimethylformamide gave the 5′-mercapto-5′-deoxyuridine [as a cyclo form, (237)] along with (236).[260,498] Cleavage of (197) with methanolic ammonia gave 2-O-methyluridine derivative first, which was subsequently converted to the isocytidine derivative (238).[477] Treatment of (197) with acetyl chloride[540] or benzoyl bromide[541] afforded the 5′-deoxy-5′-halogeno derivatives (239). Methanolysis of (197) in the presence of p-toluenesulfonic acid afforded 5′-O-methyluridine (**240**) in high yield.[542] In this case the protonation of (**197**) at the N-3 facilitated the substitution at the C-5′ position, and by the same principle (**197**) gave uridine 5′-diphenyl phosphate by treatment with boron trifluoride and diphenyl hydrogen phosphate. An attempted cleavage of (**197**) with benzyl-magnesium bromide resulted in the formation of the 5′-bromo derivative with no C-alkylation being observed.[542] The 5,6-dihydro derivative of (**197**) underwent solvolyses with hydrogen sulfide or methanol in a similar fashion.[373] Methanolysis in the presence of hydrochloric acid resulted in a cleavage of the N^3—C^4 linkage.[373]

185 $\longrightarrow$

241   242

Treatment of 2′,3′-O-isopropylidene-O^2,5′-cyclocytidine tosylate (**185**) with methanolic ammonia afforded the ribosyl-2,4-diaminopyrimidinium salt (**241**) in high yield.[543] Treatment of (**185**) with liquid hydrogen sulfide gave the 5′-deoxy-5′-mercapto-2′,3′-O-isopropylidene-4-thiouridine as the S^6,5′-epithio form (**242**).[376] In this conversion, a rapid alkyl–oxygen fission followed by sulfhydrolysis of the 4-amino group took place. This was in direct contrast with the result observed for the cleavage of the corresponding O^2,5′-cyclouridine to (**236**).[539] Treatment of (**185**) with sodium hydrogen sulfide gave a low yield of the 2-thiocytidine derivative.[543]

The cleavage of O^6,5′-cyclouridine with aqueous acid gave the ribosyl–barbituric acid[331] and with ammonia[544] or methylamine[545] afforded the 6-amino- and 6-methylaminouridines (**243**), respectively. These were further converted by the appropriate chemical reactions to a purine 3-riboside (**244**) and a thiazolopyrimidine nucleoside (**245**), respectively. The 6-methylaminouridine was sulfhydrolyzed to give the 6-thiobarbituric acid nucleoside (**246**) and a further conversion of (**246**) to afford 5-propyluridine (**247**) and 5-acyluridines (**248**) was accomplished.[546] Treatment of (**121**) with Vilsmeier reagents (POCl₃ or POBr₃–DMF) gave the 5-dimethylaminomethylenebarbituric acid nucleosides with a 5′-halogeno group. The reaction was initiated by the attack of the halide ion at the 5′-carbon. In the case of O^2,2′- or O^2,5′cyclouridines the 2′- or 5′-halogeno nucleosides were obtained as expected.[547] However, with 2′,3′-O-isopropylidene-O^6,5′-cyclocytidine, the attack of ethylmercaptide or cyanide ion gave the 5′-substituted derivatives while an attack with methoxide or ammonia gave both the 6-substituted and the 5′-substituted derivatives.[548]

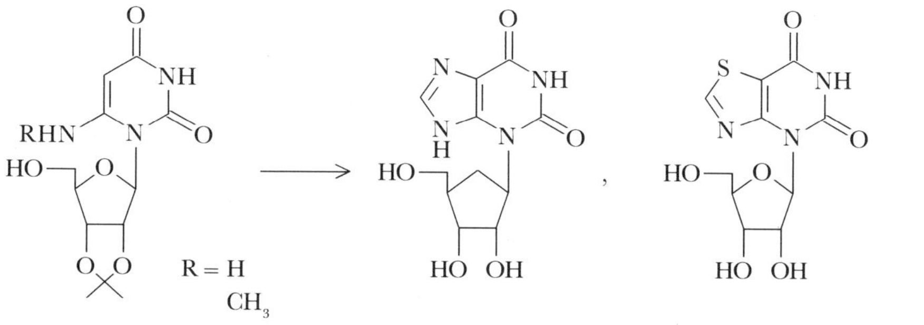

Since the introduction of a substituent into the 5′ position of pyrimidine nucleosides can be accomplished by a direct S_N2 displacement of the 5′-halogeno or 5′-O-sulfonyl derivatives, there does not seem to be any special advantage in using the $O^2,5′$-cyclonucleosides for 5′ substitution.

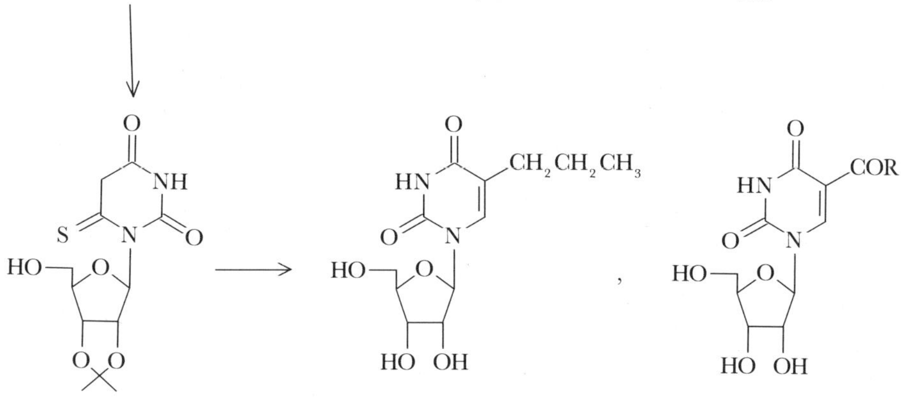

Cleavage of (**197**) with the carbon nucleophile, dimethyloxosulfonium methylide, occurred by aryl–oxygen fission to give a 2-methylenesulfonium ylide (**249**). Desulfurization of (**249**) with Ni catalyst gave the 2-methylpyrimidin-4-one derivative (**250**). Photoirradiation of (**249**) gave a novel type of carbon bridged 2,2′-methylene cyclonucleoside (**251**) as the result of a generation and subsequent insertion of the 2-carbene intermediate.[549] Formation of similar sulfonium ylides from a $O^2,2′$-cyclouridine, $O^2,3′$-cyclothymidine, and $O^2,3′$-cyclo-2′-deoxyuridine were also observed. Nickel hydrogenolysis of these compounds also gave the 2-methylpyrimidinones. Photoreaction of the 2-ylide of the arabinosyl derivative afforded a 2-formyl derivative that exists as the hemiacetal with the 2′-hydroxyl group.[549]

It would appear that some generalization, as to the relationship between the type of nucleophiles and the site of cleavage of the cyclo bond, could be formulated from the accumulated examples. The site of cleavage is not primarily dependent on the nucleophile used, as shown in the sulfhydrolysis of a $O^2,5′$-cyclouridine, and although it must depend on the reaction conditions such as the

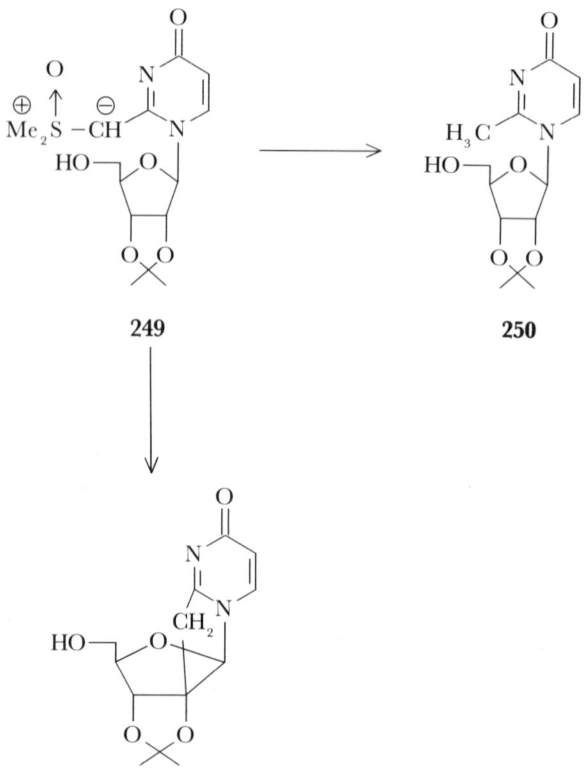

249   250

251

ionic species of the cyclonucleosides and presence or absence of blocking groups on the sugar portion, the nucleophiles can be roughly divided into the two following categories: (1) *nucleophiles bearing active hydrogen(s) to be removed after an addition to the C-2 (or C-6) tend to give aryl–oxygen fission products; and* (2) *nucleophiles of anionic species having no active hydrogens tend to give the alkyl–oxygen fission products.*

Thus, ammonia, monoalkylamines, alcohols, and hydrogen sulfide generally give 2-substituted derivatives in the reaction of, for example, $O^2,2'$- and $O^2,5'$cyclouridines. In the case of hydrogen sulfide, which dissociates even in a neutral solution to furnish more nuclephilic sulfhydryl ion, the possibility of an attack at the 5′ primary carbon increases in the reaction with $O^2,5'$-cyclouridine. Most anions such as RS^-, $RCOO^-$, X^-, $RCOS^-$, and N_3^- give sugar-substituted fission products. As stated above, the cleavage of cyclouridines with halide ions is enhanced in acidic media as a direct result of the aglycon being protonated, which makes the heterocyclic moiety a stronger leaving group in the alkyl–oxygen fission.[503,504] Of course, the above classification of nucleophiles does not exclude the possibility of an attack by a nucleophile of the anionic species at C-2 (or C-6). However, even if it occurs, the product would reversibly regenerate the cyclonucleoside again, and with the progress of the reaction at the sugar carbon (alkyl–oxygen fission) the final product would be the sugar-substituted derivative, since there would be no reverse reaction under these latter conditions.

For example, treatment of 2′,3′-O-isopropylidene-O^6,5′-cyclocytidine (252) with sodium methoxide in methanol at 40°C furnished 6-methoxycytidine (253), which reached a maximum concentration after 10 hr and then began to decrease with the concomitant formation of a second product, the 5′-O-methyl derivative (254). The first product (253), when treated with *t*-butoxide, afforded the O^6,5′-cyclo compound (252) as expected. Therefore, the 6-methoxy derivative, formed by the aryl–oxygen fission, was the kinetically controlled product while 5′-O-methyl-6-oxocytidine (254) was the thermodynamically more stable final product.[548]

254   252   253

In the synthesis of 2′,3′-O-isopropylidene-O^6,5′-cyclouridine (121) from (120), by treatment with sodium ethoxide in ethanol, the alkaline condition would lead to a dissociation of the N-3 proton of (121) and thus stabilize the cyclo linkage from further attack of ethoxide ion. The use of the stronger base, to effect the cleavage of the cyclo linkage, would result in an elimination leading to unsaturated sugar nucleosides. This reaction is treated in more detail in Section 5.5.

4.3. Cyclonucleosides Bridged by Sulfur and Nitrogen

The cyclonucleosides discussed, *vide infra*, have been oxygen-bridged O-cyclonucleosides. In studies on the stereochemistry of nucleosides and nucleotides, especially conformational analyses of the glycosyl torsion angles, it was apparent that the cyclonucleosides bridged by atoms other than oxygen would also be very useful. The different reactivities of the cyclo linkages would also be expected. The carbon-bridged cyclonucleosides seem to be the most suitable conformationally fixed models around the glycosyl linkages.

The first synthesis of an S-cyclonucleoside was accomplished by Shaw and Warrener[550] during the transformation of ribofuranosyl-2-thiothymine to thymidine *via* S^2,2′-cyclo-2-thiothymidine (255). Later, S^2,3′-cyclo-2-thiothymidine (256) was prepared from 3′-O-mesyl-O^2, 5′-cyclothymidine *via* a methanolysis followed by sulfhydryl ion attack. Cleavage of the S^2,3′-cyclo linkage with 1 N NaOH solution afforded 2′,3′-dideoxy-3′-mercapto-1-(β-D-xylofuranosyl)-thymine.[551] Since 2-thiouridine is now readily accessible from uridine[539] or by reported condensation methods,[13,552] the synthesis of S^2,2′-, S^2,3′-, and

$S^2,5'$-cyclo-2-thiouridines have been achieved by an application of procedures similar to those used in O-cyclonucleoside synthesis.

5'-*O*-Tosyl-2',3'-*O*-isopropylidene-2-thiouridine was cyclized readily, on a brief treatment with triethylamine in dioxane, to afford $S^2,5'$-cyclo-2-thiouridine **(258)**.[539] As described, the triphenylphosphine–diethyl azodicarboxylate procedure gave unprotected $S^2,5'$-cyclo-2-thiouridine **(257)** directly from 2-thiouridine.[483] Treatment of 2-thiouridine with diphenyl carbonate gives the $S^2,2'$-cyclo derivative **(258)**.[552] Compound **(258)** has also been prepared from the 2'-*O*-mesyl derivative of arabinosyl-2-thiouracil by treatment of the mesyl compound with base.[539] Treatment of **(258)** with 1 N NaOH at room temperature gave 2'-deoxy-2'-mercapto-1-(β-D-arabinofuranosyl)uracil, which exists in the 5,6-dihydro-6,2'-epithio form.[511b] Treatment of **(258)** with methyl iodide

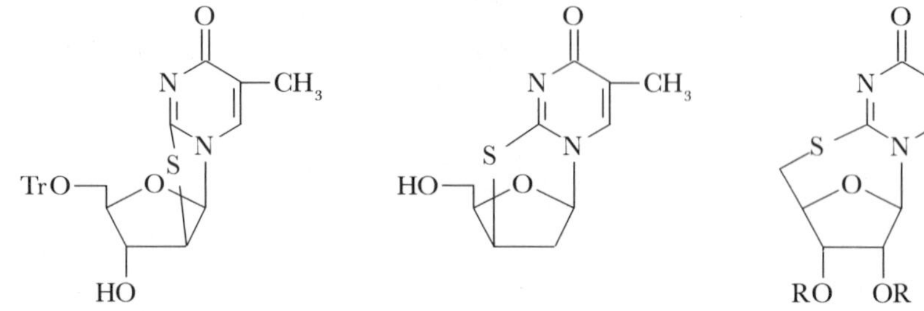

255   256   257

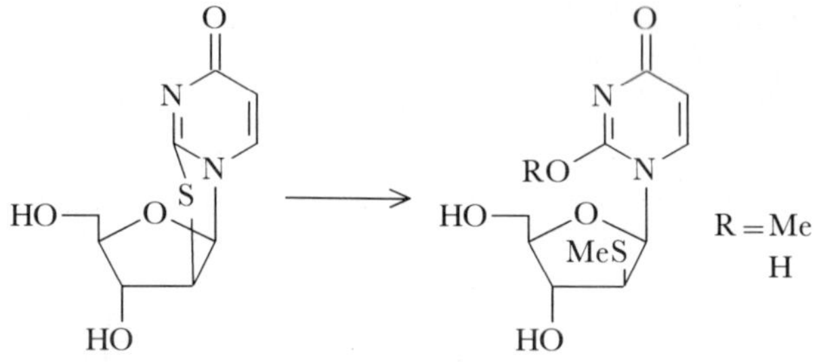

258   259

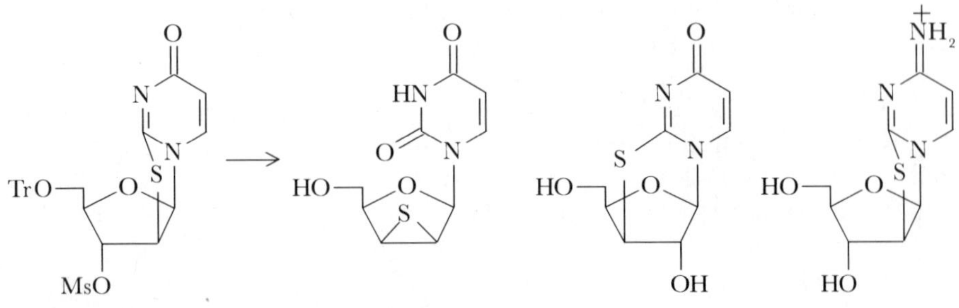

260   261   262   263

and sodium methoxide in methanol gave the 2'-methylthioarabinosyl derivative of both uracil and 2-O-methyluracil [(**259**), R = H, Me], which was then converted to the 2'-methylthio-araC by the conventional method.[553] Treatment of 3'-O-mesyl-5'-O-trityl-S^2,2'-cyclo-2-thiouridine (**260**) with dilute hydrochloric acid resulted in the formation of 2',3'-dideoxy-2',3'-epithio-1-(β-D-lyxofuranosyl)uracil (**261**) in good yield.[554] S^2,3'-Cyclo-2-thiouridine (**262**) was prepared from 2-thiouridine by the standard method used for the preparation of the O^2,3'-cyclo derivative.[539]

Treatment of 2-thiocytidine[164] and 5-fluoro-2-thiocytidine[523] with 2-acetoxyisobutyryl chloride in acetonitrile afforded the S^2,2'-cyclo-2-thiocytidine (**263**). Attempted cleavage of the S-cyclo linkage with sulfhydryl ion resulted only in defluorination.[523] 2-Selenocytidine was also converted to the Se^2,2'-cyclo derivative by similar treatment with the acyl chloride.[555]

S^6,5'-Cyclo-6-thiouridine was prepared from the readily accessible 5'-acetylthio-5'-deoxy-2',3'-O-isopropylidene-5-bromouridine by treatment with sodium methoxide in methanol followed by deacetonation in 50% formic acid.[556] A conversion of S^2,2'-cyclouridine to S^6,2'-cyclouridine (**264B**) was performed by the following rout. S^2,2'-Cyclo-2-thiouridine was brominated to afford the 5-bromo derivative (**264A**), which was treated with methoxide in methanol to furnish the S^6,2'-cyclo derivative of 2-methoxy-6-thiopyrimidin-4-one nucleoside [(**264B**), R = OMe] *via* the 2'-sulfide intermediate. The S^6,2'-cyclo compound can be converted to the corresponding cyclo derivatives of uracil, 2-thiouracil, and isocytosine [(**264B**), R = OH, SH, and NH$_2$].[557] In a similar fashion, S^2,5'-cyclo-2-thiouridine was converted to the S^6,5'-cyclouridine derivative.

264A   264B

The N^2,3'-cyclo derivative (**265**) of thymidine has been prepared by the treatment of 3'-O-mesyl-O^2,5'-cyclothymidine with methanolic methoxide followed by ammonolysis. The use of other amines such as methylamine, hydroxylamine, and hydrazine gave the N-methyl, N-hydroxy, and N-amino derivatives of (**265**).[558,559] Treatment of N^2-acetylisocytidine,[560] 5-bromo-isocytidine,[557] and 5-fluoroisocytidine[561] with diphenyl carbonate afforded the respective N^2,2'-cyclo-isocytidines (**266**). The 5-fluoro derivative was then converted to the 4-amino derivative by the standard procedure of thiation and amination. The same type of N^2,2'-cyclo derivative was also prepared by the treatment of 1-(β-D-ribofuranosyl)-2,4-diaminopyrimidinium chloride with diphenyl carbonate.[562] $N^2,5'$-Cycloisocytidine (**267**) was prepared by a direct cyclization of isocytidine with the triphenylphosphine–diethyl azodicarboxylate system[483,557] or by heating 5'-amino-5'-deoxy-2',3'-O-isopropylidene-5-bromouridine.[556] Treatment of 5'-O-benzoyl-3'-O-mesyl-$O^2,2'$-cyclouridine with ammonium azide (generated *in situ* from sodium azide and ammonium chloride) in dimethylformamide afforded the $N^2,3'$-cycloarabinosyl derivative.[563] The formation of this compound was accomplished by a cleavage of the cyclo linkage with ammonia followed by $N^2,3'$-cyclization.[564]

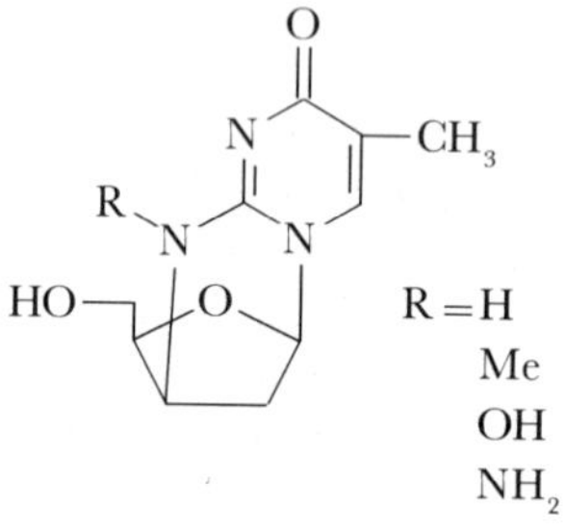

265 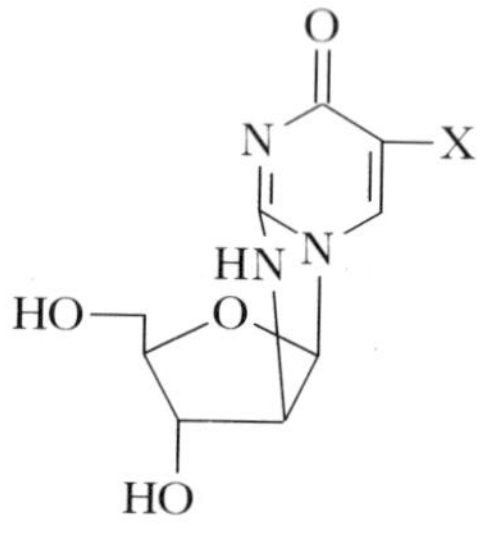   266

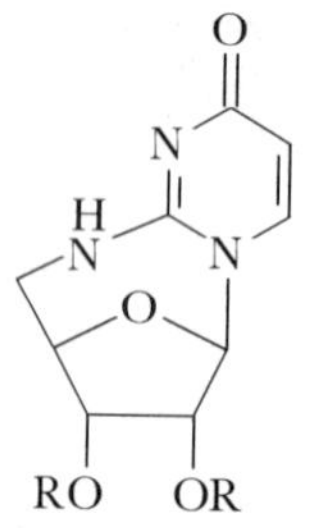

267 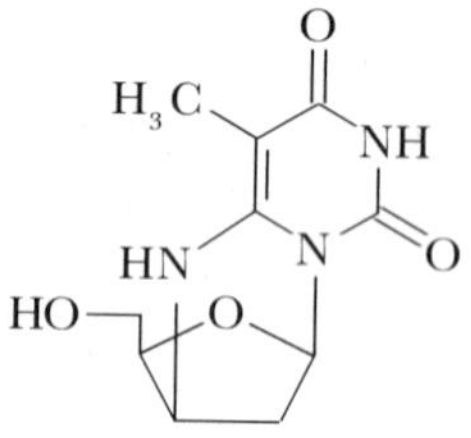   268

Treatment of 5'-O-tosyl-2',3'-O-isopropylidene-5-bromouridine with sodium azide afforded $N^9,5'$-cyclo-3-(2,3-O-isopropylidene-β-D-ribofuranosyl)-8-azaxanthine (**146**). The same compound was obtained by heating 5'-azido-5-bromouridine in dimethylformamide.[565] A 1,3-dipolar cycloaddition of the 5'-azide to the 5,6 bond was postulated in this reaction. The 5'-azido-5-bromocytidine also gave the corresponding 8-azapurine cyclonucleoside.[565] Treatment of 5'-O-trityl-3'-O-mesylthymidine with sodium azide in dimethylformamide afforded a $N^6,3'$-cyclo derivative (**268**) along with the expected 3'-azido derivative. The intermediacy of an 8-azapurine compound was postulated in the formation of $N^6,3'$-cyclothymidine.[566] 5'-Azido-5'-deoxythymidine gave the $N^6,5'$-cyclothymidine by heating.[566] The scope and mechanism of this type of reaction have been described.[567-569]

4.4. Cyclonucleosides Bridged by Carbon

In addition to using cyclonucleosides in the interconversions of pyrimidine nucleosides, the cyclonucleosides have also been used as fixed models in the conformational studies of nucleosides, nucleotides, and polynucleotides. For example, naturally occurring pyrimidine nucleosides generally exist in the anti form. In this form, the projection of the 2-carbonyl group to the plane of the sugar ring is outside the ring, and the $O^6,5'$-cyclocytidine and $O^6,5'$-cyclouridine [such as (201) and (202)] and its *S*- and *N*-isosteres are regarded as anti fixed nucleosides. In a same sense, $O^2,5'$-, $N^2,5'$-, and $S^2,5'$-cyclouridines—(199),(267), and (257)—are in the syn fixed form.

However, one serious drawback in using cyclonucleosides as models of torsion angle-fixed conformers is that the introduction of a O, S, or N linkage to the aglycon may alter the electronic structure of the original nucleobase. Perhaps the most suitable alternative for fixing the base-sugar orientation can be achieved by forming a carbon–carbon bridge, for example, a C-cyclonucleoside. For further discussions of the conformational studies utilizing cyclonucleosides, excellent reviews are available.[570,571]

Photolysis of 2',3'-*O*-isopropylidene-5'-deoxycobalamin afforded the 6,5'-cyclo-5'-deoxy-5,6-dihydrouridine derivative (269) in 40% yield, in which a direct bond between C-6 and C-5' has been generated, although the configuration at C-6 was not established.[572] This type of *C*-cyclouridine has now been obtained readily by the treatment of 5'-deoxy-5'-iodo-2',3'-*O*-isopropylideneuridine with the dropwise addition of a mixture of tributyltin hydride and azobisisobutyronitrile (AIBN) to give the *C*-6(*R*) diastereomer (269) selectively.[573] Similarly, treatment of 5'-halogeno-5'-deoxy-5-halogenouridines, with these reagents, afforded the 5-halo derivatives of cyclodihydrouridine (270) in which the cis addition was observed. Treatment of (270) with base furnished, after deprotection, 5'-deoxy-6,5'-cyclouridine (271).[574,575] A similar cyclization, starting from N^4-acetyl-2',3'-*O*-isopropylidene-5'-deoxy-5'-iodocytidine or 5-bromocytidine, gave the cyclocytidine derivative (272), although the yield of the cyclization was rather poor.[575]

Fox and co-workers have reported an interesting intramolecular aldol addition on a 5-hydroxyuridine 5'-aldehyde (273). Treatment of (273) in a sodium bicarbonate solution furnished 6,5'(*S*)-cyclo-5-hydroxyuridine [(274),

R = OH].[576] The 5-hydroxyl group in (274) was removed *via* hydrogenation of
the 5-mesyloxy derivative to give 6,5′(*S*)-cyclouridine [(274), R = H].[577] The
5′-hydroxyl group of (274) can readily be epimerized by simply warming the
5′-*O*-acetylate in pyridine. The 5′ position of (274) (R = H) was oxidized to the
5′-keto derivative (275) with pyridine sulfur trioxide and then readily reduced by
sodium borohydride. Treatment of 5′-keto-6,5′-cyclouridine (275) in alkaline
solution resulted in an epimerization of the 3′-hydroxyl group through a retro-
aldol intermediate to give the xylosyl derivative (276).[578] Reduction of the
5′-keto function of (276) with sodium cyanoborohydride gave an epimeric mixture
of the 5′-hydroxyl derivative. Ring expansion of the 5′-keto-6,5′-cyclouridine

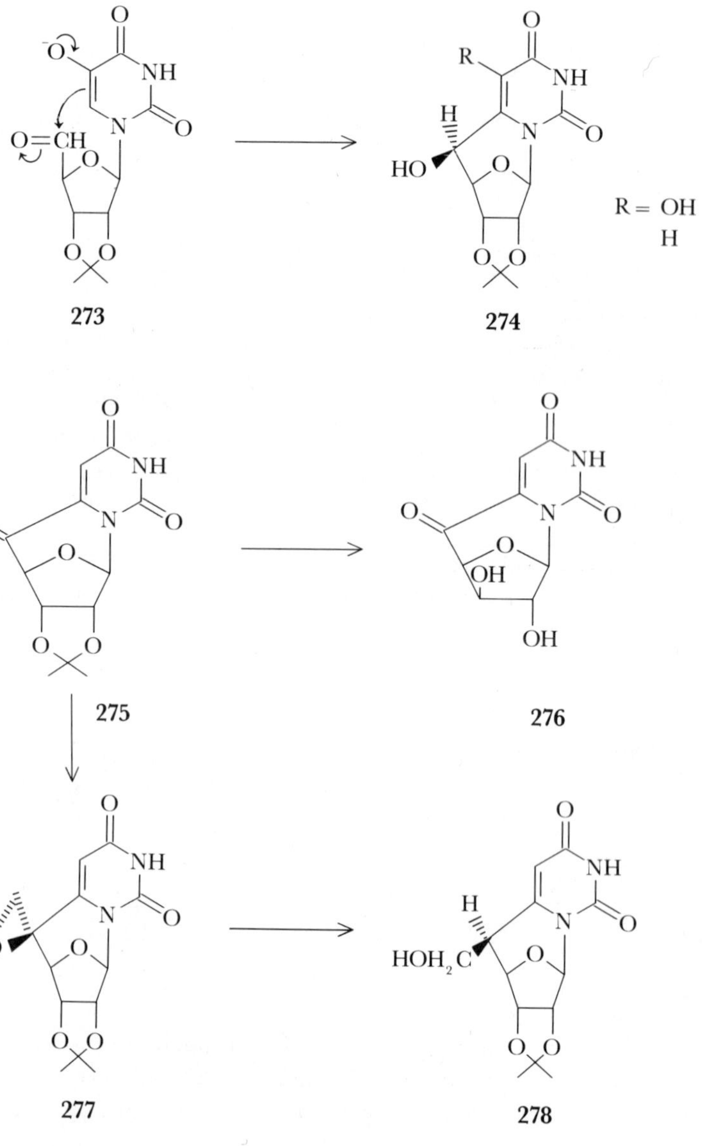

273 → 274

R = OH
H

275 → 276

277 → 278

derivative (275), with diazomethane, afforded the 5'-keto-6,6'-cyclo-3-methyl-uridine, which was then reduced to the 5'(S)-hydroxyl derivative.[579] The reaction of dimethylsulfonium methylide with the 5'-keto group of (275) gave the 5'-spiro-epoxide (277), which was then reductively cleaved to furnish the 5'-hydroxymethyl derivative (278) of 6,5'-cyclouridine. The stereochemistry of the epoxy formation and cleavage has been described.[580]

The intramolecular cyclization, using the 5'-methylene radical generation, can be adapted for the synthesis of various C-cyclonucleosides. Treatment of 5'-iodo-5'-deoxy-2',3'-O-isopropylidene-6-cyanouridine with tributyltin hydride and AIBN gave the 6'-keto-6,6'-cyclouridine (279), which was converted to the 6'-alcohol, 5',6'-etheno, and 6'-methylene (280) derivatives, respectively.[574] Radical cyclization of 5'-O-benzoyl-2',3'-dideoxy-3'-iodomethyl-5-chlorouridine gave 6,3'-methano-cyclouridine (281).[581] In the same manner, 6,2'-ethano-cyclouridine (282) was prepared from the 2'-iodoethyl-5-chlorouridine derivative.[574] 2'-Deoxy-6,2'-methano-cyclouridine (283) was prepared from the 2'-deoxy-2'-ethoxycarbonylmethyl-5-bromouridine derivative by an ionic conden-sation in the presence of strong base followed by de-ethoxycarbonylation.[582] The ribo analogue of (283) was also prepared starting from a 2'-methyleneuridine derivative.[582] An attack of the carbanion of the sugar portion to the C-6 is also exemplified in the case of the nitromethylene carbanion at the 2'position[583] or at the 5' position.[584]

As already described, attack of the C-2 carbene at the C-2' position furnished 2,2'-methylene-cyclonucleoside (251).[549]

Studies on the interaction of these C-cyclonucleosides and their phosphates, with enzymes utilizing nucleotides, is of current interest.

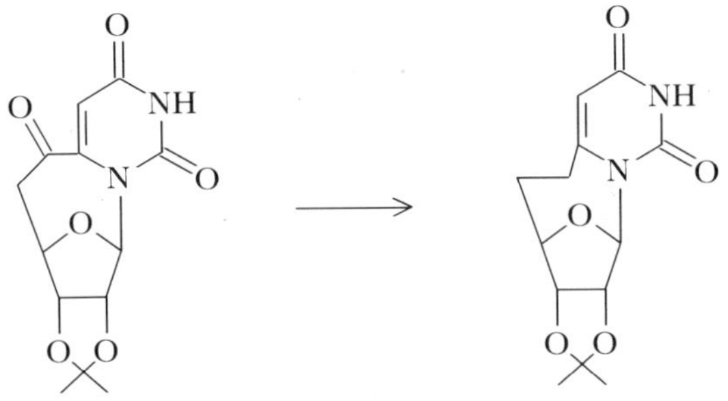

279   280

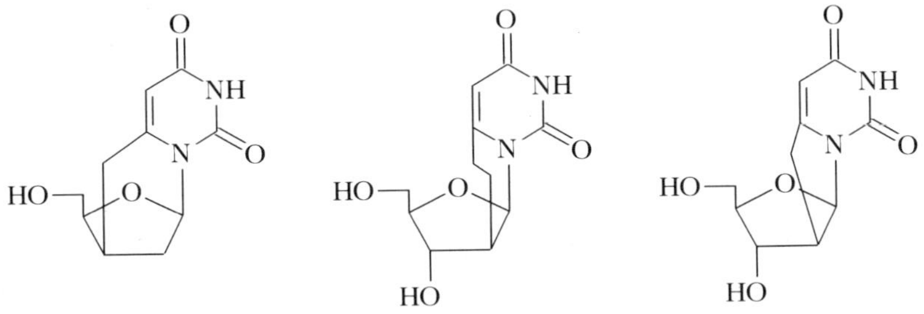

281   282   283

5. Reactions of Pyrimidine Nucleosides in the Sugar Moiety

As described in Section 4, a unique transformation of the sugar moiety in pyrimidine nucleosides often involves cyclonucleoside formation. Although the transformations of the sugar moiety, without participation of the aglycon, are regarded as general reactions in furanose chemistry, it is still worthwhile to discuss and cover this area. Numerous studies have been reported on the protection, such as acylation, acetalation and ketalation, and alkylsilylation, of the hydroxylic functions. The primary purpose of these studies has involved the synthesis of mono- and oligonucleotides; this is not treated in this chapter. Excellent reviews on this, in relation to oligonucleotide synthesis, are available.[585–588]

5.1. Alkylation of Sugar Hydroxyl Groups

A number of 2'-O-methylated nucleosides, as well as the N-methylated derivatives, have been detected as minor constituents of tRNAs. Therefore, studies on specific alkylations of the sugar portion have been an important research area. As already described, the usual procedures alkylate the nitrogen atoms of the aglycons under neutral or mild basic conditions. In strong alkaline solution, where the sugar hydroxyls dissociate, alkylation with alkyl halide occurs on the sugar hydroxyls. Treatment of 2',3'-O-isopropylideneuridine or 5'-O-trityluridine[589] and di-O-trityluridine[590] with benzyl chloride and potassium hydroxide in dioxane gave the respective O'-benzylated derivatives without any concomitant benzylation at N-3. Treatment of the fully protected uridine, except the 2'-hydroxyl, with silver oxide and methyl iodide gave the 2'-O-methyluridine derivative. The use of 2,2,2-trichloro-t-butoxycarbonyl for the N-3 protection and tetraisopropyl-1,3-disiloxanediyl for the 3',5' protection was noticeable in the 2'-O-methylation of uridine by methyl iodide and silver oxide.[591]

With cytidine derivatives, the reaction of dimethyl (or diethyl) sulfate in 1 N KOH solution gave all the possible mono-, di-, and tri-O-methylated cytidines, in which 2'-O-methylcytidine was predominant among the monomethyl derivatives.[592] Arabinosylcytosine was methylated in a similar manner.[593,594] The 5'-O-methylated araC was resistant to a cytidine deaminase.[595] Benzylation of N4-benzoylcytidines, with benzyl halide and sodium hydride, gave the O'-benzyl derivatives.[596] Cytidine itself gave the 2'-O-benzyl derivative along with some of the N-3 benzyl derivative by treatment with benzyl bromide and sodium hydride in dimethylformamide.[597] The use of p-methoxybenzyl bromide was reported recently in this reaction, for the primary purpose of oligonucleotide synthesis, since debenzylation is possible by using triphenylmethyl tetrafluoroborate.[598] 5'-O-Carboxymethylated derivatives of cytidine, araC,[599] and 2'-deoxyuridines[600] were prepared from the suitably protected starting materials. 3'-O-Carboxymethylthymidine was also prepared.[601]

Diazomethane does not usually methylate the alcoholic hydroxyls. However, the reaction of diazomethane with cytidine in aqueous 1,2-dimethoxyethane gave both the 2'- and 3'-O-methylcytidines (in a ratio of 3 : 1).[602] A similar O'-methylation was reported with ribofuranosyl-4-methoxypyrimidin-2-one, and the products were converted to the corresponding cytidine or uridine counterparts.[603] The methylation by diazomethane in the presence of stannous chloride proceeds much better to give, for example, the 2'- and 3'-O-methyl derivatives of cytidine—(285) and (286)—with the 2'-isomer being the major

product in either case.[604,605] The same reaction with adenosine afforded the 3'-O-methyl derivative as the major product. As found later, the reaction with unprotedted uridine does not alkylate the N-3 but instead gives the 2'- and 3'-O-methyl derivatives in 58% and 28% yield, respectively.[186,605] The 5'-O-trityl derivatives of uridine and N^4-benzoylcytidine improved the yields of 2'-O-methylation, when the reaction was done at 0°C.[606] Various metal salts, other than stannous chloride, are also effective as the catalyst.[607] The formation of the complex (284) of $SnCl_2$ with the 2',3'-cis-diol system would be essential for producing the 2'(3')-O-methyl derivatives but not the 5'-O-methyl derivative. Various diazoalkanes have been used in this stannous-chloride-catalyzed alkylation. These include diazoethane,[608] phenyldiazomethane,[609] o-nitrophenyldiazomethane,[610] and 1-oxido-2-picolyldiazomethane.[611] The latter two benzyl functionalities are removable by photoirradiation or acetic anhydride treatment, respectively.

Moffatt and co-workers reported the use of dibutyltin oxide for specific 2'(3')-O-alkylation of nucleosides.[612] Treatment of a ribonucleoside with dibutyltin oxide in methanol gave the 2',3'-O-stannylene derivative (287) as an isolable product.[612] Methylation of this product, with methyl iodide, afforded the 2'(3')-O-methyl derivatives. Benzyl bromide[612] and 4-methoxybenzyl bromide[598] afforded the corresponding benzyl ether. Although cytidine gives the 2',3'-O-stannylene derivative by the similar treatment, the alkylation of this compound occurred predominantly at the N-3 position.[612] Acetylation and tosylation of (287) are possible to afford the corresponding 2'(3')-O-substituted derivatives.[612] In the tosylation of cytidine or its 5'-phosphate by this procedure, the products were O^2,2'-cyclocytidine and araC derivatives.[613] The reaction of o-nitrobenzyl bromide with the stannylene nucleoside gave the 2'- and 3'-O-nitrobenzyl derivatives, which were used in the synthesis of UpU and UpA,[614] since the photolytic deblocking of the o-nitrobenzyl group is possible.

Studies of methylation, using trimethylsulfonium hydroxide in the presence of acetylacetonate of transition metals, has been reported and in the case of cytidine the 2'(3')-O-methyl derivatives were obtained.[615]

Although it is not a simple alkylation, the glycosylation of the 5' position of uridine[616] and 5-fluorouridine[617] by the use of acetobromoglucose has been reported.

5.2. Conversion of Sugar Hydroxyls to Halogeno Groups

For a derivatization of the sugar hydroxyl groups in nucleosides, the arylsulfonylation is the simplest approach. The most susceptible is the primary hydroxyl group (5'-OH), which can readily be converted to the 5'-O-tosyl or -mesyl group. In the 2'-deoxynucleosides, the selective 5'-O-sulfonation is possible.[160] The arylsulfonyl group is an effective leaving group for an S_N2 reaction, although the halogeno group has additional properties. Especially for the deoxygenation, the catalytic hydrogenation of a halogeno group should be most suitable, as in the preparation of 2',3'-dideoxyuridine and 2',3'-dideoxythymidine from their 3'-iodo derivatives[618] and 2',5'-dideoxyuridine[160] and 2',5'-dideoxycytidine[619] from their 5'-iodo derivatives. Tributyltin hydride is the current choice of reagent for the reduction of these halogeno groups.

The treatment of uridine with an excess of tosyl chloride in pyridine at 0°C gave 2',3'-di-O-tosyl-5'-chloro-5'-deoxyuridine.[620] However, a similar treatment of uridine with mesyl chloride afforded the tri-O-mesyluridine,[439a] which was then converted to the 5'-iodo derivative by treatment with sodium iodide. 2',3'-O-Isopropylidene-5'-O-tosyluridine under similar conditions gave the 5'-iodo derivative,[477,621] which was then reduced to the 5'-deoxyuridine by tributyltin hydride.[622] 5'-Deoxy-5-fluorouridine has been reported to be an effective antitumor agent.[623] 3',5'-Di-O-mesylthymidine was converted to the corresponding dihalides by treatment with lithium bromide[470] or lithium chloride.[624]

Procedures for a direct substitution of the 5'-hydroxyl group to the halogeno group have been reported. Treatment of 2',3'-O-isopropylideneuridine[625,626] and 2',3'-O-isopropylidene-N^4-acetylcytidine[627] with methyltriphenoxyphosphonium iodide in dimethylformamide at room temperature gave the respective 5'-deoxy-5'-iodo derivatives. The preferential reaction at the 5'-hydroxyls is evident since uridine, thymidine, or 5-fluorouridine on treatment with 1.1 equivalents of the reagent, also gave the corresponding 5'-iodo derivatives.[623,626] $O^2,2'$-Cyclouridine is converted to the 5'-iodo derivative without cleavage of the cyclo linkage.[626] The secondary hydroxyls will also react, but under somewhat stronger conditions. Ttreatment of 2',5'-di-O-trityluridine with this reagent gave the 3'-deoxy-3'-iodo derivative with inversion of configuration at C-3. This 3'-iodo compound was converted to 3'-deoxyuridine by hydrogenolysis.[628,629] In the case of thymidine, the product was the 3',5'-diiodo derivative with a retention of the configuration at C-3'. In this case, the intermediary formation of the $O^2,3'$-cyclothymidine is postulated. The mechanism in this conversion is fully described by Verheyden and Moffatt.[628]

The introduction of either chlorine or bromine atoms at C-5' was accomplished by the use of triphenylphosphine and carbon tetrachloride or carbon

R = H, OH
X = Cl, Br

288

tetrabromide in dimethylformamide at room temperature. Thus, uridine, thymidine, N^4-acetylcytidine, and 2',3'-O-isopropylideneuridine gave the respective 5'-deoxy-5'-halogeno derivatives (**288**).[630] A prolonged period of time (23°C, for 5 days) for the reaction of uridine or N^4-acetylcytidine under these conditions gave the 2',5'-dichloro derivatives [(**288**), R = X = Cl] as the major products, presumably *via* the O^2,2'-cyclo intermediates. In the case of thymidine, the configuration at C-3' was inverted.[630] A report[484] that claimed the formation of 3'-chloro-3'-deoxyuridine from uridine by treatment with arsenic trichloride seemed to be ambiguous.[630] The product was subsequently found to be 2'-chloro-2'-deoxyuridine.

Another approach that yields the halogeno compounds directly has been reported. Treatment of uridine with halomethylene(dimethylammonium) halide, Vilsmeier reagent, which was produced with arsenic trichloride or bromide in dimethylformamide, gave the appropriate 5'-halo derivative.[484,631] Treatment of cytidine with the Vilmeier reagent (thionyl chloride or phosphoryl chloride in dimethylformamide) gave O^2,2'-cyclocytidine as discussed previously.[460] However, treatment of cytidine with this reagent, in hexamethylphosphorotriamide at room temperature, gave the 5'-deoxy-5'-halogenocytidines in high yield.[632] Similarly, 3'-O-acetylthymidine was converted to the 5'-chloro derivative by this system.[633] The reaction of thionyl chloride with cytidine in acetonitrile gave the 2',3'-cyclic sulfinate of 5'-chloro-5'-deoxycytidinium hydrochloride, which was then converted to 5'-chloro-5'-deoxycytidine, 5'-deoxycytidine, or 5'-chloro derivative of araC.[634] Treatment of 2',3'-O-isopropylideneuridine with excess thionyl chloride afforded the 5',5'-bis-sulfite ester with a trace of the 5'chloro derivative.[635]

Various acyl halides are now used for the preparation of 2'-deoxy-2'-halogenouridine derivatives directly from uridine and uridine derivatives. Treatment of uridine with 2-acetoxyisobutyryl chloride gave the 3',5'-protected 2'-chlorouridine (**289**).[636] In the case of purine nucleosides, the products were 3'-halogeno-xylo and 2'-halogeno-arabinonucleosides.[637] In place of this rather complex acyl halide, acetyl or propionyl bromide (or chloride) in ethyl acetate–dimethylformamide is effective and provides the 2'-halogeno derivatives (**289**).[507,638] The reaction sequence most likely includes the initial formation of a 2',3'-acyloxonium ion, which is then converted to the O^2,2'-cyclouridine derivative followed by the attack of halide ion to furnish the product. Aspirin chloride was also used for the preparation of 2'-chloro-2'-deoxyuridine.[463] However, treatment of 5'-O-benzoyluridine with aspirin chloride afforded both the 2'- and 3'-chloro derivatives in a comparable ratio. The latter compound gave 3'-deoxyuridine by a reduction with tributyltin hydride.[639] The bulkiness of the

289   290

5′-protecting group may affect the mode of attack of the nucleophile to the 2′,3′-acyloxonium intermediate. The reaction of tetrachlorosilane and acetic acid with uridine and 5-bromouridine also gave the 2′-chloro derivatives. In the case of 5-bromouridine, formation of the 3′-chloro derivative was again observed.[464] Aspirin bromide also gave the di-O-acetate of 2′-bromo-2′-deoxyuridine.[640]

Whereas the treatment of cytidine with acetyl bromide gives the $O^2,2′$-cyclocytidine, N^4-acetylcytidine under similar conditions gave 3′-bromo-3′-deoxy-$N^4,2′$-$O,5′$-O-triacetylxylosylcytosine (**290**).[638] This compound gave 3′-deoxycytidine and some 2′,3′-dideoxycytidine on catalytic hydrogenation and deacetylation. With N^4-acetylcytidine, the most stable product from the acetoxonium ion should be the 3′-bromo-xylosyl derivative, (**290**), rather than the cyclocytidine or 2′-bromocytidine derivatives under these reaction conditions. The forced reaction of cytidine with the Vilsmeier reagent was reported to give 2′,5′-dichloro-2′,5′-dideoxycytidine, through the $O^2,2′$-cyclo derivative.[460]

Other reactions involving the 2′,3′-acyloxonium ion formation have been reported. Treatment of 2′,3′-O-benzylideneuridine with N-bromosuccinimide gave $O^2,2′$-cyclouridine and the 2′-bromo-2′-deoxy-5-bromouridine derivative.[641] The mechanism in this conversion is the radical abstraction of an acetal hydrogen, producing the 2′,3′-benzoxonium ion intermediate followed by the conversion already described. The use of N-bromosuccinimide and triphenylphosphine gave the 5′-bromo derivative,[642] probably by a mechanism similar to that observed in the reaction with methyltriphenoxyphosphonium iodide. Treatment of 2′,3′-O-methoxymethylideneuridine with chloro(or bromo)-trimethylsilane also produced an acetoxonium ion that gave both the cyclouridine and the 2′-halogenouridine.[643]

A radical reduction of the halosugar nucleosides to the deoxynucleosides is carried out by tributyltin hydride as described in many instances. Some additional examples are recorded.[633,644,645] The Barton reduction[646] and some modified procedures have been applied successfully to the deoxygenation of the 2′-hydroxyl group of ribonucleosides.[647-650]

Electrolysis of 3′,5′-di-O-ethoxyethyl-2′-bromo-2′-deoxyuridine or 5′-deoxy-5′-iodo-2′,3′-O-isopropylideneuridine in the presence of Cr^{2+} and butanethiol gave the respective deoxynucleosides in good yield.[651] The cathodic reduction of 3′,5′-dideoxy-3′,5′-diiodothymidine was reported to give the cyclopropanosugar nucleosides (**291**).[652]

291

The introduction of a fluoro group has been reported in some instances. 2′-Deoxy-2′-fluorouridine was prepared from $O^2,2′$-cyclouridine as described.[503] Cleavage of $O^2,3′$-cyclothymidine with hydrogen fluoride in dioxane also gave the 3′-fluorothymidine.[653] The use of aluminum fluoride–hydrofluoric acid improved the yield of the above reaction,[654] which was also applied to the 3′-fluorination of $O^2,3'$-cyclo-2'-deoxy-5-fluorouridine,[655] while treatment of 5'-O-tosyl-2',3'-O-isopropylideneuridine with potassium fluoride in ethyleneglycol resulted in the formation of the 5'-(2-hydroxyethyl) derivative along with the 5'-fluoro derivative in low yield.[656] Treatment of 5'-O-tosylates of thymidine, 2'-deoxyuridine, 5-fluorouridine, and N^4-acetylcytidine, with tetrabutylammonium fluoride, gave the 5'-fluoro derivatives in better yields.[657] 5'-Deoxy-5'-fluorouridine and 5'-deoxy-5'-fluorothymidine can be converted to their cytosine derivatives by using the standard thiation–amination procedures.[658]

5.3. Conversion of 5'(and 3')-Hydroxyl Groups to Other Functional Groups

The 5'-O-sulfonyl group, or halogeno group, can be further subjected to a substitution by a direct S_N2 displacement or *via* the $O^2,5'$-cyclonucleosides as intermediates. For example, 5'-azido-5'-deoxyuridine or 5'-azido-5'-deoxythymidine (**292**) was prepared from the 5'-O-tosyl derivative by treatment with lithium azide, which was then hydrogenated to furnish the 5'-amino derivative.[659] Likewise, the 5'-amino derivatives of thymidine,[659] araC,[660] and 5-trifluoromethyl-2'-deoxyuridine[661] were prepared by this method. Treatment of 3'-O-acetyl-5'-azido-5'-deoxythymidine with triphenyl- or trimethylphosphite gave the diester of the phosphoramidate, which can be hydrolyzed to afford 5'-amino-5'-deoxythymidine (**293**).[662] Similarly, a reduction of the azido to amino

292   293

group of thymidine was performed by treatment with triphenylphosphine followed by alkaline hydrolysis.[663] This method was adapted in the preparation of the 5'-amino derivatives of 5-halo-2'-deoxyuridine.[664] Tris(trimethylsilyl) phosphite is also effective for the azido to amino conversion.[665] Treatment of the azido derivative with hydrogen sulfide in pyridine also seems to be efficient as shown in the synthesis of the 5'-amino-2',3'-unsaturated derivative of uridine, thymidine,[666] and cytidine[667] from the respective 5'-azido derivatives. The one-pot synthesis of 5'-azido-5'-deoxynucleosides (**292**) from various nucleosides has been developed, which involves the combination of triphenylphosphine, carbon tetrabromide, and lithium azide.[668,669]

Photoirradiation converts the 5'-azido group into a 5'-aldehyde function, which furnishes an alternative method for the preparation of nucleoside 5'-aldehydes.[670] The reaction of dimethyl acetylenedicarboxylate with 5'-azidothymidine afforded a 5'-(1,2,3-triazol-1-yl) derivative by the cycloaddition.[671] The 5'-amino function of (**293**) was converted to the 5'-(thymin-1-yl) derivative, which was then used in the photodimerization studies as the dinucleoside phosphate models.[672] Further acylation of the amino function of 5'-amino-5'-deoxyuridine and 5'-amino-5'-deoxythymidine was performed, such as carboxy-methylation of 5'-aminouridine,[673] chloroacetylation of 5'-aminothymidine,[674] bridging by alkane dicarboxamide linkage,[675] and sulfonation.[676]

The 3'-hydroxyl group of thymidine can be converted to an azido group by the treatment of 5'-O-tritylthymidine with triphenylphosphine, carbon tetrabromide, and lithium azide to give the 3'-azido-xylo derivative with some concomitant formation of the $O^2,3'$-cyclothymidine. The xylosylthymine gave the 3'-azido derivative in the ribo configuration and again the $O^2,3'$-cyclonucleoside.[669] Treatment of 5'-O-tritylthymidine and 2',5'-di-O-trityluridine with triphenylphosphine and diethyl azodicarboxylate in the presence of hydrogen azide afforded the respective 3'-azido derivatives in the xylo configuration. Replacement of hydrogen azide with methyl iodide in the above reaction with thymidine gave the 3'-deoxy-3'-iodo-xylo derivative with the predominant formation of the $O^2,3'$-cyclothymidine.[677] The 1,3-dipolar cyclo-addition of the 3'-xylo-azide of thymidine, on heating, has been already described.[566]

Substitution of the 5'-O-tosyl group in pyrimidine nucleosides by various nucleophiles has been accomplished for a variety of purposes. Indole and carbazole, as the nucleophile, gave the 5'-N-indolyl or 5'-N-carbazoyl derivatives of uridine and thymidine.[678] 5'-O-p-Cyanophenyluridine was prepared and the cyano function was converted to the amidino or iminoester as the linker to protein.[679] Similarly, the reaction of p-nitrophenoxide with 5'-O-tosylated uridine, cytidine, and 2'-deoxycytidine gave the respective 5'-O-nitrophenyl derivatives, which after hydrogenation to the 5'-aminophenyl derivatives were utilized for azocoupling to various biopolymers such as serum albumin.[680] The reaction of sufur nucleophiles with 5'-deoxy-5'-iodo-2',3'-O-isopropylideneuridine gave the 5'-methylthio,[259] 5'-thioacetyl,[259,681] 5'-mercapto, and 5'-thiocyanato derivatives.[259] The 5'-mercapto derivatives of 2'-deoxyuridine and its 5-fluoro derivatives have also been prepared.[261] Homocysteine was also connected to the 5'-position of uridine.[682] The direct method for the preparation of the 5'-alkylthio derivatives from protected or unprotected uridine and other nucleosides includes the action of 2-mercaptopyrimidine and dimethylformamide di(neopentylacetal) on heating to give the 5'-pyrimidinylthioderivative.[683] Tributylphosphine and dialkyl disulfides convert nucleosides to various 5'-alkylthio-5'-deoxynucleosides in a one-pot reaction.[684] Treatment of 2',3'-O-isopropylidene-N^3 (or 2-O-)-methyluridine with 2,6-di(t-butyl)-4-nitrophenol in the presence of triphenyl-phosphine and diethyl azodicarboxylate gave the aci-nitro ester (294), which is thermally decomposed to give the 5'-aldehyde as shown in the coupling with Wittig reagents.[685]

In an attempt to extend the carbon chain of the sugar portion, 5'-iodo-5'-deoxythymidine was reacted with sodium cyanide in dimethylsulfoxide to give the 5'-cyano derivative.[686] The use of the 5'-O-tosyl derivative resulted in a formation of the $O^2,5'$-cyclo derivative.[687] Hydrogenation of the cyano group followed by deamination gave "homothymidine" (295).[686] Yields for the introduction of the cyano group were improved by using [18]-crown-6 in acetonitrile or dioxane to give, for example, 5'-cyanouridine[688] and

5′-caynothymidine[689] from the respective 5′-*O*-tosyl derivatives or from 5′-iodothymidine.[690] The 5′-aminomethylthymidine was further converted to other reactive functionalities such as 6′-bromoacetamido derivatives.[690]

294   295

Treatment of 5′-*O*-tosylthymidine with methoxide in methanol afforded a 60% yield of 5′-*O*-methylthymidine.[691] This is a rather rare case of tosyloxy to methoxy conversion without any elimination being observed.

A large number of reports have described the introduction of functional groups into the 2′ or 3′ positions. Most of these involved the reaction of cyclonucleoside cleavage, which has already been described.

5.4. Epoxides of Pyrimidine Nucleosides

The 2′, 3′-epoxy-lyxosyluracil (**296**) has been prepared by the treatment of 3′-*O*-mesyl-arabinofuranosyluracil, which was derived from 2′,3′,5′-tri-*O*-mesyluridine *via* O^2,2′-cyclonucleoside formation, under basic conditions.[692] Ammonolysis of (**296**) afforded 3′-amino-3′-deoxyarabinosyluracil (**297**), which was then converted to the corresponding uridine compound by an epimerization of the 2′ position[692]. By starting from (**297**), 3′-amino-2′-mercapto-2′,3′-dideoxyuridine has ben derived by the participation of the 3′-dithiocarbamate intermediate.[693] From 2′,3′-di-*O*-tosyluridine, (**296**) and the isocytosine derivatives were prepared.[694] The latter was eventually converted to the 3′-halogeno-arabino-nucleosides.[502] Cleavage of the epoxide bond of the 5′-*O*-benzoate of (**296**) with hydrogen fluoride afforded both the 3′- and 2′-fluoro derivative in a ratio of 2 : 1. The 3′-fluoroarabinosyl compound was then converted to the cytidine derivative by a standard thiation–amination.[695] On the contrary, a recent report showed the formation of 3′-deoxy-3′-fluorouridine by treatment of (**296**) with hydrogen fluoride, in addition to the expected 3′-fluoro-arabinosyl derivative, and uracil. Further treatment of the 3′-fluoroarabinosyl compound gave the same products along with uracil.[696] The structure of this compound was identified by a preparation from O^2,5′-cyclo-2′,5′-di-*O*-trityluridine with hydrogen fluoride. The mechanism for the formation of this compound from (**296**) is puzzling. Treatment of 5′-*O*-benzoate of (**296**) (and its N^4-benzoylcytosine derivative) with boron trifluoride in acetonitrile afforded the 3′-acetamidoarabino derivatives. In this conversion, boron trifluoride activates the epoxide ring and the nitrogen of acetonitrile becomes a nucleophile to furnish the products.[697]

The 2′,3′-lyxo-epoxide of cytidine has been prepared from 5′-O-trityl-2′,3′-di-O-mesylcytidine *via* the O^2,2′-cyclocytidine intermediate.[697,698] Cleavage of the 2′,3′-epoxy-lyxosylcytosine derivative or (**296**) with Br^-, I^-, N_3^-, and SCN^- afforded the respective 3′-substituted araC and araU (**298**).[699] Formation of 2′,3′-epoxy-lyxo derivative of 5′-benzamido-5′-deoxyuridine from the 2′,3′-di-O-mesyl derivative was also reported.[700]

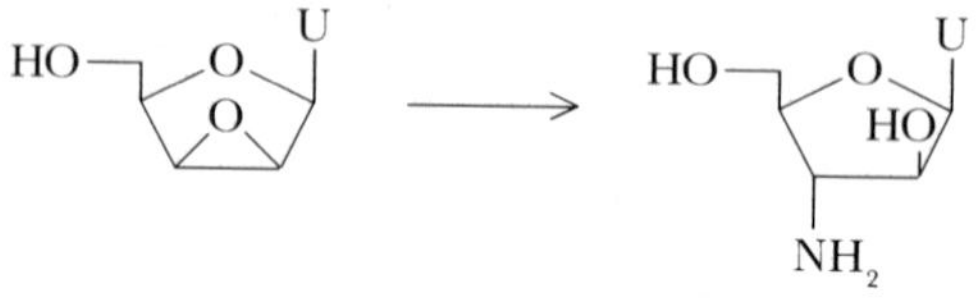

296   297

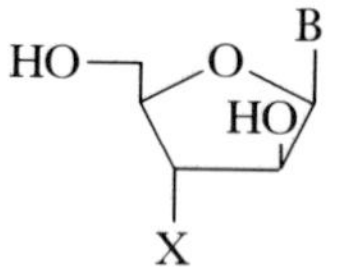

B = U, C
X = Br, N_3, SCN

298

Although the existence of 2′,3′-epoxy-ribofuranosyluracil has been postulated on many occasions, the isolation of this compound has been unsuccessful. For example, treatment of 3′-chloro-3′-deoxy-xylosyluracil with sodium methoxide afforded O^2,2′-cyclouridine through the 2′,3′-ribo-epoxide.[464] However, careful treatment of O^2,2′-cyclouridine (**187**) with 1 equivalent of sodium methoxide in methanol gave the ribo-epoxide, which was then treated with methyl iodide in DMSO to give N^3-methyluridine 2′,3′-epoxide (**299**), isolated in a crystalline form.[701] This compound is in fact very sensitive to hydrolysis. In the case of O^2,2′-cyclo-5-bromouridine, the base treatment followed by methylation gave the O^6,5′-cyclouridine 2′,3′-epoxide. The structures of these 2′,3′-epoxides are confirmed by the x-ray diffractions. A detailed investigation of the methylation of some related 5-bromouridines was recently presented, including the isolation of N^3,5,5′-O-trimethyluridine 2′,3′-epoxide.[702]

The 3′,5′-xylo-epoxide (**300**) and 2′,5′-arabino-epoxide derivative (**301**) of uridine were derived from the 5′-O-mesylates of xylosyl- and arabinosyl-uracils,[534] and 2′,5′- and 3′,5′-epoxy-lyxosyluracils were likewise derived from 5′-O-mesyl-lyxosyluracil by base treatment or from the 2′-O-mesyl derivative of the 3′,5′-epoxide (**300**), respectively.[535] Perhaps the easiest way to prepare (**301**) is by the treatment of uridine with the Vilsmeier reagent to obtain the 2′,5′-dichloro derivative and the successive treatment with alkali as recently reported.[703] The 3′,5′-epoxide of thymidine was prepared by Horwitz and co-workers[704] by the treatment of 3′,5′-di-O-mesylthymidine with 1 N NaOH solution.

The 2′,3′-epithio derivative of lyxosyluracil (**302**) was prepared from 5′-O-trityl-3′-O-mesyl-S^2,·2′-cyclo-2-thiouridine by acid treatment.[554] The riboepithio-epimer (303) was synthesized by a cleavage of the 5'-O-benzoate of (296) with ammonium thiocyanate, followed by 2'-O-mesylation and base treatment.[554] It is to be noted that, whereas the ribo-epoxide of uridine is too labile to exist, compound (303) is fairly stable.

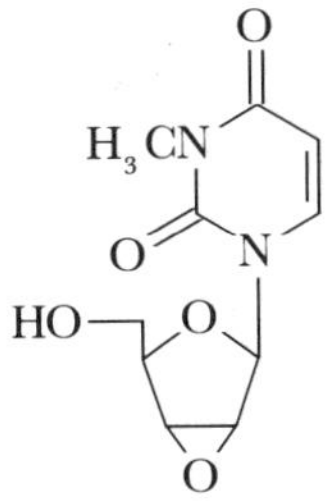

299 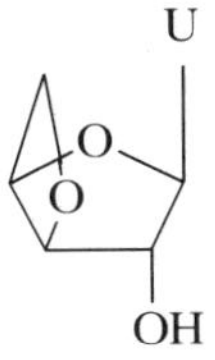   300

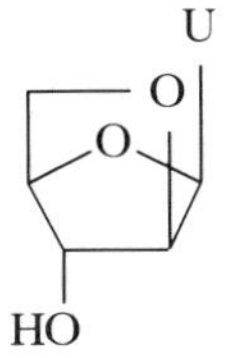

301 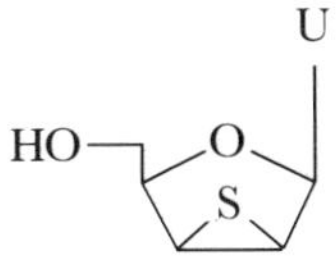   302 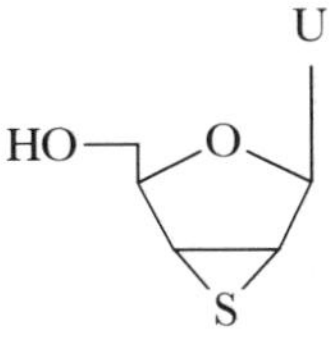   303

5.5. Introduction of Double Bonds in the Sugar Moiety

Among the various nucleoside antibiotics, blasticidin S,[705] angustmycin A,[706] and mildiomycin[707] are the three unsaturated sugar nucleosides that have so far been isolated. This is of considerable interest since the unsaturated sugar nucleosides provide potential intermediates for nucleoside transformations. In principle, four pentenofuranosyl nucleosides are possible—(304)–(307). As is

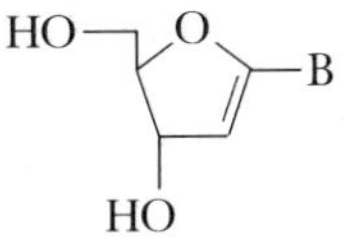

304 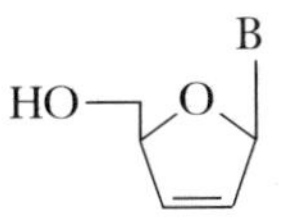   305 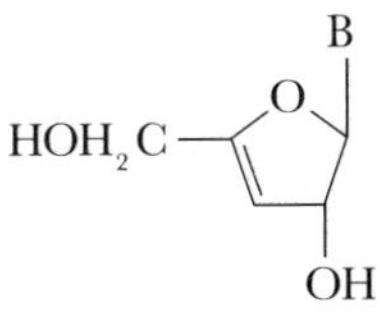   306 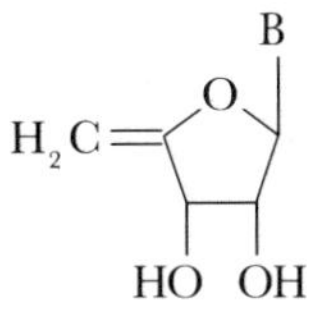   307

expected from the dihydrofurane chemistry, these compounds are not very stable and tend to aromatize by further elimination to afford a furane compound. The first pentenofuranosyl nucleoside was prepared by Horwitz and co-workers,[708–710] who obtained a 2',3'-unsaturated nucleoside (305) starting from the various intermediates by base-catalyzed elimination reactions.

5.5.1. *2′,3′-Unsaturated Nucleosides*

Treatment of 3′-*O*-mesyl-5′-*O*-trityl-2′-deoxylyxosylthymine (**308**) with a strong base at room temperature afforded the 2′,3′-unsaturated thymidine (**309**). Detritylation of (**309**) with hydrochloric acid in chloroform gave the product, 1-(2,3-dideoxy-β-D-*glycero*-pent-2-enofuranosyl)thymine (**305**). Starting from the more readily accessible 3′-*O*-mesyl-5′-*O*-tritylthymidine, *via* the O^2,3′-cyclo derivative (**310**), also gave the product (**305**).[708–710] This method was also used for the synthesis of 2′,3′-unsaturated uridine[710] and 5-fluorouridine.[711] The hydrogenation of (**305**) gave the known 3′-deoxythymidine.[470,710,712] 3′,5′-Di-*O*-mesylthymidine was also converted to (**305**) *via* the 3′,5′-epoxy intermediate (**311**).[710] The 2′-pentenofuranosylthymine was also prepared from 3′-chloro (or fluoro)-3′-deoxythymidine by *t*-butoxide treatment.[653] The cytidine derivative of (**305**) was likewise prepared by this method from the respective 2′-deoxynucleoside by way of the 3′,5′-di-*O*-mesylate, O^2,3′-cyclonucleoside, and the 3′,5′-epoxide. Hydrogenation of the double bond furnishes 2′,3′-dideoxycytidine.[713]

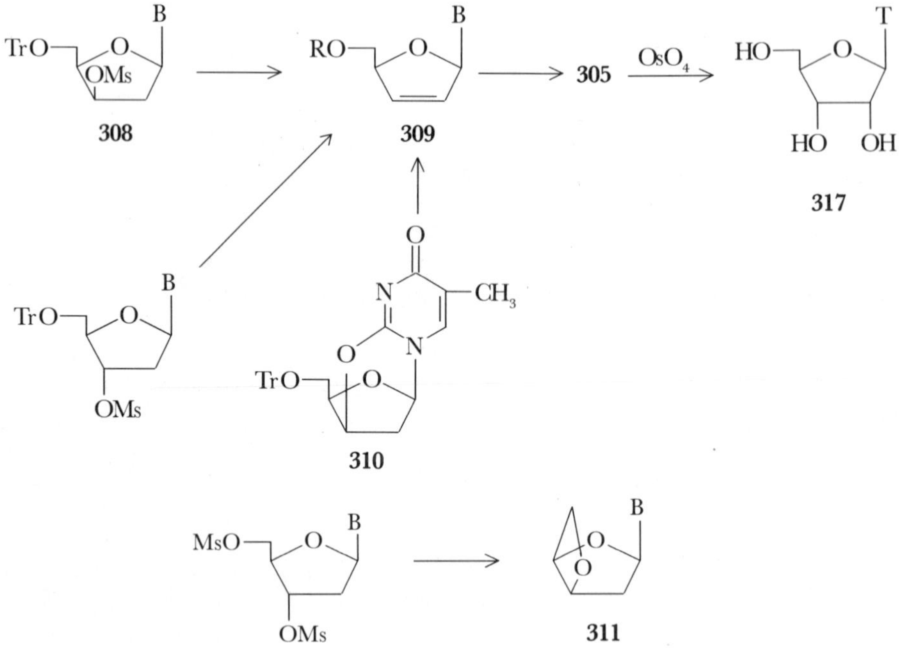

Treatment of 5′-*O*-mesyl-O^2,3′-cyclothymidine with base yielded a thyminylfurane derivative (**314**) by a two-step [(**312**) → (**313**)] elimination.[710] Attempts to prepare the O^2,5′-cyclothymidine possessing a 2′,3′-unsaturated function, starting from the 5′-iodo derivative, also resulted in the formation of the furane derivative (**314**).[714] Treatment of 5′-deoxy-5′-fluoro-3′-*O*-mesylthymidine with *t*-butoxide gave the requisite product, along with a major by-product (**314**) and a 3′,4′-unsaturated compound.[715] The 2′,4′-diene of thymidine (**313**), a postulated intermediate to (**314**), was reported to be formed by the elimination of

$3',5'$-dideoxy-$3',5'$-diiodothymidine with silver fluoride or of $5'$-iodo-$2',3',5'$-trideoxypent-2-enofuranosylthymine with DBN.[627] More recently, the $2',4'$-dienes (**313**) were prepared from the $3',5'$-dichloro derivatives of deoxyuridine, deoxycytidine, and thymidine by treatment with dilute alkali. They were then isomerized to the endodienes (**314**) by treatment with a stronger base.[716]

312   313   314

The $2',3'$-unsaturated nucleosides can also be derived from the ribonucleosides. Treatment of $5'$-O-benzoyl-$3'$-deoxy-$3'$-iodo-$2'$-O-mesylarabinosyluracil (**315**) with sodium iodide gave the desired $2',3'$-dehydrouridine (**316**).[710]

315   316

Chromous acetate was found to be effective for the preparation of (**316**) as exemplified in the treatment of $3'$-O-acetyl-$2'$-chloro-$2'$-deoxyuridine.[717] Since the reaction proceeds by a radical mechanism, the generated C-$2'$ radical may result largely in an elimination of uracil. On the other hand, elimination from $3'$-deoxy-$3'$-halogeno-xylosyladenines proceeds smoothly to give the $2'$-ene derivative.[717] An elimination of the halogeno group by electrolysis was reported to be effective in giving the $2',3'$-unsaturated uridine and cytidine from a $2'$-bromo-$2'$-deoxyuridine and a $3'$-bromo-xylosyl-N^4-acetylcytosine, respectively.[718] Desulfurization of the $2',3'$-episulfide of uridine (**302**) with triphenylphosphine afforded (**316**) in a high yield.[554] Platinum-catalyzed hydrogenation of $2'$-bromo-$2'$-deoxy-$3'$-O-mesyl-$5'$-O-benzoyluridine yielded the $2',3'$-dideoxyuridine. This was explained by assuming that the initial formation of a $2',3'$-enebromide was the result of an elimination of methanesulfonic acid and successive hydrogenation of this function.[719] The elimination reaction of $2',3'$-di-O-mesyl derivatives of lyxosyluracil, on treatment with sodium benzoate at elevated temperature, gave the $2'$-keto-$3'$-deoxyuridine.[720,721] The borohydride reduction yielded the $3'$-deoxyarabinosyluracil.[722] Similarly, treatment of $5'$-O-trityl-$2'$-$3'$-di-O-mesyl-lyxosylcytosine afforded $2'$-keto-$3'$-deoxycytidine.[723] It seems uncertain whether the elimination of the methanesulfonate was regiospecific or not, since the other expected product, a $2'$-deoxy-$3'$-keto derivative, would be very unstable[724] and most likely would not be isolated under these conditions.

Oxidation of 2′,3′-unsaturated thymidine with osmium tetroxide gave ribosylthymine (317). This reaction illustrates the conversion of a 2′-deoxyriboside to a riboside *via* the 2′,3′-unsaturated intermediate.[725]

5.5.2. 4′,5′-Unsaturated Nucleosides

A unique structural feature of the nucleoside antibiotic angustmycin A[706] is the presence of a 4′,5′-unsaturated bond. A synthesis of this nucleoside has been accomplished[726] from another closely related antibiotic—angustmycin C (psicofuranin). A similar sequence was used in the synthesis of a 4′,5′-unsaturated uridine (318a) by the reaction of 5′-O-tosyl-2′,3′-O-isopropylideneuridine with *t*-butoxide.[727] However, an attempt to remove the isopropylidene protecting group of (318a) was unsuccessful. This was overcome by the treatment of 2′,3′-di-O-acetyl-5′-deoxy-5′-iodouridine with silver fluoride in pyridine followed by deacetylation with ammonia to give (318b).[627,728] By a similar procedure, 4′,5′-dehydrothymidine[627,657] and 4′,5′-dehydrocytidine[627] were prepared. The 5′-bromo compound was also converted to the 4′,5′-dehydro product by *t*-butoxide treatment.

Treatment of (318a) with an equivalent amount of *N*-bromosuccinimide gave a low yield of a new type of cyclonucleoside, O^2,4′-cyclo-5′-bromo-5′-deoxy-

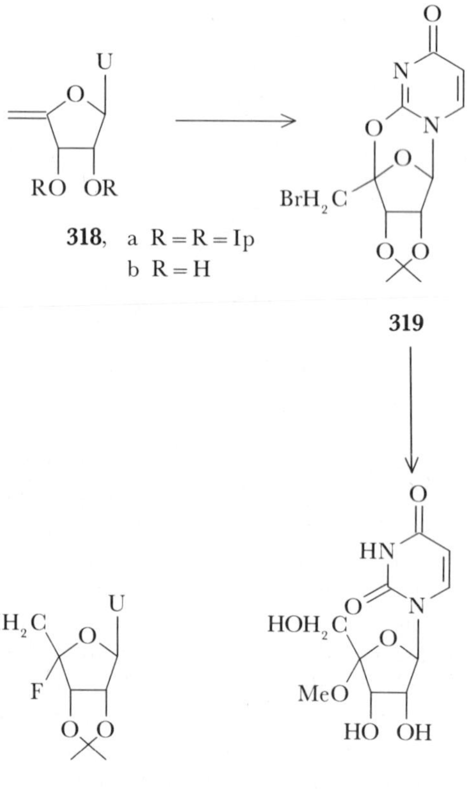

318, a R = R = Ip
       b R = H

319

321

320

uridine (**319**), along with other products.[729] Further treatment of this compound with aqueous methanol afforded 4'-methoxyuridine (**320**). The reaction of *t*-butyl hypochlorite with the 4',5'-dehydro derivatives of 2',3'-*O*-anisylideneuridine and 2',3'-*O*-anisilydene-N^4-benzoylcytidine gave the respective O^2,4'-cyclo-5'-chloro derivatives in better yields.[730] The reaction of several 4',5'-dehydrouridines with iodine in methanol has been extensively studied by Verheyden and Moffatt.[731,732] These reactions are all initiated by the electrophilic attack of iodine at the 5' position leaving the 4'-carbonium ion, which is converted to the β-D-ribo and α-L-lyxo derivatives or O^2,4'-cyclouridine derivatives. The 5'-iodo-4'-methoxy-β-D-ribo derivative was converted to 4'-methoxyuridine (**320**) by treatment with lithium benzoate followed by debenzoylation with methanolic ammonia.[732] Starting from (**318a**), the reaction with iodine fluoride afforded 5'-deoxy-4'-fluoro-5'-iodo-2',3'-*O*-isopropylideneuridine (**321**). The 5'-iodo function was then converted to various 5'-substituted 4'-fluorouridines, including 4'-fluoroudine and its O^2,5'-cyclonucleoside.[733]

A O^2,2'-cyclo derivative of 4',5'-dehydrouridine was prepared by the treatment of (**318b**) with diphenyl carbonate. The 3'-hydroxyl group of 4',5'-dehydrothymidine was converted to the 3'-iodo compound by treatment with a phosphonium iodide.[734]

5.5.3. 3',4'-Unsaturated Nucleosides

The 5'-carboxylic acid derivatives of thymidine, deoxyuridine, or their 3-methyl or 5-fluoro derivatives were converted to the 3'-*O*-mesyl-5'-ethyl esters (**322**). On mild base treatment, such as triethylamine or sodium benzoate, these compounds were converted to the 3',4'-dehydro derivatives (**323**). The 5'-ester group of (**323**) was reduced by sodium bis(ethoxymethoxy)aluminum hydride to yield the 3',4'-dehydro nucleosides (**324**).[735,736] Treatment of thymidine 5'-carboxylate with acetic anhydride under reflux, followed by esterification, also afforded (**323**).[735,736] Preparation of the 3',4'-dehydroribonucleosides from pyrimidine ribonucleoside carboxylates was rather unsuccessful.[735]

322   323   324   325

Catalytic hydrogenation of (**324**) (B = thymine) seems to be stereospecific, giving 3'-deoxythymidine (**325**) as the sole product.[735] In contrast, the hydrogenation of 3',4'-dehydroadenosine proceeded without stereoselectivity.[737] The presence or absence of the 2'-hydroxyl group might be a controlling factor in the site of hydrogen attack. Although the synthetic utilities or biochemically distinct properties of these 3',4'-dehydronucleosides have not been realized, the polymerization of their 5'-phosphates, if possible, could open a new area of

polydeoxynucleotide synthesis. The Wittig reaction between 2′,3′-O-isopropyl-
ideneuridine-5′-aldehyde and methoxycarbonylmethylene-triphenylphosphorane
gave the 4′,5′- and 3′,4′-unsaturated nucleosides instead of the expected
5′,6′-unsaturated derivatives.[738] More recent results on this aspect are described
later.

Treatment of thymidine 5′-carboxylic acid with dimethylformamide
di(neopentylacetal) afforded 2-(thymin-1-yl)-2,3-dihydrofurane, which was
hydrogenated to give the tetrahydrofuranylthymine.[739]

5.5.4. 1′,2′-Unsaturated Nucleosides

The synthesis of a 1′,2′-dehydrated nucleoside has been reported by Robins
and Trip.[740] Treatment of O^2,2′-cyclouridine with 2-methoxypropene gave a
3′,5′-di-O-protected pent-1-enofuranosyluracil. Palladium-catalyzed hydrogenation and successive acid-catalyzed deprotection gave α- and β-2′-deoxyuridine.
The free 1′,2′-dehydrouridine (327) was obtained from 3′,5′-di-O-(*t*-butyl-
dimethylsilyl)-O^2,2′-cyclouridine by the above reaction followed by desilylation of
the product. Compound (327) is reported to be very unstable in acid.

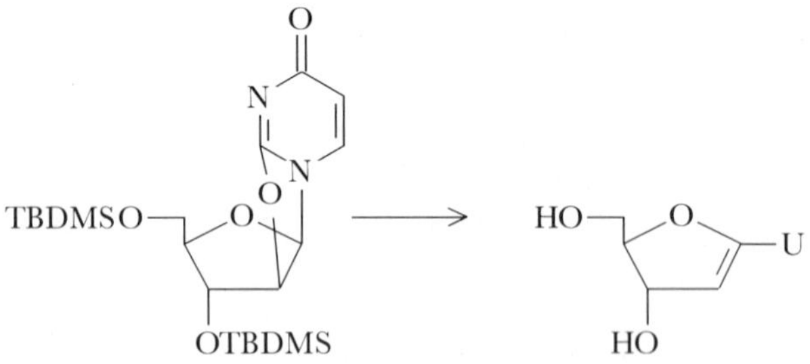

326   327

5.6. Oxidation of the Sugar Moiety

Oxidation of the primary hydroxyl group of pyrimidine nucleosides into a
5′-carboxylic acid has been accomplished by air oxidation catalyzed by platinum
at pH 9. Thus, uridine, thymidine, and purine nucleosides were converted to their
5′-carboxylates (328).[741] This procedure was adapted to other nucleosides
including 2′-deoxy-5-halouridines and arabinosyluracil and arabinosylcytosine.[134,742–744] Oxidation of thymidine and 2′-deoxycytidine, with chromic
acid in pyridine, also gave the respective 5′-carboxylates.[742] The permanganate
oxidation of 2′,3′-O-isopropylideneadenosine to furnish the 5′-carboxylic acid
seems to be restricted to the purine nucleosides.[745] Oxidation of thymidine
3′-phosphate with hydrogen peroxide in the presence of platinum catalyst gave
the 5′-carboxylic acid.[746] This and the foregoing air oxidation method were
originally attempted for the study of a stepwise degradation of polynucleotide
chains by elimination reaction, which could be induced by the introduction of the
electron attracting 5′-carboxylic acid moiety at the 5′ end of the nucleotide

sequence. The 5'-carboxylic acid function can be converted to the cyano function *via* the ester and amido intermediate. The 5'-nitrile compounds were condensed with azide ion to afford the 5'-tetrazolo derivative as reported in the conversion of 2'-deoxyuridine and 2'-deoxy-5-fluorouridine.[744] The anodic oxidation of 2',3'-O-isopropylideneuridine-5'-carboxylic acid in methanol gave the 4'-aldehydic product, which was isolated as the 4'-hemiacetal (329) and the O^2-4'-cyclonucleoside (330).[747]

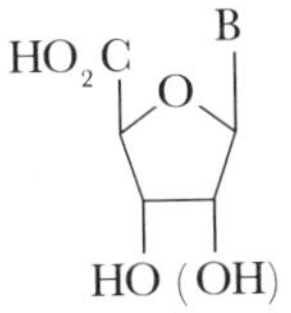

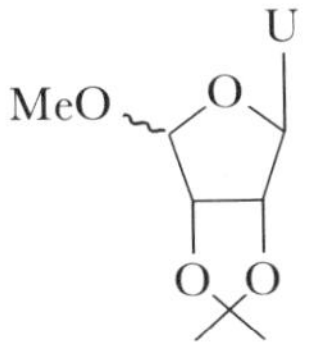

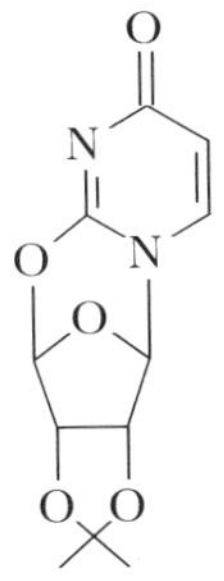

An extremely mild, yet efficient, oxidation of both primary and secondary alcohols to aldehydes and ketones was developed by Pfitzner and Moffatt.[724,748] Treatment of 3'-O-acetylthymidine with phosphoric acid and dicyclohexylcarbodiimide (DCC) in dimethylsulfoxide gave the 5'-aldehyde (331). Further oxidation of this aldehyde with hypoiodous acid gave the 5'-carboxylate.[724,748] Treatment of 2',3'-O-cyclohexylideneuridine with diisopropylcarbodiimide and dimethylsulfoxide in the presence of dichloroacetic acid gave the 5'-aldehyde (332), which was purified by making the N,N'-diphenylethylenediamine adduct.[749] Similarly, oxidation of 2',3'-O-isopropylideneuridine with DCC, trifluoroacetic acid, and dimethylsulfoxide gave the 5'-aldehyde in a crystalline form as the methyl hemiacetal.[750]

Several studies on an extension of the sugar chain by the use of the 4,5'-aldehydes were reported. Treatment of (332) with sodium cyanide, followed by peroxide, gave the epimeric hydroxyamide (333). The talo-epimer was mesylated, the mesyloxy group was replaced with azide, and acid hydrolysis afforded the

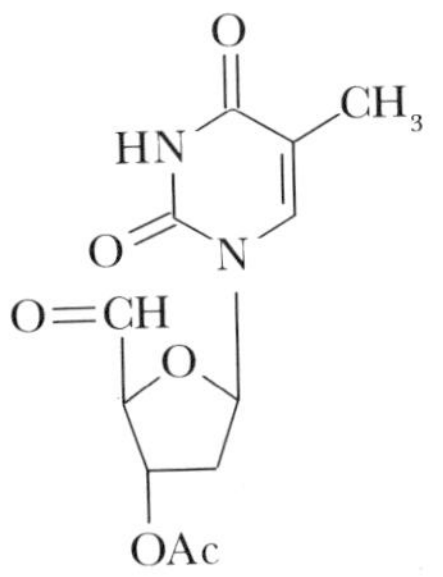

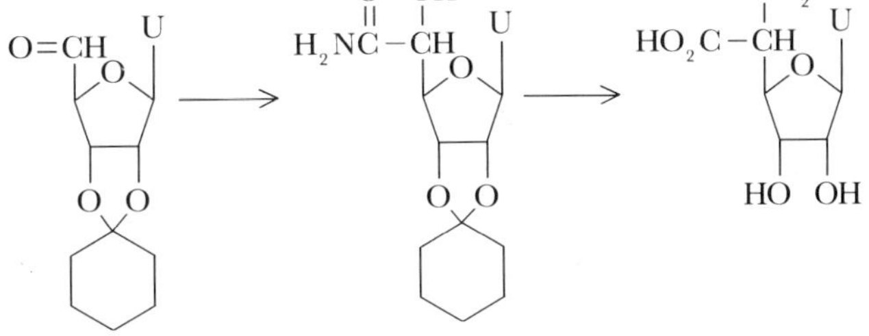

5'-azido-allofuranosyluracil-uronate. Palladium-catalyzed hydrogenation afforded the 5'-amino derivative (334),[749] the skeleton of the antibiotic polyoxins.[751] Treatment of (332) with acetic anhydride and potassium carbonate afforded the enol acetate (335), which on treatment with *N*-iodosuccinimide gave the $O^2,4'$-cyclouridine derivative (336).[752] Various reactions, starting from the enol acetate (335), were described in this report. The uridine 4',5'-enamine (337) was also prepared from (332),[753] which on treatment with allyl bromides gave the 4'-propenyl derivatives (338). These were further converte to the 4'-alkyl and 4'-hydroxyalkyl derivatives of uridine.[753] In contrast to the previous report,[738] the Wittig reaction of 2',3'-*O*-isopropylideneuridine-5'-aldehyde with acetyl-methylenetriphenylphosphorane gave the expected product, which was then deacetonized to furnish the 6'-acetyl-5',6'-dehydro derivative of uridine (339).[754]

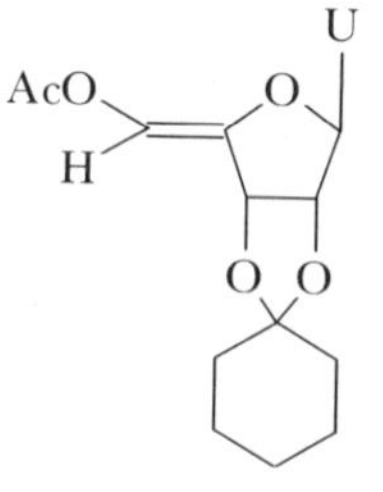

335

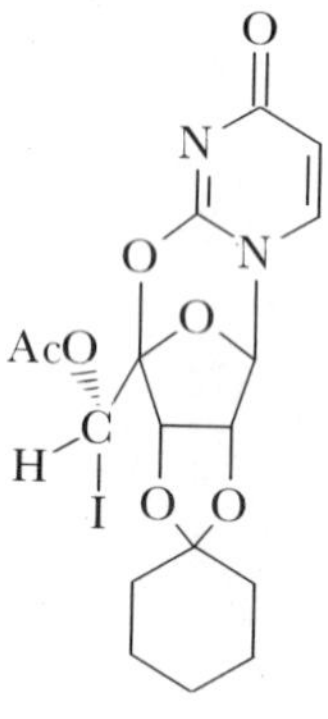

336

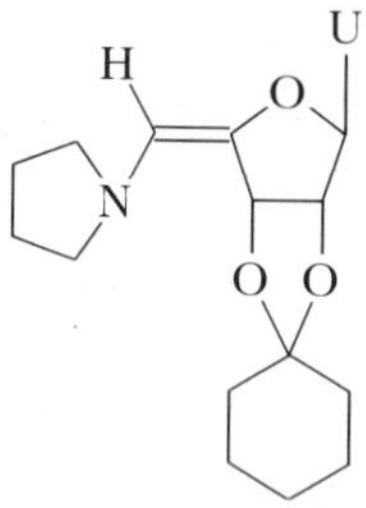

337

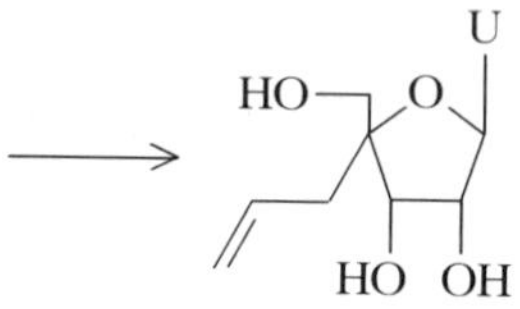

338

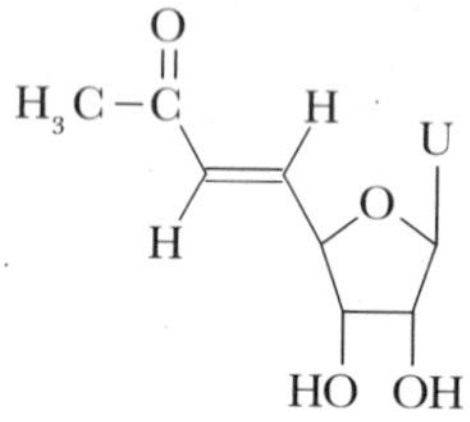

339

A Wittig reaction of the same 5'-aldehyde with diphenylphosphomethyl-enetriphenylphosphorane, giving finally a homouridine phosphonate, has been reported.[755] An aldol addition of nitromethane to 3'-*O*-acetylthymidine-5'-aldehyde afforded the expected nitroalcohol. This compound was then reduced to the 5'-aminomethyl derivative of thymidine.[756] Adaptation of the azolactone method to the 5'-aldehyde of uridine afforded the aminohepturonic acid nucleoside.[757]

An interesting crossed reaction of (332) (and other nucleoside 5'-aldehydes) with formaldehyde was reported.[758,759] Treatment of (332) with excess formaldehyde in aqueous sodium hydroxide afforded the 4′-hydroxymethyluridine (**340**). Borohydride treatment of the initially formed 4′-hydroxymethyluridine-5′-aldehyde gave the desired product in much better yields. When this reaction was accomplished with unprotected uridine-5′-aldehyde (prepared from 2′,3′-*O*-anisylideneuridine), the product was a mixture of 4′-hydroxymethyl nucleosides that were epimeric at C-3′.[759] This may be due to a retro-aldol reaction after the 4′-hydroxymethylation step. 3′-*O*-Benzoylthymidine 5′-aldehyde and 2′,3′-*O*-isopropylidene-*N*⁴-benzoylcytidine also gave the respective 4′-hydroxymethyl-thymidine and 4′-hydroxymethylcytidine by similar treatment, after suitable deprotection.[759] 4′-Hydroxymethylation of the sugar 5-aldehyde was also carried out first and the sugar component was then condensed with various purines and pyrimidines to give the 4′-hydroxymethylated nucleosides.[760]

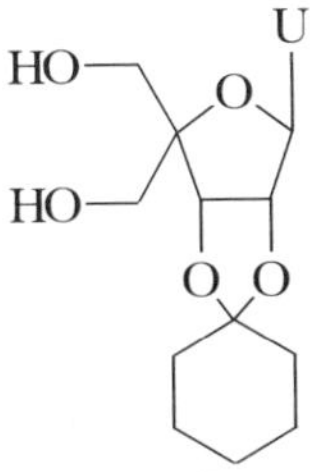

340

2′-Keto-3′-deoxyuridine was prepared by the Pfitzner–Moffatt oxidation of 3′-deoxyarabinosyluracil.[761]

Although the oxidation of the 3′-hydroxy group of 2′-deoxyribonucleosides gives the 3′-keto nucleosides, these were unstable and resulted in a rapid cleavage of the glycosyl bond.[724,748] However, photolysis of 3′-*O*-pyruvyl-5′-*O*-trityl-thymidine gave the 3′-keto derivative (**341**). This compound is unstable but was

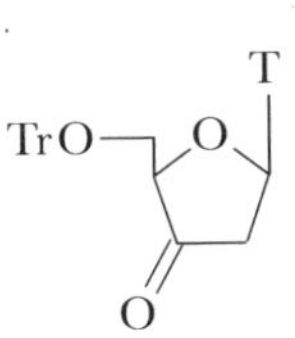

341 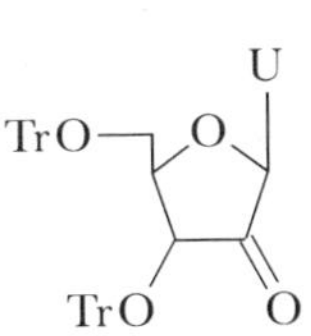   **342** 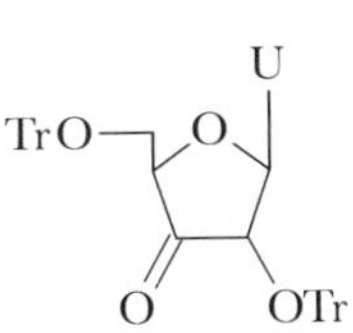   **343**

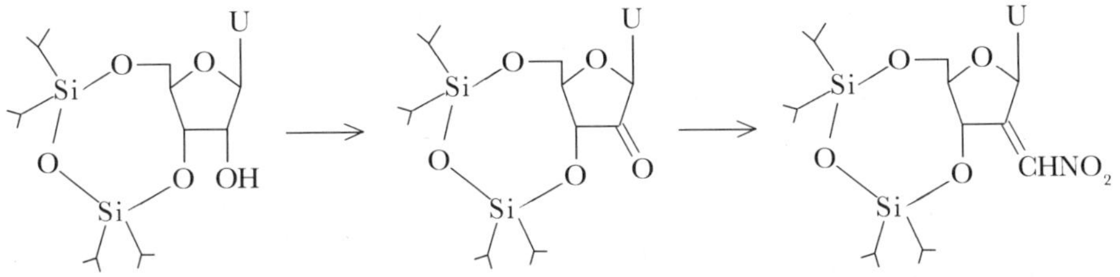

344   **345**   **346**

converted to the xylosylthymine by borohydride reduction.[762] Recently, 5'-*O*-trityl-3'-ketothymidine was prepared by an oxidation with a complex of chromium trioxide–pyridine–acetic anhydride.[763,764] The borohydride reduction of the product gave the xylosylthymine as the major product along with some thymidine.[764] The Moffatt oxidation of 2'(3'),5'-di-*O*-trityluridine[765] and 2'(3'),5'-di-*O*-tritylcytidine[766] proceeded to give the respective 2'- and 3'-keto derivatives of uridine and cytidine, (342) and (343). The compounds were stable enough to be detritylated by hydrochloric acid in chloroform. The oxidation of 3',5'-*O*-(tetraisopropyldisiloxane-1,3-diyl)uridine (344), a 3',5'-bis-protected derivative recently introduced by Markiewicz,[767] with DCC, dimethylsulfoxide, pyridine, and trifluoroacetic acid in benzene afforded the 2'-ketouridine derivative (345) in high yield.[574,583] This keto function was reacted with various Wittig reagents or carbon nucleophiles, such as nitromethane, to give the respective 2'-carbon-substituted derivatives [e.g., (346)]. Some of these compounds were used in the preparation of C-cyclonucleosides.[574,583] The chromium trioxide reagent was also used in the oxidation of the same 3',5'-silylated nucleosides and 2'(3'),5'-di-*O*-(*t*-butyldimethylsilyl) nucleosides.[764]

The oxidative cleavage of the 2',3' bond of uridine (and other ribonucleosides as well) with sodium periodate gives a nucleoside dialdehyde (347), which was then recyclized with nitromethane to give the 3'-deoxy-3'-nitro-β-D-gluco-pyranosyluracyl (348).[768,769] Formation of some galacto-epimer was recently noted in this cyclization.[770,771] A reduction of the nitro group furnished 3'-amino-3'-deoxyglucosyluracil (349).[768] The nitrite deamination of (349) gave, other than uracil, 3'(*R*)-formyl-3'-deoxyuridine (350) in a hemiacetal form, which was then reduced to give 3'-deoxy-3'(*R*)-hydroxymethyluridine (351).[771] The 2',3'-dideoxy-3'(*R* and *S*)-hydroxymethyluridines were also prepared by a multistep conversion of 3'-deoxy-3'-nitroglucosyluracil.[771]

The periodate cleavage and successive cyclization with nitromethane were also performed with cytidine to give the 3'-deoxy-3'-nitroglucosylcytosine as the major product.[769,772] Cyclization of (347) with nitroethane also gave the 3'-methyl-3'-nitroglucosyluracil.[773] The periodate oxidation of glucopyranosyl-

347

348

349

350

351

thymine was carried out to give the dialdehyde, which was again cyclized with nitromethane.[774] The chemical conversion of naturally occurring ribonucleosides to the hexopyranosyl nucleosides is of some interest in view of the fact that cytosine nucleoside antibiotics often contain the hexopyranosyl component, for example, gougerotin, blasticidin S, amicetins, and mildiomycin.

As is expected, the nucleoside dialdehydes do not exist as the free aldehyde form. The nature of the structure of ribonucleoside dialdehydes has been discused in detail.[775] The nucleoside dialdehydes have shown some antitumor activities,[776,777] which may be due to the general toxicities caused by cross-linking of proteins with the dialdehyde function.

6. Reactions Involving Cleavage of Glycosyl Bond and Anomerization

An intramolecular glycosyl migration is observed in the acid treatment of $O^2,2'$-cyclouridine derivatives. Treatment of di-O-benzoyluridine with poly-phosphoric acid at 60°C gave, after debenzoylation, a mixture of 5'-nucleotides including 3-ribosyluracil and 3-arabinosyluracil. The formation of these compounds involved an initial formation of the $O^2,2'$-cyclouridine derivative in which the glycosyl bond had migrated from N-1 to N-3.[778] In an attempt to introduce a bromo group at the 2' position, $O^2,2'$-cyclouridine-6-carboxamide (**352**) was treated with hydrobromic acid in trifluoroacetic acid at 0°C. However, the product was the $O^2,2'$-cyclo derivative of the 3-arabinoside derivative (**353**).

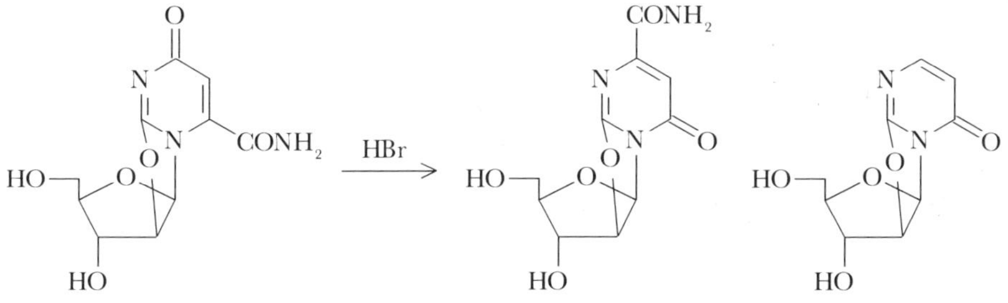

The bulkiness of the 6-carboxamide group would induce the glycosyl bond cleavage and then recyclization would furnish the more stable 3-ribosyl isomer (**353**).[779,780] Similarly, treatment of $O^2,2'$-cyclouridine (**187**) with hydrofluoric acid at 80°C gave a mixture in which the ratio of $O^2,2'$-cyclo-3-β-D-arabinosyl-uracil (**354**) to the starting cyclouridine was approximately 4 : 1.[780] No indication of glycosyl bond migration was obtained in the solvolysis of $O^2,3'$- or $O^2,5'$-cyclouridines with hydrofluoric acid.[780]

Since anomerizations require prior cleavage of the lactol ring, the dihydropyrimidine nucleosides should be more susceptible toward anomerization, providing that the degree of anomerization is parallel to the degree of ease of glycosyl bond cleavage. It has been reported that in the alkaline or acidic hydrolyzates of commercial RNA very small amounts of α-(2′ and 3′)-cytidylic acids were detected.[781] It is not obvious what caused this anomerization, but since no anomerization was detected by base or acid treatment of β-cytidylic acid,[781] it must have occurred at the polynucleotide level. On the other hand, photoirradiation of CMP and 2′-deoxytidine was reported to induce anomerization to give the α-cytidines.[782] During photoirradiation, hydration is known to occur, which implies that the anomerization will take place rather readily with the 5,6-dihydropyrimidine nucleosides and nucleotides. The anomerization (and a lactol ring isomerization) of 2′-deoxyuridine by treatment with bromine in water[783] seems to be such a case. The 5,6-dihydro-6-sulfonate derived from thymidine and 2′-deoxyuridine also anomerized and isomerized in acidic medium, 1 M HBr at 45°C for 30 min, and over 20% of α-deoxyuridine along with the α- and β-deoxyribopyranosyl derivatives were formed.[784] During a cleavage of the 2′,5′-epoxide (**301**) of uridine with hydrobromic acid, the formation of the α-arabinoside (**355**) was observed in addition to the 5′-bromo-5′-deoxyarabinofuranosyluracil (**356**).[785]

The base-catalyzed anomerization of 5-formyluridine was reported, and the mechanism for this anomerization was presented.[786] This mechanism suggests that pyrimidine nucleosides having an electron-withdrawing substituent at position 5 generally anomerize by base treatment.

One useful physical method of determining the anomeric configurations of nucleosides was reported by Imbach and co-workers. This method is based on a difference of the NMR chemical shifts for the 2′,3′-O-isopropylidene methyl groups in both anomers of ribonucleosides. Whereas the chemical shift differences of the two methyl groups in the β-ribosides (Δδ) are greater than 0.18 ppm, those of the α-anomer are under 0.15 ppm.[787] The mode of multiplicity of the 4′ protons is different and this can also be used to distinguish anomeric configurations of ribo- and deoxyribo-nucleosides and -nucleotides, with the α-anomers showing a triplet and β-anomers a multiplet.[788]

Recently, the transglycosylation of peracylated pyrimidine nucleosides with suitable purine bases giving purine nucleosides has been reported.[789–795] Similarly, perbenzoylated 3′-deoxyadenosine (cordycepin) was converted to 3′-deoxyuridine by a stannic-chloride-catalyzed transglycosylation.[796] As these results show, this method is especially useful for the preparation of base-modified nucleosides from the sugar-modified nucleoside antibiotics without isolating the sugar part of the nucleoside antibiotics.

Studies on the mechanism of acidic hydrolysis of pyrimidine nucleosides have been performed kinetically. Shapiro and Danzig have reached the conclusion,[797] by a determination of the pH-rate profile, that the acidic hydrolysis of 2'-deoxy-uridine and 2'-deoxycytidine did not involve the previously proposed[798,799] lactol ring protonation followed by Schiff base formation. They have concluded that it most likely involved a direct attack to the mono- or diprotonated form of the nucleosides with water. 2'-Deoxypyrimidine nucleosides undergo slow hydrolysis between pH 3 and 7, probably by an S_N1 mechanism.[800] The influence of a 5-halogeno group on the thermal bond cleavage of 2'-deoxyuridines was measured and the rate of thermolysis was found to parallel the Hammett *meta*-substituent constants of the 5-substituents.[801]

The glycosyl bond is also cleaved with strong alkali in nonaqueous media, with the elimination being initiated by the dissociated sugar hydroxyl group. The fully protected (on the sugar portion) nucleosides are therefore rather stable.[802] The bisulfite–oxygen system was reported to accelerate the cleavage of glycosyl bonds of pyrimidine ribonucleosides, while those of 2'-deoxynucleosides are resistant.[803] High-intensity UV irradiation is also reported to cause glycosyl bond cleavage in thymidine, 2'-deoxyadenosine, and adenosine.[804] 2-Thiopyrimidine nucleosides, when alkylated at S-2, tend to be hydrolyzed readily because of the formation of the pyrimidinium structure.[805,806]

7. References

1. J. J. Fox and I. Wempen, *Adv. Carbohydr. Chem.* **14**, 283 (1959).
2. A. M. Michelson, *The Chemistry of Nucleosides and Nucleotides*, Academic Press, New York, 1963.
3. T. Ueda and J. J. Fox, *Adv. Carbohydr. Chem.* **22**, 307 (1967).
4. R. H. Hall, *The Modified Nucleosides in Nucleic Acids*, Columbia University Press, New York, 1971.
5. N. K. Kotchetkov and E. I. Budowski, eds., *Organic Chemistry of Nucleic Acids*, Vols. 1 and 2, Plenum Press, New York, 1971.
6. L. Goodman, in *Basic Principles in Nucleic Acid Chemistry*, Vol. 1, p. 93 (P. O. P. Ts'o, ed.), Academic Press, New York, 1974.
7. R. J. Suhadolnik, *Prog. Nucleic Acids Res. Mol. Biol.* **22**, 193 (1979).
8. R. J. Suhadolnik, *Nucleosides as Biological Probes*, Wiley, New York, 1979.
9. R. T. Walker, E. De Clercq, and F. Eckstein, eds., *Nucleoside Analogues. Chemistry, Biology and Medical Applications*, Plenum Press, New York, 1979.
10. K. H. Scheit, *Nucleotide Analogs. Synthesis and Biological Functions*, Wiley, New York, 1980.
11. (a) G. H. Hilbert and T. B. Johnson, *J. Am. Chem. Soc.* **52**, 2001, 4489 (1930); (b) G. A. Howard, B. Lythgoe, and A. R. Todd, *J. Chem. Soc.*, 1052 (1947); (c) J. Piml and M. Prystas, *Adv. Heterocycl. Chem.* **8**, 115 (1967); (d) M. Prystas, *Coll. Czech. Chem. Commun.* **40**, 1786 (1975).
12. (a) T. Nishimura, B. Shimizu, and I. Iwai, *Chem. Pharm. Bull. (Tokyo)* **11**, 1470 (1963); **12**, 1471 (1964); (b) E. Wittenburg, *Chem. Ber.* **101**, 1095, 1614, 2132 (1968).
13. U. Niedballa and H. Vorbrüggen, *J. Org. Chem.* **39**, 3654 (1974).
14. U. Niedballa and H. Vorbrüggen, *J. Org. Chem.* **39**, 3660 (1974).
15. U. Niedballa and H. Vorbrüggen, *J. Org. Chem.* **39**, 3672 (1974).
16. U. Niedballa and H. Vorbrüggen, *J. Org. Chem.* **39**, 3668 (1974).
17. U. Niedballa and H. Vorbrüggen, *J. Org. Chem.* **39**, 3664 (1974).
18. U. Niedballa and H. Vorbrüggen, *J. Org. Chem.* **41**, 2084 (1976).
19. (a) H. Vorbrüggen and K. Krolikiewicz, *Angew. Chem. Int. Ed.* **14**, 421 (1975); (b) H. Vorbrüggen, K. Krolikiewicz, and B. Bennua, *Chem. Ber.* **114**, 1234 (1981).
20. H. Vorbrüggen and G. Hofle, *Chem. Ber.* **114**, 1256 (1981).
21. R. L. Shone, *Tetrahedron Lett.*, 993 (1977).
22. (a) H. Vorbrüggen and B. Bennua, *Tetrahedron Lett.*, 1339 (1978); (b) H. Vorbrüggen and B. Nennua, *Chem. Ber.* **114**, 1279 (1981).

23. Z. Tocik, R. A. Earl, and J. Beranek, *Nucleic Acids Res.* **8**, 4755 (1980).
24. A. Matsuda, Y. Kurasawa, and K. A. Watanabe, *Synthesis*, 748 (1981).
25. J. Kiss and R. D'Souza, *J. Carbohydr. Nucleosides Nucleotides* **7**, 141 (1980).
26. C. Chavis, F. Dumont, R. H. Wightman, J. C. Ziegler, and J. L. Imbach, *J. Org. Chem.* **47**, 202 (1982).
27. A. J. Hubbard, A. S. Jones, and R. T. Walker, *Nucleic Acids Res.* **12**, 6827 (1984).
28. J. J. Fox, N. Yung, J. Davoll, and G. B. Brown, *J. Am. Chem. Soc.* **78**, 2117 (1956).
29. J. J. Fox, N. C. Yung, and D. Van Praag, *J. Org. Chem.* **26**, 526 (1961).
30. K. A. Watanabe and J. J. Fox, *J. Heterocycl. Chem.* **6**, 109 (1969).
31. (a) G. T. Rogers and T. L. V. Ulbricht, *J. Chem. Soc. D*, 508 (1969); (b) G. T. Rogers and T. L. V. Ulbricht, *J. Chem. Soc. C*, 1109 (1970).
32. T. Ueda and H. Tanaka, *Chem. Pharm. Bull. (Tokyo)* **18**, 1491 (1970).
33. (a) M. W. Winkley and R. K. Robins, *J. Org. Chem.* **33**, 2822 (1968); (b) M. Prystas and F. Sorm, *Coll. Czech. Chem. Commun.* **34**, 2316 (1969).
34. R. S. Klein and J. J. Fox, *J. Org. Chem.* **37**, 4831 (1972).
35. K. A. Watanabe, D. H. Hollenberg, and J. J. Fox, *J. Carbohydr. Nucleosides Nucleotides* **1**, 1 (1974).
36. G. Shaw, R. N. Warrener, M. H. Maguire, and R. K. Ralph, *J. Chem. Soc.*, 2294 (1958).
37. N. J. Cusack, B. J. Hildick, D. H. Robinson, P. W. Rugg, and G. Shaw, *J. Chem. Soc., Perkin Trans. I*, 1720 (1973).
38. T. Naito, M. Hirata, T. Kawakami, and M. Sano, *Chem. Pharm. Bull. (Tokyo)* **9**, 703 (1961).
39. M. Sano, *Chem. Pharm. Bull. (Tokyo)* **10**, 308, 313 (1962).
40. R. Lohrmann, J. M. Lagowski, and H. S. Forrest, *J. Chem. Soc.*, 451 (1964).
41. (a) T. Ukita, Y. Yoshida, A. Hamada, and Y. Kato, *Chem. Pharm. Bull. (Tokyo)* **12**, 459 (1964); (b) T. Ukita, A. Hamada, and M. Yoshida, *Chem. Pharm. Bull. (Tokyo)* **12**, 454 (1964).
42. A. Piskala and F. Sorm, *Coll. Czech. Chem. Commun.* **29**, 2062 (1964).
43. H. Ogura, H. Takahashi, and O. Sato, *Nucleic Acids Res. Symp. Ser.* **8**, s1 (1980).
44. R. A. Sanchez and L. E. Orgel, *J. Mol. Biol.* **47**, 531 (1970).
45. D. H. Schannahoff and R. A. Sanchez, *J. Org. Chem.* **38**, 593 (1973).
46. (a) D. T. Gish, G. L. Neil, and W. J. Wechter, *J. Med. Chem.* **14**, 882 (1971); (b) R. L. Tolman and R. K. Robins, *J. Med. Chem.* **14**, 1112 (1971).
47. A. Holý, *Nucleic Acids Res.* **1**, 289 (1974).
48. C. A. Dekker, *Annu. Rev. Biochem.* **29**, 453 (1960).
49. T. Ueda and J. J. Fox, *J. Am. Chem. Soc.* **85**, 4024 (1963).
50. R. Wagner and W. von Philipsborn, *Helv. Chim. Acta* **53**, 299 (1970).
51. B. Singer, *Prog. Nucleic Acid Res. Mol. Biol.* **15**, 219 (1975).
52. P. Brookes and P. D. Lawley, *J. Chem. Soc.*, 1348 (1962).
53. P. Brookes, A. Dipple, and P. D. Lawley, *J. Chem. Soc. C*, 2026 (1968).
54. L. Sun and B. Singer, *Biochemistry* **13**, 1905 (1974).
55. R. Shapiro and S.-J. Shiuey, *J. Org. Chem.* **41**, 1597 (1976).
56. H. Hayashi, K. Yamauchi, and M. Kinoshita, *Bull. Chem. Soc. Jpn.* **53**, 277 (1980).
57. T. Ueda, Y. Nomoto, and A. Matsuda, *Chem. Pharm. Bull. (Tokyo)* **33**, 3263 (1985).
58. H. T. Miles, *J. Am. Chem. Soc.* **79**, 2565 (1957).
59. J. A. Haines, C. B. Reese, and Lord Todd, *J. Chem. Soc.*, 1406 (1964).
60. H. T. Miles, *Biochim. Biophys. Acta* **22**, 247 (1956).
61. T. Ueda, *Chem. Pharm. Bull. (Tokyo)* **10**, 788 (1962).
62. D. B. Farmer, A. B. Foster, M. Jarman, and M. J. Tisdale, *Biochem. J.* **135**, 203 (1973).
63. J. J. Fox, N. C. Yung, and R. J. Cushley, *Tetrahedron Lett.*, 4927 (1966).
64. (a) J. T. Kusmierek and B. Singer, *Nucleic Acids Res.* **3**, 989 (1976); (b) B. Singer, M. Kroger, and M. Carrano, *Biochemistry* **17**, 1246 (1978).
65. B. Singer, *FEBS Lett.* **63**, 85 (1976).
66. A. M. Serebryanyi and R. M. Mnatsakanyan, *FEBS Ltt.* **28**, 191 (1972).
67. Y. Mizuno, T. Endo, T. Miaoka, and K. Ikeda, *J. Org. Chem.* **39**, 1250 (1974).
68. Y. Kanaoka, E. Sato, M. Aiura, O. Yonemitsu, and Y. Mizuno, *Tetrahedron Lett.*, 3361 (1969).
69. N. Imura, T. Tsuruo, and T. Ukita, *Chem. Pharm. Bull. (Tokyo)* **16**, 1105 (1968).
70. H. Seliger and F. Cramer, *Angew. Chem. Int. Ed.* **8**, 609 (1969).

71. Y. Furukawa, K. Kobayashi, Y. Kanai, and M. Honjo, *Chem. Pharm. Bull. (Tokyo)* **13**, 1273 (1965).
72. T. A. Khwaja and C. Heidelberger, *J. Med. Chem.* **13**, 64 (1970).
73. J. F. Codington, R. J. Cushley, and J. J. Fox, *J. Org. Chem.* **33**, 466 (1968).
74. K. Tsuchida, Y. Mizuno, and K. Ikeda, *Heterocycles* **15**, 883 (1981).
75. K. Yamauchi, K. Nakamura, and M. Kinoshita, *J. Org. Chem.* **43**, 1593 (1978).
76. K. Yamauchi and M. Kinoshita, *J. Chem. Soc., Perkin Trans. I*, 762 (1978).
77. Y. Hisanaga, T. Tanabe, and K. Yamauchi, *Bull. Soc. Chem. Jpn.* **54**, 1509 (1981).
78. K. K. Ogilvie, S. L. Beaucage, M. F. Gillen, and D. W. Entwistle, *Nucleic Acids Res.* **6**, 2261 (1979).
79. K. K. Ogilvie, S. L. Beaucage, and M. F. Gillen, *Tetrahedron Lett.*, 3203 (1978).
80. (a) K. K. Ogilvie, S. L. Beaucage, and M. F. Gillen, *Tetrahedron Lett.*, 1663 (1978); (b) K. K. Ogilvie, S. L. Beaucage, M. F. Gillen, D. Entwistle, and M. Quilliam, *Nucleic Acids Res.* **6**, 1695 (1979).
81. (a) R. T. Markiw and E. S. Canellakis, *Tetrahedron Lett.*, 657 (1968); (b) R. T. Markiw and E. S. Canellakis, *J. Org. Chem.* **34**, 3707 (1969).
82. (a) J. Zemlicka, *Coll. Czech. Chem. Commun.* **28**, 1060 (1963); (b) J. Zemlicka, *Coll. Czech. Chem. Commun.* **32**, 3572 (1970).
83. J. Zemlicka and A. Holý, *Coll. Czech. Chem. Commun.* **32**, 3159 (1967).
84. A. Holý, R. W. Bald, and Ng D. Hong, *Coll. Czech. Chem. Commun.* **36**, 2659 (1971).
85. T. Ukita, H. Okuyama, and H. Hayatsu, *Chem. Pharm. Bull. (Tokyo)* **11**, 1399 (1963).
86. H. Mizuno, H. Okuyama, H. Hayatsu, and T. Ukita, *Chem. Pharm. Bull. (Tokyo)* **12**, 1240 (1964).
87. R. W. Chambers, *Biochemistry* **4**, 219 (1965).
88. (a) M. Yoshida and T. Ukita, *J. Biochem.* **57**, 818 (1965); (b) M. Yoshida and T. Ukita, *J. Biochem.* **58**, 191 (1965).
89. F. Johnson, K. M. R. Pillai, A. P. Grollman, L. Tseng, and M. Takeshita, *J. Med. Chem.* **27**, 954 (1984).
90. N. K. Kotchetkov, V. N. Shibaev, and A. A. Kost, *Tetrahedron Lett.*, 1993 (1971).
91. (a) J. R. Barrio, J. A. Secrist III, and N. J. Leonard, *Biochem. Biophys. Res. Commun.* **46**, 597 (1972); (b) J. A. Secrist III, J. R. Barrio, N. J. Leonard, and G. Weber, *Biochemistry* **11**, 3499 (1972).
92. J. Bierndt, J. Cieciolka, P. Gornicki, R. W. Adamiak, W. J. Kryzosiak, and M. Wiewiorowski, *Nucleic Acids Res.* **5**, 789 (1978).
93. K. Koyasuga-Mikado, T. Hashimoto, T. Negishi, and H. Hayatsu, *Chem. Pharm. Bull. (Tokyo)* **28**, 932 (1980).
94. (a) E. Zbiral and H. Hugl, *Tetrahedron Lett.*, 439 (1972); (b) H. Hugl, G. Schulz, and E. Zbiral, *Ann. der Chemie*, 287 (1973); (c) C. Ivancsics and E. Zbiral, *Ann. der Chemie*, 1934 (1975).
95. R. S. Hosmane and N. J. Leonard, *J. Org. Chem.* **46**, 1457 (1981).
96. V. Nair, R. J. Offerman, and G. A. Turner, *J. Org. Chem.* **49**, 4021 (1984).
97. V. Nair, G. A. Turner, and R. J. Offerman, *J. Am. Chem. Soc.* **106**, 3370 (1984).
98. D. Cech, R. Wohlfeil, and G. Etzold, *Nucleic Acids Res.* **2**, 2183 (1975).
99. R. P. Panzica, L. B. Townsend, D. L. Von Minden, M. S. Wilson, and J. A. McCloskey, *Biochim. Biophys. Acta* **331**, 147 (1973).
100. A. Holý, *J. Carbohydr. Nucleosides Nucleotides* **5**, 487 (1978).
101. E. Ochiai, *Aromatic Amine Oxides*, Elsevier, Amsterdam, 1967.
102. M. A. Stevens, D. I. Magrath, H. W. Smith, and G. B. Brown, *J. Am. Chem. Soc.* **80**, 2755 (1958).
103. T. J. Delia, M. J. Olsen, and G. B. Brown, *J. Org. Chem.* **30**, 2766 (1965).
104. F. Cramer and H. Seidel, *Biochim. Biophys. Acta* **72**, 157 (1963).
105. L. B. Townsend, R. P. Panzica, and R. K. Robins, *J. Med. Chem.* **14**, 259 (1971).
106. M. Araki, M. Maeda, and Y. Kawazoe, *Tetrahedron* **32**, 337 (1976).
107. (a) G. B. Brown, G. Levin, and S. Murphy, *Biochemistry* **3**, 880 (1964); (b) T. Fukui, T. Itaya, C. C. Wu, and F. Tanaka, *Tetrahedron* **27**, 2415 (1971); (c) T. Ueda K. Miura, and K. Kasai, *Chem. Pharm. Bull. (Tokyo)* **26**, 2122 (1978).
108. L. R. Subbaraman, J. Subbaraman, and E. J. Behrman, *Biochemistry* **8**, 3059 (1967).
109. A. D. Broom and R. K. Robins, *J. Org. Chem.* **34**, 1025 (1969).

110. G.-F. Huang, T. Okamoto, M. Maeda, and Y. Kawazoe, *Tetrahedron Lett.*, 4541 (1973).
111. M. Maeda and Y. Kawazoe, *Chem. Pharm. Bull. (Tokyo)* **23**, 844 (1975).
112. M. Maeda and Y. Kawazoe, *Tetrahedron Lett.*, 2751 (1973).
113. P. A. Levene and F. B. LaForge, *Chem Ber.* **45**, 608 (1912).
114. I. Wempen, I. L. Doerr, L. Kaplan, and J. J. Fox, *J. Am. Chem. Soc.* **82**, 1624 (1960).
115. R. A. Long, T. R. Matthews, and R. K. Robins, *J. Med. Chem.* **19**, 1072 (1976).
116. P. F. Torrence, G.-F. Huang, M. W. Edwards, B. Bhooshan, J. Descamps, and E. DeClercq, *J. Med. Chem.* **22**, 316 (1979).
117. J. J. Fox and D. Van Praag, *J. Org. Chem.* **26**, 526 (1961).
118. G.-F. Huang and P. F. Torrence, *J. Org. Chem.* **42**, 3821 (1977).
119. E. DeClercq, J. Descamps, G.-F. Huang, and P. F. Torrence, *Mol. Pharmacol.* **14**, 422 (1978).
120. G.-F. Huang and P. F. Torrence, *J. Carbohydr. Nucleosides Nucleotides* **5**, 317 (1978).
121. R. E. Kline, R. M. Fink, and K. Fink, *J. Am. Chem. Soc.* **81**, 2521 (1959).
122. A. Giner-Sorolla and L. Medreck, *J. Med. Chem.* **9**, 97 (1966).
123. (a) K. H. Scheit, *Chem. Ber.* **99**, 3884 (1966); (b) K. H. Scheit, *Tetrahedron Lett.*, 1031 (1965).
124. B. R. Baker, T. J. Schwan, and D. V. Santi, *J. Med. Chem.* **9**, 66 (1966).
125. G. Herrmann, D. Cech, G. Kowollik, and P. Langen, *Z. Chem.* **19**, 422 (1979).
126. E. I. Budowsky, V. N. Shibaev, and G. I. Eliseeva, in *Synthetic Procedures in Nucleic Acid Chemistry*, Vol. 1, p. 436 (W. W. Zorbach and R. S. Tipson, eds.), Interscience, New York, 1968.
127. D. V. Santi and C. F. Brewer, *J. Am. Chem. Soc.* **90**, 6236 (1968).
128. S. A. Salisburg and D. M. Brown, *J. Chem. Soc. Chem. Commun.*, 656 (1979).
129. K. Ikeda, T. Takeda, and Y. Mizuno, *Nucleic Acids Res. Symp. Ser.* **6**, s1 (1979).
130. K. Ikeda, S. Tanaka, and Y. Mizuno, *Chem. Pharm. Bull. (Tokyo)* **23**, 2958 (1975).
131. H. R. Rackwitz and K. H. Scheit, *J. Carbohydr. Nucleosides Nucleotides* **2**, 407 (1975).
132. (a) C. B. Reese and Y. S. Sanghvi, *J. Chem. Soc. Chem. Commun.*, 62 (1984); (b) A. Malkiewicz, E. Sochacka, A. F. Sayed Ahmed, and S. Yassin, *Tetrahedron Lett.* **24**, 5395 (1983).
133. S. S. Jones, C. B. Reese, and A. Ubasawa, *Synthesis*, 259 (1982).
134. K. Imai and M. Honjo, *Chem. Pharm. Bull. (Tokyo)* **13**, (1965).
135. V. W. Armstrong and F. Eckstein, *Nucleic Acids Res. Spec. Publ.* **1**, s97 (1975).
136. (a) M. P. Mertes and M. T. Shipchandler, *J. Hterocycl. Chem.* **8**, 133 (1971); (b) A. Kampf, C. T. Pillar, W. J. Woodford, and M. P. Mertes, *J. Med. Chem.* **19**, 909 (1976).
137. J. S. Park, C. T.-C. Chang, C. L. Schmidt, Y. Golander, E. DeClercq, J. Descamps, and M. P. Mertes, *J. Med. Chem.* **23**, 661 (1980).
138. K. Ikeda, A. Hisaka, Y. Misu, and Y. Mizuno, *Abstr. 103rd Annu. Meeting Pharm. Soc. Jpn.*, 191 (1983).
139. J. Farkas and F. Sorm, *Coll. Czech. Chem. Commun.* **28**, 1620 (1963).
140. H. Guglielmi and B. Athen, *Hoppe-Seyler's Z. Physiol. Chem.* **350**, 710 (1969).
141. H. Guglielmi and B. Athen., *Hoppe-Seyler's Z. Physiol. Chem.* **350**, 809 (1969).
142. D. V. Santi, *J. Heterocycl. Chem.* **4**, 475 (1967).
143. E. DeClercq, J. Descamps, C. L. Schmidt, and M. P. Mertes, *Biochem. Pharmacol.* **28**, 3249 (1979).
144. C. L. Schmidt, C. T.-C. Chung, E. DeClercq, J. Descamps, and M. P. Mertes, *J. Med. Chem.* **23**, 252 (1980).
145. H. Lonnberg, *Tetrahedron* **38**, 1517 (1982).
146. A. H. Alegria, *Biochim. Biophys. Acta* **149**, 317 (1967).
147. H. Hayatsu and M. Shiragami, *Biochemistry* **18**, 632 (1979).
148. K. Ikeda, T. Takeda, H. Inoue, T. Ueda, and Y. Mizuno, *Abstr. 100th Annu. Meeting Pharm. Soc. Jpn.*, 82 (1980).
149. R. Shapiro and S. C. Agarwal, *J. Am. Chem. Soc.*, **90**, 474 (1968).
150. R. E. Beltz and D. W. Visser, *J. Am. Chem. Soc.* **77**, 736 (1955).
151. T. J. Bardos, G. M. Levin, R. R. Herr, and H. L. Gordon, *J. Am. Chem. Soc.* **77**, 4279 (1955).
152. M. Friedland and D. W. Visser, *Biochim. Biophys. Acta* **51**, 148 (1961).
153. P. C. Srivastava and K. L. Nagpal, *Experientia* **26**, 220 (1970).
154. T. Ueda, unpublished experiments.
155. T. K. Fukuhara and D. W. Visser, *J. Biol. Chem.* **190**, 95 (1951).
156. D. W. Visser, D. M. Fritsch, and B. Huang, *Biochem. Pharmacol.* **5**, 157 (1960).
157. K. Kikugawa, I. Kuwada, and M. Ichino, *Chem. Pharm. Bull. (Tokyo)* **23**, 35 (1975).

158. W. H. Prusoff, W. L. Holmes, and A. D. Welch, *Cancer Res.* **13**, 221 (1953).
159. W. H. Prusoff, *Biochim. Biophys. Acta* **32**, 295 (1959).
160. T.-S. Lin and W. H. Prusoff, *J. Carbohydr. Nucleosides Nucleotides* **2**, 309 (1975).
161. D. Lipkin, F. B. Howard, D. Novotony, and M. Sano, *J. Biol. Chem.* **238**, 2249 (1963).
162. F. Ascoli and F. M. Kahan, *J. Biol. Chem.* **241**, 428 (1966).
163. M. J. Robins, P. J. Barr, and J. Giziewicz, *Can. J. Chem.* **60**, 554 (1982).
164. A. F. Russell, M. Prystasz, E. K. Hamamura, J. P. H. Verheyden, and J. G. Moffatt, *J. Org. Chem.* **39**, 2182 (1974).
165. A. Matsuda, H. Inoue, and T. Ueda, *Chem. Pharm. Bull. (Tokyo)* **26**, 2340 (1978).
166. T. K. Fukuhara and D. W. Visser, *J. Am. Chem. Soc.* **77**, 2393 (1955).
167. D. M. Frisch and D. W. Visser, *J. Am. Chem. Soc.* **81**, 1756 (1959).
168. H. Yoshida, J. Duval, and J. P. Ebel, *Biochim. Biophys. Acta* **161**, 13 (1968).
169. H. Hayatsu, S.-K. Pan, and T. Ukita, *Chem. Pharm. Bull. (Tokyo)* **19**, 2189 (1971).
170. E. K. Ryu and M. MacCoss, *J. Org. Chem.* **46**, 2819 (1981).
171. P. K. Chang, *J. Org. Chem.* **30**, 3913 (1965).
172. P. K. Chang and A. D. Welch, *J. Med. Chem.* **6**, 428 (1963).
173. M. Honjo, Y. Furukawa, M. Nishikawa, K. Kamiya, and Y. Yoshida, *Chem. Pharm. Bull. (Tokyo)* **15**, 1076 (1967).
174. R. Duchinsky, T. Gabriel, W. Tautz, A. Nussbaum, M. Hoffer, E. Grunberg, J. H. Burchenal, and J. J. Fox, *J. Med. Chem.* **10**, 47 (1967).
175. H. Taguchi and S. Y. Wang, *J. Org. Chem.* **44**, 4386 (1979).
176. D. Lipkin and J. A. Rabi, *J. Am. Chem. Soc.* **93**, 3309 (1971).
177. R. Teoule, B. Fougue, and J. Cadet, *Nucleic Acids Res.* **2**, 487 (1975).
178. R. Ducolomb, J. Cadet, C. Taieb, and R. Teoule, *Biochim. Biophys. Acta* **432**, 18 (1976).
179. R. Ducolomb and R. Teoule, *Tetrahedron* **33**, 1603 (1977).
180. C. Heidelberger, *Prog. Nucleic Acid Res. Mol. Biol.* **4**, 1 (1965).
181. D. H. R. Barton, R. H. Hesse, H. T. Toh, and M. M. Pechet, *J. Org. Chem.* **37**, 329 (1972).
182. M. J. Robins and S. R. Naik, *J. Am. Chem. Soc.* **93**, 5277 (1971).
183. M. J. Robins and S. R. Naik, *J. Chem. Soc. Chem. Commun.*, 18 (1972).
184. J. O. Folayan and D. W. Hutchinson, *Biochim. Biophys. Acta* **340**, 194 (1974).
185. M. J. Robins, G. Ramani, and M. MacCoss, *Can. J. Chem.* **53**, 1302 (1975).
186. M. J. Robins, *Ann. N.Y. Acad. Sci.* **255**, 104 (1975).
187. M. J. Robins, M. MacCoss, S. R. Naik, and G. Ramani, *J. Am. Chem. Soc.* **98**, 7381 (1976).
188. R. A. Scharma, M. Bobek, and A. Bloch, *J. Med. Chem.* **17**, 466 (1974).
189. H. Meinert and D. Cech, *Z. Chem.* **12**, 292 (1972).
190. D. Cech, G. Herrmann, and A. Holý, *Nucleic Acids Res.* **4**, 3259 (1977).
191. D. Cech and A. Holý, *Coll. Czech. Chem. Commun.* **41**, 3335 (1976).
192. S. Stovber and M. Zupan, *J. Chem. Soc. Chem. Commun.*, 563 (1983).
193. A. Holý and J. Philip, *Nucleic Acids Res.* **1**, 1209 (1974).
194. T. Nagamachi, P. F. Torrence, J. A. Waters, and B. Witkop, *J. Chem. Soc. Chem. Commun.*, 1025 (1972).
195. T. Nagamachi, J.-L. Forrey, P. F. Torrence, J. A. Waters, and B. Witkop, *J. Med. Chem.* **17**, 403 (1974).
196. T. Ueda, S. Watanabe, and A. Matsuda, *J. Carbohydr. Nucleosides Nucleotides* **5**, 523 (1978).
197. F. J. Dinan and T. J. Bardos, *J. Med. Chem.* **23**, 569 (1980).
198. M. Roberts and D. W. Visser, *J. Am. Chem. Soc.* **74**, 668 (1952).
199. T. Ueda, *Chem. Pharm. Bull. (Tokyo)* **8**, 455 (1960).
200. T. Kanai, M. Ichino, and T. Kojima, *Chem. Pharm. Bull. (Tokyo)* **17**, 650 (1969).
201. E. G. Podrebarac and C. C. Cheng, *Synthetic Procedures Nucleic Acid Chem.* **1**, 412 (1968).
202. T. K. Fukuhara and D. W. Visser, *Biochemistry* **1**, 563 (1962).
203. T. Szabo, T. I. Kalman, and T. J. Bardos, *J. Org. Chem.* **35**, 1434 (1970).
204. S. Choi, T. I. Kalman, and T. J. Bardos, *J. Med. Chem.* **22**, 618 (1979).
205. V. C. Solan, G. L. Szekeres, E. K. Ryu, H. Kung, Y. K. Ho and T. J. Bardos, *Nucleosides Nucleotides* **2**, 419 (1983).
206. B. A. Otter, A. Taube, and J. J. Fox, *J. Org. Chem.* **36**, 1251 (1971).
207. K. Murao, T. Hasegawa, and H. Ishikura, *Nucleic Acids Res.* **3**, 2851 (1976).

208. P. F. Torrence, J. W. Spencer, and A. M. Bobst, *J. Med. Chem.* **21**, 228 (1978).
209. G.-F. Huang, M. Okada, E. DeClercq, and P. F. Torrence, *J. Med. Chem.* **24**, 390 (1981).
210. K. Ikeda and Y. Mizuno, *Chem. Pharm. Bull. (Tokyo)* **19**, 564 (1971).
211. P. A. Levene, *J. Biol. Chem.* **63**, 653 (1925).
212. K. Ikeda, T. Sumi, K. Yokoi, and Y. Mizuno, *Chem. Pharm. Bull. (Tokyo)* **21**, 1327 (1973).
213. A. M. Bobst, A. J. Ozinskas, and E. DeClercq, *Helv. Chim. Acta* **66**, 534 (1983).
214. D. Bärwolff and P. Langen, *Nucleic Acis Res. Spec. Publ.* **1**, s29 (1975).
215. D. Bärwolff and P. Langen, in *Nucleic Acid Chemistry*, Part I, p. 75 (L. B. Townsend and R. S. Tipson, eds.), Wiley, New York, 1978.
216. G. T. Shiau, R. F. Schinazi, M. S. Cheu, and W. H. Prusoff, *J. Med. Chem.* **23**, 127 (1980).
217. D. E. Bergstrom, *Nucleosides Nucleotides* **1**, 1 (1982).
218. R. M. K. Dale, E. Martin, D. C. Livingstone, and D. C. Ward, *Biochemistry* **14**, 2447 (1975).
219. R. M. K. Dale, D. C. Livingstone, and D. C. Ward, *Proc. Natl. Acad. Sci. U.S.A.* **70**, 2238 (1973).
220. R. M. K. Dale and D. C. Ward, *Biochemistry* **14**, 2485 (1975).
221. R. M. K. Dale, D. C. Ward, D. C. Livingstone, and E. Martin, *Nucleic Acids Res.* **2**, 915 (1975).
222. D. E. Bergstrom and J. L. Ruth, *J. Carbohydr. Nucleosides Nucleotides* **4**, 257 (1977).
223. D. E. Bergstrom and J. L. Ruth, *J. Am. Chem. Soc.* **98**, 1587 (1976).
224. D. E. Bergstrom and N. K. Ogawa, *J. Am. Chem. Soc.* **100**, 8106 (1978).
225. J. L. Ruth and D. E. Bergstrom, *J. Org. Chem.* **43**, 2870 (1978).
226. D. E. Bergstrom, J. L. Ruth, and P. Warwick, *J. Org. Chem.* **46**, 1432 (1981).
227. D. E. Bergstrom, J. L. Ruth, P. A. Reddy, and E. DeClercq, *J. Med. Chem.* **27**, 279 (1984).
228. (a) C. F. Bigge, P. Kalaritis, and M. P. Mertes, *Tetrahedron Lett.*, 1653 (1979); (b) C. F. Bigge, P. Kalaritis, J. R. Deck, and M. P. Mertes, *J. Am. Chem. Soc.* **102**, 2033 (1980); (c) C. F. Bigge, K. E. Lizzote, J. S. Panek, and M. P. Mertes, *J. Carbohydr. Nucleosides Nucleotides* **8**, 295 (1981).
229. J. S. Park, C. F. Bigge, M. E. Hassan, L. Maggiora, and M. P. Mertes, *J. Chem. Soc. Chem. Commun.*, 553 (1984).
230. D. E. Bergstrom, H. Inoue, and P. A. Reddy, *J. Org. Chem.* **47**, 2174 (1982).
231. E. DeClercq, J. Descamps, P. DeSomer, P. J. Barr, A. S. Jones, and R. T. Walker, *Proc. Natl. Acad. Sci. U.S.A.* **76**, 2947 (1979).
232. A. S. Jones, G. Verhelst, and R. T. Walker, *Tetrahedron Lett.*, 4415 (1979).
233. A. S. Jones, S. G. Rahim, and R. T. Walker, *J. Med. Chem.* **24**, 759 (1981).
234. S. G. Rahim, M. J. H. Duggan, R. T. Walker, A. S. Jones, R. L. Dyer, J. Balzarini, and E. DeClercq, *Nucleic Acids Res.* **10**, 5285 (1982).
235. C. F. Bigge and M. P. Mertes, *J. Org. Chem.* **46**, 1994 (1981).
236. Y. Kobayashi, I. Kumadaki, and K. Yamamoto, *J. Chem. Soc. Chem. Commun.*, 536 (1977).
237. Y. Kobayashi, K. Yamamoto, T. Asai, M. Nakano, and I. Kumadaki, *J. Chem. Soc., Perkin Trans.* **I**, 2755 (1980).
238. R. C. Bleakley, A. S. Jones, and R. T. Walker, *Nucleic Acids Res.* **2**, 683 (1975).
239. (a) M. J. Robins and P. J. Barr, *Tetrahedron Let.* **22**, 421 (1981); (b) M. J. Robins and P. J. Barr, *J. Org. Chem.* **48**, 1854 (1983).
240. P. J. Barr, P. Chananont, T. A. Hamor, A. S. Jones, M. K. O'Leary, and R. T. Walker, *Tetrahedron* **36**, 1269 (1980).
241. I. Saito, T. Ito, T. Shinmura, and T. Matsuura, *Tetrahedron Lett.* **21**, 2813 (1980).
242. (a) S. Ito, I. Saito, and T. Matsuura, *J. Am. Chem. Soc.* **102**, 7535 (1980); (b) I. Saito, S. Ito, and T. Matsuura, *J. Am. Chem. Soc.* **100**, 2901 (1978); (c) I. Saito, S. Ito, T. Matsuura, and C. Helene, *Photochem. Photobiol.* **33**, 15 (1981).
243. A. Wexler, R. J. Balchunis, and J. S. Swenton, *J. Chem. Soc. Chem. Commun.*, 601 (1975).
244. P. Vincent, J. P. Beaucourt, and L. Pichat, *Tetrahedron Lett.* **23**, 63 (1982).
245. P. Vincent, J. P. Beaucourt, and L. Pichat, *Tetrahedron Lett.* **25**, 201 (1984).
246. I. Saito, H. Ikehira, and T. Matsuura, *Tetrahedron Lett.* **26**, 1993 (1985).
247. L. Pichat and J.-P. Gilbert, *C. R. Acd. Sci. Paris Ser. C* **277**, 1157 (1973).
248. L. Pichat, B. Masse, J. Descamps, and P. Dufay, *C. R. Acid. Sci. Paris Ser. C* **268**, 197 (1969).
249. L. Pichat, B. Masse, J. Descamps, and P. Dufay, *Bull. Soc. Chim. Fr.*, 2102 (1971).
250. P. L. Coe, M. R. Harnden, A. S. Jones, S. A. Noble, and R. T. Walker, *J. Med. Chem.* **25**, 1329 (1982).
251. L. Pichat, J. Godbillon, and M. Herbert, *Bull. Soc. Chim. Fr.*, 2715 (1973).

252. (a) H. Tanaka, I. Nasu, and T. Miyasaka, *Tetrahedron Lett.*, 4755 (1979); (b) H. Tanaka, I. Nasu, H. Hayakawa, and T. Miyasaka, *Nucleic Acids Res. Symp. Ser.* **8**, 33 (1980).
253. H. Tanaka, H. Hayakawa, and T. Miyasaka, *Chem. Pharm. Bull. (Tokyo)* **29**, 3565 (1981).
254. H. Tanaka, S. Iijima, A. Matsuda, H. Hayakawa, T. Miyasaka, and T. Ueda, *Chem. Pharm. Bull. (Tokyo)* **31**, 1222 (1983).
255. H. Tanaka, A. Matsuda, S. Iijima, H. Hayakawa, and T. Miyasaka, *Chem. Pharm. Bull. (Tokyo)* **31**, 2164 (1983).
256. H. Tanaka, H. Hayakawa, and T. Miyasaka, *Tetrahedron* **38**, 2635 (1982).
257. H. Hayakawa, H. Tanaka, and T. Miyasaka, *Chem. Pharm. Bull. (Tokyo)* **30**, 4589 (1982).
258. H. Ikehira, T. Matsuura, and I. Saito, *Tetrahedron Lett.* **26**, 1743 (1985).
259. B. Bannister and F. Kagan, *J. Am. Chem. Soc.* **82**, 3363 (1960).
260. R. W. Chambers and V. Kurkov, *J. Am. Chem. Soc.* **85**, 2160 (1963).
261. E. J. Reist, A. Benitz, and L. Goodman, *J. Org. Chem.* **29**, 554 (1964).
262. K. Isono and T. Azuma, *Chem. Pharm. Bull. (Tokyo)* **20**, 193 (1972).
263. J.-L. Fourrey and P. Jouin, *Tetrahedron Lett.*, 3393 (1977).
264. J. Cadet, L.-S. Kan, and S. Y. Wang, *J. Am. Chem. Soc.* **100**, 6715 (1978).
265. A. D. McLaren and D. Shugar, *Photochemistry of Proteins and Nucleic Acids*, Pergamon Press, Oxford, 1964.
266. J. G. Burr, in *Advances in Photochemistry*, Vol. 6, p. 193 (W. A. Noyes Jr., G. S. Hammond, and J. N. Pitts Jr., eds.), Interscience New York, 1968.
267. D. V. Santi and C. F. Brewer, *J. Am. Chem. Soc.* **90**, 6236 (1968).
268. Y. Kondo, J.-L. Fourrey, and B. Witkop, *J. Am. Chem. Soc.* **93**, 3527 (1972).
269. D. Shugar and J. J. Fox, *Biochim. Biophys. Acta* **9**, 199 (1952).
270. R. Shapiro, R. E. Servis, and M. Welcher, *J. Am. Chem. Soc.* **92**, 422 (1970).
271. (a) H. Hayatsu, Y. Wataya, and K. Kai, *J. Am. Chem. Soc.* **92**, 724 (1970); (b) H. Hayatsu, Y. Wataya, K. Kai, and S. Iida, *Biochemistry* **9**, 2858 (1970).
272. R. Shapiro, M. Welcher, V. Nelson, and V. Difate, *Biochim. Biophys. Acta* **425**, 115 (1976).
273. (a) E. G. Sander and C. L. Deyrup, *Arch. Biochem. Biophys.* **150**, 600 (1972); (b) F. A. Sedor and E. G. Sander, *Arch. Biochem. Biophys.* **161**, 632 (1974).
274. J.-L. Fourrey, *Bull. Soc. Chim. Fr.*, 4580 (1972).
275. H. Hayatsu, T. Chikuma, and K. Negishi, *J. Org. Chem.* **40**, 3862 (1975).
276. J. W. Triplett, G. A. Digenis, W. J. Layton, and S. L. Smith, *J. Org. Chem.* **43**, 4411 (1977).
277. K. Isono, S. Suzuki, M. Tanaka, T. Nanbata, and K. Shibuya, *Tetrahedron Lett.*, 425 (1970).
278. R. P. Singhal, *Biochemistry* **13**, 2924 (1974).
279. S. Slae and R. Shapiro, *J. Org. Chem.* **43**, 4197 (1978).
280. H. Taniguchi and S. Y. Wang, *J. Org. Chem.* **42**, 2028 (1977).
281. E. D. Sverdrov, G. S. Monastyrskaya, N. S. Tarabakina, and E. I. Budowsky, *FEBS Lett.* **62**, 212 (1976).
282. R. Shapiro and J. M. Weisgras, *Biochem. Biophys. Res. Commun.* **40**, 839 (1970).
283. (a) E. I. Budowski, M. F. Turchinsky, V. D. Domkin, A. G. Pogorelov, V. N. Pisaenko, K. S. Kusova, and N. K. Kotchetkov, *Biochim. Biophys. Acta* **277**, 421 (1972); (b) E. I. Budowski, E. D. Sverdlov, and G. S. Monastyrskaya, *FEBS Lett.* **25**, 201 (1972).
284. H. Hayatsu, *Biochemistry* **15**, 2677 (1976).
285. K. Negishi, C. Harada, Y. Ohara, K. Ohara, N. Nitta, and H. Hayatsu, *Nucleic Acids Res.* **11**, 5223 (1983).
286. D. Cech and A. Holý, *Coll. Czech. Chem. Commun.* **42**, 2246 (1977).
287. H. Hayatsu, *Prog. Nucleic Acid Res. Mol. Biol.* **16**, 75 (1976).
288. R. Y.-H. Wang, C. W. Gehrke, and M. Ehrlich, *Nucleic Acids Res.* **8**, 4777 (1980).
289. H. Hayatsu and M. Inoue, *J. Am. Chem. Soc.* **93**, 2301 (1971).
290. E. I. Budowsky, *Prog. Nucleic Acid Res. Mol. Biol.* **16**, 125 (1976).
291. P. D. Lawley, *J. Mol. Biol.* **24**, 75 (1967).
292. (a) D. M. Brown and J. H. Philips, *J. Chem. Soc.*, 208 (1965); (b) D. M. Brown and J. H. Philips, *J. Mol. Biol.* **11**, 666 (1965).
293. (a) C. Janion and D. Shugar, *Biochem. Biophys. Res. Commun.* **18**, 617 (1965); (b) C. Janion and D. Shugar, *Acta Biochim. Pol.* **12**, 337 (1965).
294. N. K. Kotchetkov, E. I. Budowsky, and R. P. Shibaeva, *Biochim. Biophys. Acta* **68**, 493 (1963).

295. P. M. Schalke and C. D. Hall, *J. Chem. Soc., Perkin Trans. I*, 2417 (1975).

296. N. A. Simukova, D. Yu. Yakovlov, and E. I. Budowsky, *Nucleic Acids Res.* **2**, 2269 (1975).

297. (a) D. W. Verwoerd, H. Kohlhage, and W. Zillig, *Nature* **192**, 1038 (1961); (b) D. W. Vervoerd, W. Zillig, and H. Kohlhage, *Z. Physiol. Chem.* **332**, 184 (1963).

298. F. Baron and D. M. Brown, *J. Chem. Soc.*, 2855 (1955).

299. F. Lingens and H. Schneider-Bernloehr, *Biochim. Biophys. Acta* **123**, 611 (1966).

300. A. M. Maxam and W. Gilbert, *Proc. Natl. Acad. Sci. U.S.A.* **74**, 560 (1977).

301. D. A. Peattie, *Proc. Natl. Acad. Sci. U.S.A.* **76**, 1760 (1979).

302. N. K. Kotchetkov, E. I. Budowsky, and R. P. Shibaeva, *Biochim. Biophys. Acta* **87**, 515 (1964).

303. C. Janion and D. Shugar, *Acta Biochim. Pol.* **14**, 293 (1967).

304. D. H. Marrian, V. L. Spicer, M. E. Balis, and G. B. Brown, *J. Biol. Chem.* **189**, 533 (1951).

305. A. S. Jones, A. M. Mian, and R. T. Walker, *J. Chem. Soc. C*, 1748 (1966).

306. R. E. Notari, M. L. Chin, and A. Cardini, *Tetrahedron Lett.*, 5103 (1967).

307. C. W. Whitehead and J. J. Traverso, *J. Am. Chem. Soc.* **82**, 3971 (1960).

308. N. Miller and J. J. Fox, *J. Org. Chem.* **29**, 1772 (1964).

309. (a) H. Hayatsu and T. Ukita, *Biochem. Biophys. Res. Commun.* **14**, 198 (1964); (b) H. Hayatsu, K. Takeishi, and T. Ukita, *Biochim. Biophys. Acta* **123**, 445 (1966).

310. K. Kikugawa, H. Hayatsu, and T. Ukita, *Biochim. Biophys. Acta* **134**, 221 (1967).

311. A. B. Patel and H. D. Brown, *Nature* **214**, 402 (1967).

312. K. Kikugawa, A. Muto, H. Hayatsu, K. Imura, and T. Ukita, *Biochim. Biophys. Acta* **134**, 232 (1967).

313. L. Gal-Or, J. E. Mellema, E. N. Moundrianakis, and M. Beer, *Biochemistry* **6**, 1909 (1967).

314. J. R. Barrio and N. J. Leonard, *J. Am. Chem. Soc.* **95**, 1323 (1973).

315. T. Kanai, M. Ichino, A. Hoshi, F. Kanzawa, and K. Kuretani, *J. Med. Chem.* **17**, 1076 (1974).

316. R. Shapiro and R. S. Klein, *Biochemistry* **6**, 3576 (1967).

317. (a) A. M. Moore, *Can. J. Chem.* **36**, 281 (1958); (b) N. Miller and P. Cerutti, *Proc. Natl. Acad. Sci. U.S.A.* **59**, 34 (1968); (c) W. J. Wechter and K. C. Smith, *Biochemistry* **7**, 4064 (1968).

318. F.-T. Liu and N. C. Yang, *Biochemistry* **17**, 4877 (1978).

319. I. Saito, H. Sugiyama, S. Ito, N. Furukawa, and T. Matsuura, *J. Am. Chem. Soc.* **103**, 1598 (1981).

320. S. C. Shim, H. Y. Koh, and D. Y. Chi, *Photochem. Photobiol.* **34**, 177 (1981).

321. J. Cadet, L. Voituriez, F. Gaboriau, P. Vigny, and S. D. Negra, *Photochem. Photobiol.* **37**, 363 (1983).

322. (a) T. Kunieda and B. Witkop, *J. Am. Chem. Soc.* **91**, 7751 (1969); (b) T. Kunieda and B. Witkop, *J. Am. Chem. Soc.* **93**, 3478 (1971).

323. J.-L. Fourrey, G. Henry, and P. Jouin, *Tetrahedron Lett.*, 951 (1979).

324. A. Haron, J. Sperling, and D. Elad, *Nucleic Acids Res.* **3**, 1715 (1976).

325. (a) H. P. M. Thiellier, G.-J. Koomen, and U. K. Pandit, *Tetrahedron Lett.* **33**, 1493 (1977); (b) H. P. M. Thiellier, G.-J. Koomen, and U. K. Pandit, *Tetrahedron* **33**, 2609 (1977); (c) H. P. M. Thiellier, A. M. van den Burg, G.-J. Koomen, and U. K. Pandit, *Heterocycles* **2**, 467 (1974); (d) H. P. M. Thiellier, G.-J. Koomen, and U. K. Pandit, *Heterocycles* **5**, 19 (1976).

326. D. W. Visser, S. Kabet, and M. Lieb, *Biochim. Biophys. Acta* **76**, 463 (1963).

327. T. Y. Shen, J. F. McPherson, and B. O. Linn, *J. Med. Chem.* **9**, 366 (1966).

328. G. Herrmann, D. Cech, G. Kowollik, and P. Langen, *Z. Chem.* **19**, 376 (1979).

329. T.-S. Lin and W. H. Prusoff, *J. Carbohydr. Nucleosides Nucleotides* **5**, 15 (1978).

330. (a) B. A. Otter and J. J. Fox, *J. Am. Chem. Soc.* **89**, 3663 (1967); (b) B. A. Otter, E. A. Falco, and J. J. Fox, *J. Org. Chem.* **33**, 3593 (1968).

331. (a) B. A. Otter, E. A. Falco, and J. J. Fox, *Tetrahedron Lett.*, 2967 (1968); (b) B. A. Otter, E. A. Falco, and J. J. Fox, *J. Org. Chem.* **34**, 1390 (1969).

332. B. A. Otter, E. A. Falco, and J. J. Fox, *J. Org. Chem.* **34**, 2636 (1969).

333. S. Y. Wang, *J. Am. Chem. Soc.* **81**, 3786 (1969).

334. L. Szabo, T. I. Kalman, and T. J. Bardos, *J. Org. Chem.* **35**, 1434 (1970).

335. Y. Wataya, K. Negishi, and H. Hayatsu, *Biochemistry* **12**, 3992 (1973).

336. B. C. Pal, *J. Am. Chem. Soc.* **100**, 5170 (1978).

337. T. Chikuma, K. Negishi, and H. Hayatsu, *Chem. Pharm. Bull. (Tokyo)* **26**, 1746 (1978).

338. (a) D. Lipkin, C. C. Cori, and M. Sano, *Tetrahedron Lett.*, 5993 (1968); (b) E. G. Lovett and D. Lipkin, *J. Am. Chem. Soc.* **95**, 2312 (1973).

339. H. Inoue, S. Tomita, and T. Ueda, *Chem. Pharm. Bull. (Tokyo)* **23**, 2614 (1975).
340. H. Inoue and T. Ueda, *Chem. Pharm. Bull. (Tokyo)* **19**, 1743 (1971).
341. T. Ueda, H. Inoue, and A. Matsuda, *Ann. N.Y. Acad. Sci.* **255**, 121 (1975).
342. H. Inoue and T. Ueda, *Chem. Pharm. Bull. (Tokyo)* **26**, 2657 (1978).
343. A. Holý, *Coll. Czech. Chem. Commun.* **40**, 738 (1975).
344. T. Ueda, M. Yamamoto, A. Yamane, M. Imazawa, and H. Inoue, *J. Carbohydr. Nucleosides Nucleotides* **5**, 261 (1978).
345. I. Saito, K. Shimazono, and T. Matsuura, *J. Am. Chem. Soc.* **102**, 3948 (1980).
346. A. Matsuda, H. Inoue, and T. Ueda, *Chem. Pharm. Bull. (Tokyo)* **26**, 2340 (1978).
347. A. Matsuda, K. Akimoto, and T. Ueda, unpublished experiments.
348. D. Goldman and T. I. Kalman, *Nucleosides Nucleotides* **2**, 175 (1983).
349. (a) T. C. Thurber and L. B. Townsend, *J. Heterocycl. Chem.* **9**, 629 (1972); (b) T. C. Thurber and L. B. Townsend, *J. Heterocycl. Chem.* **12**, 711 (1975).
350. (a) T. C. Thurber and L. B. Townsend, *J. Am. Chem. Soc.* **95**, 3081 (1973); (b) T. C. Thurber and L. B. Townsend, *J. Org. Chem.* **41**, 1041 (1976).
351. (a) H. U. Blank and J. J. Fox, *J. Am. Chem. Soc.* **90**, 7175 (1968); (b) H. U. Blank, I. Wempen, and J. J. Fox, *J. Org. Chem.* **35**, 1131 (1970).
352. T. Sasaki, K. Minamoto, and T. Mizuno, *J. Org. Chem.* **41**, 1100 (1976).
353. J. A. Rabi and J. J. Fox, *J. Am. Chem. Soc.* **95**, 1628 (1973).
354. M. Maeda and Y. Kawazoe, *Tetrahedron Lett.*, 1643 (1975).
355. I. Wempen, G. B. Brown, T. Ueda, and J. J. Fox, *Biochemistry* **4**, 54 (1965).
356. E. R. Garret and J. Tsau, *J. Pharm. Sci.* **61**, 1052 (1972).
357. R. S. Klein and J. J. Fox, *J. Org. Chem.* **37**, 4381 (1972).
358. J. J. Fox, D. V. Praag, I. Wempen, I. L. Doerr, L. Cheong, J. E. Knoll, M. L. Eidinoff, A. Bendich, and G. B. Brown, *J. Am. Chem. Soc.* **81**, 178 (1959).
359. I. Wempen, R. Duschinsky, L. Kaplan, and J. J. Fox, *J. Am. Chem. Soc.* **83**, 4755 (1961).
360. I. Wempen, N. Miller, E. A. Falco, and J. J. Fox, *J. Med. Chem.* **11**, 144 (1968).
361. R. A. Sharma, A. Bloch, and M. Bobek, *J. Heterocycl. Chem.* **19**, 1153 (1982).
362. J. Brokes and J. Beranek, *Coll. Czech. Chem. Commun.* **39**, 3100 (1974).
363. T. Kulikowski, B. Zmudzka, and D. Shugar, *Acta Biochim. Pol.* **16**, 201 (1969).
364. M. Ikehara, T. Ueda, and K. Ikeda, *Chem. Pharm. Bull. (Tokyo)* **10**, 767 (1962).
365. T. Ueda and J. J. Fox, *J. Med. Chem.* **6**, 697 (1963).
366. W. V. Ruyle and T. Y. Shen, *J. Med. Chem.* **10**, 331 (1967).
367. P. Faerber and K. H. Scheit, *Chem. Ber.* **103**, 1307 (1970).
368. M. Saneyoshi and F. Sawada, *Chem. Pharm. Bull. (Tokyo)* **17**, 181 (1969).
369. M. Saneyoshi, *Chem. Pharm. Bull. (Tokyo)* **23**, 1146 (1975).
370. E. A. Falco, B. A. Otter, and J. J. Fox, *J. Org. Chem.* **35**, 2326 (1970).
371. R. S. Klein, I. Wempen, K. A. Watanabe, and J. J. Fox, *J. Org. Chem.* **35**, 2330 (1970).
372. (a) T. Ueda, Y. Iida, K. Ikeda, and Y. Mizuno, *Chem. Pharm. Bull. (Tokyo)* **14**, 666 (1966); (b) T. Ueda, Y. Iida, K. Ikeda, and Y. Mizuno, *Chem. Pharm. Bull. (Tokyo)* **16**, 1788 (1968).
373. V. Skaric, B. Gaspert, and M. Hohnjec, *J. Chem. Soc. C*, 1444 (1970).
374. V. Slaric, B. Gaspert, M. Hohnjec, and G. Lacan, *J. Chem. Soc., Perkin Trans. I*, 267 (1974).
375. E. A. Faclo and J. J. Fox, *J. Med. Chem.* **11**, 148 (1968).
376. T. Ueda, M. Imazawa, K. Miura, R. Iwata, and K. Odajima, *Tetrahedron Lett.*, 2507 (1971).
377. T. Ueda, K. Miura, M. Imazawa, and K. Odajima, *Chem. Pharm. Bull. (Tokyo)* **22**, 2377 (1974).
378. K. Miura, M. Shiga, and T. Ueda, *J. Biochem.* **73**, 1279 (1973).
379. K. Miura, S. Tsuda, T. Iwano, T. Ueda, F. Harada, and N. Kato, *Biochim. Biophys. Acta* **739**, 181 (1983).
380. C.-Y. Shiue and S.-H. Chu, *J. Org. Chem.* **40**, 2971 (1975).
381. D. S. Wise and L. B. Townsend, *J. Heterocycl. Chem.* **9**, 1461 (1973).
382. (a) M. Saneyoshi and S. Nishimura, *Biochim. Biophys. Acta* **145**, 208 (1967); (b) M. Saneyoshi and S. Nishimura, *Biochim. Biophys. Acta* **204**, 389 (1970).
383. R. T. Walker, *Tetrahedron Lett.*, 2145 (1971).
384. K. H. Scheit, *Biochim. Biophys. Acta* **195**, 294 (1969).
385. B. R. Reid, *Biochemistry* **9**, 2852 (1970).
386. P. Faerber and W. Vizethum, *J. Carbohydr. Nucleosides Nucleotides* **3**, 15 (1976).

387. E. Sato, M. Machida, and Y. Kanaoka, *Chem. Pharm. Bull. (Tokyo)* **28**, 1722 (1980).

388. C.-Y. Shiue and S.-H. Chu, *J. Org. Chem.* **41**, 1847 (1976).

389. W. J. Kryzyzosiak, J. Biernat, J. Ciesiolka, P. Gornicki, and M. Wiewiorowski, *Nucleic Acids Res.* **8**, 861 (1980).

390. (a) H. Hayatsu and M. Yano, *Tetrahedron Lett.*, 755 (1969); (b) M. Yano and H. Hayatsu, *Biochim. Biophys. Acta* **199**, 303 (1970).

391. E. B. Ziff and J. R. Fresco, *J. Am. Chem. Soc.* **90**, 7338 (1968).

392. K. H. Scheit and P. Faerber, *J. Carbohydr. Nucleosides Nucleotides* **1**, 375 (1974).

393. J. Hobbs, H. Sternbach, M. Sprinzl, and F. Eckstein, *Biochemistry* **11**, 4336 (1972).

394. S. Iida, K. C. Chung, and H. Hayatsu, *Biochim. Biophys. Acta* **308**, 198 (1973).

395. C. Fombert, J.-L. Fourrey, P. Jouin, and J. Moron, *Tetrahedron Lett.*, 3007 (1974).

396. H. Vorbrüggen and K. Krolikiewicz, *Angew. Chem.* **88**, 724 (1976).

397. A. Yamane, H. Inoue, and T. Ueda, *Chem. Pharm. Bull. (Tokyo)* **28**, 157 (1980).

398. T. Ogihara and M. Mitsunobu, *Chem. Lett.*, 1621 (1982).

399. M. Mikotajczyk and J. Luczak, *J. Org. Chem.* **43**, 2132 (1978).

400. M. Kaneko and B. Shimizu, *Chem. Pharm. Bull. (Tokyo)* **20**, 1050 (1972).

401. J. Brokes and J. Beranek, *Coll. Czech. Chem. Commun.* **39**, 3100 (1974).

402. J. Zemlicka and F. Sorm, *Coll. Czech. Chem. Commun.* **30**, 2052 (1965).

403. (a) H. Vorbrüggen and U. Niedballa, *Angew. Chem.* **83**, 729 (1975); (b) H. Vorbrüggen, K. Krolikiewicz, and U. Niedballa, *Ann. der Chemie*, 988 (1975).

404. C. B. Reese and A. Ubasawa, *Tetrahedron Lett.* **21**, 2265 (1980).

405. K. J. Divakar and C. B. Reese, *J. Chem. Soc., Perkin Trans. I*, 1171 (1982).

406. S. S. Jones, B. Rayner, C. B. Reese, A. Ubasawa, and M. Ubasawa, *Tetrahedron* **36**, 3075 (1980).

407. (a) H. Hayatsu and T. Ukita, *Biochem. Biophys. Res. Commun.* **29**, 556 (1967); (b) H. Hayatsu and D. Iida, *Tetrahedron Lett.*, 1031 (1969); (c) S. Iida and H. Hayatsu, *Biochim. Biophys. Acta* **228**, 1 (1971).

408. J. Cadet and R. Teoule, *Bull. Soc. Chim. Fr.*, 1565 (1974).

409. (a) K. Burton and W. T. Riley, *Biochem. J.* **98**, 70 (1966); (b) K. Burton, *Biochem. J.* **104**, 686 (1967).

410. P. Howgate, A. S. Jones, and J. R. Tittensor, *J. Chem. Soc. B*, 275 (1968).

411. (a) L. R. Subbaraman, J. Subbaraman, and E. J. Behrman, *Bioinorg. Chem.* **1**, 35 (1971); *Chem. Abstr.* **76**, 3053z (1972); (b) F. B. Daniel and E. J. Berhman, *J. Am. Chem. Soc.* **97**, 7352 (1975).

412. L. R. Subbaraman, J. Subbaraman, and E. J. Berhman, *J. Org. Chem.* **36**, 1256 (1971).

413. S. Harayama, K. Katsumata, F. Yoneda, T. Taga, K. Osaki, and T. Nagamatsu, *Abstr. 104th Annu. Meeting Pharm. Soc. Jpn.*, 221 (1984).

414. R. S. Goody, A. S. Jones, and R. T. Walker, *Tetrahedron* **27**, 65 (1971).

415. (a) K. Ishizaki, N. Shinriki, A. Ikehata, and T. Ueda, *Chem. Pharm. Bull.* **29**, 868 (1981); (b) K. Ishizaki, N. Shinriki, and T. Ueda, *Chem. Pharm. Bull. (Tokyo)* **32**, 3601 (1984).

416. K. J. Kolonko, R. H. Shapiro, R. M. Barkley, and R. E. Sievers, *J. Org. Chem.* **44**, 3769 (1979).

417. K. Ikeda, A. Hisaka, and Y. Misu, *Abstr. 104th Annu. Meeting Pharm. Soc. Jpn.*, 221 (1984).

418. K. Kameyama, M. Sakamuki, K. Hirota, and K. Maki, *Abstr. 104th Annu. Meeting Pharm. Soc. Jpn.*, 223 (1984).

419. R. Ducolomb, J. Cadet, and R. Teoule, *Z. Naturforsch.* **29C**, 643 (1974).

420. W. E. Cohn and D. G. Doherty, *J. Am. Chem. Soc.* **78**, 2863 (1956).

421. M. Green and S. S. Cohen, *J. Biol. Chem.* **225**, 405 (1957).

422. J. T. Madison and R. W. Holley, *Biochem. Biophys. Res. Commun.* **18**, 153 (1965).

423. Y. Kondo and B. Witkop, *J. Am. Chem. Soc.* **90**, 764 (1968).

424. V. Skaric and J. Matulic-Adamic, *Helv. Chim. Acta* **66**, 687 (1983).

425. P. Cerutti, K. Ikeda, and B. Witkop, *J. Am. Chem. Soc.* **87**, 2505 (1965).

426. P. Cerutti, Y. Kondo, W. R. Landis, and B. Witkop, *J. Am. Chem. Soc.* **90**, 771 (1968).

427. E. Sato, K. Koyamna, and Y. Kanaoka, *Chem. Pharm. Bull. (Tokyo)* **19**, 837 (1971).

428. (a) V. Skaric and M. Honjec, *J. Chem. Soc., Chem. Commun.*, 495 (1973); (b) V. Skaric, B. Gaspert, and M. Honjec, *J. Chem. Soc., C*, 2444 (1970).

429. V. Skaric, G. Lacan, and D. Skaric, *J. Chem. Soc., Perkin Trans. I*, 757 (197).

430. H. Iwasaki, *Yakugaku Zasshi (Tokyo)* **82** 1368 (1962).

431. A. R. Hanze, *J. Am. Chem. Soc.* **89**, 6720 (1967).

432. J. J. Fox and D. V. Praag, *J. Am. Chem. Soc.* **82**, 486 (1960).
433. P. S. Liu, V. E. Marquez, J. S. Driscoll, R. W. Fuller, and J. J. McCormack, *J. Med. Chem.* **24**, 662 (1981).
434. S. G. Laland and G. Serck-Hanssen, *Biochem. J.* **90**, 76 (1964).
435. A. S. Jones, G. P. Stephenson, and R. T. Walker, *Tetrahedron* **35**, 1125 (1979).
436. V. M. Clark, A. R. Todd, and J. Zussman, *J. Chem. Soc.*, 2952 (1951).
437. A. Hoshi, F. Kanzawa, K. Kuretani, M. Saneyoshi, and Y. Arai, *Gann (Tokyo)* **62**, 145 (1971).
438. J. J. Fox, *Pure Appl. Chem.* **18**, 223 (1969).
439. (a) J. F. Codington, R. Fecher, and J. J. Fox, *J. Am. Chem. Soc.* **82**, 3794 (1960); (b) J. F. Codington, R. Fecher, and J. J. Fox, *J. Am. Chem. Soc.* **83**, 1889 (1961).
440. J. J. Fox and I. Wempen, *Tetrahedron Lett.*, 643 (1965).
441. J. J. Fox, N. Miller, and I. Wempen, *J. Med. Chem.* **9**, 101 (1966).
442. I. L. Doerr, J. F. Codington, and J. J. Fox, *J. Med. Chem.* **10**, 247 (1967).
443. W. Y. Ruyle, T. Y. Shen, and A. A. Pachet, *J. Org. Chem.* **30**, 4353 (1965).
444. P. Drasar, L. Hein, and J. Beranek, *Coll. Czech. Chem. Commun.* **41**, 2110 (1976).
445. L. Hein, P. Drasar, and J. Beranek, *Nucleic Acids Res.* **3**, 1125 (1976).
446. R. Marumoto and M. Honjo, *Annu. Rep. Takeda Res. Lab. (Osaka)* **26**, 21 (1967).
447. G. R. Naiz and C. B. Reese, *J. Chem. Soc., Chem. Commun.*, 552 (1968).
448. A. Hampton and A. W. Nichol, *Biochemistry* **5**, 2076 (1966).
449. J. P. H. Verheyden, D. Wagner, and J. G. Moffatt, *J. Org. Chem.* **36**, 250 (1971).
450. (a) R. L. Letsinger and K. K. Ogilvie, *J. Org. Chem.* **32**, 296 (1967); (b) K. K. Ogilvie and D. Iwacha, *Can. J. Chem.* **47**, 495 (1969).
451. T. Sowa and K. Tsunoda, *Bull. Chem. Soc. Jpn.* **48**, 505 (1975).
452. Y. Furukawa and M. Honjo, *Chem. Pharm. Bull. (Tokyo)* **16**, 2286 (1968).
453. N. C. Yung and J. J. Fox, *J. Am. Chem. Soc.* **83**, 3060 (1961).
454. F. R. Walwick, W. K. Roberts, and C. A. Dekker, *Proc. Chem. Soc.*, 84 (1958).
455. W. K. Roberts and C. A. Dekker, *J. Org. Chem.* **32**, 816 (1969).
456. R. E. Notari, D. T. Witiak, J. L. DeYoung, and A. J. Lin, *J. Med. Chem.* **15**, 1207 (1972).
457. T. Kanai, T. Kojima, O. Maruyama, and M. Ichino, *Chem. Pharm. Bull. (Tokyo)* **18**, 2569 (1970).
458. T. Kanai, M. Ichino, A. Hoshi, F. Kanzawa, and K. Kuretani, *J. Med. Chem.* **15**, 1218 (1972).
459. K. Kikugawa and M. Ichino, *Tetrahedron Lett.*, 867 (1970).
460. K. Kikugawa and M. Ichino, *J. Org. Chem.* **37**, 284 (1972).
461. J. J. Fox, E. A. Falco, I. Wempen, D. Pomeroy, M. D. Dowing, and J. H. Burchenal, *Cancer Res.* **32**, 2269 (1972).
462. E. K. Hamamura, M. Prystasz, J. P. H. Verheyden, J. G. Moffatt, K. Yamaguchi, N. Uchida, K. Sato, A. Nomura, O. Shiratori, S. Takase, and K. Katagiri, *J. Med. Chem.* **19**, 654, 663, 667 (1976).
463. U. Reichman, C. K. Chu, D. H. Hollenberg, K. A. Watanabe, and J. J. Fox, *Synthesis*, 533 (1976).
464. K. Kondo, T. Adachi, and I. Inoue, *J. Org. Chem.* **41**, 2995 (1976).
465. K. Kondo, T. Nagura, Y. Arai, and I. Inoue, *J. Med. Chem.* **22**, 639 (1979).
466. K. Kondo and I. Inoue, *J. Org. Chem.* **45**, 1577 (1980).
467. K. Kondo and I. Inoue, *J. Org. Chem.* **42**, 2809 (1977).
468. K. Kikugawa, I. Kawada, and M. Ichino, *Chem. Pharm. Bull. (Tokyo)* **23**, 3154 (1975).
469. I. L. Doerr and J. J. Fox, *J. Org. Chem.* **32**, 1462 (1967).
470. A. M. Michelson and A. R. Todd., *J. Chem. Soc.*, 816 (1955).
471. N. Miller and J. J. Fox, *J. Org. Chem.* **29**, 1772 (1964).
472. R. Letters and A. M. Michelson, *J. Chem. Soc.*, 1410 (1961).
473. G. Kowollik, K. Gaertner, and P. Langen, *Tetrahedron Lett.*, 3863 (1969).
474. R. Duschinsky and U. Eppenberger, *Tetrahedron Lett.*, 5013 (1967).
475. D. A. Schuman, M. J. Robins, and R. K. Robins, *J. Am. Chem. Soc.* **92**, 3434 (1970).
476. T. Kanai and M. Ichino, *Chem. Pharm. Bull. (Tokyo)* **16**, 1848 (1968).
477. D. M. Brown, A. R. Todd, and S. Varadarajan, *J. Chem. Soc.*, 868 (1957).
478. J. Nagyvary, *Biochemistry* **5**, 1316 (1966).
479. J. A. Secrist III, *Carbohydr. Res.* **42**, 379 (1975).
480. S. S. Tang and J. S. Roth, *Tetrahedron Lett.*, 2123 (1968).

481. M. Wada and O. Mitsunobu, *Tetrahedron Lett.*, 1279 (1972).
482. J. Kimura, Y. Fujisawa, T. Sawada, and O. Mitsunobu, *Chem. Lett.*, 691 (1974).
483. S. Shibuya, A. Kuninaka, and H. Yoshino, *Chem. Pharm. Bull. (Tokyo)* **22**, 719 (1974).
484. R. F. Dods and J. S. Roth, *J. Org. Chem.* **34**, 1627 (1969).
485. A. M. Michelson and W. E. Cohn, *Biochemistry* **1**, 490 (1962).
486. J. Zemlicka, *Coll. Czech. Chem. Commun.* **32**, 1646 (1967).
487. M. W. Winkley, *J. Chem. Soc. C*, 1365 (1970).
488. M. W. Winkley, *J. Chem. Soc. C*, 1869 (1970).
489. T. Kanai and O. Maruyama, *J. Carbohydr. Nucleosides Nucleotides* **3**, 25 (1976).
490. P. W. Wigler and H.-J. Lee, *Biochemistry* **8**, 1344 (1969).
491. K. V. B. Rao, V. E. Marquez, J. A. Kelley, and M. T. Corcoran, *J. Chem. Soc., Perkin Trans. I*, 127 (1983).
492. V. E. Marquez, K. V. B. Rao, J. V. Silverton, and J. A. Kelley, *J. Org. Chem.* **49**, 912 (1984).
493. T. Maruyama, S. Sato, and M. Honjo, *Chem. Pharm. Bull. (Tokyo)* **30**, 2688 (1982).
494. T. Maruyama, S. Kimura, Y. Sato, and M. Honjo, *J. Org. Chem.* **48**, 2719 (1983).
495. M. Honjo, T. Maruyama, Y. Wada, and K. Kamiya, *Chem. Pharm. Bull. (Tokyo)* **32**, 2073 (1984).
496. S. G. Zavgorodony, *Tetrahedron Lett.* **22**, 3003 (1981).
497. D. M. Brown, A. R. Todd, and S. Varadarajan, *J. Chem. Soc.*, 2388 (1956).
498. D. M. Brown, D. B. Parihar, A. R. Todd, and S. Varadarajan, *J. Chem. Soc.*, 3028 (1958).
499. T. J. Delia and J. Beranek, *J. Carbohydr. Nucleosides Nucleotides* **4**, 349 (1977).
500. T. Sekiya and T. Ukita, *Chem. Pharm. Bull. (Tokyo)* **15**, 1498 (1967).
501. W. V. Ruyle and T. Y. Shen, *J. Med. Chem.* **10**, 331 (1967).
502. M. Hirata, *Chem. Pharm. Bull. (Tokyo)* **16**, 437 (1968).
503. J. F. Codington, I. L. Doerr, D. V. Praag, A. Bendich, and J. J. Fox, *J. Am. Chem. Soc.* **83**, 5030 (1961).
504. (a) J. F. Codington, I. L. Doerr, and J. J. Fox, *J. Org. Chem.* **29**, 558 (1964); (b) J. F. Codington, I. L. Doerr, and J. J. Fox, *J. Org. Chem.* **29**, 564 (1964).
505. A. Holý and D. Cech, *Coll. Czech. Chem. Commun.* **39**, 3157 (1974).
506. A. Holý, *Coll. Czech. Chem. Commun.* **38**, 3912 (1973).
507. J. Brokes, H. Hrebabecky, and J. Beranek, *Coll. Czech. Chem. Commun.* **44**, 439 (1979).
508. (a) D. M. Brown, D. B. Parihar, C. B. Reese, and A. R. Todd, *J. Chem. Soc.*, 3038 (1958); (b) D. M. Brown, D. B. Parihar, and A. R. Todd, *J. Chem. Soc.*, 4242 (1958).
509. J. Hobbs and F. Eckstein, *Nucleic Acids Res.* **2**, 1987 (1975).
510. R. A. Sharma, M. Bobek, and A. Bloch, *J. Med. Chem.* **18**, 955 (1975).
511. (a) M. Imazawa, T. Ueda, and T. Ukita, *Tetrahedron Lett.*, 4807 (1970); (b) M. Imazawa, T. Ueda, and T. Ukita, *Chem. Pharm. Bull. (Tokyo)* **23**, 604 (1975).
512. M. Imazawa, K. Kudo, H. Kikutome, and T. Ueda, *Abstr. 3rd Symp. Nucleic Acid Chem.*, 1 (1975).
513. K. J. Divakar and C. B. Reese, *J. Chem. Soc., Perkin Trans. I*, 1625 (1982).
514. A. Matsuda and T. Miyasaka, *Heterocycles* **20**, 55 (1983).
515. A. Holý, *Coll. Czech. Chem. Commun.* **38**, 423 (1973).
516. A. Holý, *Coll. Czech. Chem. Commun.* **40**, 738 (1975).
517. A. Rosenthal and R. H. Dodd, *Carbohydr. Res.* **78**, 33 (1980).
518. M. Meresz, G. Horvath, P. Sohar, and J. Kuszmann, *Tetrahedron* **31**, 1873 (1975).
519. T. C. Chuwang, A. Fridland, and T. L. Avery, *J. Med. Chem.* **26**, 280 (1983).
520. J. J. Fox and B. A. Otter, *Ann. N.Y. Acad. Sci.* **255**, 59 (1975).
521. K. Kikugawa, *J. Pharm. Sci.* **64**, 424 (1975).
522. K. Kondo, A. Kojima, and I. Inoue, *Chem. Pharm. Bull. (Tokyo)* **26**, 3244 (1978).
523. A. F. Cook and M. J. Holman, *J. Org. Chem.* **43**, 4200 (1978).
524. D. H. Shannahoff and R. A. Sanchez, *J. Org. Chem.* **38**, 593 (1973).
525. R. Mengel and W. Guschlbauer, *Angew. Chem. Int. Ed.* **17**, 525 (1978).
526. A. D. Patel, W. H. Schrier, and J. Nagyvary, *J. Org. Chem.* **45**, 4830 (1980).
527. R. P. Glinski, M. S. Khan, R. L. Kalman, and M. B. Sporn, *J. Org. Chem.* **38**, 4299 (1973).
528. J. P. Horwitz, J. Chua, and M. Noel, *J. Org. Chem.* **29**, 2076 (1964).
529. K. Kikugawa and T. Ukita, *Chem. Pharm. Bull. (Tokyo)* **17**, 775 (1969).
530. K. Kikugawa, M. Ichino, and T. Ukita, *Chem. Pharm. Bull. (Tokyo)* **17**, 785 (1969).
531. K. Kikugawa, M. Ichino, T. Kusama, and T. Ukita, *Chem. Pharm. Bull. (Tokyo)* **17**, 798 (1969).
532. G. Kowollik, K. Gaertner, and P. Langen, *J. Carbohydr. Nucleosides Nucleotides* **2**, 191 (1975).

533. S. Ajimera, A. R. Bapat, K. D. Danenberg, and P. V. Danenberg, *J. Med. Chem.* **27**, 11 (1984).
534. I. L. Doerr, J. F. Codington, and J. J. Fox, *J. Org. Chem.* **30**, 467 (1965).
535. J. F. Codington, I. L. Doerr, and J. J. Fox, *J. Org. Chem.* **30**, 476 (1965).
536. M. Hirata, *Chem. Pharm. Bull. (Tokyo)* **16**, 291 (1968).
537. M. Hirata, T. Kobayashi, and T. Naito, *Chem. Pharm. Bull. (Tokyo)* **17**, 1188 (1969).
538. J. J. Fox and K. A. Watanabe, *Chem. Pharm. Bull. (Tokyo)* **17**, 211 (1969).
539. (a) T. Ueda and S. Shibuya, *Chem. Pharm. Bull. (Tokyo)* **18**, 1076 (1970); (b) T. Ueda and S. Shibuya, *Chem. Pharm. Bull. (Tokyo)* **22**, 930 (1974).
540. O. Mitsunobu, *J. Chem. Soc. Chem. Commun.*, 201 (1973).
541. J. Kimura, K. Yagi, H. Suzuki, and O. Mitsunobu, *Bull. Soc. Chem. Jpn.* **53**, 3670 (1980).
542. J. Kimura and O. Mitsunobu, *Bull. Soc. Chem. Jpn.* **51**, 1903 (1978).
543. H. Hayashi, M. Imazawa, and T. Ueda, *Abstr. 93rd Annu. Meeting Pharm. Soc. Jpn. Part II*, 122 (1973).
544. D. Lipkin, C. T. Cori, and J. A. Rabi, *J. Heterocycl. Chem.* **6**, 995 (1969).
545. Y. Mizuno, Y. Watanabe, and K. Ikeda, *Chem. Pharm. Bull. (Tokyo)* **22**, 1194 (1974).
546. H. Inoue, M. Takada, M. Takahashi, and T. Ueda, *Heterocycles* **8**, 427 (1977).
547. K. Hirota, Y. Kitade, F. Iwanami, and S. Senda, *Chem. Pharm. Bull. (Tokyo)* **32**, 2591 (1984).
548. A. Matsuda, H. Inoue, and T. Ueda, *Abstr. 1st Symp. Nucleic Acid Chem.*, 85 (1973).
549. T. Kunieda and B. Witkop, *J. Am. Chem. Soc.* **93**, 3487 (1971).
550. (a) G. Shaw and R. N. Warrener, *Proc. Chem. Soc.*, 81 (1958); (b) G. Shaw and R. N. Warrener, *J. Chem. Soc.*, 50 (1959).
551. I. Wempen and J. J. Fox, *J. Org. Chem.* **34**, 1020 (1969).
552. T. Ueda and H. Tanaka, *Chem. Pharm. Bull. (Tokyo)* **18**, 1491 (1970).
553. S. Shibuya and T. Ueda, *J. Carbohydr. Nucleosides Nucleotides* **7**, 49 (1980).
554. T. Ueda, T. Asano, and H. Inoue, *J. Carbohydr. Nucleosides Nucleotides* **3**, 365 (1976).
555. D. S. Wise and L. B. Townsend, *J. Chem. Soc., Chem. Commun.*, 271 (1979).
556. H. Inoue and T. Ueda, *Chem. Pharm. Bull. (Tokyo)* **26**, 2664 (1978).
557. S. Shibuya and T. Ueda, *Chem. Pharm. Bull. (Tokyo)* **28**, 939 (1980).
558. I. L. Doerr, R. J. Cushley, and J. J. Fox, *J. Org. Chem.* **33**, 1592 (1968).
559. I. L. Doerr and J. J. Fox, *J. Am. Chem. Soc.* **89**, 1760 (1967).
560. T. Ueda, S. Shibuya, and J. Yamashita, *Abstr. 1st Symp. Nucleic Acid Chem.*, 71 (1973).
561. A. F. Cook, *J. Med. Chem.* **20**, 344 (1977).
562. T. Ueda, M. Imazawa, and S. Hayashi, unpublished experiments.
563. T. Sasaki, K. Minamoto, and T. Sugiura, *J. Org. Chem.* **40**, 3498 (1975).
564. M. J. Robins and T. Kanai, *J. Org. Chem.* **41**, 1886 (1976).
565. T. Sasaki, K. Minamoto, M. Kino, and T. Mizuno, *J. Org. Chem.* **41**, 1100 (1976).
566. A. Matsuda, K. A. Watanabe, and J. J. Fox, *J. Org. Chem.* **45**, 3274 (1980).
567. T. Sasaki, K. Minamoto, T. Suzuki, and T. Sugiura, *J. Am. Chem. Soc.* **100**, 2248 (1978).
568. T. Sasaki, K. Minamoto, T. Suzuki, and T. Sugiura, *J. Org. Chem.* **44**, 1424 (1979).
569. T. Sasaki, K. Minamoto, T. Suzuki, and S. Yamashita, *Tetrahedron* **36**, 865 (1980).
570. T. L. V. Ulbricht, in *Synthetic Procedures in Nucleic Acid Chemisty, Vol.* 2, p. 177 (W. W. Zorbach and R. S. Tipson, eds.), Wiley-Interscience, New York, 1973.
571. M. Ikehara, *Acc. Chem. Res.* **2**, 47 (1969).
572. A. W. Johnson, D. Oldfield, R. Rodrigo, and N. Shaw, *J. Chem. Soc.*, 4080 (1964).
573. Y. Yamagata, S. Fujii, T. Fujiwara, K. Tomita, and T. Ueda, *Biochim. Biophys. Acta* **654**, 242 (1981).
574. T. Ueda, S. Shuto, T. Sano, H. Usui, and H. Inoue, *Nucleic Acids Res. Symp. Ser.* **11**, 5 (1982).
575. T. Ueda, H. Usui, S. Shuto, and H. Inoue, *Chem. Pharm. Bull. (Tokyo)* **32**, 3410 (1984).
576. J. A. Rabi and J. J. Fox, *J. Org. Chem.* **37**, 3898 (1972).
577. B. A. Otter, E. A. Falco, and J. J. Fox, *J. Org. Chem.* **41**, 3133 (1976).
578. B. A. Otter, E. A. Falco, and J. J. Fox, *J. Org. Chem.* **43**, 481 (1978).
579. B. A. Otter and E. A. Falco, *Tetrahedron Lett.*, 4383 (1978).
580. I. M. Sasson and B. A. Otter, *J. Org. Chem.* **46**, 1114 (1981).
581. T. Ueda and S. Shuto, *Nucleosides Nucleotides* **3**, 295 (1984).
582. (a) T. Sano, H. Inoue, and T. Ueda, *Chem. Pharm. Bull. (Tokyo)* **33**, 3595 (1985); (b) T. Sano, S. Shuto, H. Inoue, and T. Ueda, *Chem. Pharm. Bull. (Tokyo)* **33**, 3617 (1985).
583. T. Ueda, S. Shuto, and H. Inoue, *Nucleosides Nucleotides* **3**, 173 (1984).

584. G. H. Jones and J. G. Moffatt, *J. Carbohydr. Nucleosides Nucleotides* **6**, 127 (1979).
585. C. B. Reese, in *Protective Groups in Organic Chemistry* (J. F. W. McOmie, ed.), p. 95, Plenum Press, New York, 1973.
586. H. Kossel and H. Seliger, *Fortschr. Chem. Org. Naturst.* **32**, 297 (1975).
587. E. Ohtsuka, M. Ikehara, and D. Söll, *Nucleic Acids Res.* **10**, 6553 (1982).
588. S. A. Narang, *Tetrahedron* **39**, 3 (1983).
589. A. M. Michelson and A. R. Todd, *J. Chem. Soc.*, 3459 (1956).
590. (a) C. B. Reese and D. R. Trentham, *Tetrahedron Lett.*, 2459 (1965); (b) B. E. Griffin, C. B. Reese, G. F. Stephenson, and D. R. Trentham, *Tetrahedron Lett.*, 4349 (1966).
591. T. Kamimura, T. Masegi, and T. Hata, *Chem. Lett.*, 965 (1982).
592. (a) J. T. Kusmiereck, J. Giziewicz, and D. Shugar, *Biochemistry* **12**, 194 (1973); (b) J. T. Kusmiereck and D. Shugar, *Acta Biochim. Pol.* **18**, 413 (1971).
593. (a) E. Darzyniewicz, J. T. Kusmiereck, and D. Shugar, *Biochem. Biophys. Res. Commun.* **46**, 1734 (1972); (b) E. Darzyniewicz, J. T. Kusmiereck, and D. Shugar, *Acta Biochim. Pol.* **21**, 305 (1974).
594. J. Gigiewicz and D. Shugar, *Acta Biochim. Pol.* **20**, 73 (1973).
595. J. Gigiewicz, J. T. Kusmiereck, and D. Shugar, *J. Med. Chem.* **15**, 839 (1972).
596. W. Hutzenlaub and W. Pfleiderer, *Chem. Ber.* **106**, 665 (1973).
597. K. Kikugawa, F. Sato, N. Imura, and T. Ukita, *Chem. Pharm. Bull. (Tokyo)* **16**, 1110 (1968).
598. H. Takaku, K. Kamaide, and H. Tsuchiya, *J. Org. Chem.* **49**, 51 (1984).
599. A. Holý and H. Pischel, *Coll. Czech. Chem. Commun.* **39**, 3863 (1974).
600. H. Pischel, A. Holý, G. Wagner, and D. Cech, *Coll. Czech. Chem. Commun.* **40**, 2689 (1975).
601. M. D. Edge and A. S. Jones, *J. Chem. Soc. C*, 1933 (1971).
602. D. M. G. Martin, C. B. Reese, and G. F. Stephenson, *Biochemistry* **7**, 1406 (1968).
603. M. J. Robins and S. R. Naik, *Biochemistry* **10**, 3591 (1971).
604. M. J. Robins and S. R. Naik, *Biochim. Biophys. Acta* **246**, 341 (1971).
605. M. J. Robins, S. R. Naik, and A. S. K. Lee, *J. Org. Chem.* **39**, 1891 (1974).
606. J. Heikkila, V. Bjorkman, B. Oberg, and J. Chattopadhyaya, *Acta Chem. Scand. Ser. B* **36**, 715 (1982).
607. M. J. Robins, A. S. K. Lee, and F. A. Norris, *Carbohydr. Res.* **41**, 304 (1975).
608. G. H. Ransford, R. P. Glinski, and M. B. Sporn, *J. Carbohydr. Nucleosides Nucleotides* **1**, 275 (1974).
609. L. F. Christensen and A. D. Broom, *J. Org. Chem.* **37**, 2929 (1960).
610. (a) D. C. Bartholomev and A. D. Broom, *J. Chem. Soc. Chem. Commun.*, 38 (1975), (b) E. Ohtsuka, T. Wakabayashi, S. Tanaka, T. Tanaka, H. Oshie, A. Hasegawa, and M. Ikehara, *Chem. Pharm. Bull. (Tokyo)* **29**, 318 (1981).
611. Y. Mizuno, T. Endo, and K. Ikeda, *J. Org. Chem.* **40**, 1385 (1975).
612. D. Wagner, J. P. H. Verheyden, and J. G. Moffatt, *J. Org. Chem.* **39**, 24 (1974).
613. (a) M. Ikehara and S. Uesugi, *Chem. Pharm. Bul. (Tokyo)* **21**, 264 (1973); (b) M. Ikehara and S. Uesugi, *Tetrahedron Lett.*, 713 (1970).
614. E. Ohtsuka, S. Tanaka, and M. Ikehara, *Nucleic Acids Res.* **1**, 1351 (1974).
615. K. Yamauchi, T. Nakagima, and M. Kinoshita, *J. Org. Chem.* **45**, 3865 (1980).
616. F. W. Lichtenthaler, Y. Sanemitsu, and T. Nohara, *Angew. Chem. Int. Ed.* **17**, 772 (1978).
617. K. A. Watanabe, A. Matsuda, M. J. Halat, D. H. Hollenberg, J. S. Nisselbaum, and J. J. Fox, *J. Med. Chem.* **24**, 893 (1981).
618. K. E. Pfitzner and J. G. Moffatt, *J. Org. Chem.* **29**, 1508 (1964).
619. E. Benz, N. F. Elmore, and L. Goodman, *J. Org. Chem.* **30**, 3067 (1965).
620. P. A. Levene and R. S. Tipson, *J. Biol. Chem.* **105**, 419 (1934).
621. P. A. Levene and R. S. Tipson, *J. Biol. Chem.* **106**, 113 (1935).
622. B. Schwarz, D. Cech, A. Holý, and J. Skoda, *Coll. Czech. Chem. Commun.* **45**, 3217 (1980).
623. A. F. Cook, J. J. Holman, M. J. Kramer, and P. W. Trown, *J. Med. Chem.* **22**, 1330 (1979).
624. Y. Mizuno, T. Ueda, K. Ikeda, and K. Miura, *Chem. Pharm. Bull. (Tokyo)* **16**, 262 (1968).
625. J. P. H. Verheyden and J. G. Moffatt, *J. Am. Chem. Soc.* **86**, 2093 (1964).
626. J. P. H. Verheyden and J. G. Moffatt, *J. Org. Chem.* **35**, 2319 (1970).
627. J. P. H. Verheyden and J. G. Moffatt, *J. Org. Chem.* **39**, 3573 (1974).
628. J. P. H. Verheyden and J. G. Moffatt, *J. Org. Chem.* **35**, 2868 (1970).
629. G. A. R. Johnston, *Aust. J. Chem.* **21**, 513 (1968).
630. J. P. H. Verheyden and J. G. Moffatt, *J. Org. Chem.* **37**, 2289 (1972).

631. R. F. Dods and J. S. Roth, *Tetrahedron Lett.*, 165 (1969).
632. K. Kikugawa and M. Ichino, *Tetrahedron Lett.*, 87 (1971).
633. Y. Wang, H. P. C. Hogenkamp, R. A. Long, G. R. Revanker, and R. K. Robins, *Carbohydr. Res.* **59**, 449 (1977).
634. H. Hrebabecky, J. Brokes, and J. Beranek, *Coll. Czech. Chem. Commun.* **45**, 599 (1980).
635. P. C. Srivastava and R. J. Rousseau, *Carbohydr. Res.* **27**, 455 (1973).
636. S. Greenberg and J. G. Moffatt, *J. Am. Chem. Soc.* **95**, 4016 (1973).
637. A. F. Russell, S. Greenberg, and J. G. Moffatt, *J. Am. Chem. Soc.* **95**, 4025 (1973).
638. R. Marumoto and M. Honjo, *Chem. Pharm. Bull. (Tokyo)* **22**, 128 (1974).
639. A. A. Akhrem, G. V. Zaitseva, I. A. Mikhailopulo, and A. F. Abramov, *J. Carbohydr. Nucleosides Nucleotides* **4**, 43 (1977).
640. K. S. Bhat and A. S. Rao, *Indian J. Chem. Sec. B* **22**, 678 (1983).
641. M. M. Ponpipom and S. Hanessian, *Can. J. Chem.* **50**, 246, 253 (1972).
642. M. M. Ponpipom, *Carbohydr. Res.* **18**, 342 (1971).
643. M. W. Logue, *Carbohydr. Res.* **40**, c9 (1975).
644. H. Hrebabecky and J. Beranek, *Nucleic Acids Res.* **5**, 1029 (1978).
645. H. Hrebabecky and J. Beranek, *Coll. Czech. Chem. Commun.* **43**, 3268 (1978).
646. D. H. R. Barton and W. B. Motherwell, *Pure Appl. Chem.* **53**, 15 (1981).
647. R. A. Lessor and N. J. Leonard, *J. Org. Chem.* **46**, 4300 (1981).
648. M. J. Robins and J. S. Wilson, *J. Am. Chem. Soc.* **103**, 932 (1981).
649. K. Pankiewicz, A. Matsuda, and K. A. Watanabe, *J. Org. Chem.* **47**, 485 (1982).
650. (a) K. Fukukawa, T. Ueda, and T. Hirano, *Chem. Pharm. Bull. (Tokyo)* **29**, 597 (1981); (b) K. Fukukawa, T. Ueda, and T. Hirano, *Chem. Pharm. Bull. (Tokyo)* **31**, 1842 (1983).
651. J. Wellmann and E. Steckhan, *Angew. Chem. Int. Ed.* **19**, 46 (1980).
652. T. Adachi, T. Iwasaki, M. Miyoshi, and I. Inoue, *J. Chem. Soc. Chem. Commun.*, 248 (1977).
653. G. Etzold, R. Hintsche, G. Kowollik, and P. Langen, *Tetrahedron* **27**, 2463 (1971).
654. G. Kowollik, G. Etzold, M. von Janta-Lipinski, K. Gaertner, and P. Langen, *J. Prak. Chem.* **315**, 859 (1973).
655. S. Ajmera, A. R. Bapat, K. Danenberg, and P. V. Danenberg, *J. Med. Chem.* **27**, 11 (1984).
656. M. Schutt, G. Kowollik, G. Etzold, and P. Langen, *J. Prak. Chem.* **314**, 251 (1972).
657. G. Kowollik, K. Gaertner, G. Etzold, and P. Langen, *Carbohydr. Res.* **12**, 251 (1970).
658. G. Kowollik, K. Gaertner, and P. Langen, *Z. Chem.* **10**, 141 (1970).
659. J. P. Horwitz, A. J. Tomson, J. A. Urbanski, and J. Chua, *J. Org. Chem.* **27**, 3045 (1962).
660. T.-S. Lin and W. H. Prusoff, *J. Carbohydr. Nucleosides Nucleotides* **2**, 185 (1975).
661. T.-S. Lin, C. Chai, and W. H. Prusoff, *J. Med. Chem.* **19**, 915 (1976).
662. W. Freist, K. Shattka, F. Cramer, and B. Jastorff, *Chem. Ber.* **105**, 991 (1972).
663. W. S. Mungall, G. L. Gleene, G. A. Heavner, and R. L. Letsinger, *J. Org. Chem.* **40**, 1659 (1975).
664. T.-S. Lin and W. H. Prusoff, *J. Med. Chem.* **21**, 106 (1978).
665. T. Hata, I. Yamamoto, an M. Sekine, *Chem. Lett.*, 601 (1976).
666. (a) T. Adachi, Y. Yamada, I. Inoue, and M. Saneyoshi, *Carbohydr. Res.* **73**, 113 (1979); (b) T. Adachi, Y. Yamada, I. Inoue, and M. Saneyoshi, *Synthesis*, 45 (1977).
667. T. Adachi, Y. Arai, I. Inoue, and M. Saneyoshi, *Carbohydr. Res.* **78**, 67 (1980).
668. T. Hata, I. Yamamoto, and M. Sekine, *Chem. Lett.*, 977 (1975).
669. I. Yamamoto, M. Sekine, and T. Hata, *J. Chem. Soc., Perkin Trans. I*, 306 (1980).
670. D. C. Baker and D. Horton, *Carbohydr. Res.* **21**, 393 (1972).
671. J. J. Baker, P. Mellish, C. Riddle, A. R. Somerville, and J. R. Tittensor, *J. Med. Chem.* **17**, 764 (1974).
672. M. W. Logue and N. J. Leonard, *J. Am. Chem. Soc.* **94**, 2842 (1972).
673. E. J. Delaney, P. W. Morris, T. Matsushita, and D. L. Venton, *J. Carbohydr. Nucleosides Nucleotides* **8**, 445 (1981).
674. M. J. Gait, A. S. Jones, M. D. Jones, M. J. Shepherd, and R. T. Walker, *J. Chem. Soc., Perkin Trans. I*, 1389 (1979).
675. T.-S. Lin, G. T. Shiau, W. H. Prusoff, and J. P. Neenan, *J. Carbohydr. Nucleosides Nucleotides* **7**, 389 (1980).
676. W. S. Mungall, L. J. Lemmen, K. L. Lemmen, J. K. Dethmers, and L. L. Norling, *J. Med. Chem.* **21**, 704 (1978).

677. H. Loibner and E. Zbiral, *Liebigs Ann. Chem.*, 78 (1978).
678. M. N. Preobrazhenskaya, S. Yu Melnik, E. A. Utkina, E. G. Sokolova, and N. N. Suvorov, *Zh. Org. Khim.* **9**, 2207 (1973) [*Nucleic Acids Abstr.* **4**, 4N-4575 (1974)]; M. N. Preobrazhenskaya, S. Yu Melnik, E. A. Utkina, E. G. Sokolova, and N. N. Suvorov, *Zh. Org. Khim.* **10**, 863 (1974) [*Nucleic Acids Abstr.* **5**, 5N-1880 (1975)].
679. H. Pischel, A. Holý, and G. Wagner, *Col. Czech. Chem. Commun.* **46**, 933 (1981).
680. A. S. David and J.-C. Fischer, *Bull. Soc. Chim. Fr.* 3610 (1972).
681. J. Baddiley and G. A. Jamisen, *J. Chem. Soc.*, 1085 (1955).
682. J. Hildesheim, R. Hildesheim, and E. Lederer, *Biochimie Fr.* **53**, 1067 (1971) [*Chem. Abstr.* **76**, 14293d (1972)].
683. A. Holý, *Tetrahedron Lett.*, 585 (1972).
684. (a) I. Nakagawa and T. Hata, *Tetrahedron Lett.*, 1409 (1975); (b) I. Nakagawa, K. Aki, and T. Hata, *J. Chem. Soc., Perkin Trans. I*, 1315 (1983).
685. J. Kimura, T. Shimizu, and O. Mitsunobu, *Chem. Lett.*, 1793 (1981).
686. G. Kowollik and P. Langen, *J. Prak. Chem.* **37**, 311 (1968).
687. G. Etzold, G. Kowollik, and P. Langen, *J. Chem. Soc. Chem. Commun.*, 422 (1968).
688. W. Meyer, E. Bohnke, and H. Follmann, *Angew. Chem. Int. Ed.* **15**, 499 (1976).
689. W. Meyer and H. Follmann, *Chem. Ber.* **113**, 2530 (1980).
690. R. D. Elliot, R. W. Brockman, and J. A. Montgomery, *J. Med. Chem.* **24**, 350 (1981).
691. G. Kowollik, K. Gaertner, and P. Langen, *Angew. Chem.* **78**, 752 (1966).
692. J. F. Codington, R. Fecher, and J. J. Fox, *J. Org. Chem.* **27**, 163 (1962).
693. T. Sekiya and T. Ukita, *Chem. Pharm. Bull. (Tokyo)* **15**, 1503 (1967).
694. M. Hirata, *Chem. Pharm. Bull. (Tokyo)* **16**, 430 (1968).
695. G. Kowollik and P. Langen, *Z. Chem.* **15**, 147 (1975).
696. H. K. Misra, W. P. Gati, E. K. Knaus, and L. I. Wiebe, *J. Heterocycl. Chem.* **21**, 773 (1984).
697. U. Reichman, D. H. Hollenberg, C. K. Chu, K. A. Watanabe, and J. J. Fox, *J. Org. Chem.* **41**, 2042 (1976).
698. T. Kanai, C. Yamashita, and M. Ichino, *Chem. Abstr.* **76**, 14860m (1972).
699. D. H. Hollenberg, K. A. Watanabe, and J. J. Fox, *J. Med. Chem.* **20**, 113 (1977).
700. V. Skaric, D. Katalemic, D. Skaric, and I. Salaj, *J. Chem. Soc., Perkin Trans. I*, 2091 (1982).
701. M. Márton-Merész, J. Kuszmann, I. Pelczer, L. Párkányi, T. Koritsanzky, and A. Kálmán, *Tetrahedron* **39**, 275 (1983).
702. M. Márton-Merész, J. Kuszmann, J. Langó, L. Parkanyi, and A. Kálmán, *Nucleosides Nucleotides* **3**, 221 (1984).
703. K. Hirota, Y. Kitade, T. Tomishi, and Y. Maki, *J. Chem. Soc., Chem. Commun.*, 108 (1984).
704. J. P. Horwitz, J. Chua, J. A. Urbanski, and M. Noel, *J. Org. Chem.* **27**, 3045 (1962).
705. H. Yonehara and N. Otake, *Tetrahedron Lett.*, 3788 (1966).
706. H. Hoeksema, G. Slomp, and E. E. vanTamelen, *Tetrahedron Lett.*, 1787 (1964).
707. (a) S. Harada, E. Mizuta, and T. Kishi, *J. Am. Chem. Soc.* **100**, 4895 (1978); (b) S. Harada and T. Kishi, *J. Antibiotics* **31**, 519 (1978).
708. J. P. Horwitz, J. Chua, I. L. Klundt, M. A. DaRooge, and M. Noel, *J. Am. Chem. Soc.* **86**, 1806 (1964).
709. J. P. Horwitz, J. Chua, M. A. DaRooge, and M. Noel, *Tetrahedron Lett.*, 2725 (1964).
710. J. P. Horwitz, J. Chua, M. A. DaRooge, M. Noel, and I. L. Klundt, *J. Org. Chem.* **31**, 205 (1966).
711. T. A. Khwaja and C. Heidelberger, *J. Med. Chem.* **10**, 1066 (1967).
712. K. E. Pfitzner and J. G. Moffatt, *J. Org. Chem.* **29**, 1508 (1964).
713. J. P. Horwitz, J. Chua, M. Noel, and J. T. Donatti, *J. Org. Chem.* **32**, 817 (1967).
714. M. W. Winkley, *Carbohydr. Res.* **13**, 173 (1970).
715. G. Kowollik, K. Gaertner, and P. Langen, *Tetrahedron Lett.*, 1737 (1971).
716. Y. Wang and H. P. C. Hogenkamp, *J. Org. Chem.* **43**, 3324 (1978).
717. T. C. Jain, I. D. Jenkins, A. F. Russell, J. P. H. Verheyden, and J. G. Moffatt, *J. Org. Chem.* **39**, 30 (1974).
718. T. Adachi, T. Iwasaki, M. Miyoshi, and I. Inoue, *J. Org. Chem.* **44**, 1404 (1979); *Nucleic Acids Res. Spec. Publ.* **2**, s93 (1976).
719. Y. Furukawa, Y. Yoshida, K. Imai, and M. Honjo, *Chem. Pharm. Bull. (Tokyo)* **18**, 554 (1970).
720. T. Sasaki, K. Minamoto, and H. Suzuki, *J. Org. Chem.* **38**, 598 (1973).
721. T. Sasaki, K. Minamoto, and K. Hattori, *J. Org. Chem.* **38**, 1283 (1973).

722. T. Sasaki, K. Minamoto, and K. Hattori, *Tetrahedron* **30**, 2684 (1974).
723. T. Sasaki, K. Minamoto, K. Hattori, and T. Sugiura, *J. Carbohydr. Nucleosides Nucleotides* **2**, 47 (1975).
724. K. E. Pfitzner and J. G. Moffatt, *J. Am. Chem. Soc.* **85**, 3027 (1963).
725. V. Skaric and J. Matulic-Adamic, *Helv. Chim. Acta* **63**, 2179 (1980).
726. J. R. McCarthy, Jr., R. K. Robins, and M. J. Robins, *J. Am. Chem. Soc.* **90**, 4993 (1967).
727. M. J. Robins, J. R. McCarthy, Jr., and R. K. Robins, *J. Heterocycl. Chem.* **4**, 313 (1967).
728. J. P. H. Verheyden and J. G. Moffatt, *J. Am. Chem. Soc.* **88**, 5684 (1966).
729. (a) T. Sasaki, S. Kuroyanagi, K. Minamoto, and K. Hattori, *Tetrahedron Lett.*, 2731 (1973);
     (b) T. Sasaki, K. Minamoto, and K. Hattori, *J. Am. Chem. Soc.* **95**, 1350 (1973).
730. T. Sasaki, K. Minamoto, T. Asano, and M. Miyake, *J. Org. Chem.* **40**, 106 (1975).
731. J. P. H. Verheyden, I. D. Jenkins, G. R. Owen, S. D. Dimitrijevich, C. M. Richards, P. C. Srivastava, N. LeHong, and J. G. Moffatt, *Ann. N.Y. Acad. Sci.* **255**, 151 (1975).
732. J. P. H. Verheyden and J. G. Moffatt, *J. Am. Chem. Soc.* **97**, 4386 (1975).
733. G. R. Owen, J. P. H. Verheyden, and J. G. Moffatt, *J. Org. Chem.* **41**, 3010 (1976).
734. M. W. Winkley, *Carbohydr. Res.* **16**, 462 (1971).
735. J. Zemlicka, J. V. Freisler, R. Gasser, and J. P. Horwitz, *J. Org. Chem.* **38**, 990 (1973).
736. J. Zemlicka, R. Gasser, and J. P. Horwitz, *J. Am. Chem. Soc.* **92**, 4744 (1970).
737. M. J. Robins, R. Mengel, and R. A. Jones, *J. Am. Chem. Soc.* **95**, 4074 (1973).
738. P. Howgate, A. S. Jones, and J. R. Tittensor, *Carbohydr. Res.* **12**, 403 (1970).
739. J. Zemlicka, R. Gasser, and J. P. Horwitz, *J. Am. Chem. Soc.* **94**, 3213 (1972).
740. M. J. Robins and E. M. Trip, *Tetrahedron Lett.*, 3369 (1974).
741. (a) C. B. Reese, K. Schofield, R. Shapiro, and A. Todd, *Proc. Chem. Soc.*, 290 (1960);
     (b) G. P. Moss, C. B. Reese, K. Schofield, and A. Todd, *J. Chem. Soc.*, 1149 (1963).
742. (a) A. S. Jones and A. R. Williamson, *Chem. Ind.*, 1624 (1960); (b) A. S. Jones, A. R. Williamson, and M. Winkley, *Carbohydr. Res.* **1**, 187 (1965).
743. R. F. Schinazi, M. S. Chu, and W. H. Prusoff, *J. Med. Chem.* **21**, 1141 (1978).
744. J. J. Baker, A. M. Mian, and J. R. Tittensor, *Tetrahedron* **30**, 2939 (1974).
745. (a) R. R. Schmidt, U. Scholtz, and D. Schwiller, *Chem. Ber.* **101**, 590 (1968); (b) R. R. Schmidt and H. J. Fritz, *Chem. Ber.* **103**, 1867 (1970).
746. J. P. Vizsoli and G. M. Tener, *Chem. Ind.*, 2631 (1962).
747. K. Kondo and I. Inoue, *J. Org. Chem.* **44**, 4713 (1979).
748. K. E. Pfitzner and J. G. Moffatt, *J. Am. Chem. Soc.* **87**, 5661 (1965).
749. N. P. Damodara, G. H. Jones, and J. G. Moffatt, *J. Am. Chem. Soc.* **93**, 3812 (1971).
750. T. Ueda, T. Sakurai, and H. Inoue, unpublished results.
751. K. Isono, K. Asahi, and S. Suzuki, *J. Am. Chem. Soc.* **91**, 7490 (1969).
752. S. L. Cook and J. A. Secrist III, *J. Am. Chem. Soc.* **101**, 1554 (1979).
753. J. A. Secrist III and W. J. Winter, Jr., *J. Am. Chem. Soc.* **100**, 2554 (1978).
754. J. M. J. Tronchet and M. J. Valero, *Helv. Chim. Acta* **62**, 2788 (1979).
755. G. H. Jones and J. G. Moffatt, *J. Am. Chem. Soc.* **90**, 5337 (1968).
756. F. Kappler and A. Hampton, *J. Carbohydr. Nucleosides Nucleotides* **2**, 109 (1975).
757. A. Rosenthal and K. Dooley, *J. Carbohydr. Nucleosides Nucleotides* **1**, 61 (1974).
758. R. Youssefyeh, D. Tegg, J. P. H. Verheyden, G. H. Jones, and J. G. Moffatt, *Tetrahedron Lett.*, 435 (1977).
759. G. H. Jones, M. Taniguchi, D. Tegg, and J. G. Moffatt, *J. Org. Chem.* **44**, 1309 (1979).
760. R. D. Youssefyeh, J. P. H. Verheyden, and J. G. Moffatt, *J. Org. Chem.* **44**, 1301 (1979).
761. T. Sasaki, K. Minamoto, and K. Hattori, *Tetrahedron* **30**, 2689 (1974).
762. (a) R. W. Winkley, D. C. Hehemann, and W. W. Winkley, *Carbohydr. Res.* **58**, C10 (1977);
     (b) R. W. Winkley, D. G. Hehemann, and W. W. Winkley, *J. Org. Chem.* **43**, 2573 (1978).
763. F. Hansske and M. J. Robins, *Tetrahedron Lett.*, 1589 (1983).
764. F. Hansske, D. Madej, and M. J. Robins, *Tetrahedron* **40**, 125 (1984).
765. A. F. Cook and J. G. Moffatt, *J. Am. Chem. Soc.* **89**, 2697 (1967).
766. U. Brobeck and J. G. Moffatt, *J. Org. Chem.* **35**, 3552 (1970).
767. W. T. Markiewicz, *J. Chem. Res. (S)*, 24 (1979).
768. (a) K. A. Watanabe and J. J. Fox, *Chem. Pharm. Bull. (Tokyo)* **12**, 975 (1959); (b) K. A. Watanabe, J. Beranek, H. A. Friedman, and J. J. Fox, *J. Org. Chem.* **30**, 2735 (1965).
769. (a) F. W. Lichtenthaler, *Angew. Chem.* **76**, 84 (1964); (b) F. W. Lichtenthaler, H. P. Albrecht, and

G. Offermann, *Angew. Chem.* **77**, 131 (1965); (c) F. W. Lichtenthaler and H. P. Albrecht, *Chem. Ber.* **99**, 575 (1966).

770. A. Matsuda, M. W. Chun, and K. A. Watanabe, *Nucleosides Nucleotides* **1**, 173 (1982).
771. S. Shuto, T. Iwano, H. Inoue, and T. Ueda, *Nucleosides Nucleotides* **1**, 263 (1982).
772. S. Takei and Y. Kuwada, *Chem. Pharm. Bull. (Tokyo)* **16**, 944 (1968).
773. F. W. Lichtenthaler and H. Zinke, *Angew. Chem. Int. Ed.* **5**, 737 (1966).
774. F. W. Lichtenthaler and H. P. Albrecht, *Chem. Ber.* **100**, 1845 (1967).
775. A. S. Jones, A. F. Markham, and R. T. Walker, *J. Chem. Soc., Perkin Trans. I*, 1567 (1976).
776. W. Dvonch, H. Fletcher III, F. J. Gregory, E. H. Healy, G. H. Warren, and H. E. Alburn, *Cancer Res.* **26**, 2386 (1966).
777. J. J. Kinahan, Z. P. Pavelic, R. J. Leonard, A. Bloch, and G. B. Grindey, *Cancer Res.* **40**, 598 (1980).
778. M. Irie and C. A. Dekker, *Abstr. Sixth Int. Congr. Biochem. N.Y.*, 1–85 (1964).
779. C. V. Z. Smith, R. K. Robins, and R. L. Tolman, *J. Org. Chem.* **37**, 1418 (1972).
780. (a) J. O. Polazzi, D. L. Leland, and M. P. Kotick, *J. Org. Chem.* **39**, 3114 (1974); (b) J. O. Polazzi and M. P. Kotick, *Tetrahedron Lett.*, 2939 (1973).
781. H. A. Gassen and H. Witzel, *Biochim. Biophys. Acta* **95**, 244 (1965).
782. R. A. Sanchez and L. E. Orgel, *J. Mol. Biol.* **47**, 531 (1970).
783. (a) H. Quelo, J. Cadet, R. Teoule, and CR. Hebd, *C.R. Acad. Sci. Ser. C.* **275**, 1137 (1972); (b) J. Cadet, *Nucleic Acid Chemistry, Part* 1, p. 311 (L. B. Townsend and R. S. Tipson, eds.), Wiley, New York, 1978.
784. J. Cadet, *Tetrahedron Lett.*, 867 (1974).
785. H. Hrebabecky, J. Brokes, and J. Beranek, *Coll. Czech. Chem. Commun.* **47**, 2961 (1982).
786. V. W. Armstrong, J. K. Dattagupta, F. Eckstein, and W. Saenger, *Nucleic Acids Res.* **3**, 1791 (1976).
787. (a) J. L. Imbach, J. L. Barascut, B. L. Kam, B. Rayner, C. Tamby, and C. Tapiero, *J. Heterocycl. Chem.* **10**, 1066 (1973); (b) B. Rayner, C. Tapiero, and J. L. Imbach, *Carbohydr. Res.* **47**, 195 (1976).
788. M. MacCoss, M. J. Robins, B. Rayner, and J. L. Imbach, *Carbohydr. Res.* **59**, 575 (1977).
789. M. Miyaki, A. Saito, and B. Shimizu, *Chem. Pharm. Bull. (Tokyo)* **18**, 2459 (1970).
790. K. E. Norris, O. Manscher, K. Brunfeldt, and J. P. Peterson, *Nucleic Acids Res.* **2**, 1093 (1975).
791. K. Isono, P. F. Crain, and J. A. McCloskey, *J. Am. Chem. Soc.* **97**, 943 (1975).
792. (a) T. Azuma, K. Isono, P. F. Crain, and J. A. McCloskey, *Tetrahedron Lett.*, 1687 (1976); (b) T. Azuma, K. Isono, and J. A. McCloskey, *Nucleic Acids Res. Spec. Publ.* **2**, s19 (1976).
793. T. Azuma, K. Isono, D. F. Crain, and J. A. McCloskey, *J. Chem. Soc., Chem. Commun.*, 159 (1977).
794. M. Imazawa and F. Eckstein, *J. Org. Chem.* **44**, 2039 (1979).
795. M. Imazawa and F. Eckstein, *J. Org. Chem.* **43**, 3044 (1978).
796. R. J. Suhadolnik and T. Uematsu, *Carbohydr. Res.* **61**, 545 (1978).
797. R. Shapiro and M. Danzig, *Biochemistry* **11**, 23 (1972).
798. H. Venner, *Z. Physiol. Chem.* **339**, 14 (1964).
799. E. R. Garret, J. K. Seydel, and A. J. Sharpen, *J. Org. Chem.* **31**, 2219 (1966).
800. R. Shapiro and S. Kang, *Biochemistry* **8**, 1806 (1969).
801. P. G. Olafsson and A. M. Bryan, *Arch. Biochem. Biophys.* **165**, 46 (1974).
802. H. Follman, *Tetrahedron Lett.*, 397 (1973).
803. N. Kitamura and H. Hayatsu, *Nucleic Acids Res.* **1**, 75 (1974).
804. T. N. Menshonkova, N. A. Simukova, E. I. Budowsky, and L. B. Rubin, *FEBS Lett.* **112**, 299 (1980).
805. M. Kroger and F. Cramer, *Chem. Ber.* **110**, 361 (1977).
806. T. Ueda and H. Ohtsuka, *Chem. Pharm. Bull. (Tokyo)* **21**, 1530 (1973).

Chapter 2

Synthesis and Properties of Purine Nucleosides and Nucleotides

Prem C. Srivastava, Roland K. Robins,
and Rich B. Meyer, Jr.

1. Introduction

Inosinic acid was the first nucleotide obtained from nature and was reported by von Liebig in 1847. This was followed by the isolation of the first nucleoside, guanosine, by E. Schulze and E. Brosshard in 1885. Subsequently, other naturally occurring purine nucleosides and nucleotides were isolated and characterized as the degradation products of ribonucleic acids (RNAs) and deoxyribonucleic acids (DNAs). The biochemical significance of nucleosides and nucleotides became more apparent in the early 1950s when the era of modern molecular biology was introduced by the classical work of J. D. Watson and F. H. C. Crick on the structure of DNA in 1953 and by the great discovery of the "second messenger" of hormone action, adenosine cyclic 3',5'-(hydrogen)phosphate (cyclic AMP or cAMP) by Earl W. Sutherland in 1957. The synthesis of nucleosides and nucleotides then became instrumental in identifying the naturally occurring nucleosides and their metabolites and in an understanding of the biochemical and enzymatic reactions and pathways. Discoveries made in the field of purine nucleoside and nucleotide chemistry have contributed substantially to a better understanding of biology at the molecular level, which reached a pinnacle with the decoding of the genetic code and ultimate synthesis of the chemical gene by H. G. Khorana.

Many of the pioneers of nucleoside chemistry had an intense interest not only in the naturally occurring nucleosides and their biochemical phenomenon but also in the effects of synthetic nucleosides on living systems and in the synthesis of modified nucleosides as chemotherapeutic agents. Beginning in 1914, with Emil Fischer's historic reports on the first syntheses of nucleosides and nucleotides, we

Prem C. Srivastava • Nuclear Medicine Group, Health and Safety Research Division, Oak Ridge National Laboratory, Oak Ridge, Tennessee 37831. *Roland K. Robins* • Nucleic Acid Research Institute, Costa Mesa, California 92626. *Rich B. Meyer, Jr.* • MicroProbe Corporation, Bothell, Washington 98021.

have seen a marked increase in the synthesis of nucleosides that show a multitude of biological activities. The synthetic activity in this field has further been spurred by recent reports on the discovery of various nucleoside antibiotics such as puromycin, tubercidin, toyocamycin, sangivamycin, and formycin, which mimic the purine nucleosides and nucleotides such as 9-β-D-arabinofuranosyladenine (araA) and 9-[(2-hydroxyethoxy)methyl]guanine (acyclovir, Zovirax), which are clinically effective as antiviral agents. A host of modified purine nucleoside analogues and derivatives with various modifications involving all positions in the parent molecules have been synthesized.

2. Synthesis of Purine Nucleosides and Nucleotides

2.1. Direct Glycosylation of Purines

2.1.1. Synthesis Using Heavy Metal Salts of Purines

The first synthesis of nucleosides was reported by Fischer and Helferich[1] and involved a condensation of silver salts of both 2,6,8-trichloropurine and 2,8-dichloroadenine with acetobromoglucose to form glucosyl derivatives. These nucleosides were fully characterized much later.[2] The field lay dormant for several years after this classic work, primarily because of the unavailability of the sugar derivatives necessary for the preparation of ribonucleosides. Interest in the synthesis of nucleosides was rekindled when the Todd group adapted the Fischer silver salt method for the coupling of 2,8-dichloroadenine with 2,3,5-tri-*O*-acetyl-D-ribofuranosyl bromide to provide an intermediate blocked nucleoside that was converted to both adenosine and guanosine.[3]

This work led Davoll and Lowy[4] to investigate the use of the more easily prepared chloromercuri derivatives of 6-benzamidopurine and 2,6-dibenzamidopurine in a condensation with 2,3,5-tri-*O*-acetyl-D-ribofuranosyl bromide or chloride to give nucleosides that were then converted to adenosine and guanosine. The freshly prepared sugar, 2,3,5-tri-*O*-acetyl-β-D-ribofuranosyl chloride, was used as a syrup before it became available in a storable crystalline form[5] in 1974.

The introduction of 2,3,5-tri-*O*-benzoyl-D-ribofuranosyl chloride,[6] an apparently more stable activated sugar, furnished higher yields of blocked nucleosides.

The heavy metal salt method has been dealt with in general reviews, most notably those of Michelson[7] and Montgomery and Thomas.[8] This method still finds extensive use today[9–12] because of its predictability (*vide infra*). However, the possibility of mercury contamination of the products, and concurrent misleading biological test results, is a problem that should be recognized. The examples included herein are intended to show the scope of the procedure with regard to the variety of purines and carbohydrates that have been used in the condensation.

When the heavy metal salt of a heterocyclic base is condensed with a glycosyl halide, having an acyloxy group on the 2 position, the sugar forms an intermediate acyloxonium ion that then reacts with the base to give a nucleoside with

Scheme I

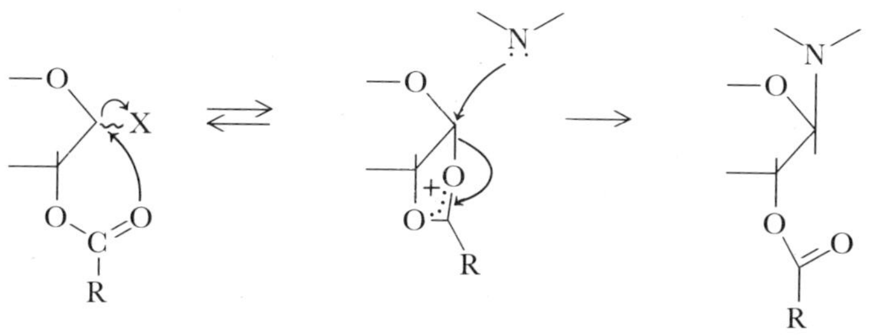

a 1′,2′-*trans* configuration. This occurs regardless of the initial configuration of the halide (Scheme I). This observation,[13] the "*trans* rule," has often been invoked in assigning the anomeric configuration of synthetic nucleosides, although *cis*-nucleosides have also been found as minor products in many cases. The position of sugar attachment for most of the common purines has been the 9 position. Although the exact mechanism may vary with the specific case cited, some particularly significant studies of Shimizu and Miyaki have shed some light on the nature of this reaction.

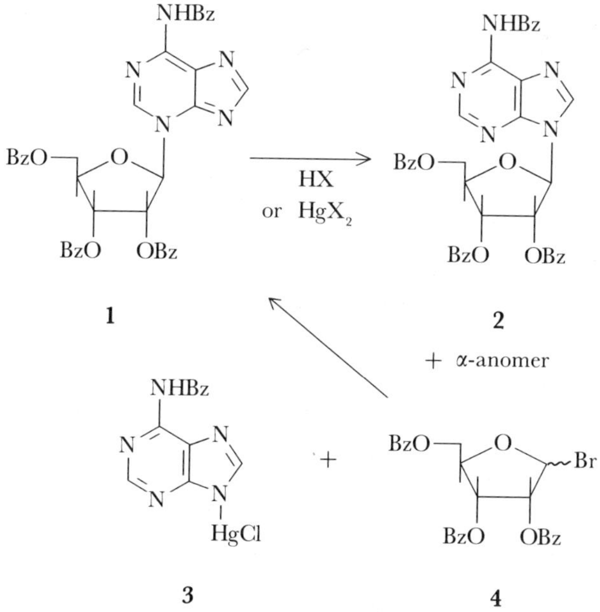

They noted that alkylation of an *N*-acyladenine in DMF gave a mixture of 3- and 9-alkyl-*N*-acyladenines (predominantly the 3-isomer) and that *N*-benzoyl-3-benzyladenine formed *N*-benzoyl-9-benzyladenine on heating.[14,15] This led to an investigation on the glycosylation of certain purines. When pure *N*-benzoyl-3-(2,3,5-tri-*O*-benzoyl-β-D-ribofuranosyl)adenine (**1**) was heated with either a hydrogen halide or mercuric halide in a nonpolar solvent, glycosyl migration occurred to give tetrabenzoyladenosine (2) in good yield.[16] The addition of labeled *N*-benzoyladenine to the mixture gave labeled tetrabenzoyladenosine (2), indicating that this was an intermolecular reaction. Furthermore, some of the 9-α-anomer was formed during the migration, with more of the α-anomer being formed with hydrogen halide than with mercuric halide catalysts. This result, coupled with the fact that appreciable quantities of 3-glycosylpurines were isolated from the reaction of chloromercuri-*N*-benzoyladenine (3) with 2,3,5-tri-*O*-benzoyl-D-ribofuranosyl bromide (4), indicates that the actual glycosylation reaction must occur *via* an initial alkylation at N-3, followed by a rapid migration to N-9 by an intermolecular attack.[16]

Further studies revealed that the direct alkylation of *N*,*N*-dimethyladenine with 2,3,5-tri-*O*-benzoylribofuranosyl bromide gave a mixture of the 3- and 9-ribosyl derivatives and that the 9-ribosyl isomer was formed by an analogous N-3 → N-9 migration.[17] Furthermore, alkylation of *N*-acetylguanine (5) with the ribofuranosyl bromide (4) occurred on N-3, and this compound then underwent a migration to give a mixture of *N*-acetyl-2′,3′,5′-tri-*O*-benzoylguanosine (6) and *N*-acetyl-7-(2,3,5-tri-*O*-benzoyl-β-D-ribofuranosyl)guanine (7).[17,18]

This suggests that, in the reaction of glycosyl halides either directly with purines or with their chloromercuric derivatives, glycosylation occurs first at N-3

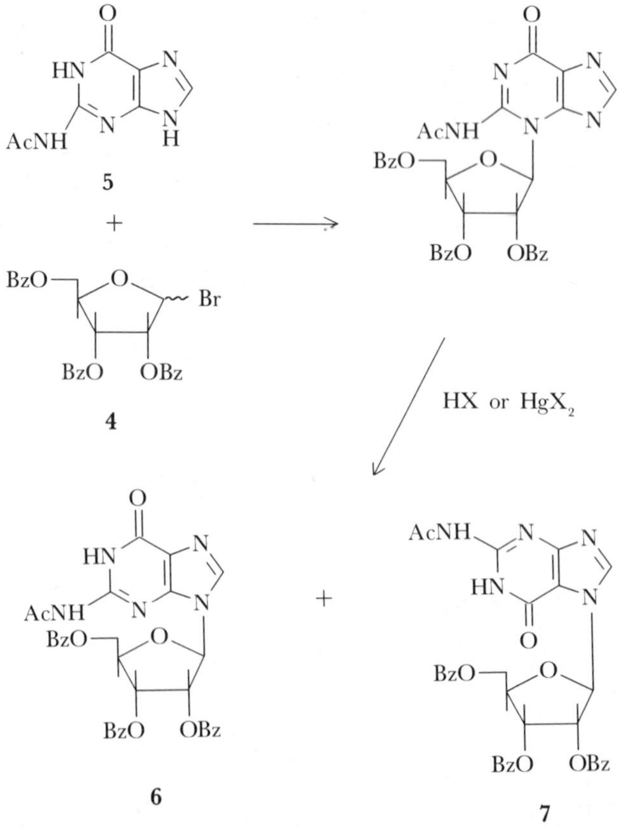

followed by an attack of another molecule of purine to give the thermodynamically more stable (9) isomer. However, in the case of theophylline, which
cannot be further alkylated on the 3 position, 7-glycosyltheophylline is formed
directly[19] (see Section 2.1.2).

Reactive glycosyl halides, which cannot form an intermediate acyloxonium
ion without acyloxy groups adjacent to the anomeric carbon, usually give
mixtures of α- and β-nucleosides. Thus, the condensation of chloromercuri-
N-benzoyladenine with either 2-deoxy-3,5-di-*O*-*p*-nitrobenzoyl-D-*erythro*-pento-
furanosyl chloride[20] or 2-chloro-2-deoxy-3,5-di-*O*-acetyl-D-arabinofuranosyl
chloride[21] (8) gave a mixture of the α- and β-nucleosides. These anomers were
separated and after appropriate transformations[22] were converted to α-deoxy-
adenosine (9) and β-deoxyadenosine (10), respectively. On the other hand,
either chloromercuri-*N*-benzoyladenine (3), 2,6-dibenzamidopurine–HgCl (11),
6-chloropurine–HgCl (12), or *N*-benzoyl-*N*-methyladenine reacted smoothly with
2,5-di-*O*-benzoyl-3-deoxy-D-ribofuranosyl bromide (13), to give the corresponding
β-nucleosides in good yield.[23,24] These compounds were used to provide cordycepin [3'-deoxyadenosine, (14)] and some substituted purine analogues after
deblocking.

Often the preparation of the halogenose required for the mercury procedure
has been difficult. As a simplification of this problem, Baker introduced an
alternate procedure that involves the formation of the halogenose *in situ* by
the addition of titanium tetrachloride (TiCl$_4$) to a solution of peracylated sugar
and the mercuripurine.[25] Thus, 1-*O*-acetyl-2,5-di-*O*-benzoyl-3-acetamido-3-
deoxy-D-ribofuranose (15), chloromercuri-*N*-benzoyladenine (3), and titanium
tetrachloride gave a 57% yield of *N*-benzoyl-9-(3-acetamido-3-deoxy-2,5-di-*O*-
benzoyl-β-D-ribofuranosyl)adenine (16), which contained some α-anomer. It was
believed that TiCl$_4$ catalyzed the anomerization of the initially formed β-
nucleoside.[25]

The syntheses of cordycepin[26] and some other 3-deoxy analogues,
9-(3-deoxy-β-D-galactofuranosyl),[27] 9-(3-deoxy-α-L-arabinofuranosyl),[27] and
9-(3-deoxy-β-D-glucofuranosyl)[28] of adenine were accomplished using the TiCl$_4$
catalyst. Titanium tetrachloride also catalyzed the condensation of chloromercuri-
N-benzoyladenine with 1,2-di-*O*-acetyl-5-*O*-benzoyl-3-*O*-methyl-D-ribofuranose
and 1-*O*-acetyl-3-benzoylthio-3-deoxy-2,5-di-*O*-benzoyl-D-ribofuranose, respectively, to provide good yields of intermediate products, which were subsequently
converted into 3'-*O*-methyladenosine[29] and 3'-thioadenosine, respectively.[30]
Several "branched-chain" nucleosides have been prepared by similar procedures
using chloromercuripurines and the appropriate sugars. Condensation of
chloromercuri-*N*-benzoyladenine (3) with 3-*C*-methyl-2,3,5-tri-*O*-benzoyl-D-
ribofuranosyl bromide (17) gave, after deblocking, 3'-*C*-methyladenosine (18) in
a 40% overall yield.[31] Treatment of 2-*O*-acetyl-3-deoxy-5-*O*-*p*-nitrobenzoyl-
3-nitromethyl-β-D-ribofuranosyl chloride (19) with chloromercuri-6-chloropurine
(12) gave a 33% yield of blocked chloropurine nucleoside (20), which, after treatment with dimethylamine followed by reduction, gave 3'-(aminomethyl)-
3'-deoxy-*N*,*N*-dimethyladenosine (21), an interesting 3'-homologue of the amino
nucleoside,[32] puromycin. Another interesting branched-chain amino nucleoside
was prepared by the reaction of 2'-*O*-acetyl-5'-*O*-benzoyl-1',3'-dideoxy-3'(*R*)-

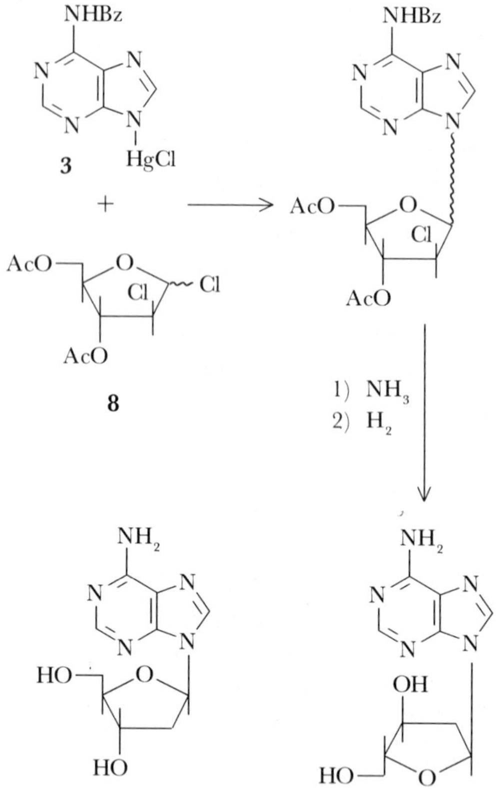

3 **8** **10** **9**

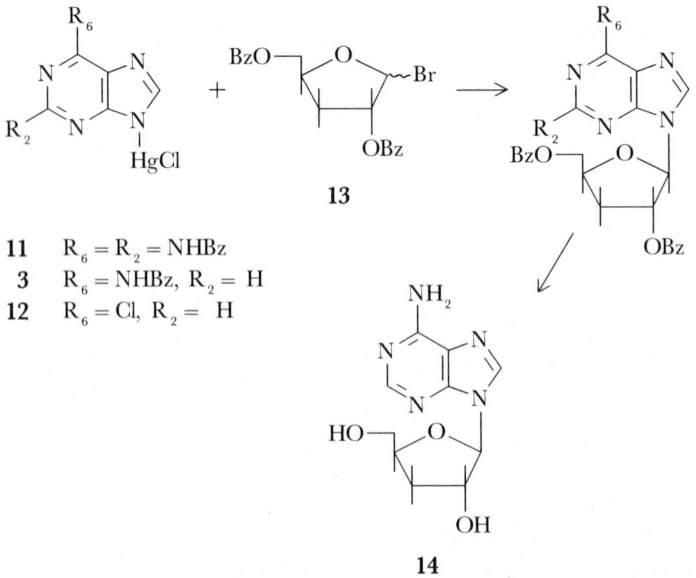

13 **14**

11 $R_6 = R_2 = NHBz$
3 $R_6 = NHBz$, $R_2 = H$
12 $R_6 = Cl$, $R_2 = H$

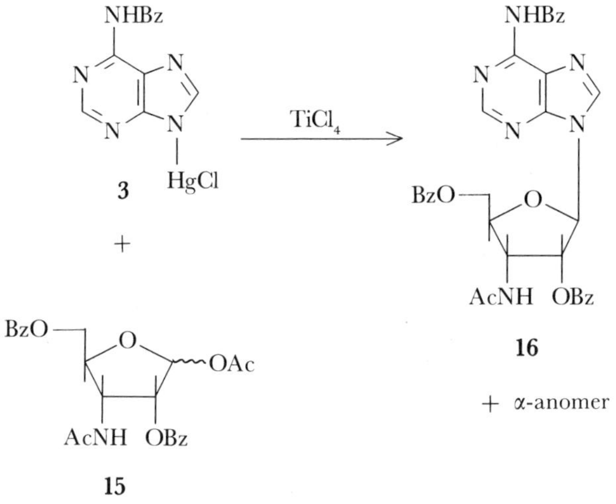
NHBz
HgCl
3
+
BzO
OAc
AcNH OBz
15
TiCl4
NHBz
BzO
AcNH OBz
16
+ α-anomer

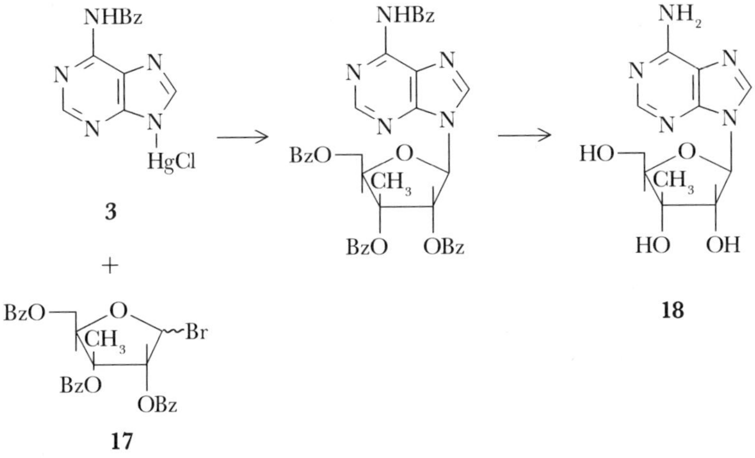
NHBz
HgCl
3
+
BzO
CH3
Br
BzO
OBz
17
NHBz
BzO
CH3
BzO OBz
NH2
HO
CH3
HO OH
18

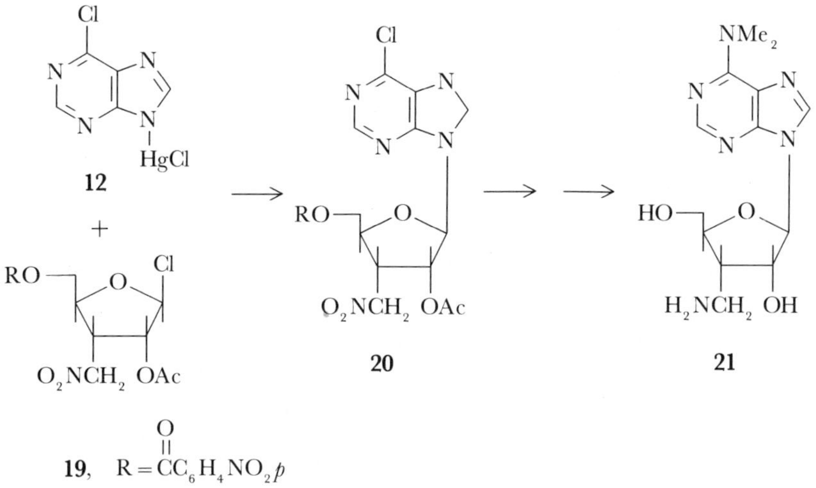
Cl
HgCl
12
+
RO
Cl
O2NCH2 OAc
19, R = C(=O)C6H4NO2p
Cl
RO
O2NCH2 OAc
20
NMe2
HO
H2NCH2 OH
21

spiro(imidazolidine-4,3'-D-*erythro*-pentofuranose)-2,5-dione (**22**) with chloromercuri-*N*-benzoyladenine (**3**). Barium hydroxide treatment gave the free amino acid nucleoside, 9-(3-amino-3-*C*-carboxy-3-deoxy-β-D-ribofuranosyl)adenine[33] (**23**).

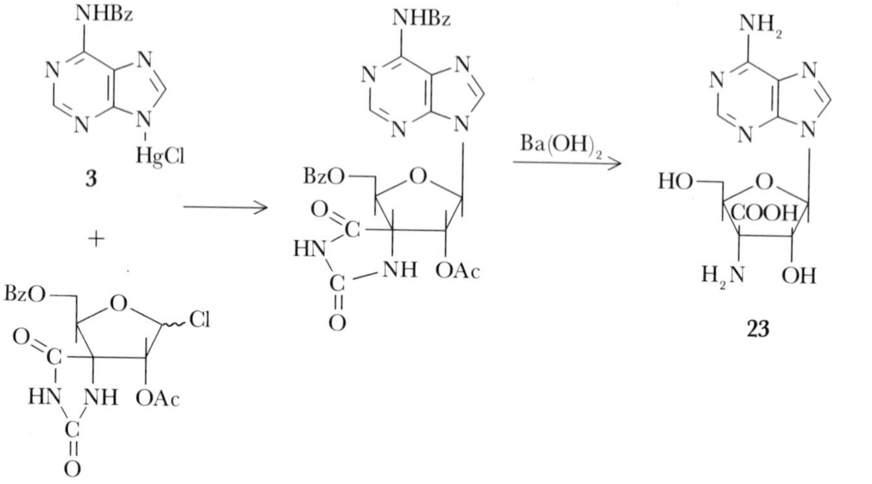

All eight possible tetrose nucleosides of adenine were prepared[34] by a condensation of D- and L-erythrose triacetate and D- and L-threose triacetate with chloromercuri-*N*-benzoyladenine in the presence of TiCl$_4$, with the expectation that this would lead to both the α- and β-anomers of each compound.[25] This indeed proved to be the case, and yields of 31–57% of the 1',2'-*trans* compounds were isolated, along with 7–13% of the *cis*-isomer. A related compound was prepared from the reaction of 1,2-di-*O*-acetyl-3-*O*-benzyl-3-*C*-benzyloxymethyl-L-*threo*-furanose (**24**) with chloromercuri-*N*-benzoyladenine (**3**) in the presence of TiCl$_4$, to yield, after sodium methoxide treatment and catalytic hydrogenolysis, 9-(3-*C*-hydroxymethyl-α-L-*threo*-furanosyl)adenine and [9-(β-d-apio-L-furanosyl)-adenine][35] (**25**). This approach was later applied to the synthesis of the corresponding 3-*C*-methyl analogue, 9-(3-*C*-methyl-α-L-*threo*-furanosyl)adenine[36] (**27**). The reaction of 3-deoxy-3-*C*-methyl-2,5-di-*O*-(*p*-nitrobenzoyl)-α-D-xylofuranosyl chloride (**28**) with chloromercuri-*N*-benzoyladenine (**3**) gave a mixture of α- and β-nucleosides (**29**) in 69% yield. These were separated after sodium methoxide treatment to yield 9-(3-deoxy-3-*C*-methyl-α(and β)-D-xylofuranosyl)adenine in a 1 : 4 ratio[37] (total yield, 69%). The unusually high amount of α-nucleoside formed was attributed to the diminished neighboring-group participation by the electron-deficient *p*-nitrobenzoyl group. This concept was tested by the use of the same sugar with benzoyl blocking groups (**30**), and in that case a 50% yield of β-nucleoside and a 3.3% yield of α-nucleoside were isolated.[37]

An unusually high proportion of α-nucleoside occurred in the condensation of chloromercuri-*N*-benzoyladenine (**3**) with 1,2-di-*O*-acetyl-3,5-dideoxy-3-trifluoro-acetamido-α-D-ribofuranose (**31**) in the presence of TiCl$_4$. In this case, a 1 : 1

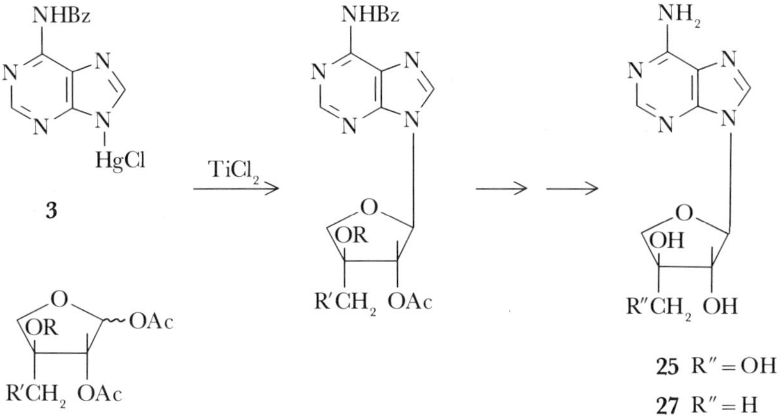

NHBz

N N

N N

HgCl

3

$\xrightarrow{\text{TiCl}_2}$

NHBz

N N

N N

O

OR

R′CH$_2$ OAc

$\longrightarrow$ $\longrightarrow$

NH$_2$

N N

N N

O

OH

R″CH$_2$ OH

25 R″ = OH

27 R″ = H

O

OR ~OAc

R′CH$_2$ OAc

24 R = CH$_2$Ph, R′ = OCH$_2$Ph
26 R = Bz, R′ = H

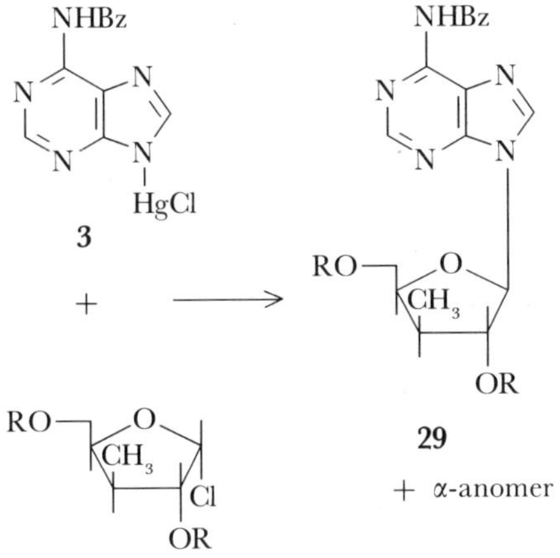

NHBz

N N

N N

HgCl

3

+

RO O

CH$_3$

Cl

OR

$\longrightarrow$

NHBz

N N

N N

RO O

CH$_3$

OR

29

+ α-anomer

28 R = $-\overset{\overset{\text{O}}{\|}}{\text{C}}-$⟨⟩$-NO_2$

30 R = $-\overset{\overset{\text{O}}{\|}}{\text{C}}-$⟨⟩

ratio of 9-(2-*O*-acetyl-3,5-dideoxy-3-trifluoroacetamido-α(and β)-D-ribofuranosyl)-*N*-benzoyladenine (**32**) as obtained in a 55% overall yield.[38] This represents an unusually high proportion of α-anomer for a 2-*O*-acetyl sugar compared to the majority of other cases. A similar TiCl$_4$-catalyzed condensation of (**3**) and 1,2-di-*O*-acetyl-xylofuranoses with a variety of blocking groups at the 3 and 5 positions has been studied.[39] The yields of the resulting nucleosides were in the order of 23, 55, and 84%, respectively, for 2,3- or 3,5-cyclic carbonate, acetyl, and benzoyl

blocking groups of xylofuranose. The appropriately blocked nucleosides were converted into 3'-*C*-hydroxymethyl- and 3'-*C*-methyl-β-D-xylofuranosyl-9-adenine,
which proved to be substrates of adenosine deaminase.[39]

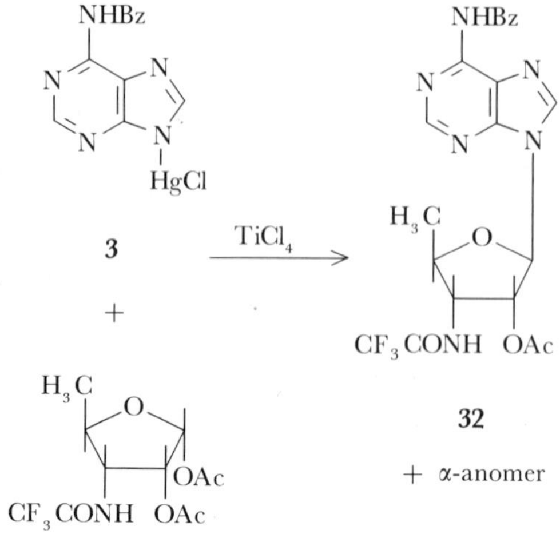

The hexofuranosyl adenine nucleosides have proved interesting as substrates
and inhibitors of calf intestinal adenosine deaminase.[40] A number of these
nucleosides have been prepared[40,41] using a TiCl$_4$ catalyst. For instance,
9-(5-deoxy-β-D-xylohexofuranosyl)adenine (**35**) and 9-(5-deoxy-α-L-arabino-
hexofuranosyl)adenine (**36**) were prepared[42] by a condensation of (**3**) with 1,2-
di-*O*-acetyl-3,6-di-*O*-benzoyl-5-deoxy-D-xylohexofuranoside (**33**) and 1-*O*-acetyl-
2,3,6-tri-*O*-benzoyl-5-deoxy-α-L-arabinohexofuranoside (**34**), respectively, in the
presence of TiCl$_4$. The *E* (*trans*) and *Z* (*cis*) isomers of 9-(5,6-dideoxy-β-*erythro*-
hex-4-enofuranosyl)adenine[43] as analogue of decoyinine and the nucleosides with
a terminal vinyl group, 9-(5,6-dideoxy-α-D-arabino-hex-5-enofuranosyl)adenine
(**37**) and 9-(5,6-dideoxy-α-L-arabino-hex-5-enofuranosyl)adenine (**38**), were
similarly prepared.[44] A convenient synthesis of guanine and thioguanine
nucleosides has been developed by a condensation of the chloromercury derivative
of 2-acetamido-6-chloropurine with 2,3,5-tri-*O*-acetyl-β-D-xylofuranosyl bromide,
followed by ammonia treatment, to give 2-amino-6-chloro-9-β-D-ribofuranosyl-
purine. The latter compound, on treatment with sodium hydrosulfide, gave
2-amino-6-mercapto-9-β-D-xylofuranosylpurine, while treatment with sodium
hydroxyethylmercaptide gave 9-β-D-xylofuranosylguanine.[45] The use of mercaptoethanol and base to convert chloropurines to hydroxypurines[46] also serves
as a convenient alternative to acidic or oxidative conditions.
    It is obvious from this discussion that the most often used method for the
preparation of new nucleosides is the coupling of an appropriate sugar and
purine. However, this route is rarely used for the synthetic preparation of
nucleotides or by nature, since in nature this occurs by the *de novo* purine
nucleotide biosynthesis pathway or by a transfer of ribose 5-phosphate to the base.

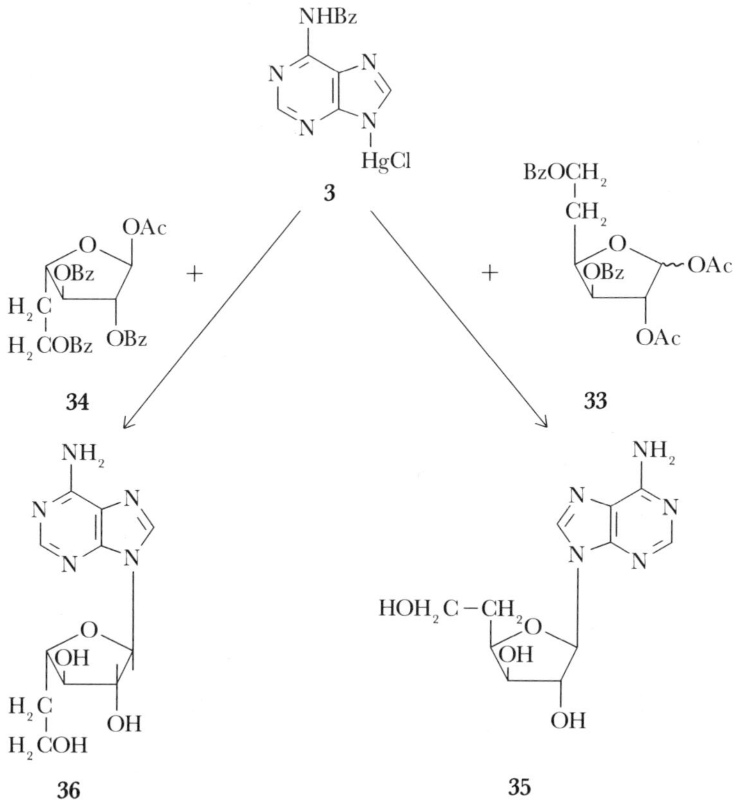

Although a few cases of direct chemical synthesis of nucleotides have been reported, these methods are infrequently used since the advent and development of more facile phosphorylation techniques.

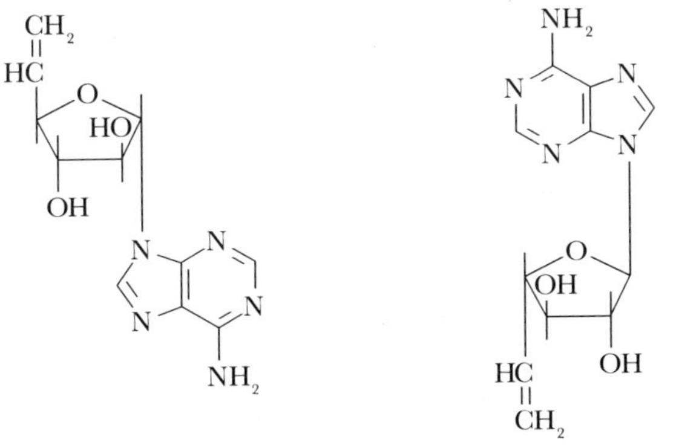

Treatment of 2,3-di-*O*-benzoyl-5-*O*-(diphenylphosphoryl)-D-ribofuranosyl bromide (**39**) with chloromercuri-*N*-benzoyladenine (**3**) gave the blocked nucleoside (**40**) in 65 % yield. Treatment of (**40**) with alkali gave the monophenyl ester of 5′-adenylic acid (41). The naturally occurring nucleotide [5′-phosphate derivative of adenosine, generally referred to as 5′-AMP (42)] was obtained on treatment of (41) with phosphodiesterase from Russell's viper venom.[47]

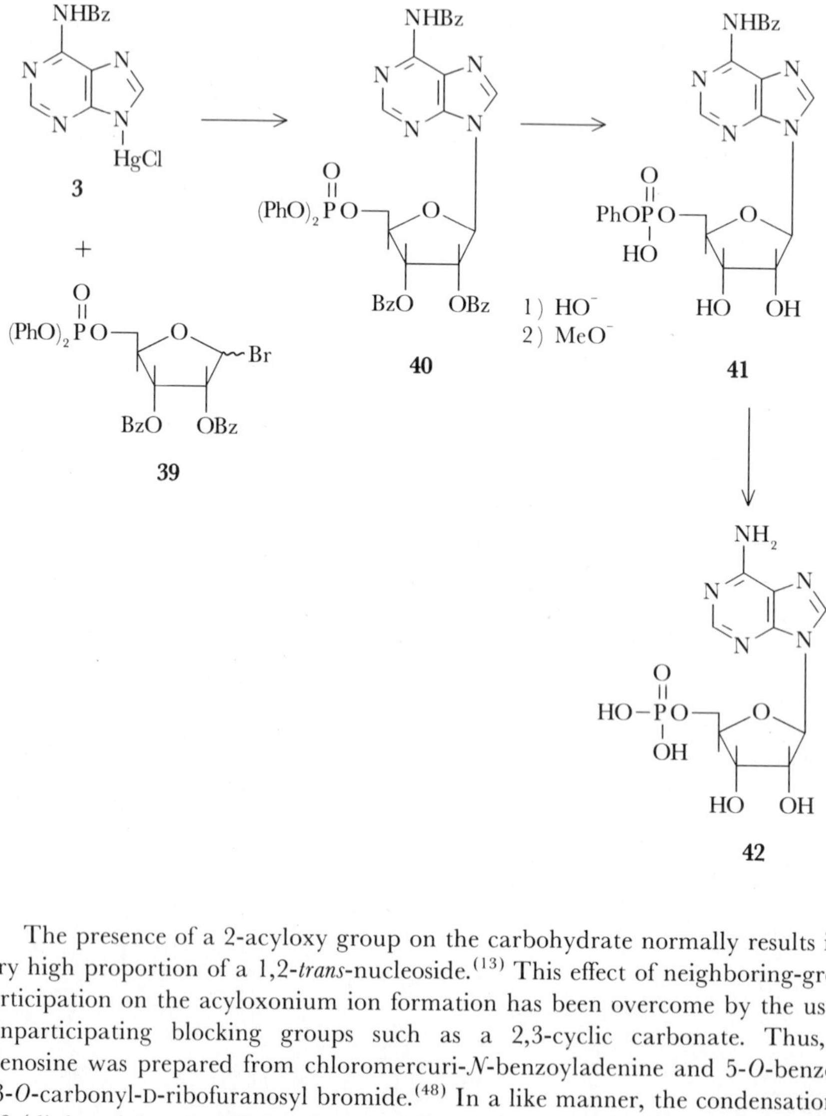

The presence of a 2-acyloxy group on the carbohydrate normally results in a very high proportion of a 1,2-*trans*-nucleoside.[13] This effect of neighboring-group participation on the acyloxonium ion formation has been overcome by the use of nonparticipating blocking groups such as a 2,3-cyclic carbonate. Thus, α-adenosine was prepared from chloromercuri-*N*-benzoyladenine and 5-*O*-benzoyl-2,3-*O*-carbonyl-D-ribofuranosyl bromide.[48] In a like manner, the condensation of 5-*O*-(diphenylphosphoryl)-2,3-*O*-carbonyl-D-ribofuranosyl bromide with chloro-mercuri-*N*-benzoyladenine (3) gave a mixture of the α- and β-nucleosides. Sodium methoxide treatment of this mixture gave the monophenyl adenosine 5′-phosphate as an anomeric mixture, which was then treated with *Trimeresurus floroviridio* venom to give β-adenosine and α-5′-AMP in a 1 : 4.5 ratio. The venom enzyme had obviously cleaved the phenyl esters and then removed the phosphate of 5′-AMP to form adenosine.[49] The isopropylidine moiety can also be considered as another nonparticipating blocking group. The reaction of chloromercuri-*N*-benzoyladenine and 2,3:5,6-di-*O*-isopropylidine-D-mannofuranosyl chloride gave crude *N*-benzoyl-9-(2,3:5,6-di-*O*-isopropylidine-D-mannofuranosyl)-adenine, which on treatment with acid followed by base gave a 38% yield of the apparently pure 9-(2,3-*O*-isopropylidine-α-D-mannofuranosyl)adenine. More rigorous acid treatment (25% aqueous acetic acid at 100°C) gave 9-α-D-mannofuranosyladenine with the α configuration being assigned on the basis of optical rotation and the assumption that steric hindrance of the fused 2,3-isopropylidine group prevented a β attack.[50]

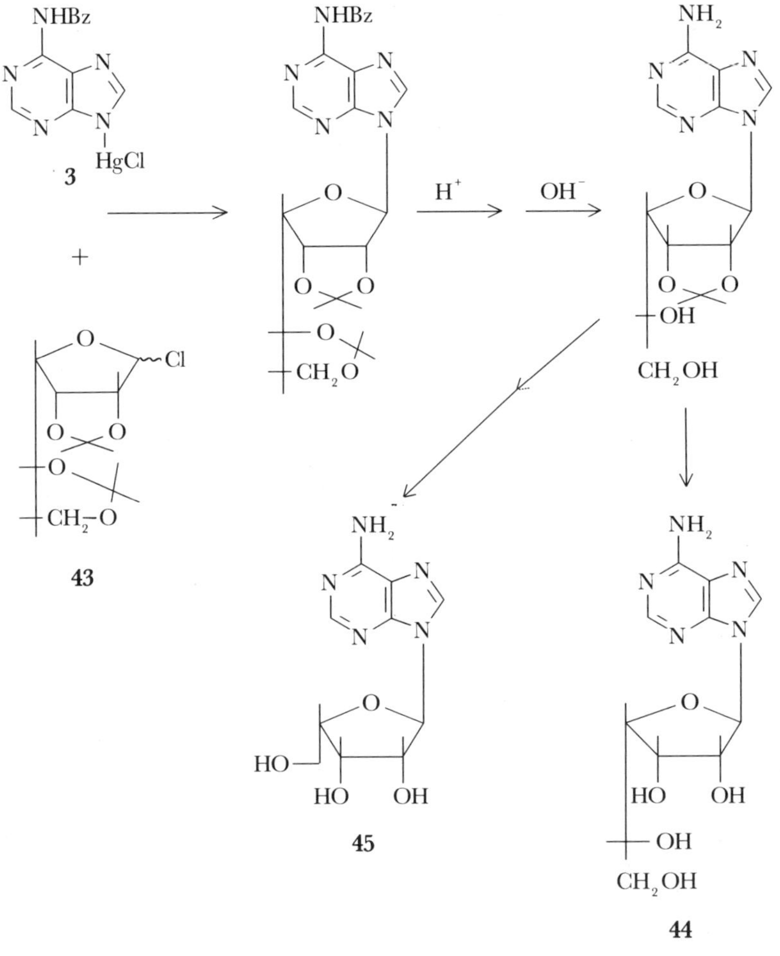

Similarly, a condensation of chloromercuri-*N*-benzoyladenine and 2,3:5,6-di-*O*-isopropylidine-D-gulofuranosyl chloride (**43**) gave, after acidic and basic deblocking, 9-β-D-gulofuranosyladenine[51] (**44**). The structure was confirmed by comparison to material synthesized from tetra-*O*-benzoyl-D-gulofuranosyl chloride[52] where the *trans* rule dictated product stereochemistry, and by oxidation and then reduction, to 9-α-L-lyxofuranosyladenine (**45**). This compound had the same properties, except for an equal and opposite rotation, as those obtained from the D-enantiopmorph.[53] A 2-amino sugar can conveniently be blocked by the electron-withdrawing 2,4-dinitrophenyl group, which not only inhibits neighboring-group participation but also renders the amino group sufficiently inert to participation in any side reactions during the glycosylation reaction. Thus, treatment of 2-deoxy-2-(2,4-dinitroanilino)-3,4,6-tri-*O*-acetyl-α-D-glucopyranosyl bromide (**47**) with chloromercuri-*N*-acetyladenine (**46**) gave α- (15%) and β- (25%) *N*-acetyl-9-[2-deoxy-2-(2,4-dinitroanilino)-3,4,6-tri-*O*-acetyl-D-glucopyranosyl)]adenine. Complete deblocking was achieved by treatment of this compound with Dowex 1 (OH⁻) to give 25 and 30% yields of 9-(2-amino-2-deoxy-β-D-glucopyranosyl)adenine (**48**) and its α-anomer,[54,55] respectively.

Sugars in which the oxygen of the furanose or pyranose ring has been replaced by sulfur or nitrogen might be expected to behave very differently in a glycosylation reaction. However, it was found that both 2,3,5-tri-*O*-(*p*-nitrobenzoyl)-4-thio-L-ribofuranosyl bromide and 2,3,5-tri-*O*-acetyl-4-thio-D-ribofuranosyl chloride (**49**) would condense with chloromercuri-*N*-benzoyladenine (**3**) to give, after deblocking, the L- and D-4′-thioadenosine (**50**) in 20 and 26% yields,[56]

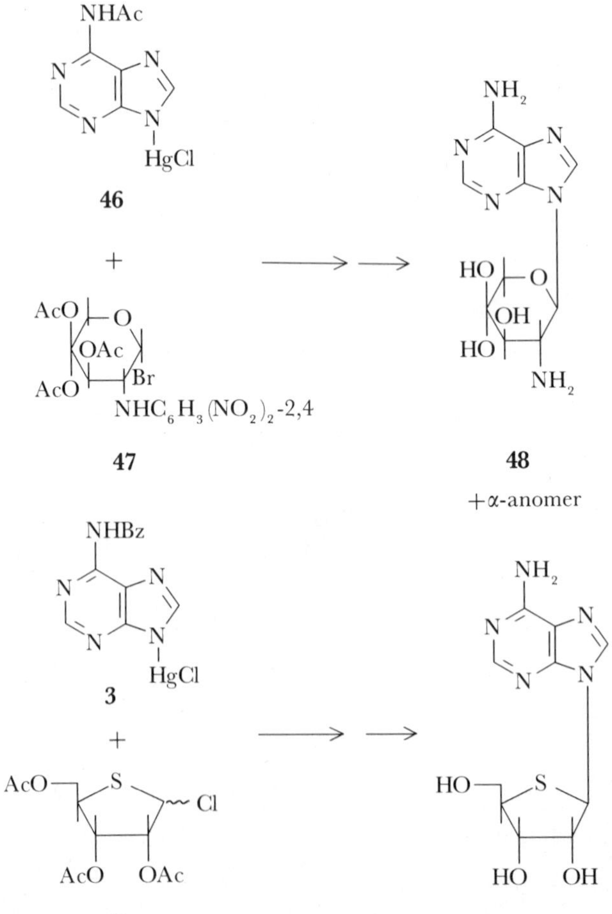

respectively. Furthermore, the same base (**3**) was reacted with 1,2,5-tri-*O*-benzoyl-3-deoxy-4-thio-α- or β-D-ribofuranose and $TiCl_4$ to give, after deblocking, excellent yields of 4′-thiocordycepin.[57] As a further demonstration of the *trans* rule, the α- and β-benzoates gave 82 and 78% yields of the β-nucleoside, respectively.

The 4-amino isostere of ribose presented some difficulties. A condensation of the apparently unstable 4-acetamido-4-deoxy-2,3,5-tri-*O*-(*p*-nitrobenzoyl)-D-ribofuranosyl bromide with chloromercuri-6-benzamidopurine gave no product. However, the same purine was condensed with 4-acetamido-5-*O*-acetyl-2,3-di-*O*-benzoyl-4-deoxy-D-ribofuranosyl chloride or with the corresponding acetate (**51**) in the presence of $TiCl_4$ to yield a nucleoside that gave, after sodium methoxide deblocking, 9-(4-acetamido-4-deoxy-β-D-ribofuranosyl)adenine (**52**) in 32 and 42% yields, respectively.[58]

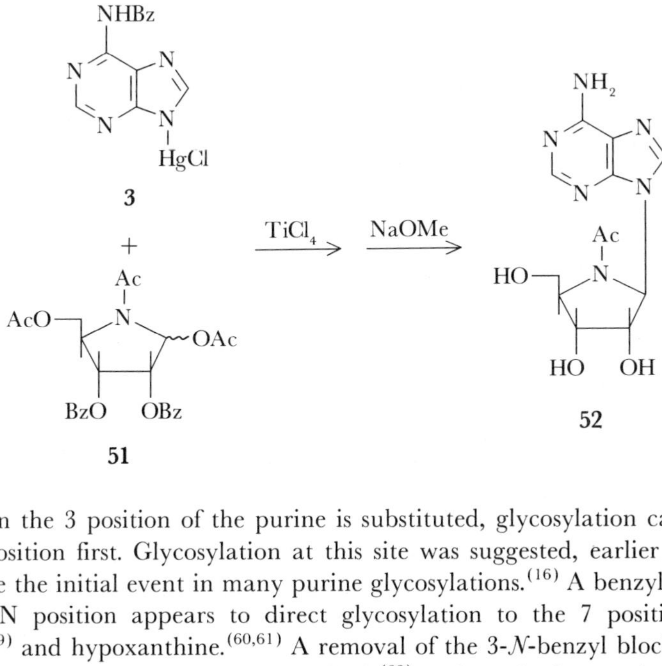

When the 3 position of the purine is substituted, glycosylation cannot occur at that position first. Glycosylation at this site was suggested, earlier in this section, to be the initial event in many purine glycosylations.[16] A benzyl substituent in the 3-N position appears to direct glycosylation to the 7 position of both adenine[59] and hypoxanthine.[60,61] A removal of the 3-*N*-benzyl blocking groups by catalytic hydrogenation is often difficult[62] and results in a reduction of the purine ring to a dihydro derivative,[62] and affords low yields of the 7-glycosyl-nucleoside. This prompted a search for more readily removable purine ring nitrogen-blocking groups. One such functionality, introduced by Mizuno, is the 2-picolyl-1-oxide group. Chloromercuri-3-(1-oxy-2-picolyl)hypoxanthine (**53**) and 2,3,5-tri-*O*-benzoyl-D-arabinofuranosyl bromide (**54**) gave 3-(1-oxy-2-picolyl)-7-(2,3,5-tri-*O*-benzoyl-α-D-arabinofuranosyl)hypoxanthine (**55**), from which the picolyl-1-oxide group was removed by treatment with acetic anhydride to give 7-(α-d-arabinofuranosyl)hypoxanthine (**56**) after debenzoylation.[63] Recent attempts[62] to introduce picolyl-1-oxide group on the 3-N position of adenine were unsuccessful and the general utility of this interesting, easily cleaved group remains questionable.

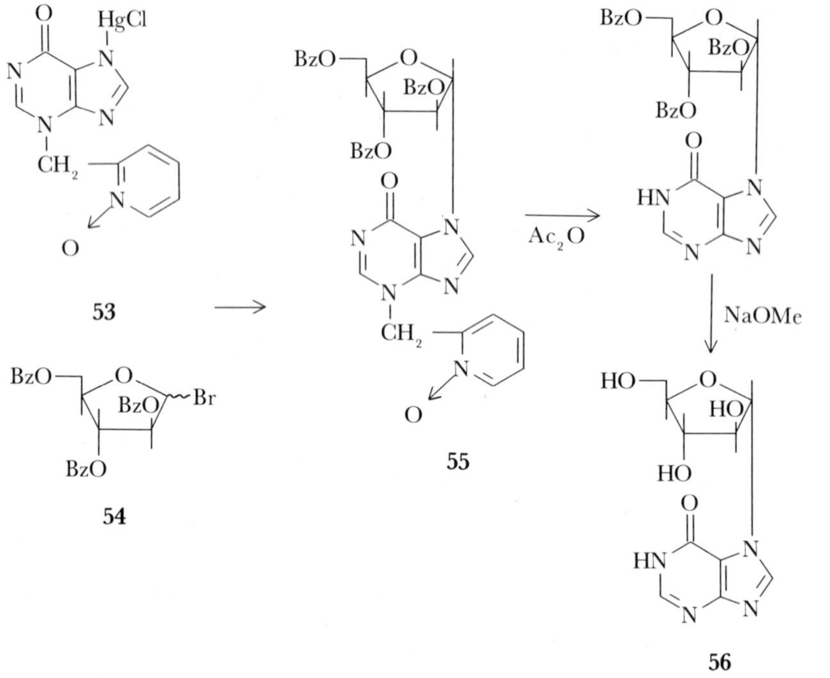

53 ⟶ 55

54

56

The preparation of mercury derivatives of the purines may occasionally present some difficulties. A method for circumventing this problem was found in the use of mercuric cyanide or mercuric acetate as a condensation catalyst.[64] Inasmuch as this method always involves the use of a glycosyl halide for coupling, it is considered with the foregoing examples of the heavy metal procedure although the mercuripurine derivative is formed *in situ*.

Acetobromoglucose was found to condense readily with 2,6-dichloro-, 6,8-dichloro-, 2,6,8-trichloro-, and 6-benzamidopurines, in addition to theophylline, in good yield when heated at reflux temperature with mercuric cyanide in nitromethane.[64] Mercuric cyanide successfully mediated the coupling of 2,3,5-tri-O-benzyl-D-arabinofuranosyl chloride with 2,6-dichloropurine to yield the α (25%) and β (11%) of 2,6-dichloro-9-(2,3,5-tri-O-benzyl-D-arabinofuranosyl)-purine.[65]

Treatment of 4-C-(acetoxymethyl)-1,2,3,5-tetra-O-acetyl-L-*threo*-pento-furanose (57) with TiCl$_4$, followed by N-benzoyladenine (58) and Hg(CN)$_2$ in nitromethane, gave, after deblocking, 9-[4-C-(hydroxymethyl)-α-L-*threo*-pento-furanosyl]adenine[66] (59). The use of mercuric cyanide also simplified the synthesis of guanine nucleosides. Treatment of 2-amino-6-chloropurine (60) with excess hexamethyldisilazine gave a trimethylsilyl derivative of undetermined structure, which was treated with 2-deoxy-3,5-di-O-p-toluoyl-D-*erythro*-pento-furanosyl chloride[67,68] and mercuric cyanide in benzene to give a mixture of 80% of pure α and β (1 : 1) 2-amino-6-chloro-9-(2-deoxy-3,5-di-O-p-toluoyl-D-*erythro*-pentofuranosyl)purine (62). Treatment of (62) with sodium 2-hydroxy-ethylmercaptide[46] provided an efficient synthesis of α- and β-2′-deoxyguanosine as an α, β mixture.[69]

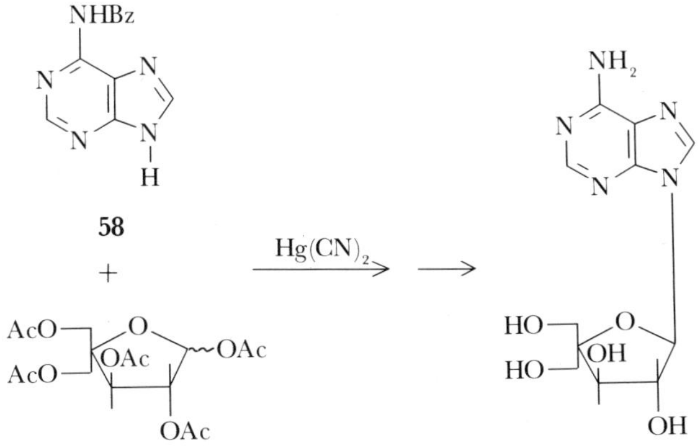

$$Tol = p—CH_3C_6H_4C—\overset{O}{\overset{\|}{}}$$

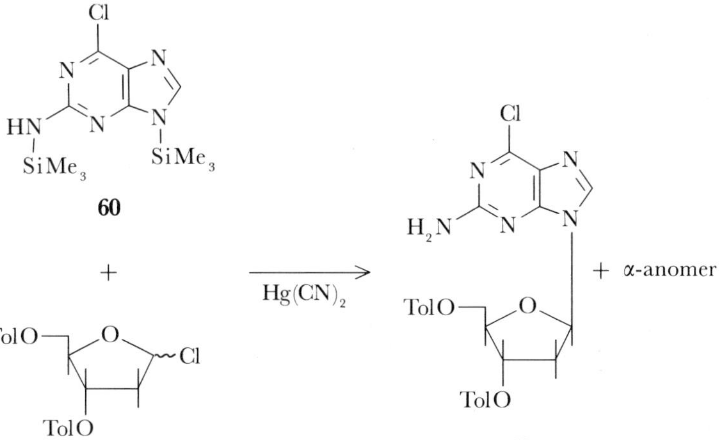

Direct glycosylation of the mercury derivative of *N*-acetylguanine gave a mixture of the 7- and 9-isomers[17,18] (*vide ante*). The treatment of tris(trimethylsilyl)-*N*-acetylguanine (**63**) with 3-deoxy-2,5-di-*O*-*p*-toluoyl-D-ribofuranosyl bromide (**64**) and mercuric acetate in acetonitrile at room temperature gave only 3′-deoxyguanosine (**65**) in 34% yield after deblocking.[70] Bis(trimethylsilyl)-*N*-benzoyladenine gave a 48% overall yield of 3′-deoxyadenosine when treated with the same halogenose and mercuric acetate in acetonitrile at room temperature. This method represents a facile preparation of the nucleosides of purines and of guanine in particular.

Further investigations, however, showed that the position of glycosylation of tris(trimethylsilyl)-*N*-acetylguanine by this method depended on the halogenose used. A 45 : 55 ratio of 7-isomers to 9-isomers was obtained in 32% yield from the reaction of this heterocycle and mercuric acetate with 1,3,4,6-tetra-*O*-*p*-toluoyl-

D-psicofuranosyl chloride, whereas 1,3,4,6-tetra-*O*-benzoyl-D-fructofuranosyl bromide and tris(trimethylsilyl)-*N*-acetylguanine gave only the 9-isomer in 88% yield in the presence of mercuric acetate.[71]

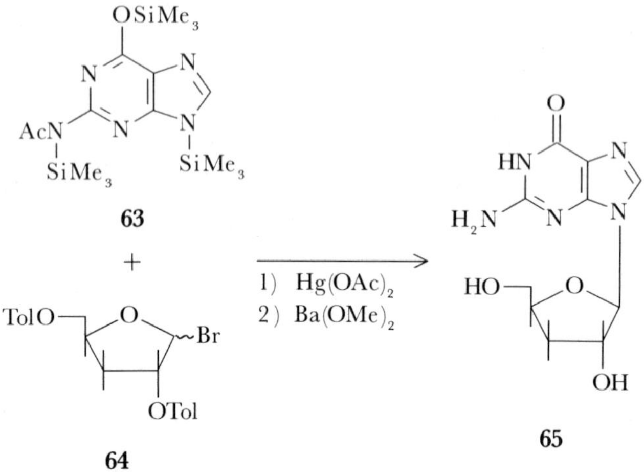

2.1.2. Synthesis Using Acid-Catalyzed, High-Temperature Fusion

High-temperature fusion, condensation by the melting of a heterocycle base, and *O*-acylated sugar mixture, usually in the presence of an acidic catalyst, proved to be a most potent, simple, and straightforward procedure for the synthesis of blocked nucleosides. The procedure was originally used by Sato and co-workers to prepare tri-*O*-acetylribofuranosyl derivatives of theophylline,[72] 2-methylthioadenine,[73] 2,6-dichloroadenine, 2,6,8-trichloropurine, and various other purines, mostly with halogen substituents.[74]

It was reported[75,76] that various purine 2′-deoxyribonucleosides could be obtained directly by the fusion of 6-chloropurine, 2,6-dichloropurine, 6-methylpurine, 6-benzamidopurine, 2,6,8-trichloropurine, and purine *per se* with the appropriate deoxyfuranose derivative in the presence of a catalytic amount of chloroacetic or dichloroacetic acid. The yields in these condensations were 20–65% as mixtures of the α- and β-anomers in approximately equal proportions. The deoxyfuranose derivatives (66) and (67) were fused smoothly with 6-benzyloxy-2-fluoropurine (68) and 2,6-dichloropurine (69) to provide a convenient route to the synthesis of 2′-deoxyguanosine (70) and 2′-deoxy-2-chloroadenosine[77] (71), respectively.

A wide variety of catalysts have been investigated, including *p*-toluenesulfonic acid, various Lewis acids,[78] polyphosphoric acid, ethyl polyphosphate,[79] bis(*p*-nitrophenyl)phosphate,[80] and iodine.[81] The fusion procedure has improved the yields obtained from the condensation of tetra-*O*-acetyl-D-ribofuranose (72) with 6-chloropurine and its 2-methylthio derivative by two to three times.[82]

Perhaps the prototype example[74,83] of the fusion method for the synthesis of

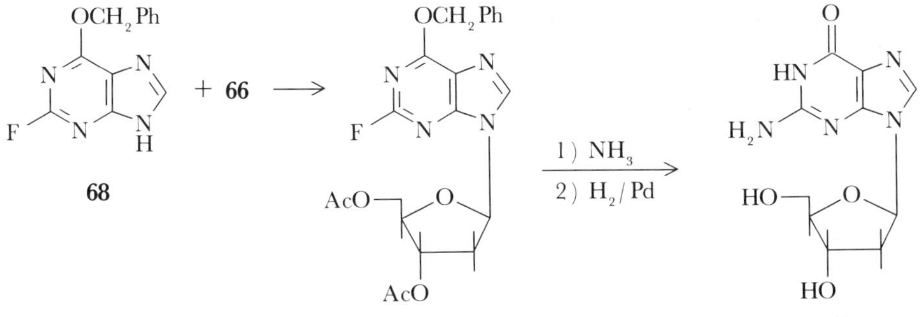

68

AcO— / AcO

+ 66 →

OCH$_2$Ph

1) NH$_3$
2) H$_2$/Pd

+ α-anomer

70

+ α-anomer

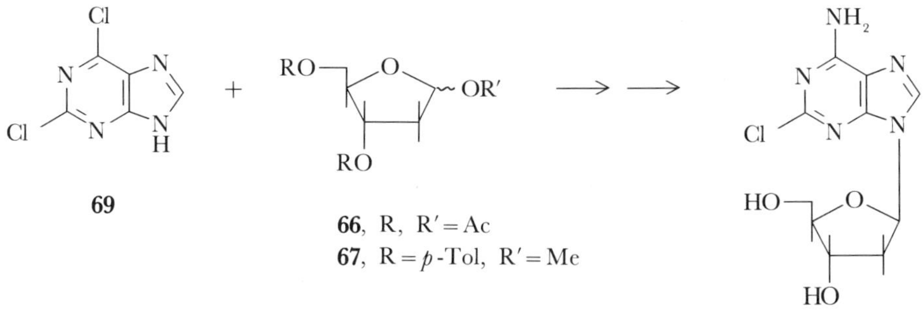

69

66, R, R′ = Ac
67, R = p-Tol, R′ = Me

71

+ α-anomer

a nucleoside has been the condensation of 2,6-dichloropurine (69) and its bromine congener,[84] (73), with 1,2,3,5-tetra-O-acetylribofuranose (72) at 160–170°C. This reaction proceeded in good yield, with or without a catalyst. In fact, it seems that the more acidic heterocycles, such as the purines with electron-withdrawing groups, are capable of undergoing the fusion reaction autocatalytically.[83] The fusion product of (73) after ammonia treatment gave 2-bromoadenosine[84] (74). It was reported that 6-methyl-, 6-ethyl-, and 2-fluoro-6-methylpurine were all good substrates of the fusion reaction with (72) and gave good yields of the corresponding 9-ribofuranosyl purines.[85] The fusion of purine (75) with (72) and other sugars in the presence of bis(p-nitrophenyl)phosphate gave mixtures of purine-7 and purine-9 nucleosides including the 2,3,5-tri-O-acetate of nebularine.[86]

Peracylated pyranose sugars have also been used successfully in the fusion procedure.[87] The fusion of acylamidopyranoses with purines in the presence of activating agents such as p-nitrophenol, 2,4-dinitrophenol, p-toluenesulfonamide and o-nitrobenzoic acid, respectively, has been studied as a general route for the synthesis of aminoglycosylnucleosides.[88] It has been observed that the fused nucleoside was obtained in improved yields when an activating agent was used in combination with p-toluenesulfonic acid as the catalyst[89] rather than using the activating agent alone.[88] Although on comparison[89] the mercury

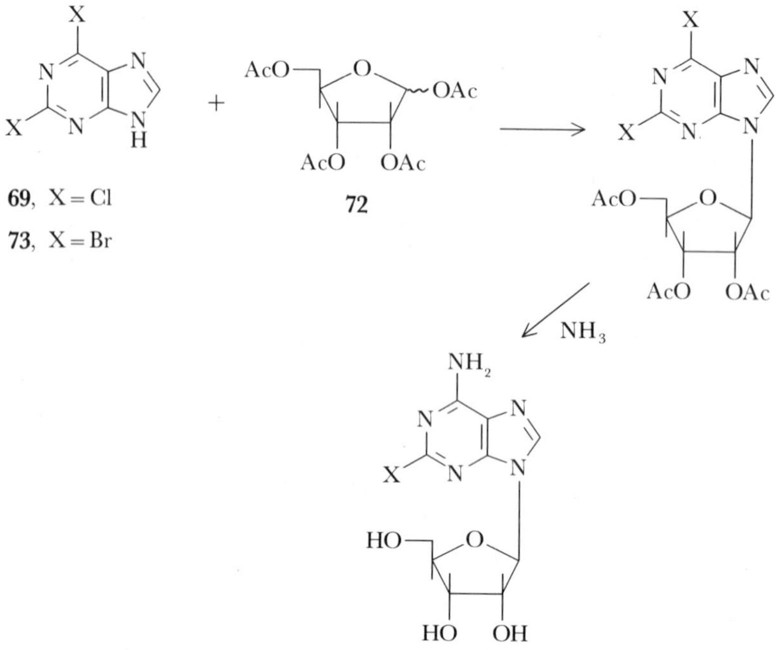

69, X = Cl

73, X = Br

72

74, X = Br

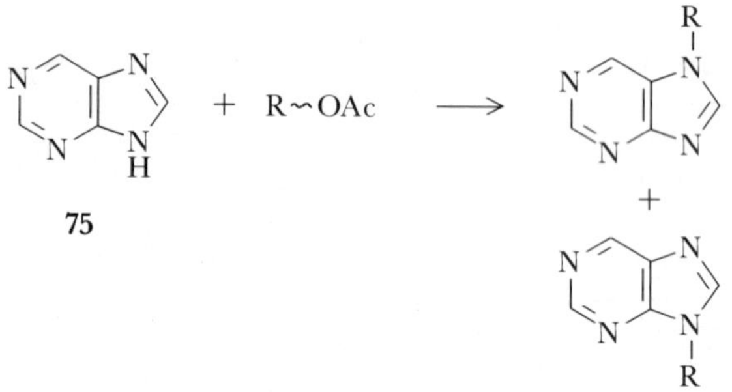

75

R = 2,3,5-tri-*O*-acetyl-D-ribofuranosyl
2,3,5-tri-*O*-acetyl-D-xylofuranosyl
2,3,4-tri-*O*-acetyl-D-ribopyranosyl

salt method was superior to the fusion method, the products by fusion method were obtained in better yields and higher purity and were easily separated. Fusion of bis(trimethylsilyl)-*N*-benzoyladenine with 2,3-di-*O*-benzoyl-5-(diphenyl-phosphoryl)-D-ribofuranosyl bromide at 100°C gave 6.2% of the α-anomer and 8.3% of the β-anomer of *N*-benzoyl-9-[2,3-di-*O*-benzoyl-5-(diphenylphosphoryl)-D-ribofuranosyl]adenine. These compounds were converted into 5′-AMP and α-5′-AMP by treatment with sodium methoxide followed by phosphodiesterase.[90] The hypoxanthine and 7-β-theophylline nucleotides were similarly synthesized.[91,92] The fusion of purines with 1,2-unsaturated sugar derivatives (glycals) has proved to be an effective route to many 2′-deoxypurine nucleosides.[93,94]

Subsequent investigations showed that acid-catalyzed fusion reactions also gave 2,3-unsaturated nucleosides[95] that were structurally related to the nucleoside antibiotic blasticidin.[96] In a typical example,[89] the fusion reaction of theophylline (76) with 2-acetamido-1,3,4,6-tetra-*O*-acetyl-2-deoxy-α-D-gluco-pyranose (77), in the presence of the acid catalyst and *p*-nitrophenol, gave a 45% yield for the fused nucleoside, 7-(2-acetamido-3,4,6-tri-*O*-acetyl-2-deoxy-β-D-glucopyranosyl)theophylline (78). Fusion of 6-chloropurine (79) and D-xylal (80) gave a mixture[95] of four compounds: 4-*O*-acetyl-3-(6-chloropurin-9-yl)-1,2,3-trideoxy-D-*erythro*-pent-1-enopyranose (81), 4-*O*-acetyl-3-(6-chloropurine-7-yl)-1,2,3-trideoxy-D-*erythro*-pent-1-enopyranose (82), and 1-(6-chloropurin-7(and 9)-yl-(4-*O*-acetyl)-2,3-dideoxy-α- and β-d-*glycero*-pent-2-enopyranose [(83) and (84)]. The lack of saturated nucleosides in the products derived from di-*O*-acetyl-D-xylal (80) was attributed[95] to the facile elimination of the acetate in the trans-substituted sugar to give the functional intermediate (85). The stereospecificity of formation of the 3-(purine-7 or 9-yl) products from the unsaturated sugar is conveniently envisaged by the assumption that a transient 3,4-acetoxonium ion is formed and then attacked by a purine derivative at the 3 (allylic) position.

The fusion of tri-*O*-acetyl-D-glucal (86) with purines was further investigated to show that, under optimal conditions, a mixture of four compounds was obtained: the anomeric (4,6-di-*O*-acetyl-2,3-dideoxy-β- or α-d-*erythro*-hex-2-enopyranosyl)purines (87) and (88) and the 4,6-di-*O*-acetyl-3-(purinyl)-1,2,3-trideoxy-D-ribo or arabino-hex-2-enopyranoses (89) and (90). The purines used, all of which gave 9-substitution, were 6-chloro-2-methylthio purine,[97] 2,6-dichloropurine,[98,99] 2-acetamido-6-chloropurine,[100] and 6-benzamidopurine.[100]

Acid-catalyzed fusion of tri-*O*-acetyl-D-glucal[98,99] and tri-*O*-acetyl-D-galactal[101] gave analogous results, with the sugar attached to the 7 position of theophylline. The hex-2-enopyranosylpurine [(87) and (88)] have been shown to be the products of kinetic control of the D-glucal fusions, and they rearranged on

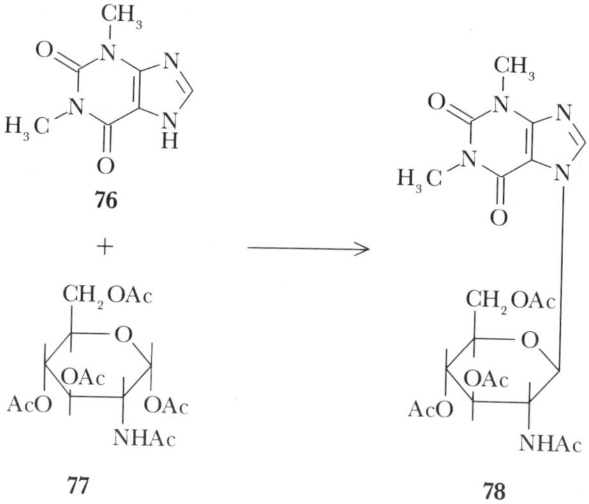

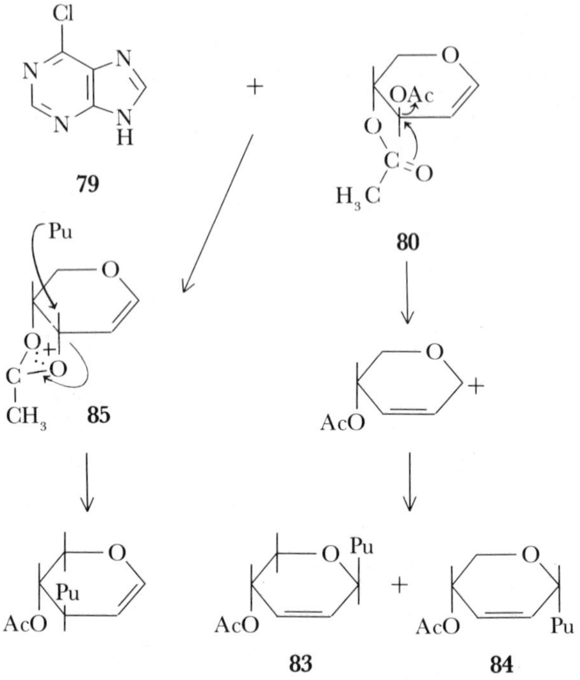

Pu = 6-chloropurin-   Pu = 6-chloropurin-9-yl
9-yl **81**
and
6-chloropurin-
7-yl **82**

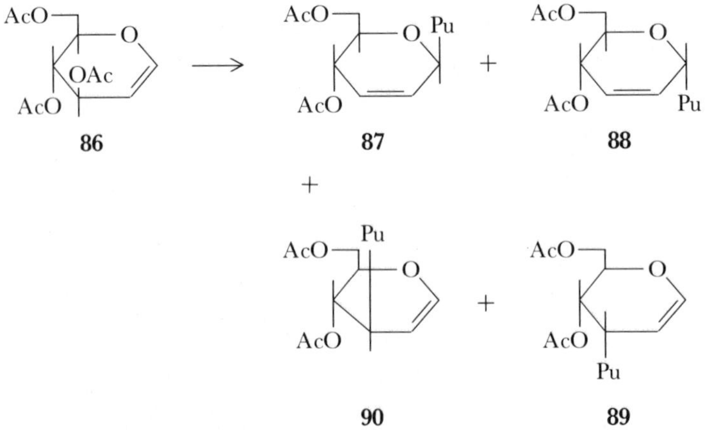

Pu = theophyllin-7-yl
2,6-dichloropurin-9-yl
2-acetamido-6-chloropurin-9-yl
6-benzamidopurin-9-yl

heating to the thermodynamically preferred 3-(purinyl)hex-1-enopyranose [(**89**) and (**90**)]. The direct synthesis of nucleosides of guanine has been made accessible *via* fusion of an *N*-acetylguanine with peracetylated sugars. Fusion of *N*-acetyl-guanine with 1,2,3,4-tetra-*O*-acetyl-β-D-ribofuranose gave the 7-β-, 9-β-, and 9-α-nucleosides in a 43 : 29 : 28 ratio, whereas 1,2,3,4-tetra-*O*-acetyl-α- or β-xylopyranose of 1,2,3,4,6-penta-*O*-acetyl-α- or β-d-glucopyranose gave 7-β- and 9-β-nucleosides in a ratio of 9 : 11. No α-anomer was observed for the pyranose sugars, except for a small amount that formed from using tetra-*O*-acetyl-β-D-ribopyranose. Bis(*p*-nitrophenyl)phosphate, *p*-toluenesulfonic acid, and ZnCl$_2$ were equally effective as catalysts for these fusion reactions. The initial anomeric configuration of the sugar appeared to have no effect on the final product ratio. The product ratio actually seemed to reflect the thermodynamic stability of the products[102] as indicated by isomerization of one of the pure products on heating. The lack of any α-anomer from xylo- and glucopyranoses may be attributed to the fact that the β position occupies the equatorial position in the all-equatorial-substituted compounds.

In contrast, the fusion of 1,2,3,5-tetra-*O*-acetyl-D-xylofuranose with *N*,7(or 9)-diacetylguanine, in the absence of catalyst, gave a mixture of α- and β-anomers of *N*-acetyl-7-D-xylofuranosylguanine with only a very small amount of the 9-isomer being formed.[103]

The affinity toward glycosidic bond formation of 6-chloropurine with some benzylthio[104] and *N*-methylacetamido[105] sugars at high temperatures further demonstrates the versatility of the fusion procedure for the synthesis of nucleosides.

In addition to a facile 1′,2′-cis glycosidic bond formation in the fusion reaction, epimerization at the 2′ position has also been reported to occur. Fusion of 2,6-dichloropurine (**69**), using *p*-toluenesulfonic acid as the catalyst, with ethyl 1,2,3-tri-*O*-acetyl-5,6-dideoxy-D-riboheptofuranuroate (**91**) at 140°C gave a 30% yield of ribonucleosides (**92**) (3β : 1α) and a 17% yield of the arabinonucleosides

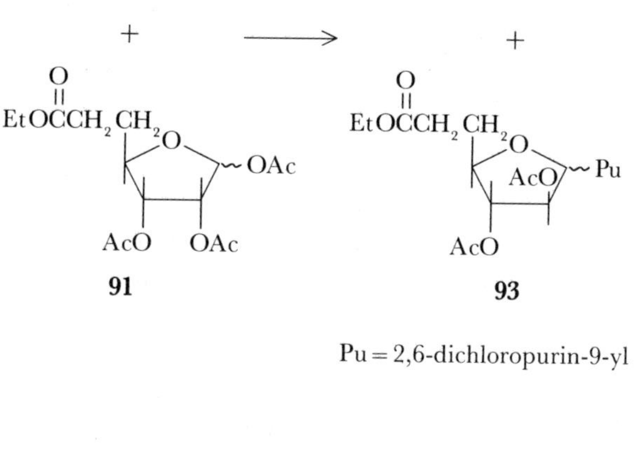

Pu = 2,6-dichloropurin-9-yl

(93) ($3\alpha : 1\beta$), which indicates epimerization of the sugar occurred under ambient reaction conditions.[96] Epimerization was also observed in the chloroacetic-acid-catalyzed[107] fusion of 3-acetamido-1,2-di-O-acetyl-3,5-dideoxy-D-ribofuranose (79) with 6-chloropurine (79) and 3-acetamido-3-deoxy-1,2,5-tri-O-acetyl-β-D-ribofuranose with 2,6-dichloropurine.[108] In the former case, the mixture of nucleosides obtained from the fusion was treated with dimethyl amine to yield 9-(3-acetamido-3,5-dideoxy-α- and β-D-ribofuranosyl)-6-dimethylaminopurine [(95), 5 and 17%, respectively] and 9-(3-acetamido-3,5-dideoxy-α- and β-D-arabinofuranosyl)-6-dimethylaminopurine [(96), 4 and 9%, respectively].[107] Thus, the need for conclusive identification of the products of a fusion reaction, especially when low yields are obtained, is obvious.

The nature of the starting sugar may also have an effect on the course of the reaction. It was found that when 1-O-acetyl-2,3,5-tri-O-benzoyl-D-arabino-furanose was fused with 2,6-dichloropurine the $\alpha(1,2$-*trans*)-anomer of the starting sugar reacted readily to give the α-nucleoside, whereas the $\beta(1,2$-*cis*)-sugar did not react.[109] The addition of an acid catalyst did not change the course of the reaction.

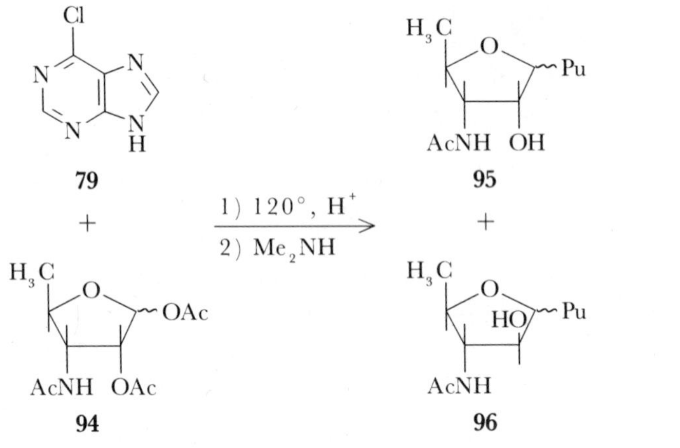

Pu = N,N-dimethyladenin-9-yl

2.1.3. Glycosylation of Purines in the Presence of Lewis Acid Catalysts

A glycosylation technique found to have widespread usage is the condensation of acylated furanoses and pyranoses with purines under the influence of a Friedel–Crafts catalyst (a Lewis acid).

Baker and co-workers found substantial utility in the use of titanium tetrachloride complexes of 1-O-acylfuranoses as reactive carbohydrate derivatives for coupling with chloromercuripurines.[25] Ishido *et al.* have also investigated numerous Lewis acids as catalysts for the high-temperature fusion method of glycosylation.[78] However, Furukawa and Honjo found that Lewis acids could efficiently catalyze a direct condensation between the aglycones and 1-O-acylfuranoses in solution in good yield.[110] Solubility in the required apolar solvents was a problem, but it was found that N-octanoyladenine (**97**) and 1-O-acetyl-2,3,5-tri-O-benzoyl-β-D-ribofuranose (**98**) could be condensed in good yield in 1,2-dichloroethane solution with 1 equivalent of $SnCl_4$ or $AlCl_3$ catalyst giving N-octanoyl-2,3,5-tri-O-benzoyladenosine (**99**) in 50–60% yield.[110] The octanoyl blocking group seemed to provide the necessary solubility. Similarly, N-palmitoyl-guanine (**100**) and 1,2,3,5-tetra-O-acetyl-D-ribofuranose (**72**) in chlorobenzene gave a 66% yield of N-palmitoyl-2′,3′,5′-tri-O-acetylguanosine (**101**) when $AlCl_3$ was used as a catalyst.[110]

Investigations of the $AlCl_3$-mediated reaction of N-nonanoyl- and N-palmitoylguanine with 1,2,3,5-tetra-O-acetyl-D-xylofuranose showed that approximately 40–60% of an $\alpha:\beta$ (1:1) mixture of N-palmitoyl-9-(2,3,5-tri-O-acetyl-D-xylofuranosyl)guanine was obtained with the palmitoylguanine, whereas up to 15% of the 7-isomer was also obtained with the nonanoyl derivative.[111]

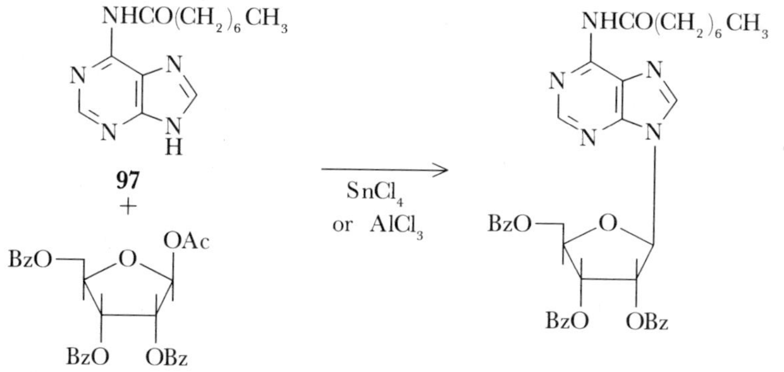

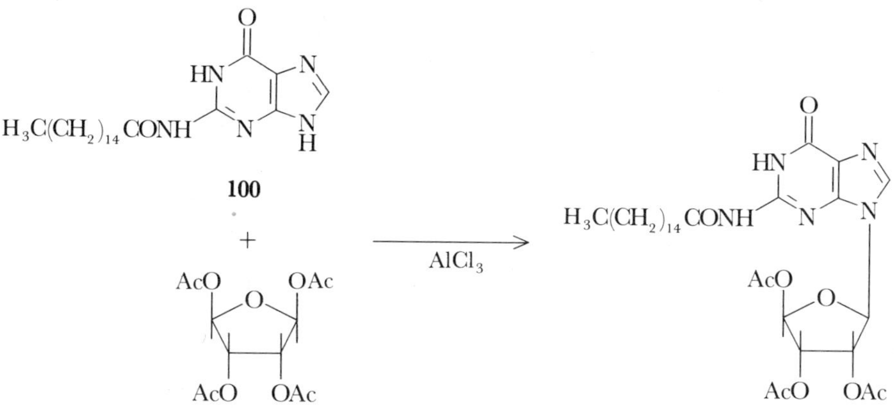

This method appears to offer advantages over the fusion procedure in that fewer side products are obtained, that is, less α- and 7-glycosylated isomers. A significant advance in the use of this method was the replacement of the ionizable hydrogens on the purine with trimethylsilyl groups. The silyl method, which is a modification of the Hilbert–Johnson method,(112) became the most used among various methods for the synthesis of nucleosides. When 6-*N*-benzoyl-*N*-9-bis(trimethylsilyl)adenine (**102**) was warmed in 1,2-dichloroethane with a variety of 1-*O*-acylglycoses (**103**) and $SnCl_4$ catalyst, the 9-*β*-nucleosides (**104**) were obtained in 41–68% yield with only small amounts of side products being reported.(113,114) The introduction of the trimethylsilyl groups greatly enhanced the solubility of the purine. Condensation of *O*,9-bis(trimethylsilyl)hypoxanthine (**105**) and 1-*O*-acetyl-2,3,5-tri-*O*-benzoyl-*β*-D-ribofuranose (**98**), with $SnCl_4$ as catalyst, gave 2′,3′,5′-tri-*O*-benzoylinosine (**106**) in 71% yield after water treatment had removed the silyl groups.(113)

A condensation of the trimethylsilyl derivative (**107**) of 8-iodoisoguanine with 1-*O*-acetyl-2,3,5-tri-*O*-benzoyl-*β*-D-ribofuranose (**98**) in 1,2-dichloroethane/$SnCl_4$ gave, after water treatment, 6-amino-8-iodo-3-(2,3,5-tri-*O*-benzoyl-

102 **103** **104**

R, R¹ = Ac or Bz

105 **98** **106**

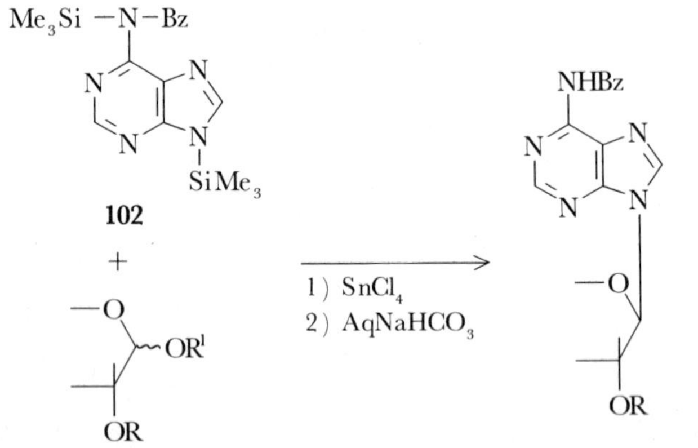

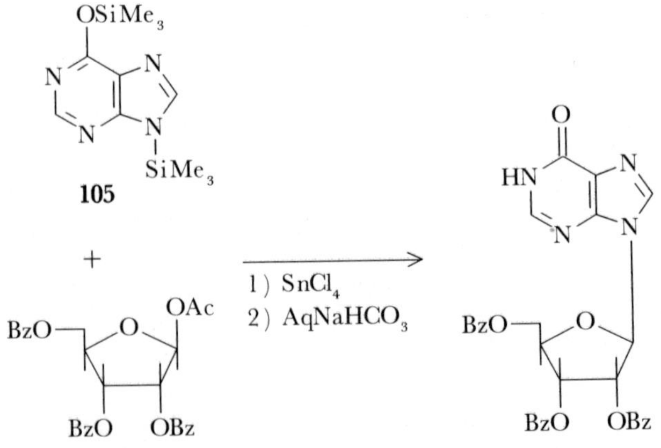

β-D-ribofuranosyl)purin-6-one **(108)** in 30% yield.[115] This indicates that the "directive effect" due to steric repulsion by a bulky 8-substituent toward glycosylation at position 9 is also in effect for this procedure.[116]

The 9- and 7-, 6-deoxy-β-D-allofuranosyl[117] and α-L-talofuranosyl[118] hypoxanthines [**(111)** and **(112)**] were prepared by treating bis(trimethylsilyl)-hypoxanthine **(110)** with 6-deoxy-2,3,5-*tris-O-(p*-nitrobenzoyl)-D-allofuranosyl bromide **(109)** and the corresponding L-talofuranose, respectively, in the presence of $SnCl_4$, followed by deblocking. The corresponding β-D-allofurano-sylguanine[117] analogues were prepared in a similar manner using *O*,9-bis (trimethylsilyl)-N^2-acetylguanine **(113)**. These nucleosides represent a modification of the 5' position, bearing a methyl and a less readily esterified secondary hydroxyl group, with proposed potential purine nucleoside phosphorylase (PNPase) and antitumor activities.[119,120]

After the readily available Friedel–Crafts catalyst, tin tetrachloride (stannic chloride, $SnCl_4$), was established as a catalyst for nucleoside synthesis, several other new catalysts,[121] trimethylsilyl triflate [trimethylsilyl trifluoromethane-sulfonate, $(CH_3)_3SiOSO_2CF_3$], trimethylsilyl nonaflate [$(CH_3)_3SiOSO_2C_4F_9$], and trimethylsilyl perchlorate [$(CH_3)_3SiClO_4$], less acidic as Lewis acids than $SnCl_4$, were studied. Tin tetrachloride decomposes into tin salts in an aqueous bicarbonate–CH_2Cl_2 medium during workup, which generates an emulsion. This problem can be avoided with the use of the new "destannyl" catalysts. In addition, the new catalysts have optimum acidity to allow the formation of acyloxonium cations **(114)** and form only week σ-complexes **(115)** with the silylated bases, resulting in higher yields of the nucleosides.[121]

Purine nucleoside formation under the Friedel–Crafts conditions has been proposed[122] to occur as follows. The Lewis acid catalyst, such as trimethylsilyl triflate, reacts simultaneously with the peracylated sugar bearing a 2-acyloxy group as in **(98)** to generate *in situ* the electrophilic 1,2-acyloxonium cation **(114)** and with the silylated purine [e.g., **(102)**] to form *in situ* the trimethylsilyl σ-com-plex at N-1, the center of highest electron density in **(115)**. The σ-complex **(115)** could also exist in equilibrium with the N-1 trimethylsilyl derivative **(116)**. Either of the silylated intermediates, (102) or (116), could react with the acyloxonium cation (114) to result in a 1-, 3-, 7-, or 9-N-ribosylation of the aglycon. Under optimum reaction conditions, the 1-, 3-, and 7-isomers should rearrange[121] to afford the most thermodynamically favorable N-9-isomer (117). However, the formation of other isomers (N-3 and N-7) cannot be completely precluded.[123,124] The purine nucleosides, adenosine (81%), guanosine (66%), and xanthosine (49%) were synthesized by this procedure following saponification.[121] A simple one-pot Friedel–Crafts-catalyzed condensation of peracylated sugars and silylated bases to provide purine nucleosides has also been studied.[125]

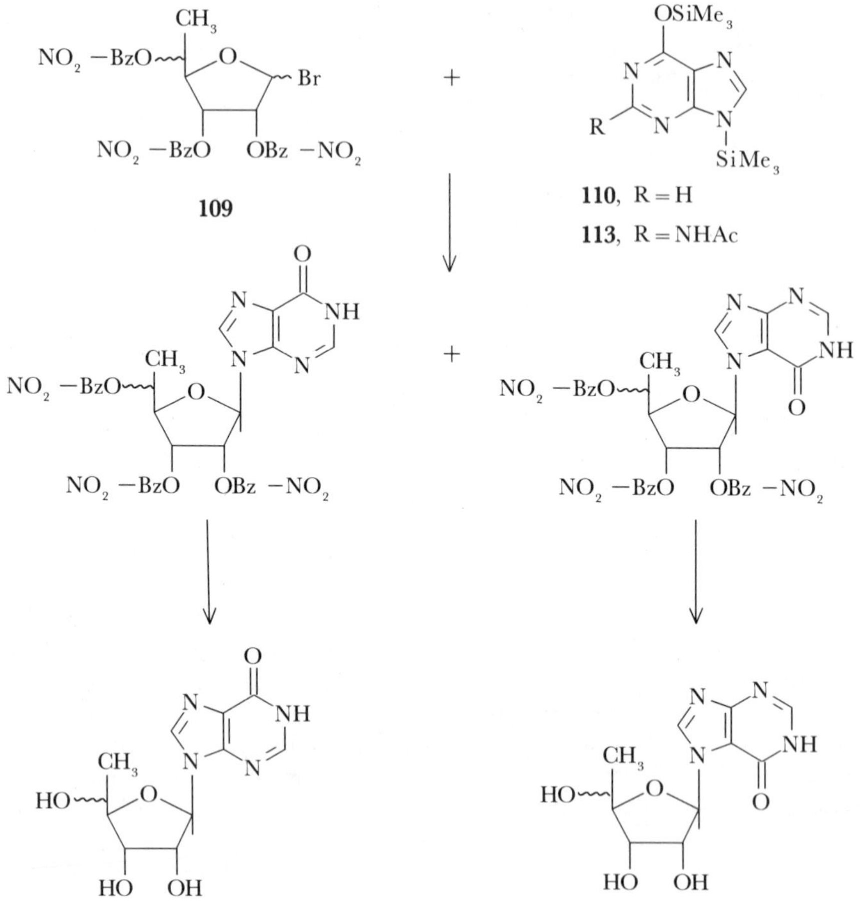

    The Friedel–Crafts condensation has also been useful for the synthesis of deoxynucleosides, 3'-azido-3'-deoxyadenosine,[126] and various analogues of aminodeoxypyranosyltheophylline.[127] These compounds were synthesized by the reaction of a persilylated purine base and an appropriate sugar in the presence of SnCl$_4$. Recently, bis(trimethylsilyl)acetamide (BSA), instead of hexamethyldisilazane, has been used[124] for the silylation and one-pot condensation of

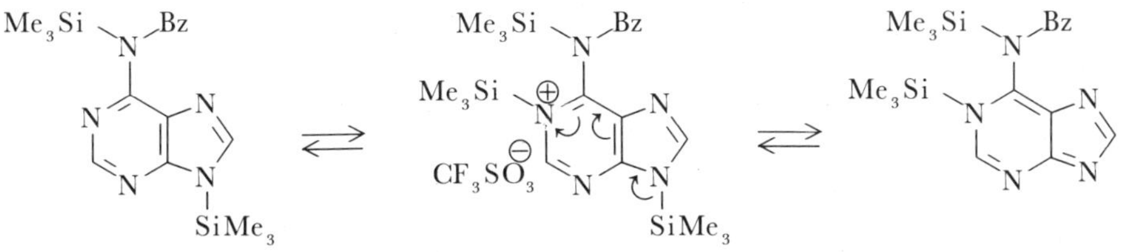

102   **115**   **116**

+

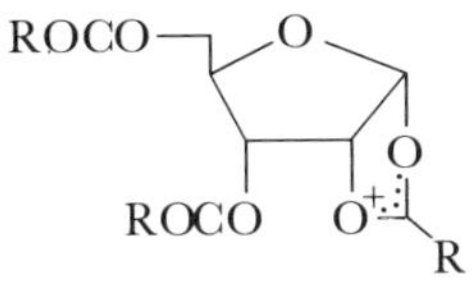

114

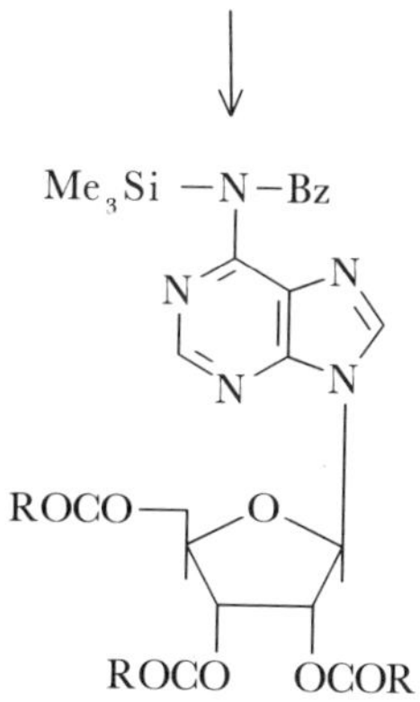

117

N^2-(p-n-butylphenyl)guanine (BUPG) with 1-chloro-3,5-di-O-p-toluoyl-2-deoxy-ribofuranose in the presence of trimethylsilyl triflate. This has provided the 7-α'-, 7-β'-, 9-α'-, and 9-β'-isomers of BUPG 2'-deoxyribonucleosides in an overall 70% yield. BSA does not introduce chemically interactive side products into the reaction mixture and was found to be superior to HMDS as the silylating agent, especially for one-pot reactions.[128]

2.1.4. Transglycosylation

The intramolecular migration of the glycosyl moiety from the N-1 or N-3 position of the pyrimidine ring to the N-9 position of the imidazole ring in a purine moiety is considered to be the normal chain of events taking place during the purine nucleoside synthesis in a Lewis acid medium (*vide ante*). Similar inter-molecular transglycosylations have also been observed.[129] The transfer of ribosyl and phosphoribosyl moieties of acylated pyrimidine nucleosides, nucleotide, and trimethylsilylated nucleotides to purines has been accomplished successfully in the presence of an acid or Lewis acid catalysts.[129] The transglycosylation method of introducing sugar in a purine base is unique and often useful, especially for the synthesis of purine nucleosides of functionalized sugars that are difficult to prepare by conventional procedures.

Ribose exchange of 2',3'-*O*-isopropylidene (118) inosine with acetobromo-glucose (119) in nitromethane occurs in the presence of mercuric cyanide. Presumably, this occurs *via* the electrophilic attack of the 1,2-acyloxonium cation of (119) at N-7 of (118) to give the quaternary intermediate (120), which then collapses into the 1,5-anhydro sugar (122) and the N-7 nucleoside, which then rearranged to furnish[130] the N-9-peracetylglucosylhypoxanthine (121).

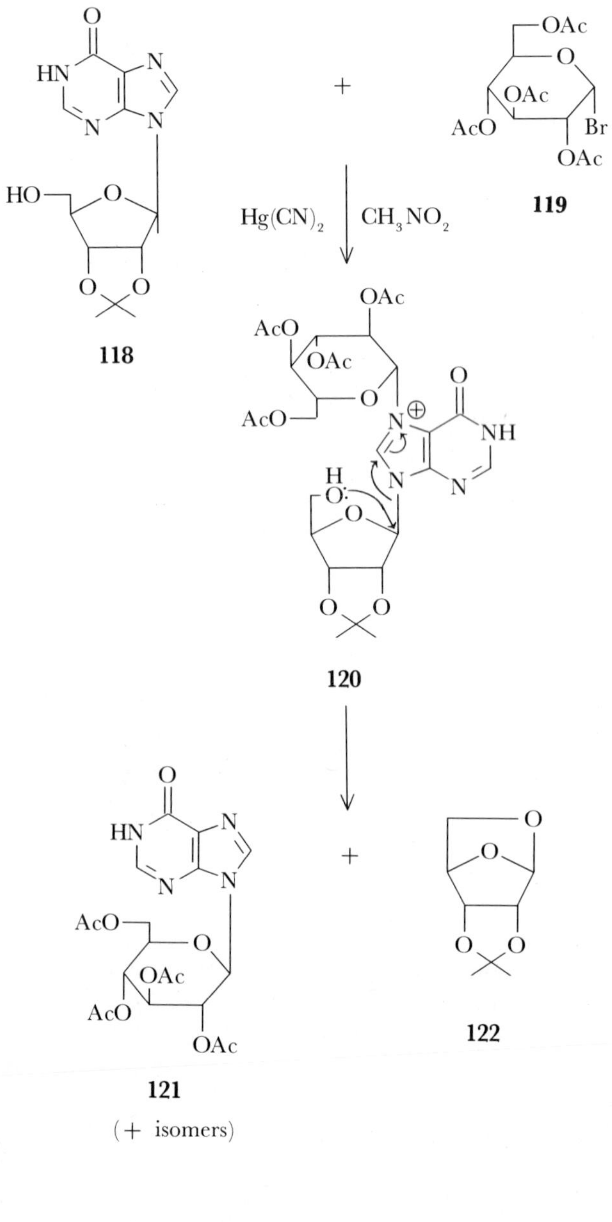

121

(+ isomers)

The transglycosylation reactions of azido-[128] and aminodeoxyribosyl-pyrimidines[131] with silylated N^6-alkanoylpurines, using trimethylsilyl triflate, have provided the azido- and aminodeoxyribosylpurines (adenine and guanine) as a mixture of α- and β-anomers.

The adenosine analogue (**124**) of the pyrimidine nucleoside antibiotic (**123**) has been prepared by a transglycosylation reaction of (**123**) with (**102**) in the presence of trimethylsilyl perchlorate.[132] Plausibly, the sugar transfer is driven by O^2-trimethylsilylation and 1′,2′-acetylcarbonium ion formation in (**123**) to yield (**124**).

The preceding examples indicate that although the transglycosylation is an excellent method for introducing functionalized sugars to purines predominantly at the 9 position, the formation of other isomers may also be observed. This problem, to some extent, has been circumvented by an enzymatic trans-glycosylation. 9-(β-D-Arabinofuranosyl)adenine [araA, (**125**)] is formed by the incubation of adenine with 1-(β-D-arabinofuranosyl)uracil (araU) in the presence of *Enterobacter aerogenes* at 60°C. Incubation at 37°C provided only the deaminated product, 9-(β-D-arabinofuranosyl)hypoxanthine[133] [araHX, (**126**)].

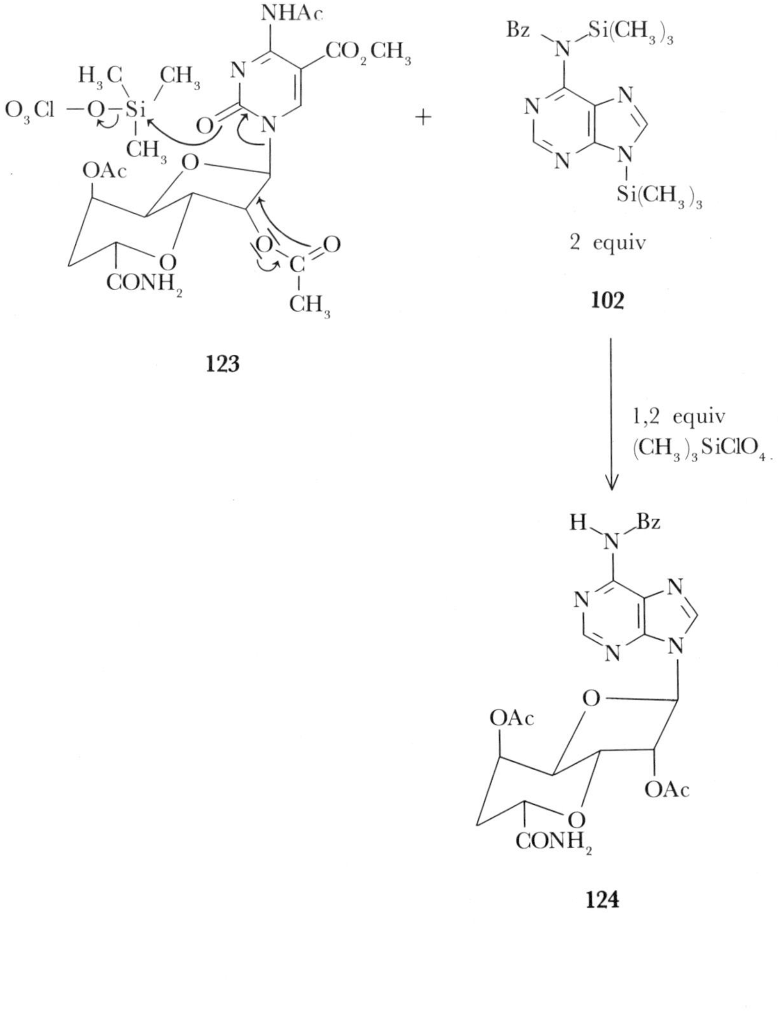

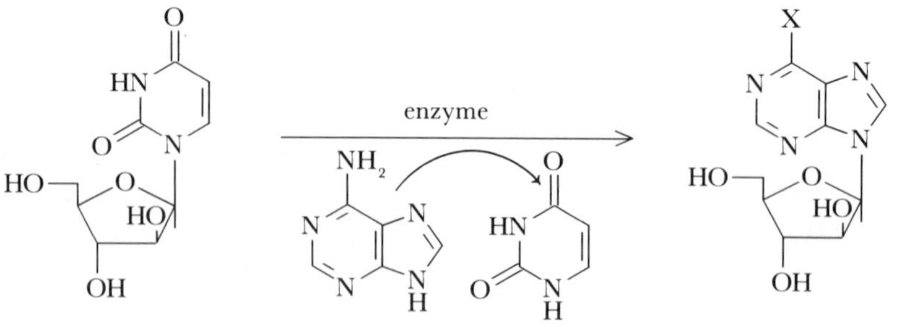

AraU

125, X=NH$_2$

126, X=OH

The 2'-amino-2'-deoxy analogues of inosine[134] and 2-chlorohypoxanthine-9-riboside[135] were similarly prepared by enzymatic transaminoribosylation using 2'-amino-2'-deoxyuridine in the presence of *Erwinia herbicola*.

An interesting uncatalyzed transglycosylation reaction was observed to occur when *N*-(2,3,5-tri-*O*-acetyl-D-ribofuranosyl)maleimide [(**127**), prepared from the silver salt of maleimide and glycosyl chloride], was heated at reflux with 6-benzamidopurine (**58**) in nitromethane. A 45% yield of adenosine (**128**) was obtained after sodium methoxide deblocking.[136]

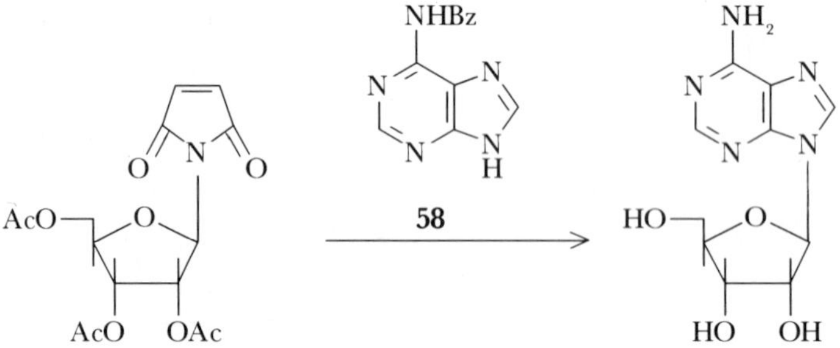

127   **58**   **128**

2.1.5. Direct, Uncatalyzed Alkylation of Purines or Their Alkali Metal Salts by Glycosyl Halides

The term "direct alkylation" is used to describe procedures for nucleoside synthesis that involve the attachment of a sugar to a heterocyclic base in the absence of a catalyst or without the formation of a heavy metal derivative of the base.

Perhaps the most fundamental reactions in this category, thus far reported, are the preparation of the ribonunucleosides of adenine, guanine, hypoxanthine, and xanthine under "prebiotic" conditions.[137] When these heterocycles were mixed with ribose and either natural or synthetic seawater and heated at 100°C, the corresponding β-ribofuranosides, in addition to other products, were obtained in yields of up to about 5%.

In a similar context, adenine was found to form adenosine in 30% yield when treated with ribose and polyphosphoric acid in moist DMF.[138] Adenine and some 6-substituted purines were condensed directly with 2-deoxy-D-ribose (**129**) in water or polar organic solvents at 100°C to give the diastereomeric 2,3-dideoxy-3-(purinyl)-D-*erythro*- (**130**) and *threo*-pentoses[139] (**131**). It was envisaged that the base-catayzed elimination of water from deoxyribose gave the 2,3-unsaturated pentose (**132**) and that the purine derivative then reacted with that species by a Michael addition.

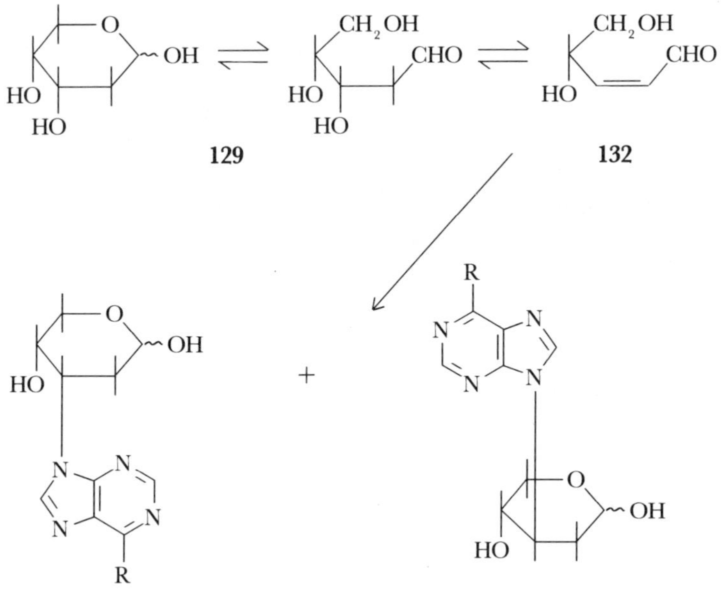

129   **132**

130   +   **131**

R = NH$_2$, NMe$_2$, SMe

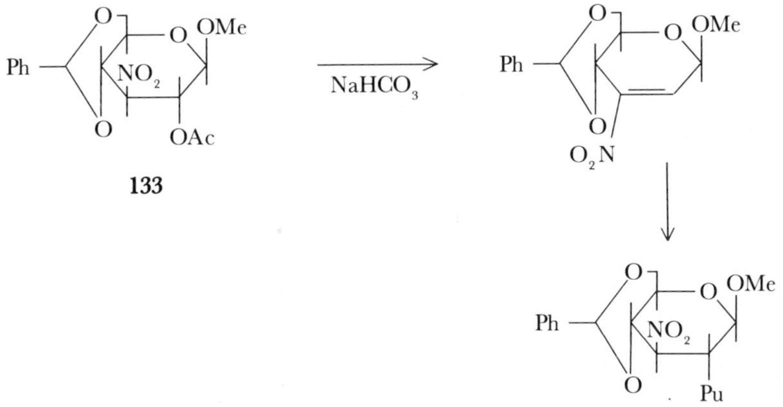

133 $\xrightarrow{\text{NaHCO}_3}$

134, Pu = theophyllin-7-yl

135, Pu = 2,6-dichloropurin-9-yl

Another example[140] of a Michael addtion was found to occur when methyl-2-*O*-acetyl-4,6-*O*-benzylidine-3-deoxy-3-nitro-*β*-D-glucopyranoside **(133)** was treated with 3,7-dihydro-1,3-dimethyl-1*H*-purin-2,6-dione(theophylline) or 2,6-dichloropurine and solid $NaHCO_3$ in tetrahydrofuran at reflux. The purine added to the activated double bond, formed by base-catalyzed elimination of acetate, to give methyl 4,6-*O*-benzylidine-2,3-dideoxy-3-nitro-2-(theophyllin-7-yl- or 2,6-dichloropurine-9-yl)-*β*-D-glucopyranoside [(**134**), 96% and (**135**), 78%, respectively].

The sodium salt of adenine has been alkylated[141] directly by 5′-*O*-trityl-1,2-anhydro-L-ribofuranose (**136**) to give a 90% yield of 5′-*O*-trityl-L-adenosine (**137**). The sodium salt of guanine was alkylated with (**136**) in a similar manner to give a 40% yield of 5′-*O*-trityl-L-guanosine.[141]

Alkylation[142] of 6-methylthiopurine (**138**) with the epoxy sugar methyl-2,3-anhydro-5-deoxy-α-D-ribofuranoside (**139**) occurred in *N*,*N*-dimethylacetamide in the presence of K_2CO_3. The two possible products of an attack on the epoxide were obtained: methyl-2,5-dideoxy-2-(6-methylthiopurin-9-yl)-α-D-arabinofuranoside [(**140**), 19%)] and methyl-3,5-dideoxy-3-(6-methylthiopurin-9-yl)-α-D-xylofuranoside [(**141**), 7%]. This reaction gave unsatisfactory results with methyl-2,3-anhydro-α-D-riboside.[142]

136 **137**

Pu = adenin-9-yl
guanin-9-yl

138 **139** **140** **141**

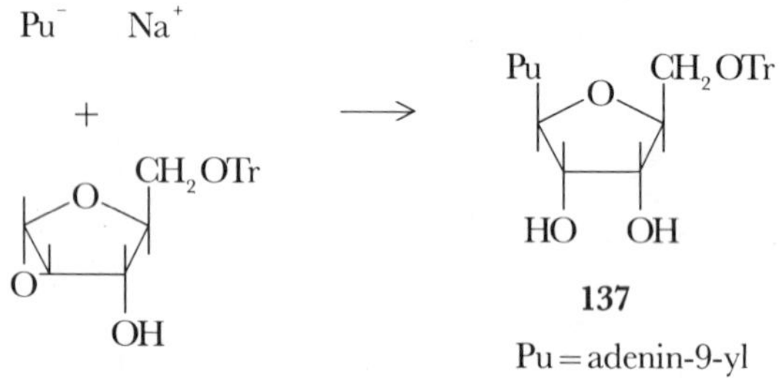

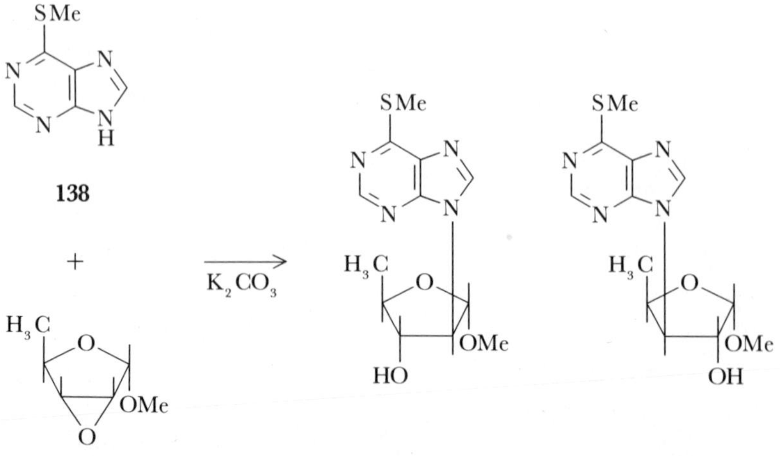

It has been possible to prepare various purine nucleosides by a direct alkylation of the purine without the presence of a catalyst or acid acceptor. Treatment of adenine **(142)** directly with 2,3,5-tri-*O*-benzoyl-D-ribofuranosyl bromide **(4)** in DMF or acetonitrile gave a 25% yield of the 3-isomer **(143)** and 18% of the 9-isomer **(144)**. Treatment of **(143)** with methanolic ammonia gave "isoadenosine," 3-β-D-ribofuranosyladenine.[143]

Treatment of adenine directly with 2,3-di-*O*-benzoyl-5-*O*-(diphenylphosphoryl)-D-ribofuranosyl bromide in acetonitrile gave a 29% yield of 3-[2,3-di-*O*-benzoyl-5-*O*-(diphenylphosphoryl)-β-D-ribofuranosyl]adenine and 23% of the 9-isomer.[144] Alkaline hydrolysis of the former compound, followed by venom phosphodiesterase treatment, gave 3-isoAMP.[144]

As discussed before,[115,116] the steric influence of a large halogen atom on the 8 position of a purine was found to hinder the glycosylation on the imidazole ring, giving a predominance of the 3-glycosylated isomer. This was also the case when 8-chloroadenine and 8-iodoadenine were directly alkylated with 2,3,5-tri-*O*-benzoyl-D-ribofuranosyl bromide **(4)**. The 8-iodoadenine as compared to its chloro congener gave better yields of the blocked nucleoside.[116] The trimethylsilylated derivative of 8-iodoisoguanine **(145)** gave, upon treatment with **(4)** followed by sodium methoxide deblocking, 8-iodo-3-β-D-ribofuranosylisoguanine[145] **(146)**.

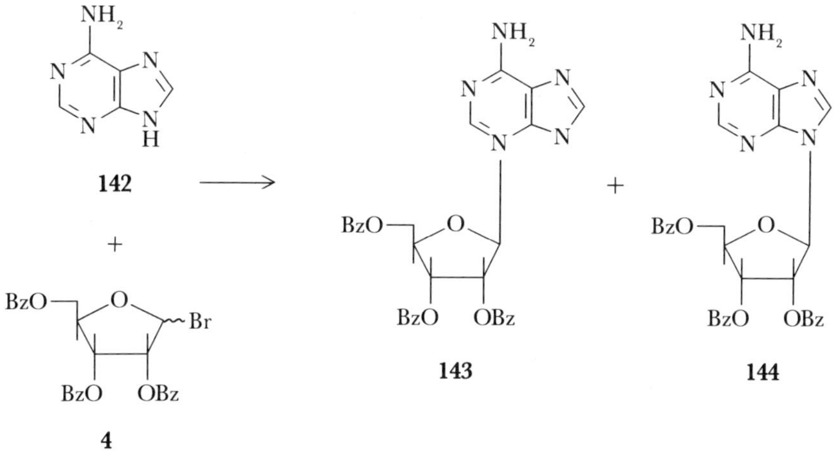

142

4

143

144

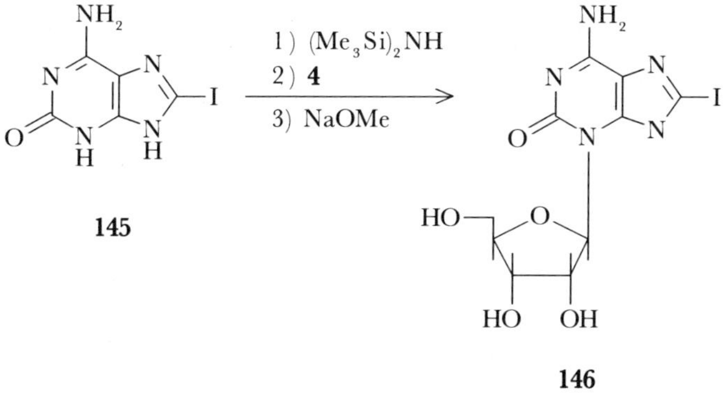

145

146

The syntheses of purine deoxyribonucleosides are elaborate, require the separation of complex anomeric mixtures, and provide poor yields. A direct deoxyribosylation procedure reported recently[146] appears to circumvent many of these problems. Thus, sodium hydride treatment of 2,6-dichloropurine (69) in acetonitrile followed by alkylation with 1-chloro-2-deoxy-3,5-di-O-p-toluoyl-α-D-*erythro*-pentofuranose (61) provided 2,6-dichloro-9-(2-deoxy-3,5-di-O-p-toluoyl-β-D-*erthyro*-pentofuranosyl)purine (147) in 59% yield. The corresponding 6-chloropurine analogue (148) was similarly prepared in a comparable yield. The 7-isomer was obtained in each case in 11–13% yields. The blocked nucleosides (147) and (148) on ammonolysis gave 2-chloro-2′-deoxyadenosine (71) and 2′-deoxyadenosine (10), respectively.[146]

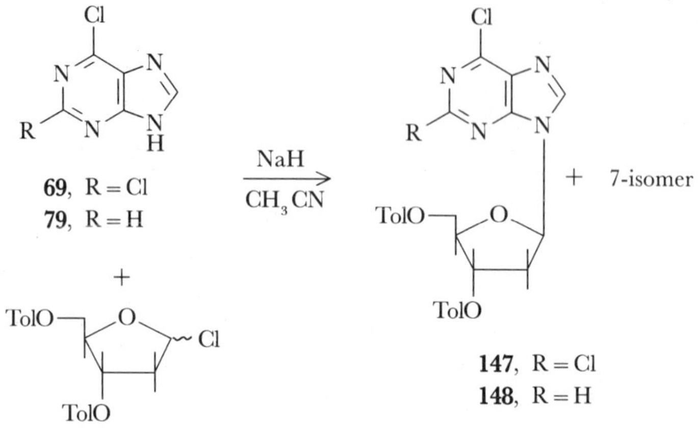

2.2. Synthesis of Purine Nucleosides and Nucleotides from Imidazole and Pyrimidine Nucleosides and Nucleotides

2.2.1. Synthesis of Purine Nucleosides and Nucleotides from Imidazole Precursors

The preparation of purine nucleosides and nucleotides from glycosylated imidazole precursors[147,148] has been used to furnish otherwise difficult to prepare types of compounds: 7-glycosylpurines, purine nucleosides with varying substituents in the 2 position, and 2-[*C]-labeled purine nucleosides important in biochemical research. As noted in Section 2.1, the glycosylation of purines most readily occurs at the 9 position of the purine ring. The only commonly used purine that has been observed to glycosylate consistently in the 7 position is theophylline.[8,93]

The chloracetic-acid-catalyzed fusion of 4(5)-bromo-5(4)-nitroimidazole (149) with 1,2,3,5-tetra-O-acetyl-β-D-ribofuranose (72) provided the intermediate nucleoside (150), which was then reacted with KI/KCN to give[149,150] the nitroimidazolecarbonitrile derivative (**151**). A catalytic reduction of the nitro group of (**151**) gave 4-amino-1-(2,3,5-tri-*O*-acetyl-β-D-ribofuranosyl)imidazole-5-carbonitrile (**152**). The ethoxymethylidine derivative (**153**) was formed with triethyl orthoformate. Treatment of (**153**) with ammonia affected both a deacetylation and ring closure[59,149,150] to give 7-β-D-ribofuranosyladenine (**154**). Substitution of triethyl-orthoacetate for triethyl-orthoformate in the former step ultimately yielded 2-methyl-7-(β-D-ribofuranosyl)adenine[150] (**155**) via (**156**). Ring closure of 4-amino-1-(β-D-ribofuranosyl)imidazole-5-carboxamide

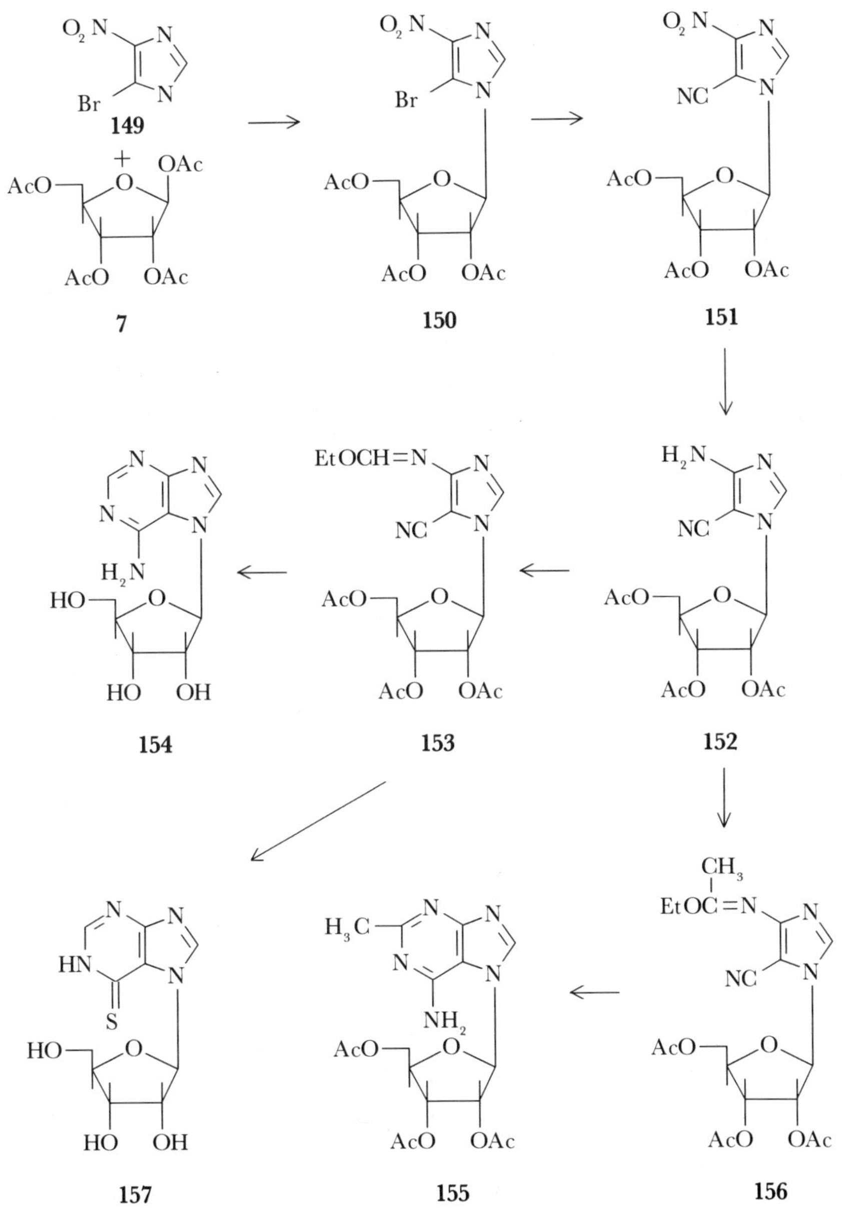

with diethoxymethyl acetate gave a good yield of 7-(β-D-ribofuranosyl) hypoxanthine.[150] A ring closure of the 4-ethoxymethylidineamino-1-(β-D-ribofuranosyl)imidazole-5-carbonitrile intermediate (153) with H_2S in pyridine gave 7-(β-D-ribofuranosyl)purin-6-thione (157), which was used to prepare other 6-substituted derivatives by selective nucleophilic displacements.[151]

Appropriately, 4,5-disubstituted imidazole nucleosides and nucleotides are the pivotal intermediates in the synthesis of purine nucleosides and nucleotides by ring annulation. Some of the first imidazole nucleoside intermediates were synthesized from inosine precursors (*vide infra*). Subsequently, 5-amino-1-(β-D-ribofuranosyl)imidazole-4-carboxamide [AICA-riboside, (158)] became available in large quantities by a fermentation process and added impetus to the synthesis of a variety of 2-substituted purine nucleosides from (158).

A classical example of the synthesis of a purine nucleoside from AICA-riboside is that of inosine (159), formed by ring annulation *via* amino-formylation.[152] Of more practical use, however, is the cyclization reaction[153] of AICA-riboside with carboxylic acid esters such as ethyl formate, diethyl carbonate, and ethyl acetate, which in the presence of a base give inosine (159), xanthosine (160), and 2-methylinosine (161), respectively. Dichlorocarbene similarly reacts with (158) to yield 5-dichloromethylamino intermediate, which cyclizes after the expulsion of hydrogen chloride to form inosine.[154]

Treatment of AICA-riboside,[155,156] or its 5'-phosphate[156] (162), with sodium methyl xanthate ($CH_3OCSSNa$) or phenylisothiocyanate[156,157] gave

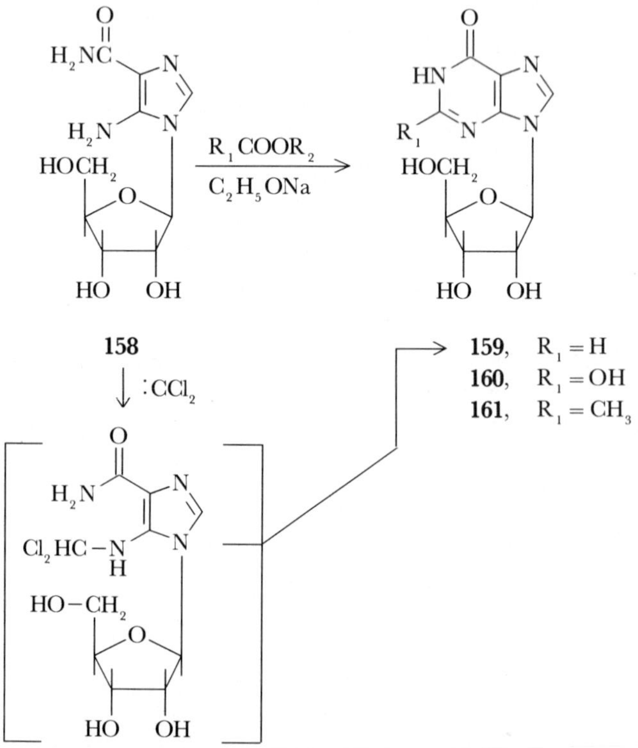

158

159, $R_1 = H$
160, $R_1 = OH$
161, $R_1 = CH_3$

158, R = H
162, R = H$_2$PO$_3$

163, R = H
164, R = H$_2$PO$_3$

2-thioinosine (**163**) and 2-thioinosinic acid (**164**), respectively. A variety of 2-alkyl- and aryl-substituted inosines have also been prepared by the sodium-ethoxide-catalyzed ring closure of AICA-riboside with appropriate esters.[158]

A 2-amino substituent was introduced for the synthesis of guanosine from 5-amino-1-(2,3-*O*-isopropylidine-β-D-ribofuranosyl)imidazole-4-carboxamide (**165**) by a four-step process. Treatment of (**165**) with benzoyl isothiocyanate, followed by methyl iodide, gave the *N*-benzoyl-*S*-methylisothiourea (**166**), and treatment with ammonia followed by hydroxide gave 2',3'-*O* isopropylidine-guanosine (**167**) in 40% overall yield.[159]

The intermediate (**166**) was found to give the cyanoaminonucleoside (**168**) on brief treatment with 0.1 N NaOH. This compound gave a 59% yield of 2',3'-*O*-isopropylidineguanosine (**167**) when treated with concentrated NaOH, or 2',3'-*O*-isopropylidineisoguanosine [(**169**), 23%)] on treatment with dilute NaOH.[160] Without the isopropylidine group, N^5-(*N'*-benzoyl-*S*-methylisothio-carbamoyl)-AICA-riboside (**170**) gave the unusual cyclonucleoside (**171**) when treated with NaOH. In addition, NaOH, NaSH, or NH$_4$OH treatment of (**170**) gave xanthosine (**160**), 2-thioinosine (**163**), and guanosine (**172**), respectively.[161] The formation of nucleosides from (**171**) can be explained by the anhydro ring opening by the nucleophile R$_1$ followed by intramolecular ring annulation.

The 5-amino-1-(β-D-ribofuranosyl)imidazole-4-carbonitrile [AICN-riboside, (**173**)] displays an interesting base-catalyzed ring-annulation reaction to form isoguanosine (**175**) analogues, possibly *via* a ureido intermediate (**174**). This method has been useful in the synthesis of *N*-methylisoguanosines of therapeutic value,[162,163] which occur in marine organisms and are difficult to synthesize by the direct methylation of isoguanosine.[164,165] The reaction[162] of 2',3',5'-tri-*O*-acetyl-AICN-riboside (**176**) with methyl isocyanate in DMF at 100°C gave, after deacetylation, the N^1-methylisoguanosine (**177**). The N^3-methylisoguanosine (**179**) was synthesized[165] by the dilute sodium hydroxide treatment of 5-(*N*-cyano-*N*-methylamino)-2',3',5'-tri-*O*-acetyl-AICN-riboside (**178**). The reaction probably proceeds by the addition of H$_2$O to the N^5-nitrile to form the 5-*N*-methylureido intermediate followed by cyclization. A variety of N^1-alkylisoguanosine[166,167] and 9-β-D-arabinofuranosyl-N^1-methylisoguanine[167]

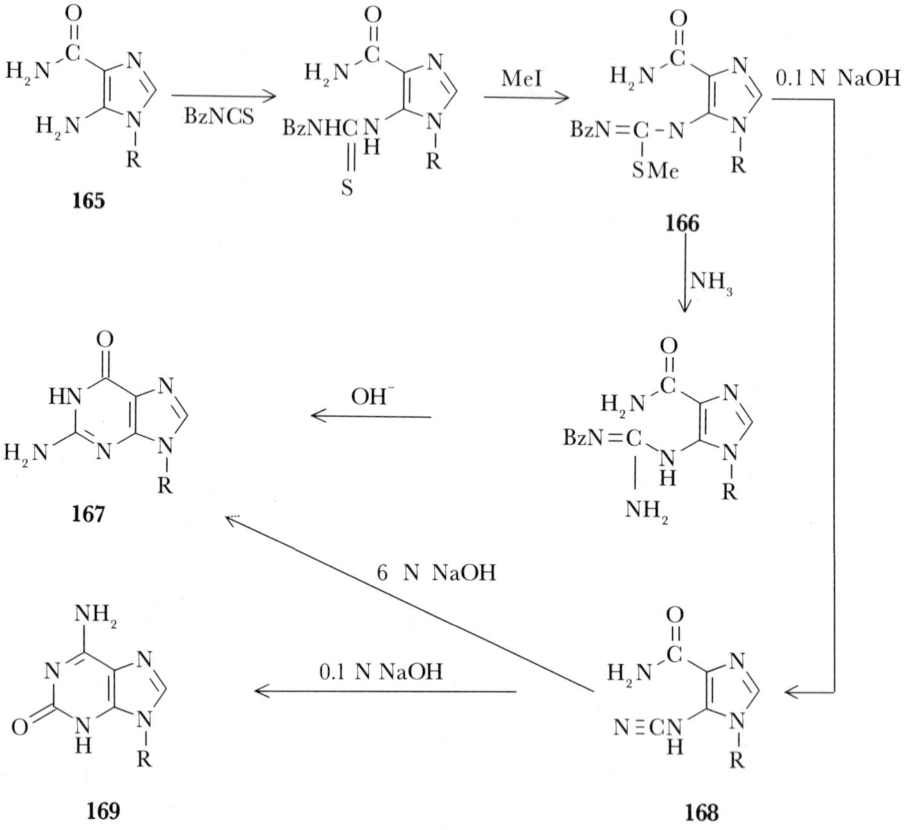

R = 2,3-*O*-isopropylidine-β-D-ribofuranosyl

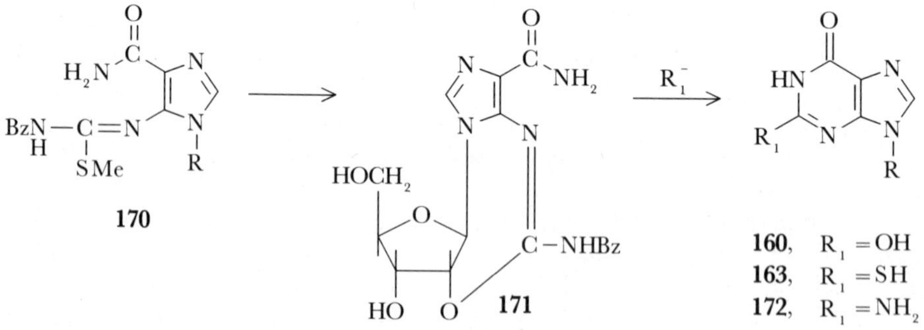

160, R$_1$ = OH
163, R$_1$ = SH
172, R$_1$ = NH$_2$

R = β-D-ribofuranosyl

analogues have been prepared from the corresponding AICN precursors. Treatment of AICN-riboside (**173**) with CS$_2$ followed by methanolic ammonia gave a good yield of 9-β-D-ribofuranosylpurine-2,6-dithione (**180**) *via* the thiazine intermediate.[168,169] The aminocyanoimidazole nucleoside (**173**) was converted to 1-methyl- and 1-phenyl-2-mercaptoadenosine with methyl- and phenylisothiocyanate, respectively, and to an extensive series of 2-alkyl- and 2-aryl-adenosines (**181**) with the appropriate alkyl or aryl nitrile in methanolic ammonia at 180°C.[168] Thus, (**173**) also represents an excellent starting material for the preparation of 2-substituted adenosines.[170]

The 1-(β-D-ribofuranosyl-), 1-(2′-deoxy-β-D-ribofuranosyl-), 1-(β-D-arabino-furanosyl-),[171] and 1-(β-D-ribofuranosyl-cyclic 3′,5′-phosphate) (**182**)[172] deriv-atives of 5-aminoimidazole-4-carboxamidine have been prepared. However, extensive use has been made only of the cyclic nucleotide (**182**),[173,174] which was

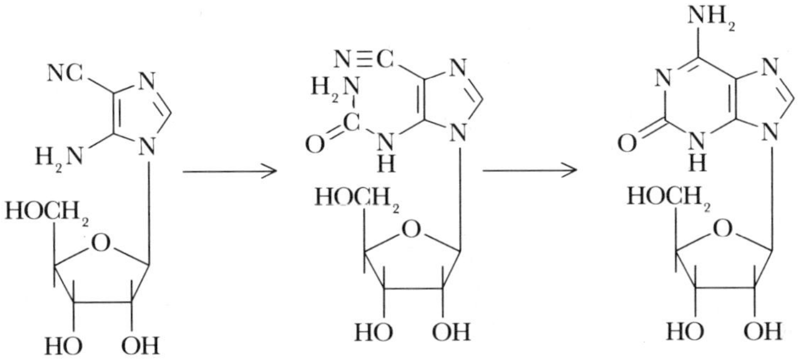

173   174   175

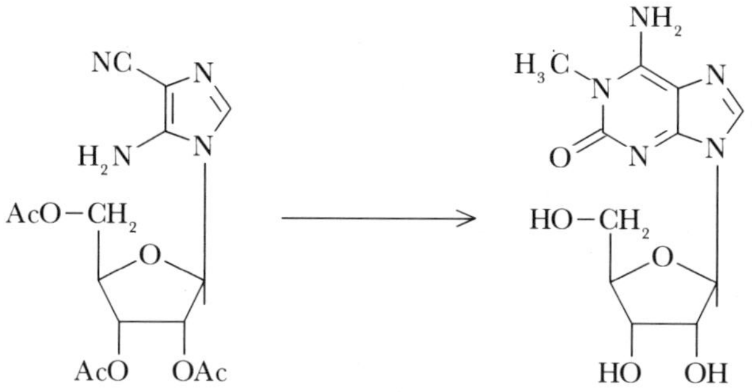

176   177

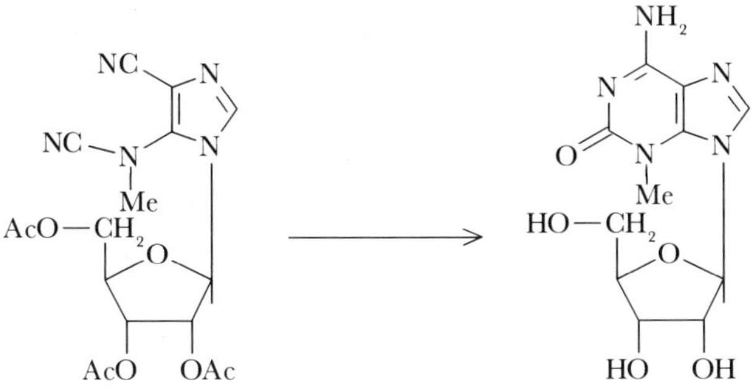

178   179

prepared from adenosine cyclic 3′,5′-phosphate (cAMP). Treatment of this versatile intermediate with carbonyl- or thiocarbonyldiimidazole gave 2-hydroxy-cAMP (183) and 2-mercapto-cAMP (184) directly.[173] The imidazole nucleotide (182) reacted with lower alkyl orthoesters such as triethyl orthoacetate and orthopropionate to give 2-methyl- (185) and 2-ethyl-cAMP (186), respectively, in good yields. With ethyl trifluoroacetate the nucleotide (182) gave 2-trifluoro-methyl-cAMP (187).[173]

A very convenient procedure for ring closure of the imidazole cyclic nucleotide (182) was the treatment with an aldehyde to yield a 2-substituted-2,3-dihydro-cAMP derivative (188). This intermediate was readily dehydrogenated in the presence of air, chloranil, or palladium on carbon to give the aromatized 2-substituted-adenosine cyclic 3′,5′-phosphate (189). By this method, a large number of 2-alkyl- and 2-aryl-cAMP derivatives were prepared[173,174] in good yields.

In an interesting reaction where one reagent served two roles, 2-mercapto-adenosine (192) was prepared directly in excellent yield from 5-amino-1-(β-D-ribofuranosyl)imidazole-4-carboxamidoxime (190) by treatment with CS_2 in pyridine/methanol.[175] It was speculated that 2-mercaptoadenosine-1-oxide (191) was initially formed,[176] followed by a sulfur-mediated deoxygenation of the

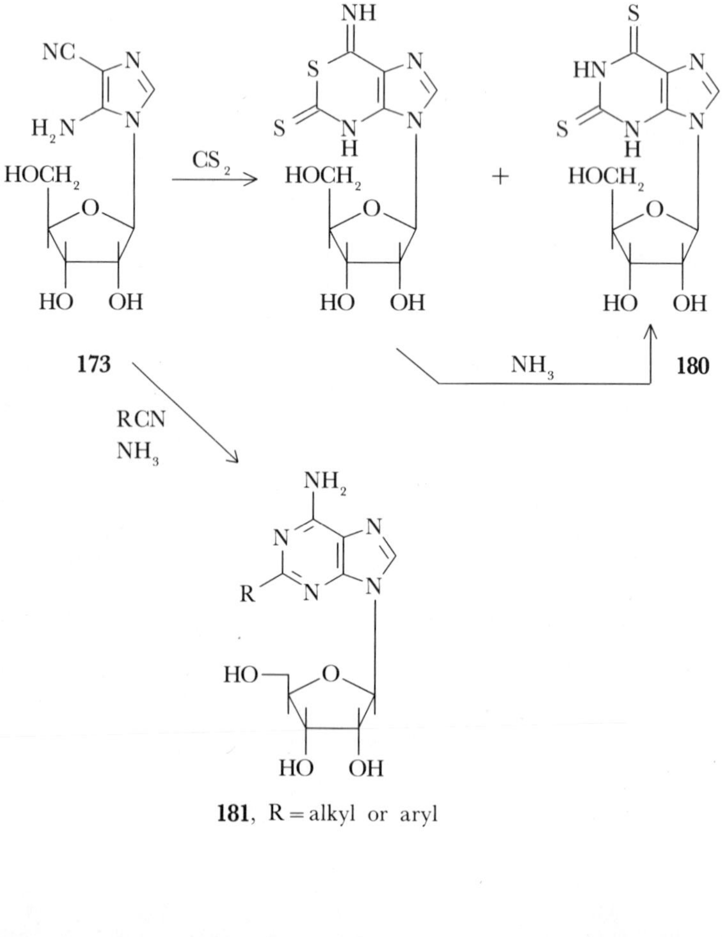

181, R = alkyl or aryl

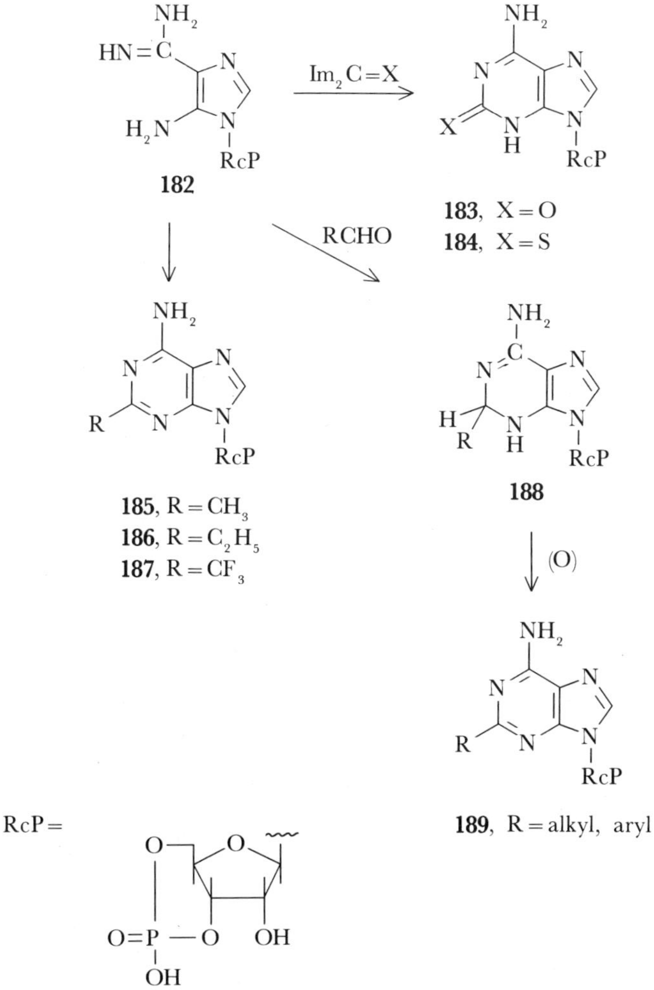

RcP =

182

183, X = O
184, X = S

185, R = CH$_3$
186, R = C$_2$H$_5$
187, R = CF$_3$

188

189, R = alkyl, aryl

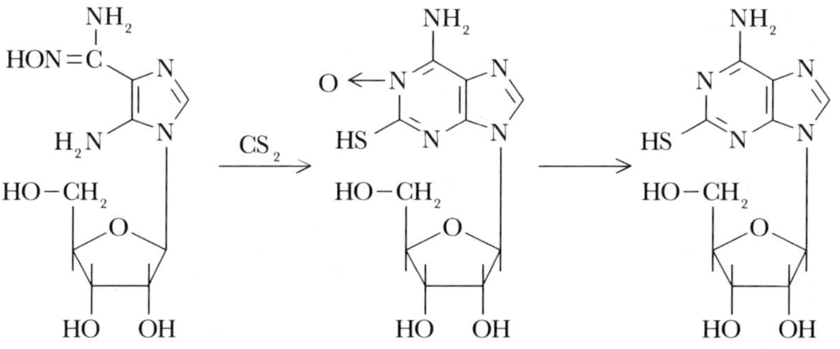

190

191

192

193   **194**

N-oxide. The further amenability of imidazole nucleosides toward ring annulation is shown by the intermolecular coupling of methyl-5-amino-1-(β-D-ribofuranosyl) imidazole-4-carboximidate (**193**) in ammonia at 100°C to provide a unique "binucleoside," 2-[5-amino-1-(β-D-ribofuranosyl)imidazol-4-yl]adenosine (**194**).[177a]

Because of the ready availability of the aminoimidazole carboxamidine nucleosides and nucleotides, the introduction of a one-carbon fragment could give an adenine nucleoside or nucleotide with labeled carbon in the 2 position. Thus, [^{13}C]formaldehyde was incorporated into the imidazole cyclic nucleotide (**182**) to yield 2-[^{13}C]cAMP, and it was also incorporated into the 5'-ribonucleotide (prepared in the same manner as the cyclic nucleotide) to give 2-[^{13}C]AMP.[177b]

2.2.2. Synthesis of Purine Nucleosides and Nucleotides from Pyrimidine Precursors

The starting material for the preparation of a purine nucleoside or nucleotide from a pyrimidine precursor is usually a 4-glycosylamino-5-aminopyrimidine, which may be cyclized to a purine by the introduction of a one-carbon fragment. This route was used extensively some years ago by the Todd group. For instance, the condensation of 4,6-diamino-2-methylthiopyrimidine (**195**) and 5-*O*-benzyl-2,3,4-tri-*O*-acetyl-D-ribose (**196**) gave, after ammonia treatment, 6-amino-2-methylthio-4-(5-*O*-benzyl-D-ribofuranosyl)aminopyrimidine (**197**). This compound was then coupled with diazotized *p*-chloroaniline and, after reduction (and desulfurization), thioformylation, and ring closure, gave adenosine[178] (**128**). A similar approach was used to prepare the 5'-deoxy-5'-(diethylphosphonate) analogue (**198**) of 2-methylthioadenosine by a condensation of diethyl-5-deoxy-2,3,4-tri-*O*-acetyl-D-ribose-5-phosphonate with 4,6-diamino-2-methylthiopyrimidine.[179]

4-Amino-6-(dimethylamino)-5-nitropyrimidine (**199**) was fused with 1-*O*-acetyl-2,3,5-tri-*O*-benzoylribofuranose (**200**) and the resultant tri-*O*-benzoyl-ribofuranosylaminopyrimidine was then reduced and treated with glyoxylic acid hemiacetal. However, in this instance, *N*,*N*-dimethyladenosine (**201**) was isolated as a side product[180] in addition to the expected 4-(dimethylamino)-8-(β-D-ribofuranosyl)pteridin-7(8*H*)-one (**202**). It (**201**) was formed presumably by cyclization of the aldehyde function, decarboxylation, and dehydration. The facile preparation of these 5-nitro-4(or 6)-ribofuranosylaminopyrimidine precursors by the fusion reaction[180,181] could prompt the future preparation of a wide variety of 8-substituted purine nucleosides by this method.

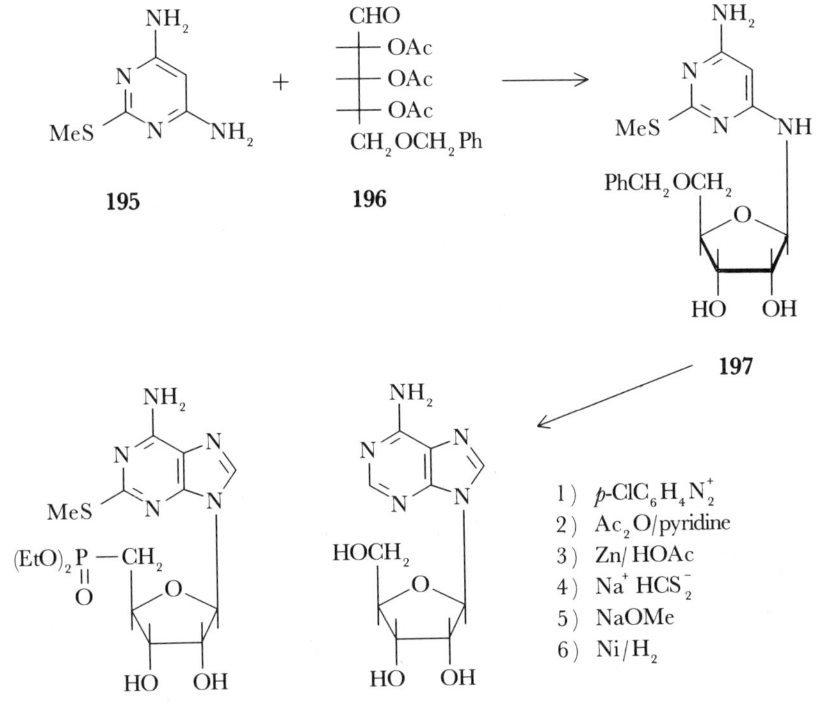

195 + 196 → 197

198   128

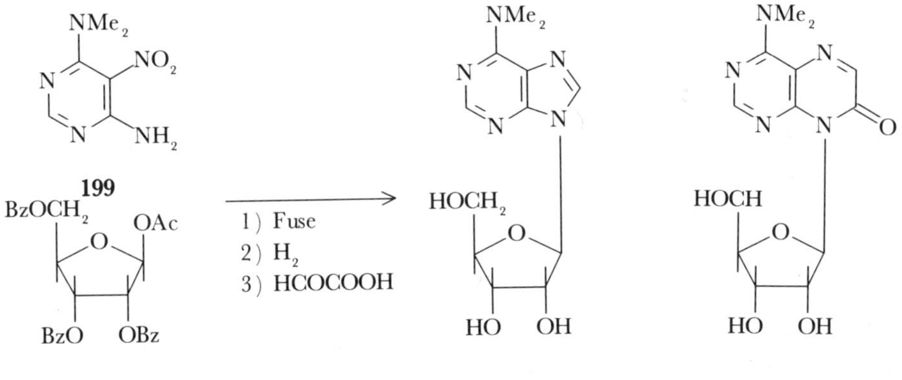

1) p-ClC$_6$H$_4$N$_2^+$
2) Ac$_2$O/pyridine
3) Zn/HOAc
4) Na$^+$HCS$_2^-$
5) NaOMe
6) Ni/H$_2$

199   200   201   202

1) Fuse
2) H$_2$
3) HCOCOOH

Treatment of 5-amino-4,6-dichloropyrimidine (**203**) with ($\pm$)-4α-amino-2α,3β-dihydroxy-1α-cyclopentanemethanol (**204**) gave the pyrimidine derivative (**205**), which was then cyclized with triethyl orthoformate and subsequently aminated to provide the xylofuranosyladenine analogue[182] (**206**). A similar treatment of 2-amino-4,6-dichloropyrimidine with ($\pm$)*trans*-3-amino-*trans*-5-(hydroxymethyl)-*cis*-1,2-cyclopentanediol gave the pyrimidine derivative, which was aminated in the 5 position by diazo coupling and reduction. This compound was then cyclized with triethyl orthoformate to furnish a carbocyclic purine analogue. Hydrolysis of the 6-chloro group gave the racemic carbocyclic analogue of guanosine ($\pm$)-9-[*trans*-2,*trans*-3-dihydroxy-*cis*-4-(hydroxymethyl)cyclopentyl] guanine.[183] Shealy and co-workers had previously used variants of this route to obtain the carbocyclic analogues of adenosine,[184] 2'- and 3'-deoxyadenosine,[185] and other analogues of purine nucleotides and nucleosides.[186]

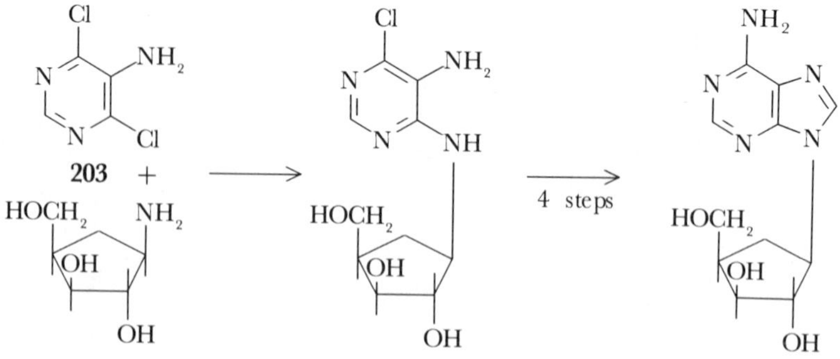

A similar route was used for the preparation of a carbocyclic analogue (**207**) of puromycin lacking the 5'-carbon.[187,188] The enantiomer of this compound, which possessed the same absolute configuration as a D-ribose analogue, was almost as effective as puromycin itself in binding to ribosomes.[189]

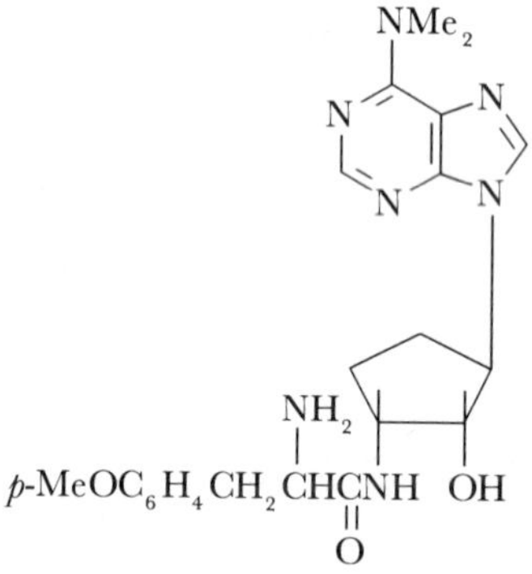

207

Ranganathan[190] followed a unique approach for the preparation of the antiviral nucleoside, 9-(β-D-arabinofuranosyl)adenine (**208**), via a treatment of the bromomercuri salt of the fused oxazolidinethione sugar derivative (**209**) with 4-amino-6-chloro-5-nitropyrimidine (**210**). Reduction of the nitro group, followed by ring closure and desulfurization, gave araA (**208**) in a 39% overall yield.

2.3. Phosphorylation of Purine Nucleosides

The phosphorylation of nucleosides has been studied extensively for some years, and numerous comprehensive reports on the many methods employed have been published. In this section, we focus only on the methods that seem to have received the most use in recent years.

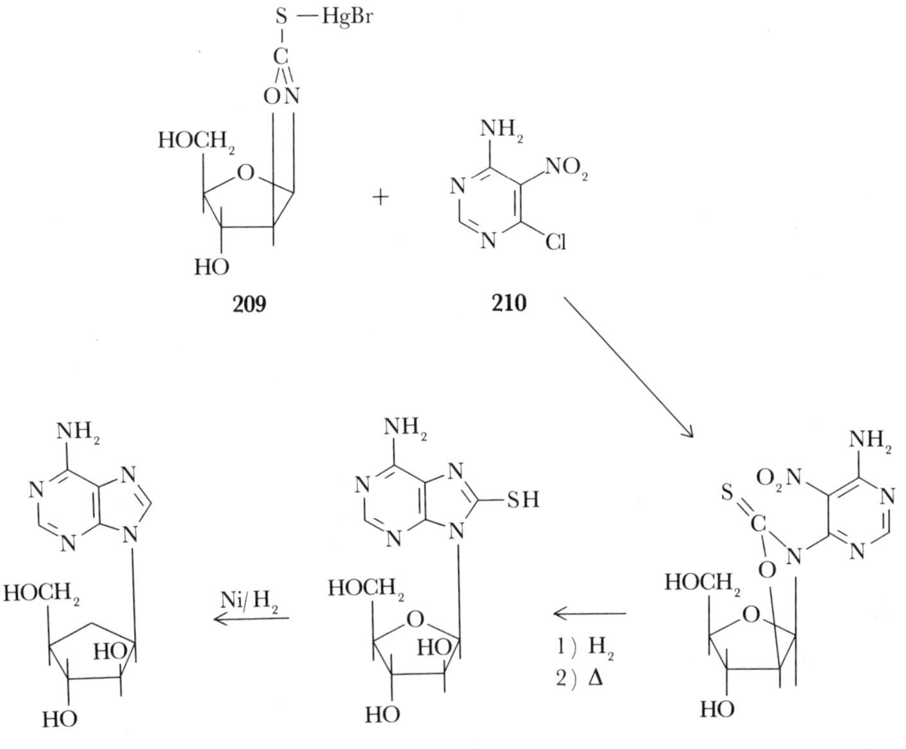

The early work of Khorana had shown that a monoalkyl phosphate (211) reacts with *N,N'*-dicyclohexylcarbodiimide [DCC, (212)] in pyridine to give an activated intermediate (213), which may then react to give an activated cyclic triphosphate (214). Either the phosphate–DCC intermediate, the cyclic triphosphate, or other pyrophosphate intermediates may then react with an alcohol to give a new dialkyl phosphate (215).[191] If R′OH in Scheme II is a blocked nucleoside, with one free hydroxyl group, and the R group in (211) is a readily removable blocking group, then it becomes apparent that this DCC-mediated coupling provides a facile method of nucleotide synthesis.

The early procedures for phosphorylation of nucleosides involved a phosphate bearing benzyl esters, for example, the use of dibenzyl phosphorochloridate[192] and *p*-nitrophenyl esters or the use of tetra(*p*-nitrophenyl)pyrophosphate.[193] For a removal of the blocking groups, hydrogenolysis was required in the former case and strong alkali or an enzymic method in the latter case. One of the most widey used of all phosphorylating reagents, *β*-cyanoethyl phosphate (216), was introduced by Tener in 1961.[194,195] The *β*-cyanoethyl blocking group may be conveniently removed with dilute alkali or ammonia by an α, β elimination of acrylonitrile from the phosphate. The reagent has been found to be very useful, especially in the preparation of nucleotides that may be sensitive to other debocking methods.[196] The search for new phosphorylation methods grew simultaneously with the need for new phosphorylated nucleosides. *O*-Phenylene

Scheme II

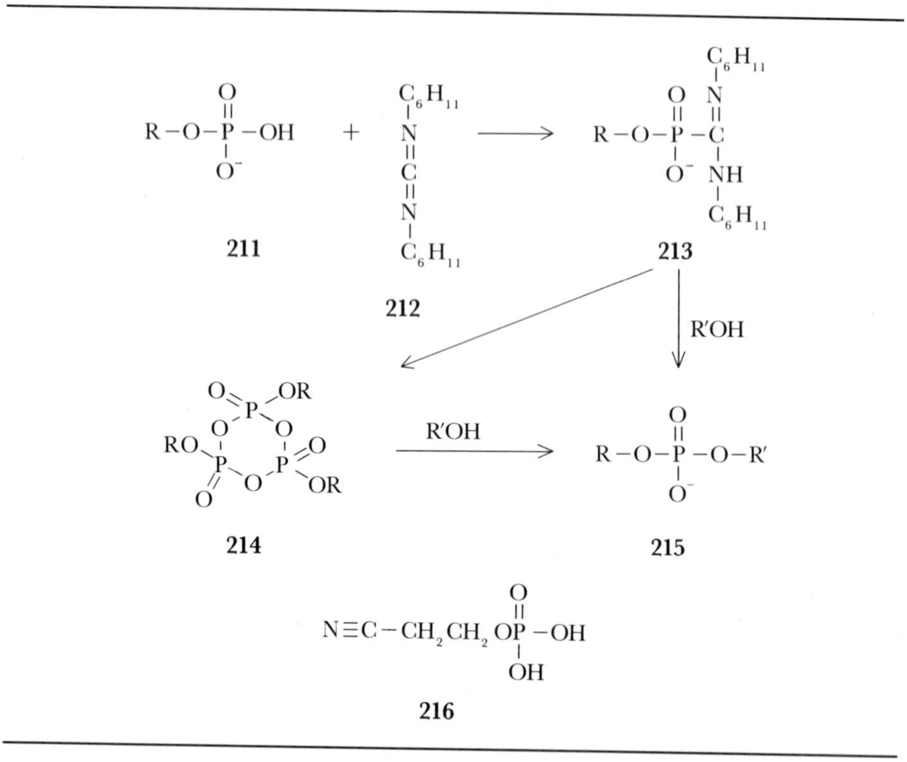

phosphorochloridate **(217)**[197] was used in acetonitrile containing lutidine for the phosphorylation of isopropylidine adenosine **(218)** and its *N,N*-dimethyl derivative. The addition of water at the termination of the reaction cleaved the phosphotriester **(219)** to give the corresponding 5'-*O*-hydroxyphenylphosphate ester **(220)**. After a removal of the isopropylidine group, oxidative cleavage of the *O*-hydroxyphenyl ester by brief treatment with bromine water furnished a good yield of adenosine 5'-phosphate **(221)** and its *N,N*-dimethyl derivative. However, these methods for the synthesis of 5'-phosphates of nucleosides are elaborate. A more convenient procedure was introduced by Yoshikawa and co-workers.[198] They found that phosphoryl chloride (neat) with a very small amount of water added was highly effective in the phosphorylation of isopropylidine nucleosides. In their development studies of this reaction, they found that the isopropylidine nucleoside, in a trialkyl phosphate solvent, was even more smoothly phosphorylated with only a slight excess of $POCl_3$.[199] However, the most crucial observation was the fact that a wide variety of unprotected ribonucleosides were selectively phosphorylated at the 5' position by $POCl_3$ in a trialkyl phosphate, particularly trimethyl or triethyl phosphate. With guanosine, for instance, 85% of 5'-GMP and only 9% bisphosphates (or higher) were obtained.[199] The addition of water, in an amount equivalent to the nucleoside, gave higher yields of the 5'-phosphate, and the competitive phosphorylation of the 2'(or 3')-hydroxyl group was strongly inhibited by *in situ* protection of the *cis*-diol.

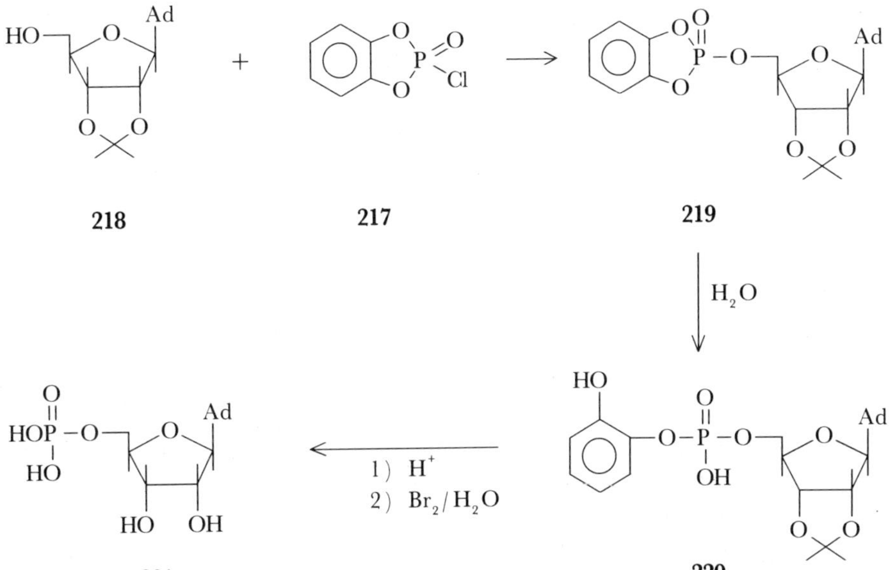

Xanthosine, which had been very difficult to phosphorylate using other techniques, smoothly gave 5'-XMP in 80% yield. Even deoxynucleosides, which do not have a *cis*-2',3'-diol and are quite acid-sensitive, were effective substrates in this reaction. Deoxyinosine gave a 73% yield of deoxy-IMP and only 6% of a bisphosphate.[199] This method for the production of 5'-nucleotides has met with widespread acceptance and is indeed elegant in its simplicity. The method has also been used for the 5'-phosphorylation of a variety of 2-substituted adenosines such as 2-azidoadenosine,[200] 2-methylthioadenosine,[201] and 2-chloroadenosine[202] in satisfactory yields.

Selective phosphorylation of a primary hydroxyl has also been achieved by other reagents. Pyrophosphoryl chloride reacts with a variety of unprotected nucleosides in organic solvents to produce good yields of 5'-nucleotides with only a minimal contamination of higher phosphates.[203] The best solvents were found to be *m*-cresol, *o*-chlorophenol, and a variety of esters and nitriles. The phenols seemed to be the most efficacious, converting inosine into IMP in quantitative yield. This method also provided a selective phosphorylation of deoxynucleosides. A modification of this procedure was introduced by Sowa and Ouchi[204] who used POCl₃ with pyridine and water in acetonitrile to generate an *in situ* species, trichloropyrophosphopyridinium chloride, which then effected a selective 5'-phosphorylation.

Occasionally, the application of phosphoryl chloride has resulted in some anomalous results. 3'-Amino-3'-deoxyadenosine gave an 80% yield of 3'-amino-5'-chloro-3',5'-dideoxyadenosine but when treated with a suspension of triethyl phosphate in POCl₃ for a shorter time at 0°C, the 5'-phosphate was obtained as the predominant product.[205] Further application of Yoshikawa's procedure has been made by using thiophosphoryl chloride for the synthesis of adenosine 5'-phosphorathioate.[206]

2.4. Synthesis of Purine Nucleoside Cyclic Phosphates

2.4.1. Cyclic 3',5'-Phosphates

The role of adenosine cyclic 3',5'-phosphate [cAMP, (**223**)] as an intracellular mediator of numerous hormones and other processes has been elucidated dramatically since its discovery in 1957.[207] This resulted in Earl Sutherland receiving the Nobel Prize in medicine in 1971. There is now a journal devoted to the study of biological events mediated by cyclic nucleotides. The chemistry of all types of cyclic esters of phosphoric acid and phosphinic acid has been reviewed.[208] Only the points pertinent to cyclic purine nucleotides are presented here.

The first synthesis of cAMP was accomplished by the barium hydroxide treatment of adenosine 5'-triphosphate.[209,210] This reaction most likely proceeds by the attack of an anion formed at the 3'-hydroxyl group of ATP on the α-phosphate, liberating pyrophosphate and forming the cyclic diester. In a systematic investigation of the chemistry of the cyclic nucleotides, Smith *et al.*[211] prepared the cyclic 3',5'-phosphates of adenosine and guanosine by treating adenosine 5'-phosphate (**221**) or guanosine 5'-phosphate with DCC in pyridine. High dilution was used in this process to avoid intermolecular esterification. The 2'-deoxy derivatives of cAMP, cGMP, and cIMP were later prepared by the same method.[212] This method has been the one most widely used for the preparation of nucleoside cyclic 3',5'-phosphates. An alternative procedure for cyclization was found *via* the treatment of an adenosine 5'-phosphate *p*-nitrophenyl ester (**222**) with potassium *tert*-butoxide.[213] This method has been used for the preparation of other cyclic nucleotides but also appears to require high dilution.

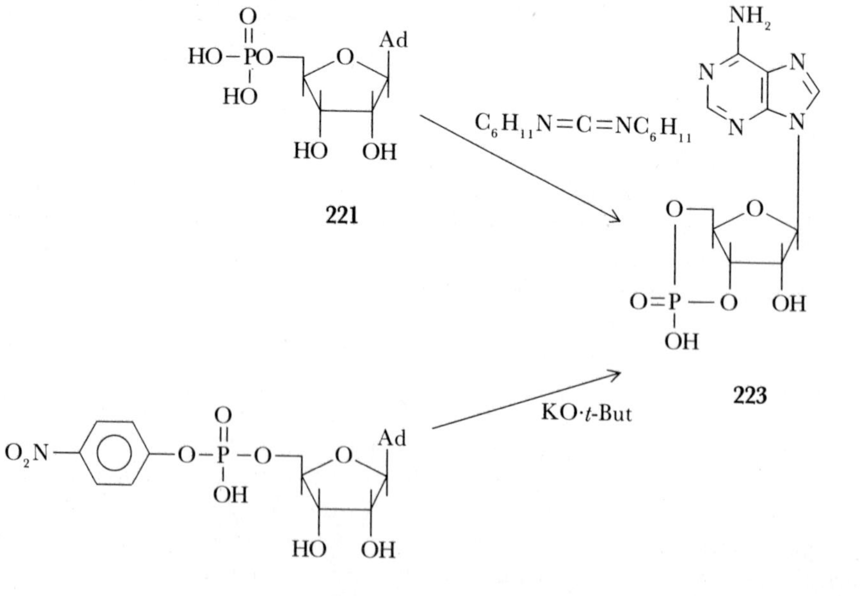

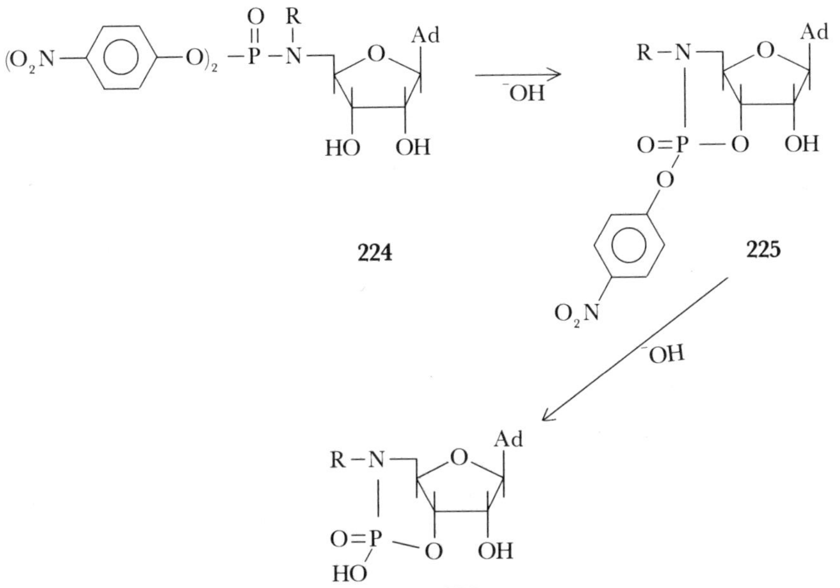

224   225

226

R = alkyl

A number of 5'-amino-5'-deoxyadenosine cyclic 3',5'-phosphoramidates have been prepared by variations of the latter method. Treatment of a series of bis(p-nitrophenyl)5'-alkylamine-5'-deoxyadenosine 5'-phosphates (224) with pyridine/ammonia/water or NaOH gave the cyclic phosphoramidates (226). This reaction probably occurred *via* the cyclic phosphoramidate p-nitrophenyl ester, which then hydrolyzed to yield (226).[214,215] The cyclic phosphoramidate with no substitutent on nitrogen, prepared by a removal of one of the p-nitrophenyl esters under mild conditions followed by cyclization with potassium *tert*-butoxide in DMSO, was found to be unstable at pH < 7 in water.[215] The cyclic diester amide (225) is formed stereospecifically. The *endo*-ester is the predominate product of the treatment of (224) in acetone with NaOH and then isomerizes to the *exo*-ester with sodium p-nitrophenolate in acetone.[216]

When bis(p-nitrophenyl)thiophosphoryl chloride was used as the phosphorylating agent for the 5'-methyl(or octyl)amino-5'-deoxyadenosines, the corresponding thiophosphonyl diester amides (227) were obtained. Treatment

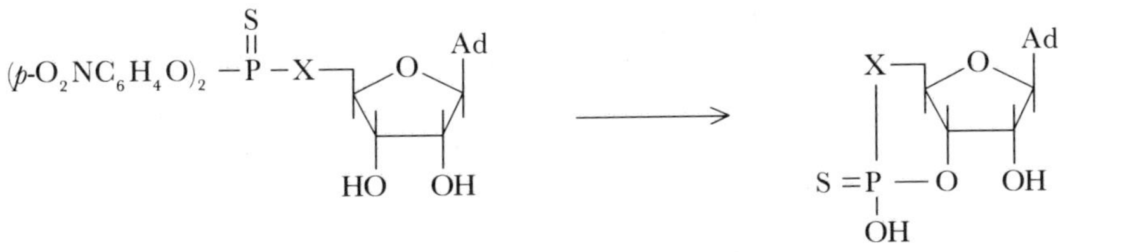

227,   X = N−CH$_3$ or C$_8$H$_{17}$

229,   X = O

228,   X = N−CH$_3$ or C$_8$H$_{17}$

230,   X = O

of (227) with base gave the 5'-methyl(or octyl)amino-5'-deoxyadenosine cyclic 3',5'-thiophosphoramidates (228).[217] The preparation of adenosine cyclic 3',5'-phosphorothioate (230) *per se* was accomplished in low yield from adenosine 5'-O,O-bis(*p*-nitrophenyl)phosphorothioate (229).[218]

The 5'-oxygen of cAMP has also been replaced by sulfur and carbon. 5'-Deoxy-5'-thioadenosine 5'-phosphorothiolate gave, upon treatment with DCC–pyridine at reflux, 5'-deoxy-5'-thioadenosine cyclic 3',5'-phosphorothiolate, albeit in low yield.[219] These conditions were also used to cyclize 5'-deoxy-5'-(dihydroxyphosphinylmethyl)adenosine.[220] Interestingly, in the latter reaction, the cyclization proceeded much more smoothly than in the case of 5'-AMP and the presence of 5% water in the pyridine did not adversely affect the high yield.

An even more facile cyclization was found to occur when 3'-amino(or benzylamino)-3'-deoxyadenosine 5'-phosphate (231) was the substrate.[221] In this case, 3'-amino(or benzylamino)-3'-deoxyadenosine cyclic 3',5'-phosphoramidate (232) was obtained in excellent yield when the water-soluble carbodiimide 1-ethyl-3-dimethyl-aminopropylcarbodiimide hydrochloride was used in water at pH 7.2–7.5 at 37°C. The increased nucleophilicity of the 3'-amino group, as compared to hydroxyl, seemed to be responsible for the ease of this cyclization. When the corresponding 2',5'-diphosphate is treated with the same reagent, the cyclic 3',5'-phosphoramidate is formed, indicating that the vicinal 3'-amino and 2'-phosphate cyclizations may be competing, but that the reaction is apparently reversible. These 3'-phosphoramides were also found to be unstable at pH < 7.[221]

R = H or benzyl

One very unusual nucleotide has been prepared by the carbodiimide method. Treatment of cordycepin (3'-deoxyadenosine)5'-phosphate (233) with DDC and pyridine gave a 6% yield of 3'-deoxyadenosine cyclic 2',5'-phosphate (234),[222] which should be a very strained compound. The same compound was also prepared by a cyclization of 8,3'-anhydro-8-mercaptoadenosine 5'-phosphate (235), followed by desulfurization.[222]

One very different alternative[223] to the above methods is exemplified by the phosphorylation of unprotected nucleosides (236) with trichloromethyl-phosphonodichloridate in triethyl phosphate. This yields the nucleoside 5'-trichloroethylphosphonates (237), which are then smoothly cyclized to the corresponding cyclic nucleotides (238) by treatment with potassium *tert*-butoxide in DMF. This procedure, when applicable, has some advantages since the nucleoside substrate does not have to be blocked, the difficult-to-remove by-products are not formed, and the reaction does not require a highly dilute medium. The mechanism would seem to be *via* an attack of the 3'-oxy anion on phosphorus, eliminating Cl_3C^-. Since a cleavage of the Cl_3C-P bond has been shown to be catalyzed by fluoride ion,[224] the substitution of KF for potassium *tert*-butoxide might provide an even milder method for the synthesis of cyclic nucleotides. However, the 3'SH adenosine analogues of (**237**) could not be cyclized when treated with potassium *tert*-butoxide.[225]

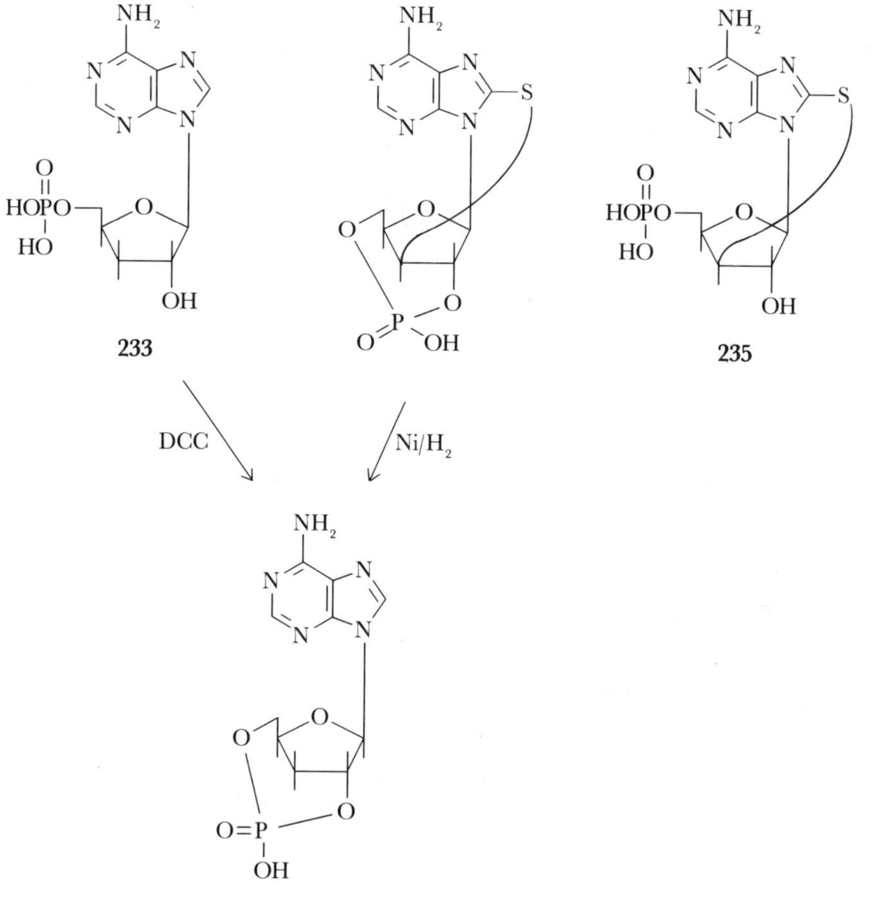

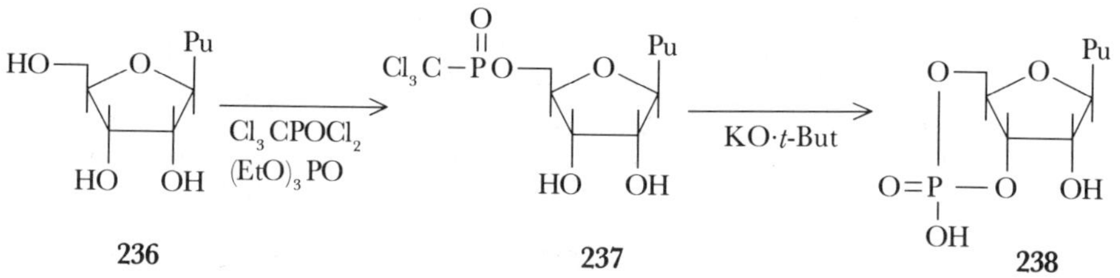

2.4.2. Cyclic 2′,3′-Phosphates

Ribonucleoside 2′,3′-cyclic phosphates are useful intermediates for chemical and enzymatic synthesis of oligoribonucleotides and are generally available by the cyclization of the corresponding 2′(3′)-phosphates.[226] The purine ribonucleosides have also been converted to 2′,3′-cyclic phosphates *via* an oxidative cyclization of the corresponding 2′(3′)-phosphite[227] and by direct 2′,3′-cyclic phosphorylation.[228]

3. Reactions of Purine Nucleosides and Nucleotides

The purine and carbohydrate moieties of nucleosides and nucleotides with various reactive centers, independently or simultaneously, play a pivotal role in the synthesis of modified nucleosides and nucleotides of biological and biochemical importance. Among the numerous possible modifications, some are conveniently grouped together as follows.

3.1. Reactions of the Carbohydrate Portion of Purine Nucleosides and Nucleotides

3.1.1. Alkylation, Alkylidination, and Acylation

The hydroxyl groups of the carbohydrate moiety of nucleosides and nucleotides undergo esterification and etherification reactions. The products have provided a "prodrug" for biologically active parent compounds and the blocked nucleosides have also been used for further chemical transformations. Various esterified and etherified nucleosides have been found in nature. Since selective blocking agents for oligonucleotide synthesis are dealt with in Chapter 3, many of those agents, like the trityl group, are not discussed in detail here.

The discovery of 2′-*O*-methyladenosine (**239**) as a naturally occurring constituent of RNA[229] stimulated interest in the preparation of *O*-alkylated nucleosides and nucleotides. The first direct preparation of (**239**) was accomplished by the treatment of adenosine (**128**) with diazomethane in aqueous dimethoxyethane.[230] This reaction was substantially selective for the alkylation of the 2′-hydroxyl. However, methylation of guanosine with diazomethane or methyl iodide in neutral solution gave 7-methylguanosine[231,232] (see Section 3.2.1). On the other hand, methylation of 2-amino-6-chloro-9-(β-D-ribofuranosyl)purine with diazomethane in aqueous dimethoxyethane did not give any nuclear alkylation but did furnish the 2′-*O*-methyl derivative. This derivative was readily converted into 2′-*O*-methylguanosine.[233] The selectivity of the reaction for the 2′-hydroxyl is remarkable and was further confirmed by the predominant 2′-*O*-methylation of adenosine using methyl iodide under alkaline conditions.[234]

The use of stannous chloride dihydrate as catalyst has provided a method for a high yield, direct methylation of nucleosides in the 2′ and 3′ positions, with the exception that the 3′-isomer predominated. Adenosine (**128**) in methanol gave, upon treatment with diazomethane and a catalytic amount of $SnCl_2 \cdot 2H_2O$, a nearly quantitative yield of a $2:3$ mixture of 2'- and 3'-O-methyl-adenosine,[235–237] **(239)** and **(240)**. Direct sugar hydroxyl methylation of guanosine was also accomplished by this method using DMF as a solvent in 50% total $(2' + 3')$ yield,[236] thus avoiding a methylation of the heterocycle.

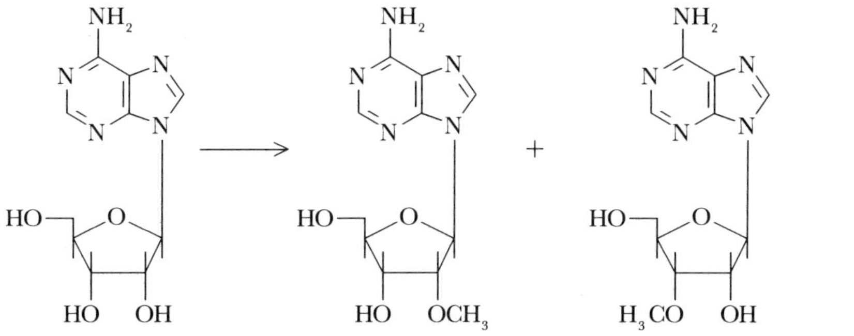

128   239   240

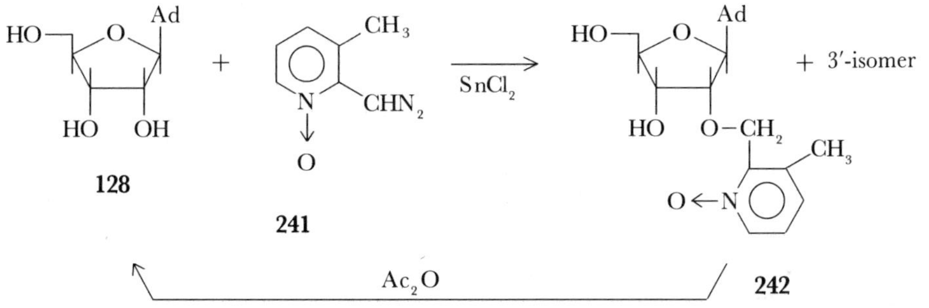

128

241

242

Stannous chloride as a catalyst has also been used with phenyl-diazomethane[238] in the preparation of 2'- and 3'-O-benzyl derivatives of adenosine, inosine, and guanosine. The photolabile *o*-nitrobenzyl group was introduced directly onto the 2'- and 3'-hydroxyl of adenosine and inosine with $SnCl_2$/*o*-nitrophenyldiazomethane.[239] The 3-methyl-2-picolyl-1-oxide protecting group, removable by excess acetic anhydride, is stable under conventional tritylation, benzoylation, and phosphorylation reaction conditions.[240] It was selectively introduced into the 2' **(242)** and 3' positions of adenosine by treatment with $SnCl_2$/1-oxido-3-methyl-2-pyridyldiazomethane **(241)**.[240]

The alkylation of nucleosides under basic conditions requires that more attention be given to the type of blocking groups on the purine moiety and the sugar. It was found that $N,5'$-O-ditrityladenosine smoothly underwent a reaction with benzyl bromide in the presence of potassium hydroxide to give an 80% yield of a mixture of 2'-O- and 3'-O-benzyl-$N,5'$-O'ditrityladenosine in a $4:1$ ratio.[241]

Selective alkylation is not required if the nucleoside is suitably blocked. One such readily available blocked adenosine derivative is adenosine cyclic 3',5'-phosphate [cAMP, **(223)**]. Treatment of cAMP with aqueous NaOH and either methyl iodide or ethyl iodide in DMF gave a 50–60% yield of the 2'-O-alkyl cyclic nucleotide (243).[242] These compounds were readily converted to the 2'-O-alkyladenosine 5'-phosphate (244) by treatment with cyclic nucleotide phosphodiesterase.

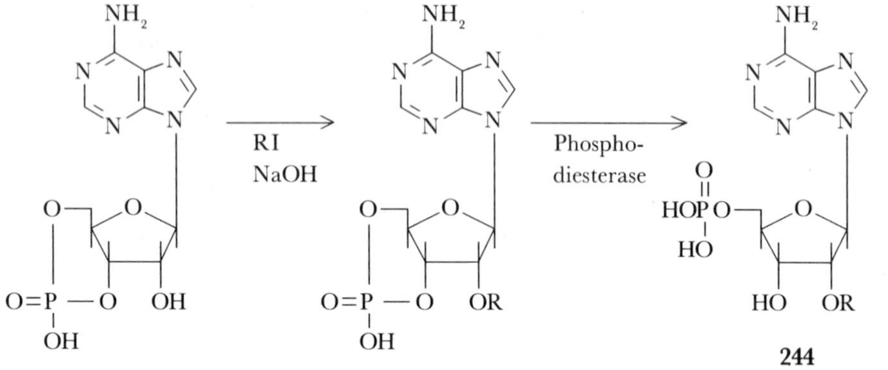

Another class of ethers that has recently experienced a surge of interest are the trialkylsilyl ethers.[243–245] These blocking groups have been used by McCloskey and co-workers for the synthesis of pertrimethylsilylated nucleosides and nucleotides, which are more volatile derivatives that can be used in gas chromatography and mass spectrometry.[246,247] The O-trialkylsilylated nucleosides have been prepared by several methods. A representative example is the preparation of 6-chloro-2-methyl-9-(2,3,5-*tris*-O-(trimethylsilyl)-β-D-ribofuranosyl-purine (246) in quantitative yield by the treatment of 6-chloro-2-methyl-9-(β-D-ribofuranosyl)purine (245) with chlorotrimethylsilane in anhydrous pyridine. The trimethylsilyl blocking groups are stable to thin-layer chromatography on silica gel but are hydrolyzed in boiling sodium carbonate solution. This has proved to be very useful in the case where a readily prepared, base-stable hydroxyl blocking group was required.[248]

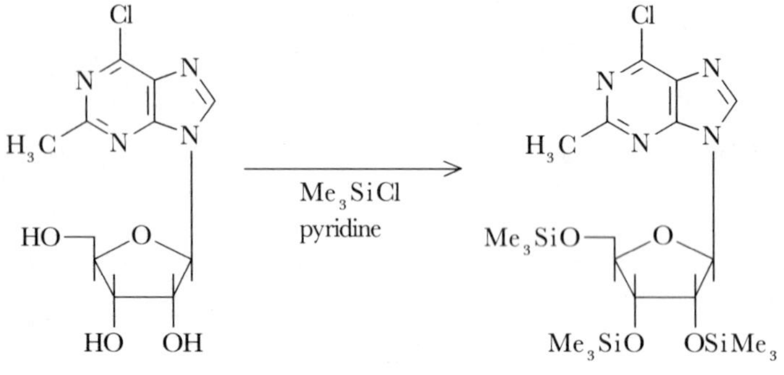

A more complete account on the use of a variety of trialkylsilyl blocking groups has been given,[243,249] with model studies for synthesis and reactivity being performed on purine and pyrimidine nucleosides. These authors found, for instance, that *tert*-butyldimethylsilyl groups required 5 hr in 80% acetic acid/water for hydrolysis from primary hydroxyls and 25 hr for hydrolysis from secondary hydroxyls at room temperature.[249] The silyl groups were stable to 15% ethanolic ammonium hydroxide but were conveniently removed by tetra-*n*-butylammonium fluoride in tetrahydrofuran in 30 min at room temperature.[249,250]

Studies on the selective blocking of ribo- and deoxyribonucleosides with *tert*-butyldimethylsilyl chloride have been performed.[251] The synthesis of 5'-hydroxyl blocked 2'-deoxyadenosine has been reported in an 80% yield using *tert*-butyldimethylsilyl chloride in DMF with imidazole as catalyst.[252] In addition to *tert*-butylammonium fluoride, this group can readily be cleaved at 100°C in 15 min by 80% acetic acid. Although silyl groups are now widey accepted for blocking the sugar hydroxyl groups of nucleosides, care should be exercised in their use for structural determinations since the silyl derivatives of ribonucleosides undergo transmigration by a first-order equilibration reaction, particularly in methanol in the presence of base.[243,253]

A convenient bifunctional silyl group, tetra(isopropyldisiloxane)-1,3-diyl (TIPDSi),[254] allows a selective and simultaneous 3',5'-hydroxyl protection of nucleosides in high yields. Thus, 1,3-dichloro-1,1,3,3-tetra(isopropyldisiloxane) (TIPDSiCl$_2$) reacts with nucleosides (**247**) initially at the 5' position and then cyclizes with the 3' position to yield the corresponding 3',5'-*O*-TIPDSi-diyl-nucleoside derivatives (**248**). The TIPDSi group has also been used for 3',5'-protection of a variety of nucleosides[255] and nucleotides.[256] However, recent reports indicate that a TIPDSi group can migrate from the 3',5' to the 2',3' position.[256]

Some other bifunctionally active carbonyl compounds such as diphenyl carbonate,[257] phosgene,[258] and carbonyldiimidazole[259] react with *cis*-hydroxyls of ribonucleosides to form nucleoside 2',3'-cyclic carbonates (**249**). These

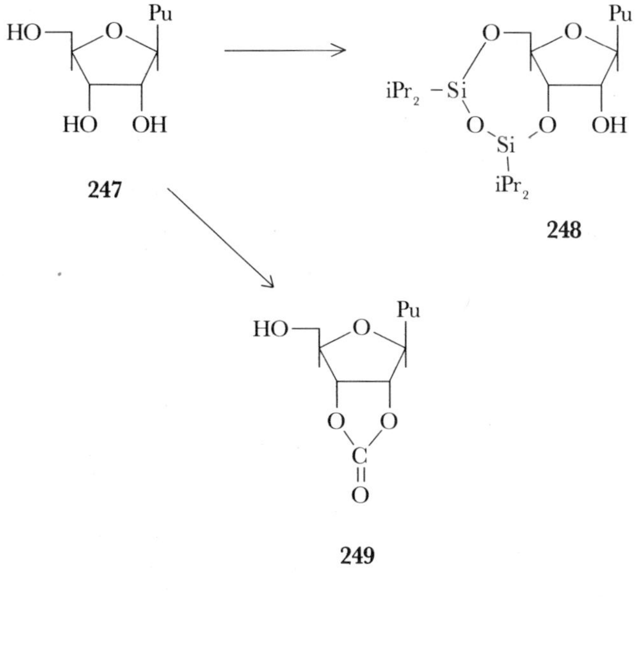

247

248

249

derivatives then serve as useful intermediates for modifications at the 5' position. The cyclic carbonate group is base labile and thus different from other frequently used 2',3' blocking groups, which are cleaved under acidic medium as described below.

The 2',3'-*cis*-hydroxyl groups of ribonucleosides are also protected by forming the 2',3'-*O*-alkylidene,[260-262] arylidene,[261-264] or orthoester[265-267] derivatives, which are all acid labile and have found considerable use in nucleoside chemistry.

The orthoester, or 2',3'-*O*-alkoxyalkylidine group, was introduced by Reese and co-workers primarily for the purposes of introducing an acyl group selectively into either the 2'- or 3'-hydroxyl. However, it has seen much wider use as a preparative tool. Adenosine (128) was treated with trimethyl orthoacetate and *o*-toluenesulfonic acid to give a 67% yield of 2',3'-*O*-methoxyethylidineadenosine (250) as a mixture of diastereomers.[268] This blocking group is remarkably unstable to acid. At pH 3, the hydrolysis of (250) is immeasurably fast while at pH 5.6 the half-life is 83 min.[268] This is about 1000 times faster than the corresponding 2',3'-*O*-methoxymethylidine derivative and the product of hydrolysis is 3'(or 2')-*O*-acetyladenosine.[269] The 3'-acetyl isomer is the more stable isomer.[270,271] If the material is allowed to crystallize slowly, the 3'-acetyl isomer is obtained predominantly owing to an acyl migration. Thus, (250) gave an 88% yield of 3'-*O*-acetyladenosine (251) after a very brief treatment with 60% acetic acid.[268]

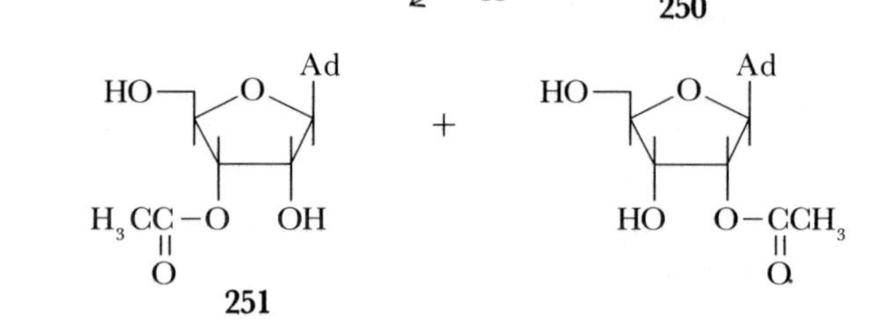

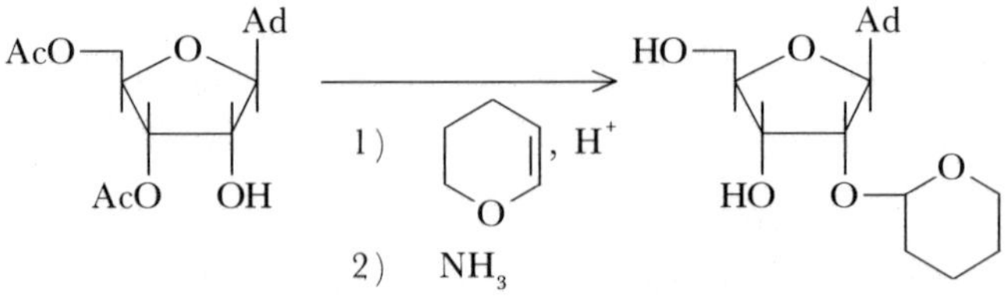

The 2-tetrahydropyranyl group has seen wide use as an acid-labile, base-stable, nonmigrating blocking group for a hydroxyl group. This group was introduced by Khorana and co-workers and found wide use in oligonucleotide synthesis.[272] Treatment of 3',5'-di-*O*-acetyladenosine (252) with dihydropyran and *p*-toluenesulfonic acid gave, after ammonia treatment to remove the acetyl groups, a 75% yield of the diastereomers of 2'-*O*-tetrahydropyran-2-yladenosine (253) in a 3 : 1 isolated ratio.[273]

The 4-methoxytetrahydropyran-2-yl functionality is an alternative to the 2'-*O*-tetrahydropyranyl group. This group was used to block the 2'-hydroxyl of 3',5'-di-*O*-acylnucleosides by treatment with 4-methoxy-5,6-dihydro-2*H*-pyran under acid catalysis.[274-276] These groups are rapidly removed at pH 4, being about 1000 times less stable than the tetrahydropyranyl group,[274] but they introduce no additional center of asymmetry into the blocked nucleosides.

Phenoxythiocarbonyl chloride (phenyl chlorothionocarbonate) has been used for the thioacylation of hindered secondary alcohols such as the 2'-OH of (248), in the presence of 4-dimethylaminopyridine catalyst. This has provided the corresponding 2'-*O*-phenoxythiocarbonyl (PTC) derivatives.[255] The 2'-*O*-PTC nucleoside derivatives undergo a free radical-initiated reductive homolytic deoxygenation with tri-*n*-butyltin hydride to provide the corresponding 2'-deoxyribonucleosides in excellent yields.[255,277]

Charged acylating reagents such as *N*-acylpyridinium chlorides, in an aprotic, nonpolar solvent such as *N*,*N*-dimethylformamide, have been found to be effective for the synthesis of 5'-*O*-acylated nucleosides (255) of 9-*β*-D-(arabinofuranosyl)adenine (254) with only minor peracylation problems.[278] However, the specific synthesis of 2'-*O*-acylated derivatives of (254) requires a specific blocking of the 3' and 5' positions.[279] The very selective reactivity of 2'-*O*-acyl groups of fuly acylated ribonucleosides toward hydrazine hydrate in an appropriate solvent results in a regioselective 2'-*O*-deacylation.[280]

The adenosine 5'-hydroxyl group reacts selectively with tri-*n*-butyl-2-nitrophenylselenophosphonium cyanide, apparently formed *in situ* by the reaction of 2-nitrophenylselenocyanate and tri-*n*-butylphosphine, to provide 5'-(2-nitrophenylseleno)-5'-deoxyadenosine (257) *via* an oxophosphonium salt (256).[281] The selenide (257) undergoes oxidation with hydrogen peroxide, followed by *syn* elimination of the selenoxide group under mild basic conditions, to furnish an exocyclic, 4',5' double bond in adenosine.[281]

254   255

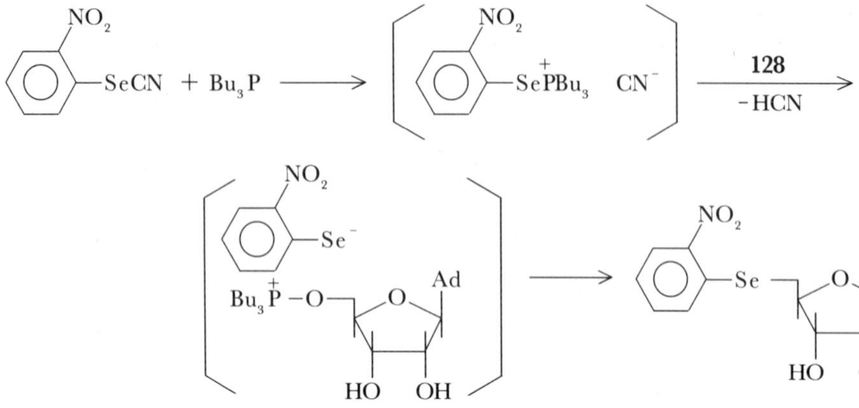

256   257

3.1.2. Oxidation of the Carbohydrate and Reactions of Oxidized Sugars in Nucleosides

Oxidation of the carbohydrate moiety has served as a convenient and versatile method for functionalizing nucleosides and nucleotides. Early work on the oxidation of nucleosides gave 5'-carboxylic acids and cleavage products of the vicinal *cis*-hydroxyls. The most useful synthetic intermediates, however, have been the aldehydes and ketones, which were very difficult to prepare until about 1963.

The most versatile nucleoside oxidation evolved from an experiment by Pfitzner and Moffatt.[282] In an effort to eliminate solubility problems inherent in *N,N'*-dicyclohexylcarbodiimide (DCC)-mediated coupling reactions of nucleotides in pyridine, they used dimethylsulfoxide (DMSO) as the solvent and observed a sulfide-like smell and a disappearance of thymidine 5'-phosphate from the reaction mixture. A closer investigation of the reaction conditions revealed that DMSO–DCC in the presence of an acid catalyst effected the oxidation of alcohols to aldehydes in good yield under quite mild conditions.[282–284] Thus, 2',3'-*O*-isopropylidineadenosine (**218**) in DMSO–DCC and dichloroacetic acid gave the 5'-aldehyde (**258**) in good yield.[282] This method represented the first readily reproducible oxidation of nucleoside hydroxyls to the corresponding aldehydes and ketones. In an improved variant of this reaction, the *N*6-benzoyl-2',3'-*O*-isopropylidineadenosine (**259**) was used and the aldehyde was isolated as its crystalline 1,3-diphenylimidazolidine derivative (**260**) (prepared by the addition of 1,2-dianilinoethane to the crude aldehyde) in 69% yield. The protecting group for the aldehyde was readily removed under mild acidic conditions, and the hydrate of the pure aldehyde (**261**) was obtained in 79% yield.[285] The free aldehyde was generated by an azeotropic removal of water.

The mechanism of this reaction was envisaged as proceeding *via* the initial acid-catalyzed formation of an adduct of DMSO and DCC, followed by attack of an alcohol on sulfur to give an alkoxysulfonium salt. This compound then rearranged by a cyclic mechanism, after loss of a proton, to give dimethyl sulfide and the carbonyl compound (Scheme III).[283]

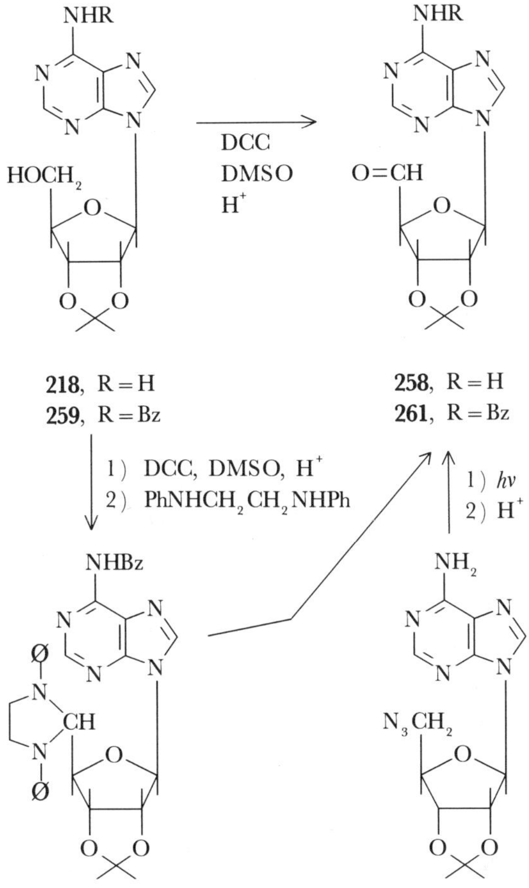
NHR
HOCH2
DCC
DMSO
H+
NHR
O=CH
1) DCC, DMSO, H+
2) PhNHCH2CH2NHPh
1) hν
2) H+
NHBz
CH
NH2
N3CH2

218, R = H
259, R = Bz

258, R = H
261, R = Bz

260

262

Acetic anhydride has been used as the dehydrating agent in place of DCC in this reaction.[286–288] These reaction conditions give more of a by-product, the methylthiomethyl ether **(264)** of the alcohol, than was obtained when DCC was used.[287–291] It was proposed[290] that this product was formed by a decomposition of the acetylated DMSO **(263)** to give a reactive sulfonium cation **(265)**, which then gave the methylthiomethyl ether **(264)** upon attack by the hydroxyl group of the alcohol.

Another method for the preparation of 5′-aldehydes of purine nucleosides is by photolysis of the corresponding 5′-deoxy-5′-azido derivatives. Irradiation (Corex filter, $\lambda > 260$ nm) of 5′-azido-5′-deoxy-2′,3′-*O*-isopropylidineadenosine **(262)** in benzene at room temperature gave a presumed aldimine intermediate, resulting from a loss of nitrogen and subsequent hydrogen migration in the intermediate nitrene. Very mild acid hydrolysis liberated the free 5′-aldehyde of isopropyldineadenosine in good yield.[292] The value of 1,2-dianilinoethane[285] should be emphasized as a "handle" for purification of these aldehydes, which in the free form may undergo many undesirable side reactions.

Scheme III

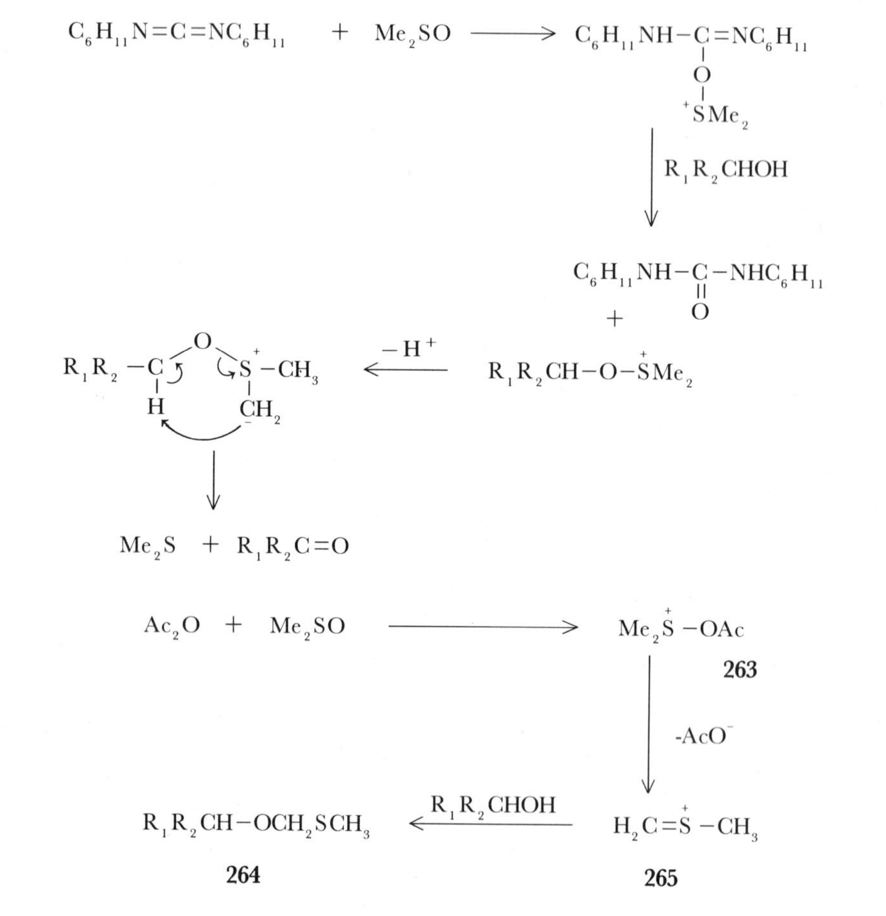

$$C_6H_{11}N=C=NC_6H_{11} \quad + \quad Me_2SO \longrightarrow C_6H_{11}NH-\underset{\underset{\overset{|}{{}^+SMe_2}}{\overset{|}{O}}}{C}=NC_6H_{11}$$

$$\Big\downarrow R_1R_2CHOH$$

$$C_6H_{11}NH-\underset{\overset{\|}{O}}{C}-NHC_6H_{11}$$
$$+$$

$$R_1R_2-\underset{H}{\overset{O}{C}}\overset{+}{\underset{CH_2}{S}}-CH_3 \quad \xleftarrow{-H^+} \quad R_1R_2CH-O-\overset{+}{S}Me_2$$

$$\Big\downarrow$$

$$Me_2S \quad + \quad R_1R_2C=O$$

$$Ac_2O \quad + \quad Me_2SO \longrightarrow Me_2\overset{+}{S}-OAc$$
$$\textbf{263}$$

$$\Big\downarrow {-AcO^-}$$

$$R_1R_2CH-OCH_2SCH_3 \quad \xleftarrow{R_1R_2CHOH} \quad H_2C=\overset{+}{S}-CH_3$$
$$\textbf{264} \qquad\qquad\qquad\qquad\qquad \textbf{265}$$

The nucleoside 5′-aldehydes have been used for a variety of transformations to give novel nucleoside and nucleotide analogues. Treatment of the 2′,3′-*O*-isopropylidineadenosine 5′-aldehyde (**258**) with sodium borodeuteride gave 2′,3′-*O*-isopropylidineadenosine-5′-*d*.[292] The same crude 5′-aldehyde (**258**) was treated *in situ* with diphenoxyl triphenylphosphoranylidinemethylphosphonate (**266**) to give the unsaturated phosphonate (**267**) in good yield. A reduction of this compound gave 5′-deoxy-5′-diphenoxyphosphinylmethyl-2′,3′-*O*-isopropylidine-adenosine. The most convenient method for a removal of the phenyl esters is by transesterification with sodium benzyloxide to give (**268**), followed by Pd/H$_2$ hydrogenolysis. Heating the resultant free acid in water removed the isopropylidine group to give the important analogue of 5′-AMP, 5′-deoxy-5′-dihydroxyphosphinylmethyladenosine (**269**).[293]

The DMSO–DCC orthophosphoric acid oxidation of 6-chloro-9-(2,3-*O*-isopropylidine-*β*-D-ribofuranosyl)purine gave the corresponding 5′-aldehyde (**270**). This (**270**) was then treated directly with ethoxycarbonylmethylidine-

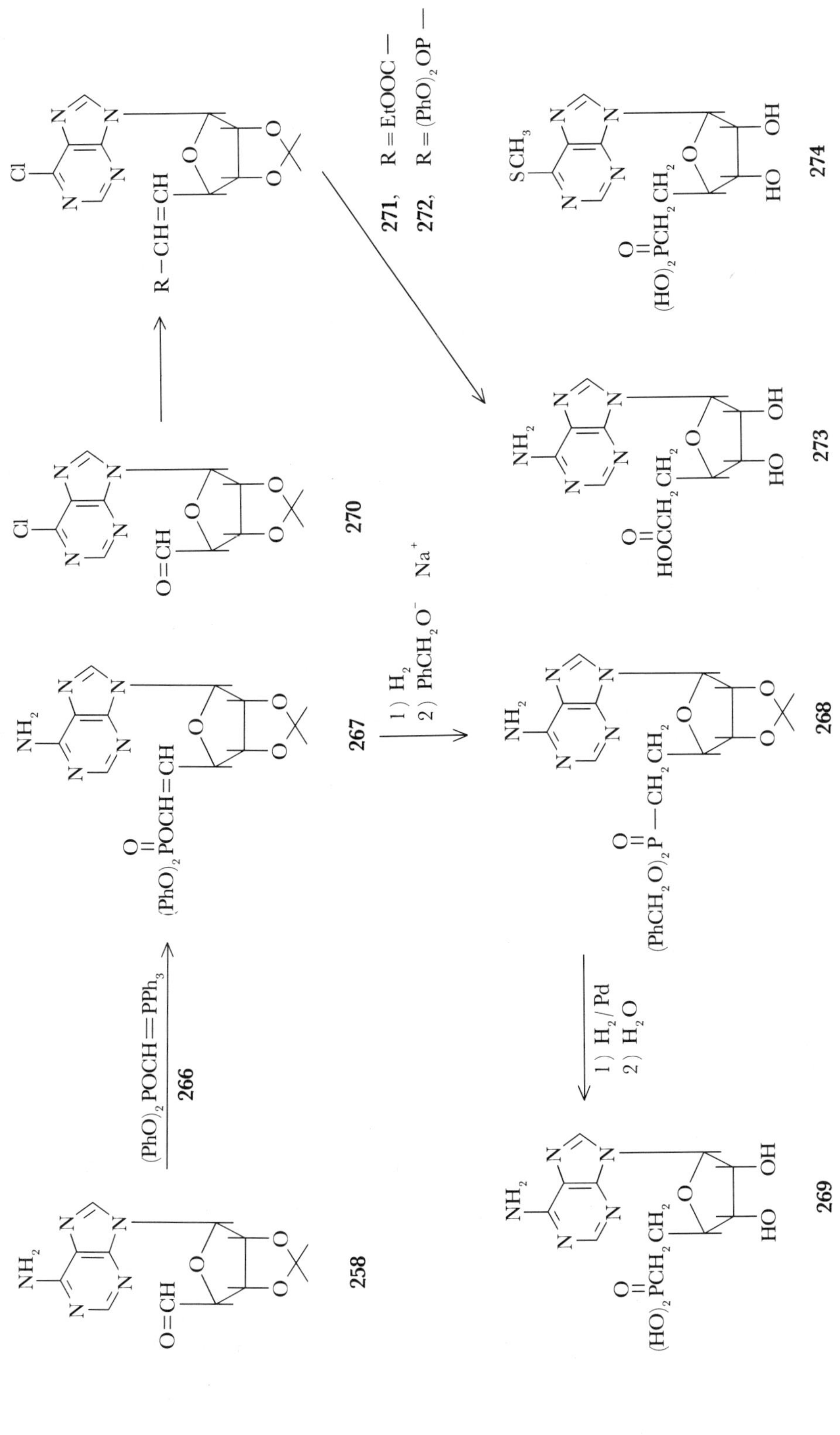

triphenylphosphorane or the Wittig reagent (266) to give the corresponding olefins (271) and (272) in good yield.[294] After treatment of (271) with sodium azide, reduction, and deisopropylidination, 5′-carboxymethyl-5′-deoxyadenosine (273) was obtained. This compound was also prepared from adenosine 5′-carboxaldehyde.[295,296] A series of similar reactions, beginning with 9-(2,3-O-isopropylidine-β-D-ribofuranosyl)-6-methylthiopurine, eventually furnished the 5′-deoxy-5′-phosphonomethyl analogue (274) and other derivatives of 6-methylthio-9-(β-D-ribofuranosyl)purine.[294]

The DMSO–DCC oxidation of arabinonucleosides has also been studied and N^6-benzoyl-9-(2,3-di-O-benzoyl-1,5-pentodialdo-β-D-arabinofuranosyl)adenine[297] (276) has been synthesized by Pfitzner–Moffatt oxidation of N^6-benzoyl-9-(2,3-di-O-benzoyl-β-D-arabinofuranosyl)adenine (275). The aldehyde (276) was condensed directly, without isolation, with (carbethoxymethylene)triphenyl-phosphorane to give, exclusively, ethyl-(E)-1-(6-benzamidopurin-9-yl)-1-5-6-trideoxy-2,3-di-O-benzoyl-β-D-arabino-hept-5-eno-1,4-furanuronate (277) in 65% yield. The arabinoheptene (277) could not be deprotected because of the carbonyl conjugation with H-4′. However, (277) was hydrogenated to afford a pivotal intermediate, which was transformed into a variety of 5′-substituted derivatives (278) of 9-(β-D-arabinofuranosyl)adenine.[298]

The two enantiomeric 5′-C-methyl analogues of adenosine were prepared by a reaction of N-benzoyl-2′,3′-O-isopropylidineadenosine 5′-aldehyde (261) with

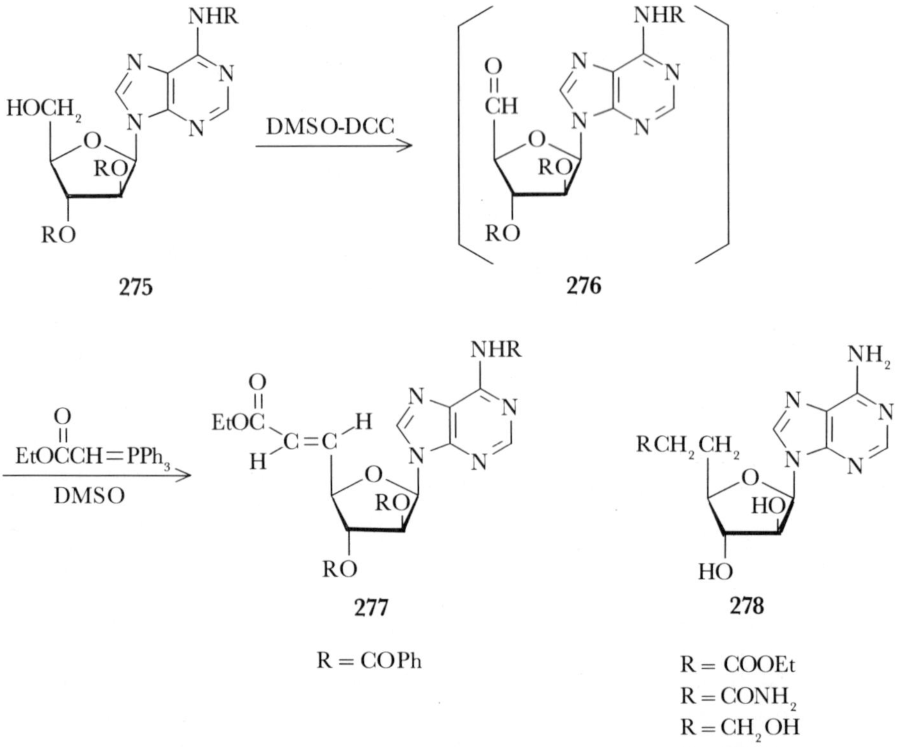

275   276

277   278

R = COPh

R = COOEt
R = CONH$_2$
R = CH$_2$OH
R = COOH

excess methylmagnesium chloride. A 34% yield of pure *N*-benzoyl-9-(6-deoxy-2,3-*O*-isopropylidine-β-D-allofuranosyl)adenine (**279**) was obtained. In addition, a small amount of the L-talo enantiomer (**280**) was also isolated.[285]

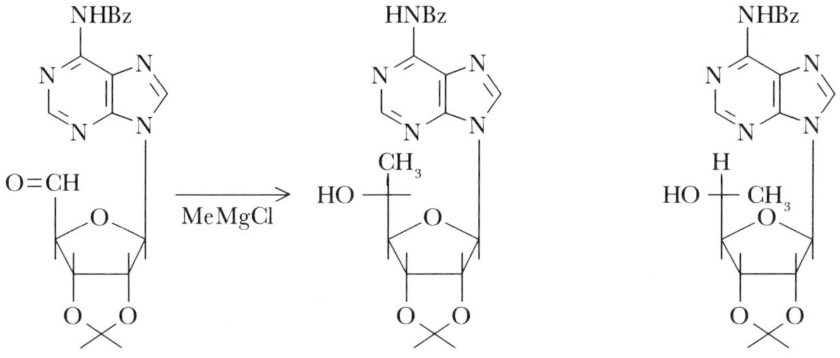

261   279   280

For the purine nucleosides, the oxidation of a secondary hydroxyl group on the sugar is not well documented. Because of a propensity for α, β elimination of the heterocycle to occur, there is a paucity of 3'-ketonucleosides of furanoses since they are rather unstable. The use of ruthenium tetroxide, in the presence of sodium metaperiodate, gives the 2'-keto derivative (**282**) from 9-(3,5-*O*-isopropylidine-β-D-xylofuranosyl)adenine (**281**). This compound readily condensed with nitromethane, with an exclusive addition on the α side of the sugar, to give, after deacetonation, 9-(2-*C*-nitromethyl-β-D-xylofuranosyl)adenine (**283**).[301]

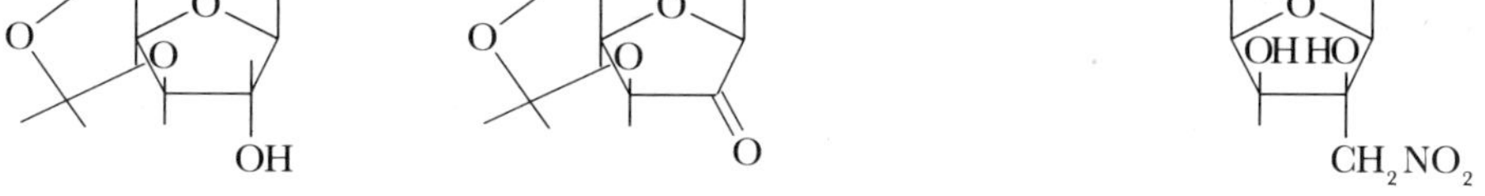

281   282   283

Several ketonucleosides containing a pyranose ring attached to theophylline and 6-chloropurine have been prepared by an oxidation with DMSO and a dehydrating reagent.[289,291,302]

More vigorous oxidation of the 5'-hydroxyl group by a number of methods has furnished nucleoside 5'-carboxylic acids. Chromium trioxide in pyridine was originally used and converted deoxyadenosine and deoxyguanosine to the corresponding 5'-carboxylic acids.[303] Adenosine and guanosine were converted to their 5'-carboxylic acid derivatives in higher than 50% yield by bubbling oxygen into a pH 8.8 solution of the nucleoside with platinum catalyst present.[304] The 5'-carboxylic acid (284) of 2',3'-O-isopropylidineadenosine has also been prepared by an oxidation with potassium permanganate in alkali.[305,307] It has also been reported that these same conditions gave the 5'-carboxylic acid derivative of 2-amino-9-(2,3-O-isopropylidine-β-D-ribofuranosyl)purine, whereas the only product to be isolated from the alkaline reaction of $KMnO_4$ and 2',3'-O-isopropylidineguanosine and 2',3'-O-isopropylidineinosine was oxalic acid.[307] The 5'-carboxylate of 2',3'-O-isopropylidineinosine was obtained by an oxidation with CrO_3 in acetic acid.[308]

The 5'-carboxylic acids have readily been esterified. Diazomethane treatment of 1-(adenin-9-yl)-2,3-O-isopropylidine-β-D-ribofuranuronic acid (284) gave the corresponding methyl ester (285) in good yield.[306] Alternatively, 1-(adenin-9-yl) and 1-(hypoxanthin-9-yl)-β-D-ribofuranuronic acid or their isopropylidine derivatives were converted into a variety of alkyl esters by treatment with either the alcohol and thionyl chloride,[309] potassium hydroxide, and an alkyl halide or sulfate,[309] or silver nitrate and an alkyl halide.[310]

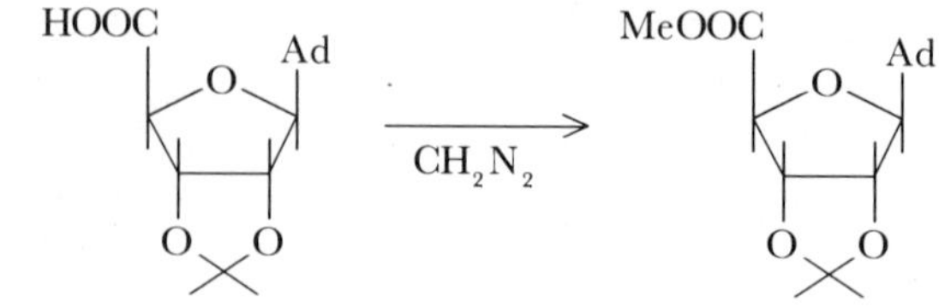

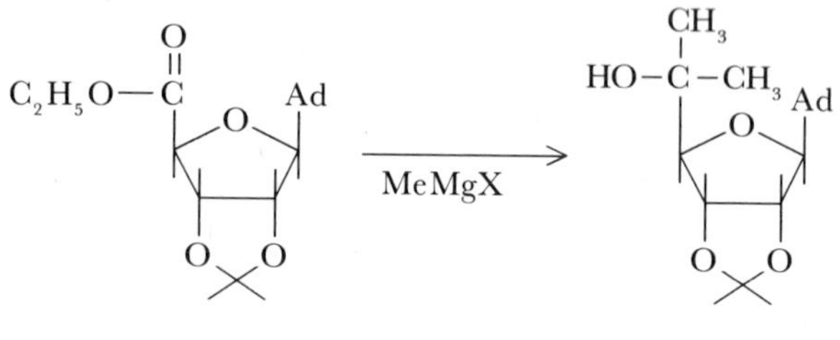

These esters have been subjected to further typical reactions of esters, such as treatment of (286) with methylmagnesium iodide to give 5',5'-di-C-methyl-2',3'-O-isopropylidineadenosine (287).[306] A superior yield (94%) was obtained by using methylmagnesium chloride.[285] Treatment of the ester (286) with ammonia and then phosphoryl chloride in cold pyridine gave the 5'-nitrile derivative (288).[311] This nitrile and sodium azide gave, after acid treatment, the unusual C-tetrazole nucleoside, 1-(adenin-9-yl)-4-(R)-C-(tetrazol-1-yl)-β-D-erythro-furanose (289).[311]

The oxidation of nucleosides and nucleotides with vicinal cis-hydroxyls to yield a dialdehyde has been amply described in previous reviews. In fact, it has at times been used as a primary method for structural elucidation. These "dialdehydes" are also useful synthetic intermediates, as demonstrated by their condensation reactions with nitroalkanes. Treatment of *N,N*-dimethyladenosine (**290**) with sodium periodate in water gave a good yield of the dialdehyde 2-*O*-[(*R*)-formyl-(6-dimethylaminopurin-9-yl)methyl]-(*R*)-glyceraldehyde (**291**). This compound was condensed with nitromethane in methanolic sodium methoxide to give 75% of a mixture of 3'-deoxy-3'-nitropyranosides. From this mixture, the *manno* (**292**), *galacto* (**293**), and *gluco* (**294**) isomers were isolated, and they were

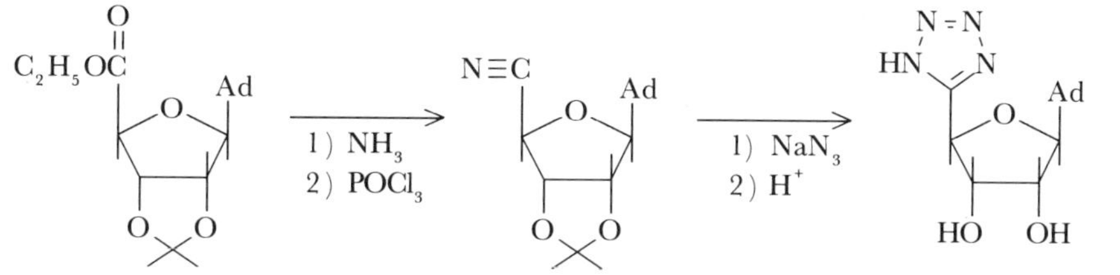

286   288   289

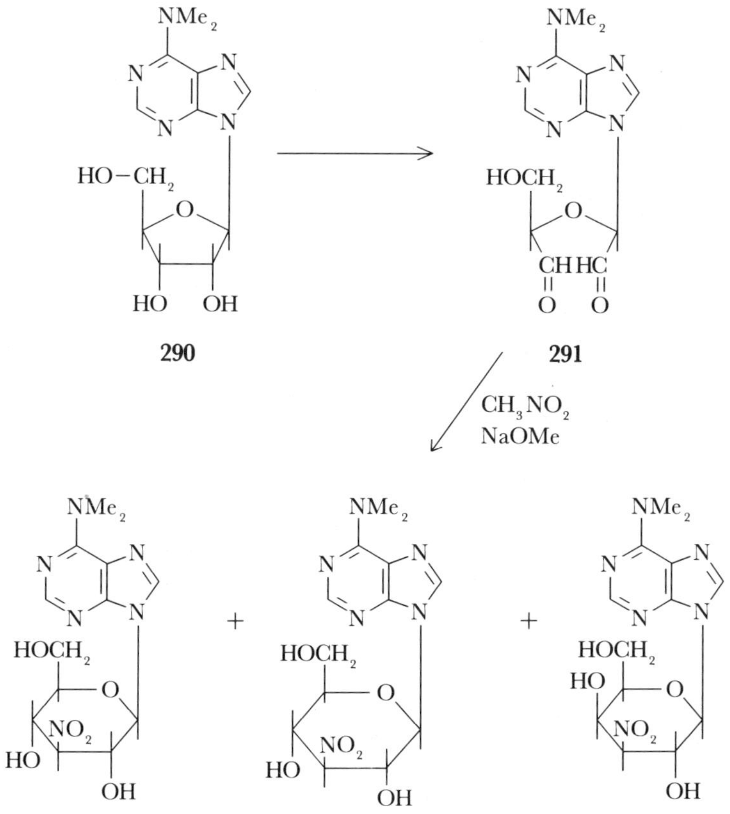

290   291

CH_3NO_2
NaOMe

294   292   293

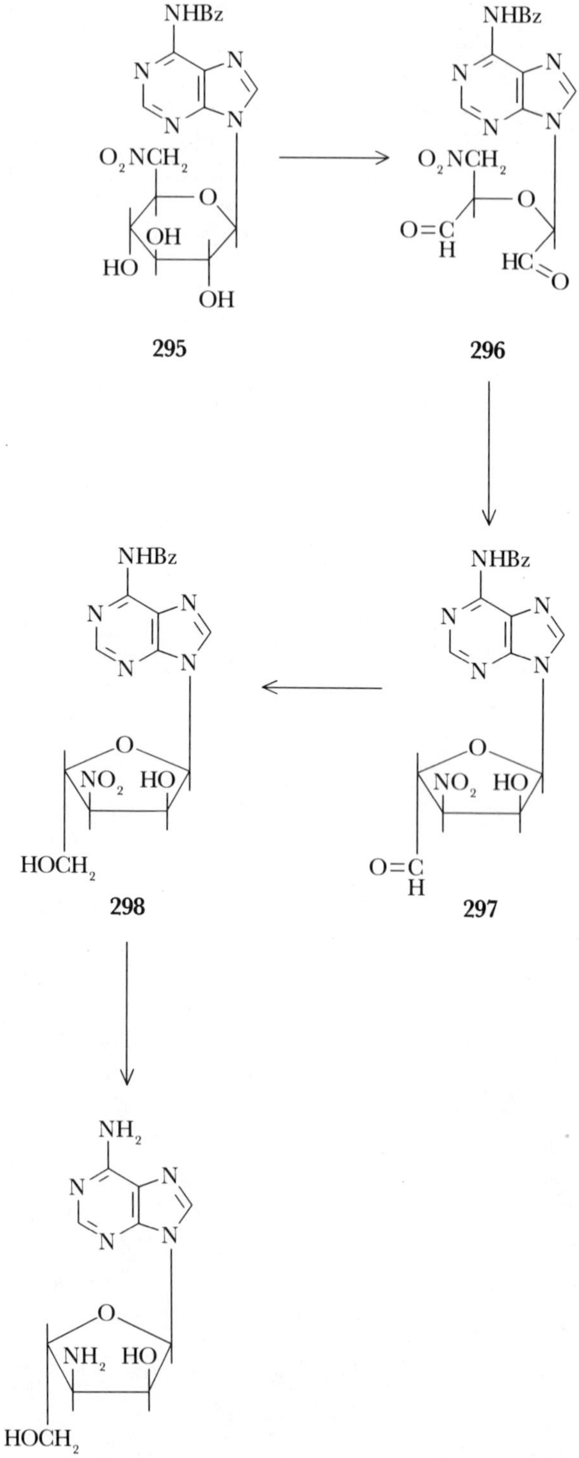

295

296

298

297

299

then reduced to the corresponding 3'-deoxy-3'-aminopyranosides.[312] A series of 3'-deoxy-3'-nitropyranosyl derivatives of hypoxanthine were prepared[313] in a similar manner.

A related ring closure reaction was performed with a *C*-nitro sugar. Oxidation of *N*-benzoyl-9-(6-deoxy-6-nitro-*β*-D-glucopyranosyl)adenine **(295)** consumed 2 equivalents of periodate to give the nitro nucleoside dialdehyde **(296)**. Treatment of this compound with sodium methoxide effected a condensation of one aldehyde with the carbanion formed next to the nitro group. An *in situ* reduction of the aldehyde **(297)**, with sodium borohydride gave *N*-benzoyl-9-(3-deoxy-3-nitro-*α*-L-ribofuranosyl)adenine **(298)**. Compound **(298)** was hydrogenated to afford the aminonucleoside, which was then treated with picric acid to remove the benzoyl group and give 9-(3-deoxy-3-amino-*α*-L-ribofuranosyl)adenine **(299)**.[314]

3.1.3. Conversion of a Sugar Hydroxyl to Halogen and Reactions of Nucleosides Containing Halogen in the Carbohydrate

The conversion of a hydroxyl group on the glycosyl moiety of purine nucleosides and nucleotides to a halogen has been of interest because of the synthetic utility of those halosugar derivatives and because of an interest in their biological properties. Many methods for these reactions are available from classical organic chemistry; our emphasis is on those methods that are most applicable to nucleoside and nucleotide chemistry.

The direct introduction of fluorine into the carbohydrate moiety of a nucleoside has seldom been accomplished. However, 9-(2,3-anhydro-*β*-D-lyxo-pyranosyl)adenine **(300)** with KHF_2 in ethylene glycol at reflux gave a 41% yield of 9-(3-fluoro-3-deoxy-*β*-D-arabinofuranosyl)adenine **(301)** as the sole nucleoside product.[315] Treatment of 9-(2,3-anhydro-5-*O*-benzoyl-*β*-D-ribofuranosyl)-*N*,*N*-dibenzoyl(or *N*-pivaloyl)adenine **(302)** with tetraethylammonium fluoride in hot acetonitrile also resulted in a nucleophilic attack at the 3' position, to give, after deacylation with sodium methoxide, 9-(3-deoxy-3-fluoro-*β*-D-xylofuranosyl) adenine **(303)** in a 63% yield.[316]

The traditional method of synthesis for a nucleoside containing halogen in the carbohydrate moiety has been accomplished by the action of a halide on an

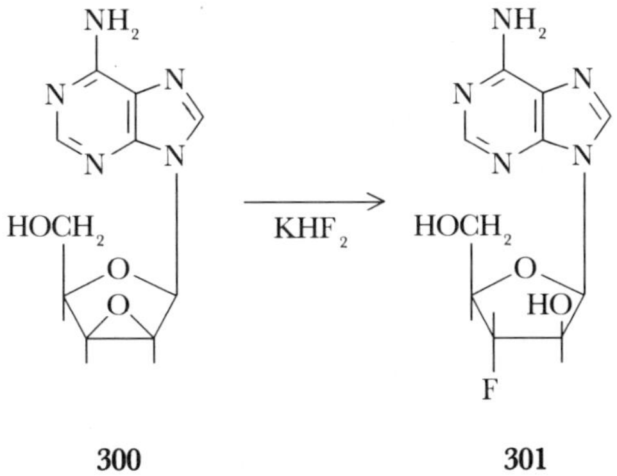

300   301

302, R$_1$, R$_2$ = Bz

or R$_1$ = H, R$_2$ = Me$_3$CCO

303

O-tosyl nucleoside.[317] Treatment of 3′-*O*-*p*-nitrobenzenesulfonyladenosine with sodium iodide in 2,4-pentandione gave a low yield of "3′-deoxy-3′-iodoadenosine" with no structure proof but which most likely had the D-xylo configuration.[318]

A reduction of this product gave the first synthesis of cordycepin.[319] Substitution of a tosyl group at the 5′ position made the 5′-carbon susceptible to an intramolecular nucleophilic attack. This resulted in a facile N^3,5′-cyclization and formation of anhydronucleosides (cyclonucleosides),[320,321] (see Section 3.3). Acetylation or formylation of the 6-amino group, by virtue of lowering the nucleophilicity of N-3 in the purine ring, inhibited N^3,5′-cyclization and allowed a direct displacement of the 5′-tosyloxy group by chloride, bromide, iodide, and azide anions to occur.[321] Sodium methylmercaptide was found to react preferentially with the 5′-tosyloxy group, at a lower temperature, without any N^3,5′-cyclization or need for formylation of the 6-amine group.[322]

Other methods and reagents have been studied for the conversion of a hydroxyl group to a halogen group and include methyltriphenoxyphosphonium iodide (Rydon reagent) (**304**).[323] Treatment of 2′,3′-*O*-isopropylidineadenosine (**218**) or 2′,3′-*O*-isopropylidineguanosine with the reagent in DMF solution rapidly gave the corresponding 3′,5′-cyclonucleoside (**306**) in high yield as their iodide salts.[324] Cyclonucleoside formation indicates the propensity of the 5′-halopurine nucleosides to undergo an intramolecular displacement of the leaving group (see Section 3.3). However, it was possible to isolate a small amount of uncyclized 5′-deoxy-2′,3′-*O*-isopropylidine-5′-iodoinosine (**308**) from the reaction of (**304**) and 2′,3′-*O*-isopropylidineinosine (**307**).[324] These reactions were presumed to occur by an intramolecular attack on the 5′-*O*-phosphonium intermediate (**305**) initially formed in the reaction. Attack of this intermediate by iodide (via path a) would give the 5′-iodo derivative, whereas attack by N-3 (via path b) would give the cyclonucleoside.

It is important to discuss the factors that minimize the undesired N^3,5′-cyclization and optimize the iodide attack to yield, selectively, the desired 5′-deoxy-5′-iodonucleoside. As already demonstrated by Jahn,[321] an acylation of the 6-amino function in adenosine reduces the electronegativity of N^3 and the tendency toward a N^3,5′-cyclization. This was also the case with $N^6,N^6,O^{2'},O^{3'}$-tetrabenzoyladenosine, which then treated with the Rydon reagent (**304**) gave the

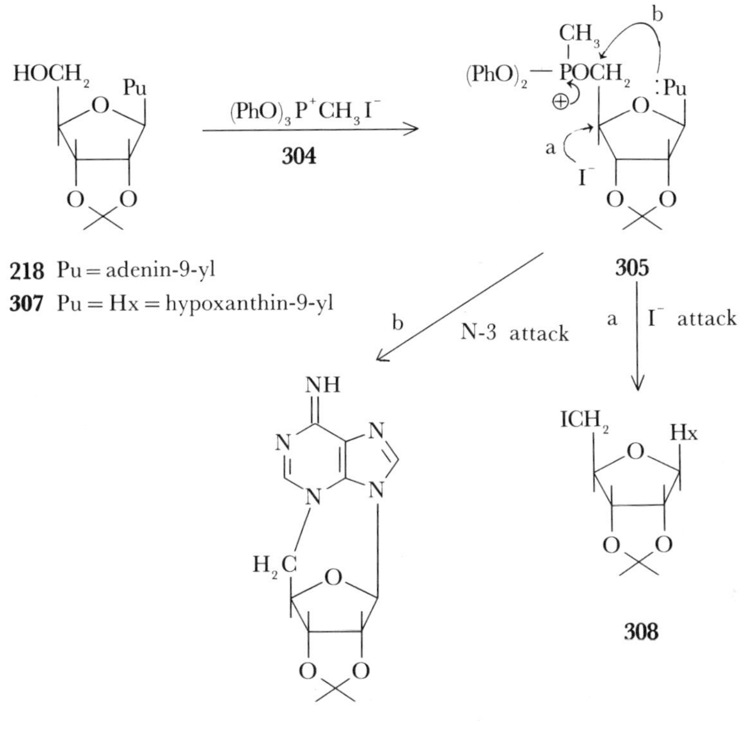

218 Pu = adenin-9-yl
307 Pu = Hx = hypoxanthin-9-yl

306

corresponding 5'-deoxy-5'-iodonucleoside in high yield.[325] In addition to heterocyclic derivatization to control the nucleophilicity of N^3, Moffatt and co-workers[258,326,327] succeeded in controlling the relative rates of paths a and b by changing the solvent and temperature. The cation and iodide anion solvation of the phosphonium intermediate (**305**) in dimethylformamide results in a preferential intramolecular ($N^3,5'$) cyclization and a decreased nucleophilicity of iodide anion, respectively.[326] The use of nonpolar solvents, such as tetrahydrofuran[326] and dichloromethane,[326,327] with the Rydon reagent gave 5'-deoxy-5'-iodopurine nucleosides in good yield.

Another group of phosphonium salts that have been used in nucleoside chemistry are the trihalomethyltriphenylphosphonium halides, prepared *in situ* from triphenylphosphine and carbon tetrahalides (CX_4). 2',3'-O-Isopropylidine-inosine gave a quantitative yield of the corresponding 5'-chloronucleoside with Ph_3P and CCl_4 in triethyl phosphate at 100°C.[328] It was also found that the adduct of Ph_3P and cyanogen bromide, bromine, or iodine could also effect a conversion of a hydroxyl group to the corresponding halogen group.[328]

The use of $Ph_3P^+CBr_3Br^-$ and $Ph_3P^+Cl_3I^-$ was also investigated by Verheyden and Moffatt.[329] A reaction of $Ph_3P^+CCl_3Cl^-$ with 2',3'-O,N,N-tetrabenzoyladenosine gave the 5'-chloro-5'-deoxy derivative, and 2',3',-O-iso-propylidineinosine with $Ph_3P^+CBr_3Br^-$ gave the 5'-bromo-5'-deoxy derivative, both in good yield.

The mechanism of reaction of this reagent is similar to that proposed for the

Scheme IV

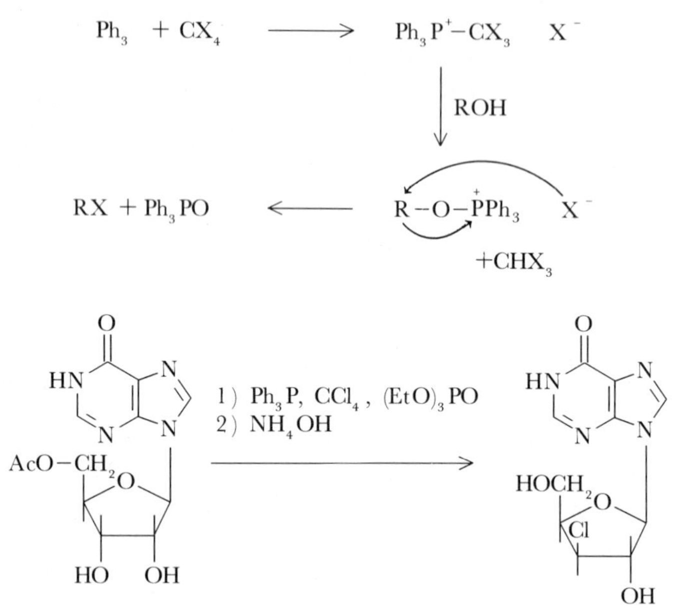

$$Ph_3 + CX_4 \longrightarrow Ph_3P^+\!-CX_3 \quad X^-$$

$$\downarrow ROH$$

$$RX + Ph_3PO \longleftarrow R-O-\overset{+}{P}Ph_3 \quad X^-$$

$$+CHX_3$$

1) Ph_3P, CCl_4, $(EtO)_3PO$
2) NH_4OH

309   310

Rydon reagent. The electrophilic phosphorus is attacked by the hydroxyl oxygen with an elimination of CHX_3. Halide then attacks the sugar carbon to eliminate the phosphine oxide (Scheme IV).[330]

This reagent was found to be somewhat more successful than (304) with secondary hydroxyls. Treatment of 5'-O-acetylinosine (309) with triphenyl-phosphine and CCl_4 gave, after deblocking, a 19% yield of 9-(3-chloro-3-deoxy-β-D-xylofuranosyl)hypoxanthine (310).[328]

Another convenient variant of this procedure is the use of *N*-bromo-succinimide and triphenylphosphine.[331] This reagent gave a good yield of 5'-bromo-5'-deoxy-2',3'-O-isopropylidineinosine from 2',3'-O-isopropylidine-inosine. Furthermore, it was shown to be quite selective for primary hydroxyls. Selective 5'-halogenation of adenosine using thionyl halide and hexamethyl-phosphoroamide has been performed in good yield.[332]

This method was primarily designed for the replacement of hydroxyl groups by halogens, but it has also been used successfully for the direct introduction of an azide group at the C-5' position of 2'-deoxy-[333] and 2',3'-unsaturated purine[334] nucleosides. A very convenient preparation of 5'-chloro-5'-deoxy-2',3'-O-iso-propylidineadenosine is by a direct treatment of 2',3'-O-isopropylidineadenosine with thionyl chloride as reported by Srivastava *et al.*[335] However, thionyl chloride (or bromide) reacts with hexamethylphosphoramide to form an adduct that readily reacts with sugar hydroxyls to give the activated intermediate and the alkoxyphosphonium salt (311). Adenosine (128) has been converted in one step[(332)] to 5'-chloro-5'-deoxyadenosine in a 75 % yield. The 2',3'-*cis*-diol system of ribonunucleosides was protected by the formation of a 2',3'-cyclic hexamethyl-phosphortriamide **(311)**,[(336)] which was then decomposed in the workup to yield only the 5'-halonucleoside **(312)**. 2'-Deoxyadenosine **(313)** reacted with $SOCl_2$/HMPT to give a 2',3',5'-trideoxy-3',5'-dichloroadenosine derivative **(314)** of unknown configuration (with *threo* being the probable), but 9-(β-D-arabino-furanosyl)adenine gave a good yield of only the 5'-chloro-5'-deoxy derivative.[(336)] Although the 2',3'-cyclic intermediate was disallowed in the latter case, steric hindrance to the attack of chloride on the 2'- and 3'-carbons of the tris-*O*-(hexamethylphosphonium)nucleoside was invoked as a probable explanation[(336)] for the selective reaction.

The reaction of adenosine **(128)** with α-acetoxyisobutyryl chloride or bromide **(315)**[(337,338)] has provided two interesting intermediates for nucleoside conversions. This reaction gave an intermediate acetoxonium ion **(316)**, formed by a reaction of the nucleoside with the α-acetoxyisobutyryl halide. This intermediate was then reacted with halide to give a mixture of 9-(2-*O*-acetyl-3-

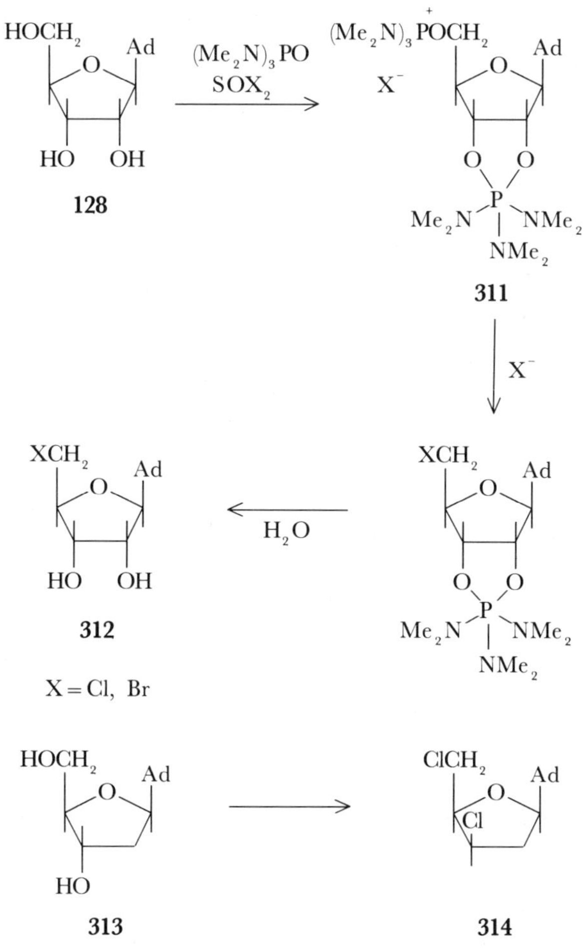

deoxy-3-halo-β-D-xylofuranosyl) and 9-(3-O-acetyl-2-deoxy-2-halo-β-D-arabino-furanosyl)adenines [(**317**) and (**318**), respectively] with a 2,5,5-trimethyl-1,3-dioxolan-4-on-2-yl ether on the 5' position.[338] The use of "moist" acetonitrile in such a bromoacetylation of adenosine inhibited the glycosyl cleavage and also avoided the 2',3',5'-tris orthoester by-product formation[339] as observed in earlier studies.[338] Treatment of these compounds *vide supra* with HCl in methanol gave the deblocked halonucleosides (**319**) and (**320**). Treatment of (**319**) and (**320**) or the blocked compounds (**317**) and (**318**) with sodium methoxide or with Amberlite IRA-400 (OH⁻) resin in dry methanol[339] gave 9-(2,3-anhydro-β-D-ribofuranosyl)adenine (**321**).

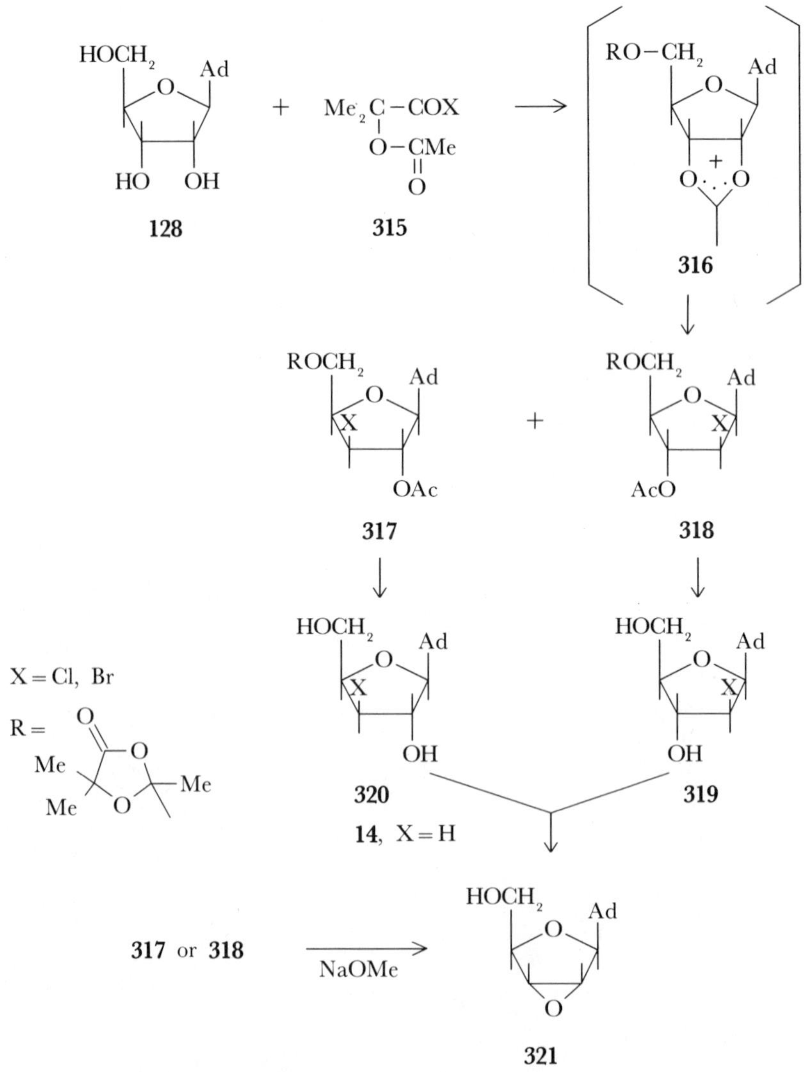

Hydrogenolysis of the 3'-bromo compound (**320**) (X = Br) gave cordycepin (**14**), whereas the corresponding 2'-O-acetyl compound (**322**) gave a mixture of cordycepin and 2′,3′-dideoxyadenosine (**323**). The dideoxynucleoside (**323**) was presumed to have arisen *via* a palladium-catalyzed elimination of acetyl hypobromite to give an olefin (**324**), which was subsequently reduced to afford (**323**).[338] Subsequently, it was shown that treatment of (**318**) with chromous acetate gave, after a base-catalyzed removal of the dioxalanone group, a 62% yield of 9-(2,3-dideoxy-β-D-*glycero*-pent-2-enofuranosyl)adenine (**324**).[340] Dehydrohalogenation of (**318**) was affected by 1,5-diazabicyclo(4,3,0)non-5-ene, which after addition of methanol gave the 3′,4′-unsaturated nucleoside, 9-(3-deoxy-β-D-*glycero*-pent-3-enofuranosyl)adenine (**325**).[340]

HOCH$_2$ Pd Ad / Br / OAc → HOCH$_2$ O Ad → HOCH$_2$ O Ad

322   **324**   **323**

318 $\xrightarrow[\text{2) MeOH}]{\text{1) DBU}}$ HOH$_2$C — O Ad / OH

325

There is another closely related procedure that has also made use of the intermediate 2′,3′-*O*-acyloxonium ions of nucleosides for the introduction of halogen into the 2′ and 3′ positions. The acyloxonium ion (**326**), generated *in situ* by treatment of 2′,3′-*O*-methoxyethylidineadenosine (**250**) with pivaloyl chloride in hot pyridine, on attack by chloride predominantly at the 3′ position, gives 9-(2-*O*-acetyl-3-chloro-3-deoxy-5-*O*-pivaloyl-β-D-xylofuranosyl)-6-pivalamidopurine (**327**).[316,341] The hypoxanthine[342] and guanine[343] nucleoside analogues of (**250**) react in a similar manner with pivaloyl chloride in pyridine to yield the corresponding 2′(or 3′),5′-blocked-3′(or 2′)-chloronucleosides. The use of sodium iodide in the reaction gives the corresponding 3′(or 2′)-iodides.

The halonucleoside (**327**) was treated with methoxide to give 2′,3′-anhydroadenosine (**321**). This epoxynucleoside was unstable and slowly formed the 3,3′-cyclonucleoside. The epoxynucleoside was then benzoylated for increased stability.[316] This blocked anhydronucleoside (**328**) was then subjected to a nucleophilic attack by fluoride, benzoate, and azide, and the attack occurred at the 3′ position to give the 3′-deoxy-3′-substituted-xylofuranosyladenines.[316]

A similar *in situ* formation of acyloxonium ion occurs when 3′,5′-di-*O*-acetyladenosine is treated with boron trifluoride etherate followed by phosphorus tribromide or when adenosine is treated with tetraacetoxysilane and phosphorus tribromide in the presence of boron trifluoride etherate. In either case, 9-(2,5-di-*O*-acetyl-3-bromo-3-deoxy-β-D-xylofuranosyl)adenine is formed as a major product *via* a bromide attack on the acyloxonium cation.[344]

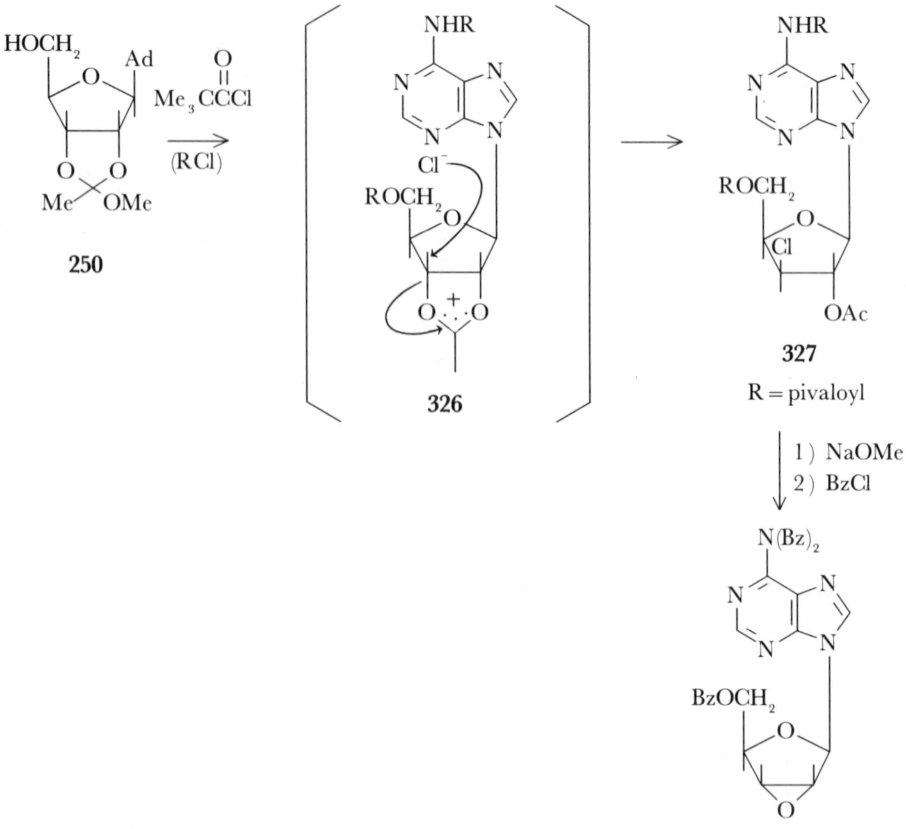

250

326

327

R = pivaloyl

1) NaOMe
2) BzCl

328

Methanesulfonyl chloride in anhydrous dimethyl formamide has been reported to effect the chlorination of a theophylline nucleoside.[345] This method has not been explored in terms of a general application. Another method for the introduction of halogens of the riboconfiguration is by halodetrifluoromethane-sulfonylation and various 2'-halogeno(F, Cl, Br, and I)-2'-deoxyguanosines have been prepared by the treatment of N^2-isobutyryl-9-(2-O-trifluoromethane-sulfonyl-3,5-di-O-tetrahydrofuranyl-β-D-arabinofuranosyl)guanine with tetra-n-butylammonium fluoride or an appropriate metal halide.[346]

Halodeoxynucleosides have normally served as precursors for the synthesis of the corresponding deoxy nucleosides by dehalogenation. A convenient and general route for such a transformation is the reduction of halodeoxynucleosides with tri-n-butylin hydride in the presence of α,α'-azobis(isobutyronitrile) as the initiator.[347,348] The 5'-chloro group in adenosine is easily replaced by a variety of nucleophiles such as homocysteine, sodium benzene selenolate, and sodium sulfite to yield biochemically useful intermediates, for example, S-adenosylhomo-cysteines,[349] 5'-phenylseleno-5'-deoxyadenosine,[350] and adenosine 5'-deoxy-5'-sulfonic acid, respectively.[351]

One of the more elegant uses of nucleosides containing a halogen in the sugar is the use of episulfonium ions to change the configuration of sugar substituents and to introduce new substituents. Treatment of 2',3'-anhydroadenosine (**321**)[352] with sodium ethylmercaptide gave 9-[3-deoxy-3-(ethylthio)-β-D-xylofuranosyl] adenine (**329**). Direct treatment of (**329**) with cold thionyl chloride gave an 86% yield of crystalline 9-[3-chloro-2,3-dideoxy-2-(ethylthio)-β-D-arabinofuranosyl] adenine (**330**). Both free hydroxyl groups of (**329**) presumably formed chlorosulfinyl esters (**331**). The sulfur then displaced the chlorosulfinate from the 2' position to give the episulfonium ion (**332**), which was attacked preferentially at the 3' position by chloride. Water hydrolysis of the chlorosulfinate (**333**) in the workup[352] then regenerated (**330**). Treatment of (**330**) with nucleophiles such as acetate,[352] azide,[353] or thiocyanate gave the corresponding 9-[3-deoxy-3-sub-stituted-2-(ethylthio)-β-D-arabinofuranosyl]adenines (**335**) *via* the episulfonium intermediate (**334**). Desulfurization of (**335**) (R = OH) gave 2'-deoxyadenosine. This was one of the first examples of a chemical synthesis of a purine 2'-deoxy-nucleoside.[354]

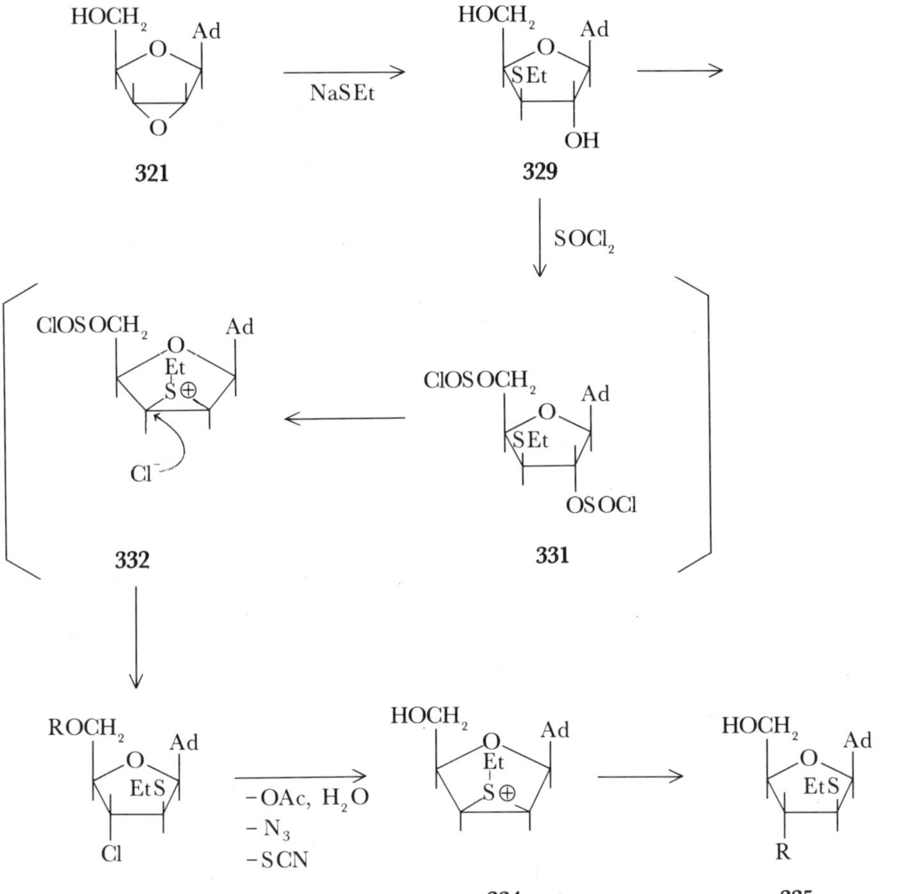

3.1.4. Preparation and Reactions of Miscellaneous Esters of the Sugar Hydroxyls

Aside from acyl groups, the most important agent used for esterification of the sugar hydroxyl groups is phosphoric acid. The synthesis and reactions of these phosphate esters are discussed in the sections on phosphorylation and reactions of the phosphate of nucleotides. Other oxyacids, notably sulfuric, nitric, and their derivatives, have been used for esterification of the sugar hydroxyls, which have then been used for a variety of synthetic and medicinal applications.

In the classical structure proof of *S*-adenosylmethionine (or "active methionine" as it was known at the time), Baddiley treated 2′,3′-*O*-isopropylidine-5′-*O*-*p*-toluenesulfonyladenosine with sodium methylmercaptide to give the intermediate 5′-methylthio derivative. This derivative was then treated with α-amino-δ-bromobutyric acid to give *S*-adenosylmethionine.[355] The first tosylation of a purine nucleoside, 2′,3′-*O*-isopropylidineinosine, was accomplished in 1935 by Levene and Tipson.[356] This sulfonylation of hydroxyl groups produced a leaving group that proved to be very useful in nucleoside and nucleotide research. Section 3.3 describes the preparations and some uses of these derivatives. Note that the usual function of sulfonylation is to convert a hydroxyl group to a function suitable for elimination or substitution. Sulfonylation accomplishes this goal with retention of configuration whereas the conversion of a hydroxyl to halogen generally proceeds with inversion. A number of analogues of *S*-adenosyl-homocysteine have recently been prepared[349] (see also Section 3.1.3), for instance, by treatment of the 2′,3′-*O*-isopropylidine-5′-*O*-*p*-toluenesulfonyl-adenosine in liquid NH_3 with homocysteine (generated from its *S*-benzyl derivative). These derivatives have included modifications in the nucleoside[357,358] and homocysteine[359,360] moieties. Replacement of a 5′-tosyl function by iodide has also been discussed.[321]

The sulfonylation of unprotected nucleosides has been investigated, however, when 8-bromoadenosine was treated with 2,4,6-triisopropylbenzenesulfonyl chloride (TPS-Cl) in pyridine at room temperature; yields of 38% of the 2′-*O*- and 44% of the 3′-*O*-2,4,6-triisopropylsulfonate derivatives were obtained.[361] The ratio was increased to 70% for the 2′-isomer and 25% for the 3′-isomer when 8-bromo-5′-*O*-trityladenosine was the substrate. This hindered sulfonyl chloride seems to offer some selectivity, especially in the presence of bulky 5′-*O*-groups. When 8-bromoguanosine was treated with methanesulfonyl chloride in pyridine, it was found that this much more reactive reagent introduced a mesyl group at both 5′ and 2′ (~40% yield).[362]

Sulfonylation of nucleotides in pyridine is unsatisfactory because of the poor solubility of the nucleotides. However, Ikehara and co-workers found that 5′-AMP in aqueous sodium hydroxide gave up to a 60% yield of 2′-*O*-*p*-toluene-sulfonyladenosine 5′-phosphate when treated with *p*-toluenesulfonyl chloride.[363] When 9-(5-deoxy-β-D-xylofuranosyl)adenine (**336**) was treated with *p*-toluene-sulfonyl chloride in pyridine, the 3′-*O*-tosylate (**337**) was formed in a 60% yield.[364] This was contrary to what had been anticipated on the grounds of relative steric hindrance. It was argued that perhaps the adenine ring had assisted in an ionization of the 3′-OH and this could be avoided by tosylation of the dianion of the nucleoside. In this latter case, the 2′-*O*-tosylate (**338**) was converted directly to 9-(2,3-anhydro-5-deoxy-β-D-lyxofuranosyl)adenine (**339**).[364]

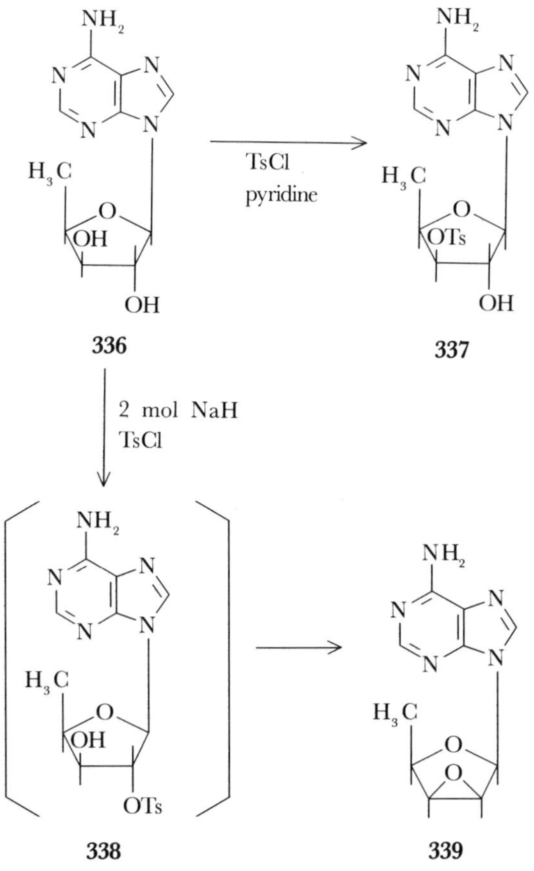

    This example of the formation of an anhydronucleoside and the methods described in Section 3.1.3 are important procedures in the inversion of hydroxyl groups on the carbohydrate moiety. In the classic synthesis of 9-(β-D-arabino-furanosyl)adenine,[365] Baker and co-workers established also that these epoxides are opened predominantly by an attack on the 3′-carbon. Thus, mesylation of 9-(3,5-*O*-isopropylidine-β-D-xylofuranosyl)adenine (**340**) gave the 2′-*O*-mesylate. Deacetonation (90% aqueous acetic acid, 100°C) followed by treatment with methanolic sodium methoxide gave 9-(2,3-anhydro-β-D-lyxofuranosyl)adenine (**341**). Treatment of this epoxide with sodium benzoate in aqueous *N*,*N*-dimethyl-formamide gave araA (**254**) in good yield,[365,366] the intermediate 3′-*O*-benzoyl group being cleaved by water in the process. A similar procedure was used for the preparation of 9-(β-D-arabinofuranosyl)guanine,[367] but 2-chloroadenine was used as the aglycon in the hydroxyl interconversion steps. Conversely, the epoxide (**341**) has been prepared in high yields by a direct treatment of araA (**254**) or xyloA with triphenylphosphinediethyl azodicarboxylate (TPA).[368] Baker had noted earlier that 9-(2,3-anhydro-β-D-lyxofuranosyl)-6-dimethylamino-2-methylthiopurine, prepared in an analogous fashion from the xyloside, gave predominantly the 3'-deoxy-3'-aminoarabinoside on treatment with ammonia.[369]

As noted in Section 3.1.3, these anhydronucleosides are amenable to attack by a variety of nucleophiles. Baker and co-workers found that 9-(2-*O*-acetyl-5-*O*-methoxycarbonyl-3-*O*-*p*-toluenesulfonyl-β-D-ribofuranosyl)adenine **(343)** gave, upon treatment with base, 2',3'-anhydroadenosine **(321)**. This epoxide underwent an attack by ethylmercaptide ion at the 3' position to give 9-(3-deoxy-3-ethylthio-β-D-xylofuranosyl)adenine **(329)**.[352]

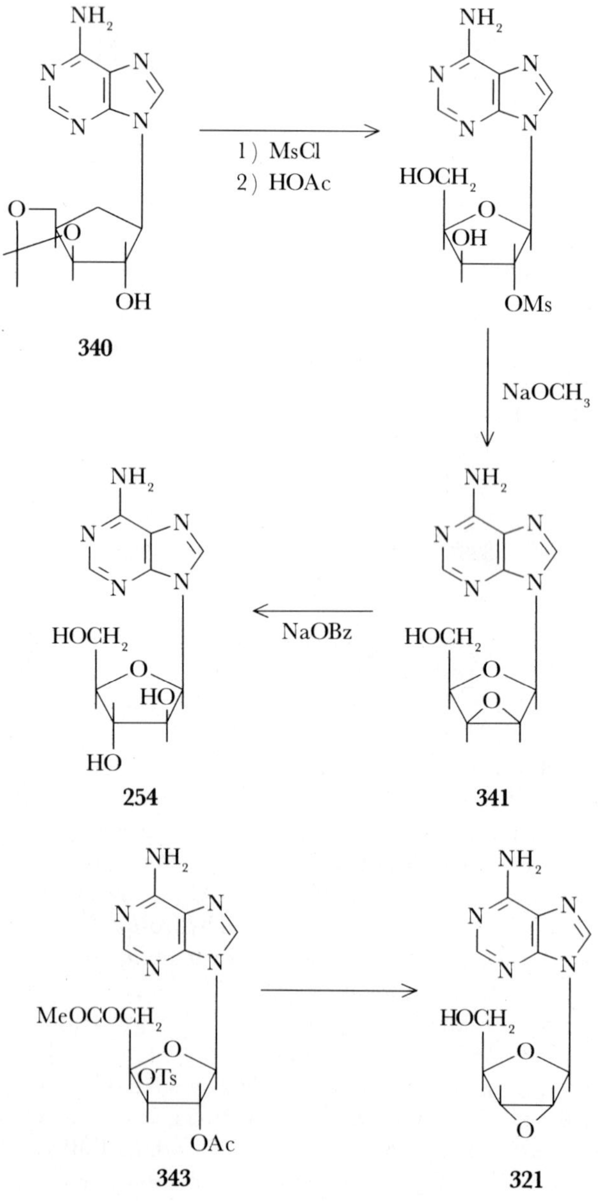

The "*trans* rule" predicts the formation of 1,2-*trans*-nucleosides from 2-*O*-acyl activated sugars (Section 2.1.1) *via* 1,2-acyloxonium ions. However, in other closely related carbohydrate reactions, the intermediate onium ions or oxazolines (from acylamido sugars) open to give vicinal *cis* configurations. For instance, 9-(3-acetamido-3-deoxy-2-*O*-methanesulfonyl-β-D-arabinofuranosyl)-6-dimethyl-amino-2-methylthiopurine (**344**) gave, on treatment with sodium acetate, the acetamidonucleoside (**345**) with the ribo configuration. Likewise, *N*-benzoyl-9-(3,5-di-*O*-benzoyl-2-*O*-methanesulfonyl-β-D-xylofuranosyl)adenine (**346**) was treated with sodium fluoride to furnish the 2′,3′-acyloxonium ion (**347**). Hydrolysis of (**347**) gave the 2′,3′-*cis* derivative, 9-(β-D-lyxofuranosyl)adenine (**348**), as the primary product.[370] The similar formation of a carboxonium type intermediate (**350**) has also been implied to occur during the reaction of 3′,5′-di-*O*-acetyl-2′-*O*-methanesulfonyladenosine (**349**) with sodium azide in DMF. The azide attack on (**350**) gives 9-(2,5-di-*O*-acetyl-3-azido-3-deoxy-β-D-xylofuranosyl)adenine (**351**).[371]

A similar class of sulfonylated nucleosides may be found in the nucleoside-*O*-sulfamates. The sulfamoyl moiety, having the ability to both donate and accept hydrogen bonds, may be the best candidate to date for a nonionic group that can stimulate the binding of a phosphate on nucleotide-binding enzymes. Nature has provided us an excellent example of this class of compounds in the nucleoside

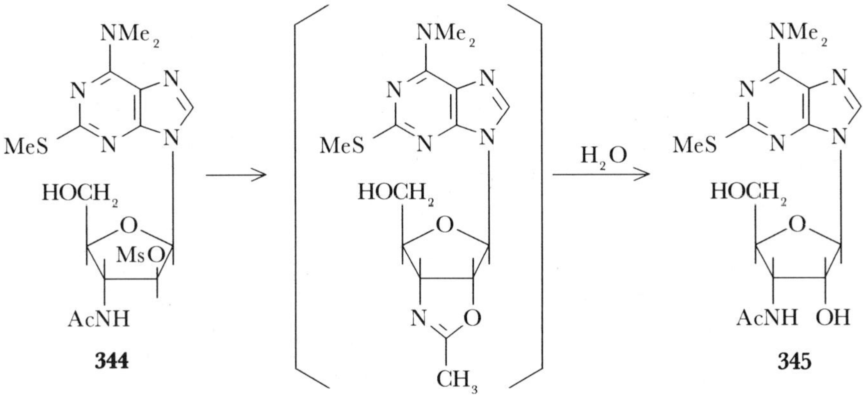

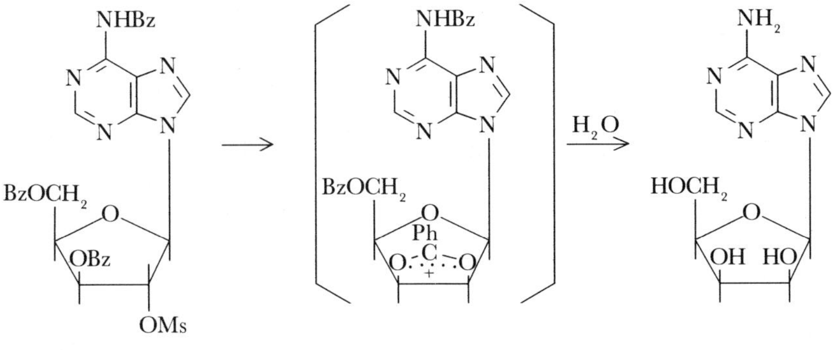

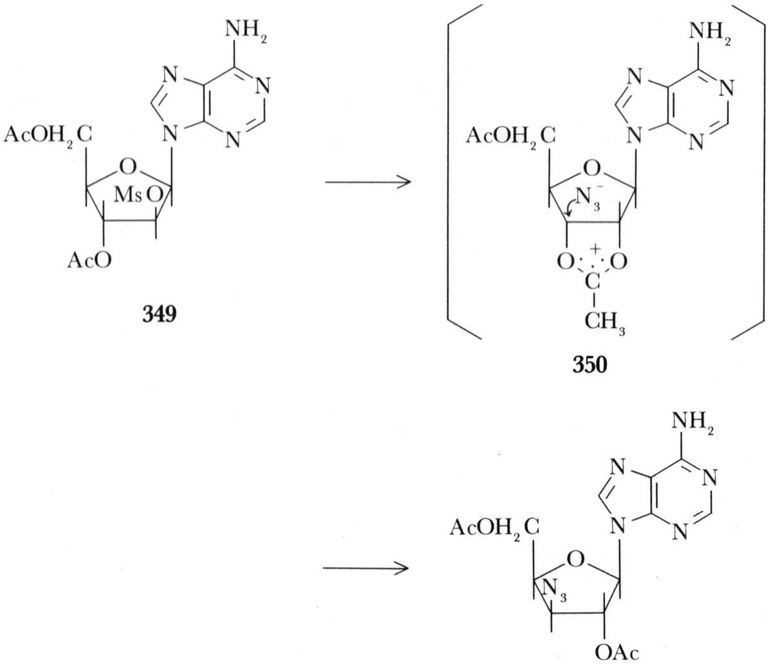

349 → **350**

351

antibiotic nucleocidin.[372] The structure of this compound was shown by spectral and indirect evidence to be 4'-fluoro-5'-O-sulfamoyladenosine **(357)**.[373,374] A series of both 5'-O- and 3'-O-nucleoside sulfamates were prepared by treatment of, for example, 2',3'-O-ethoxymethylidineadenosine **(352)** with sodium hydride, then sulfamoyl chloride.

Deblocking **(352)** with cold 5% formic acid gave 5'-O-sulfamoyladenosine **(353)**.[374] Moffatt and co-workers, in their elegant total synthesis of nucleosidin, treated 4'-fluoro-2',3'-O-isopropylidineadenosine **(354)** with hexabutyl-

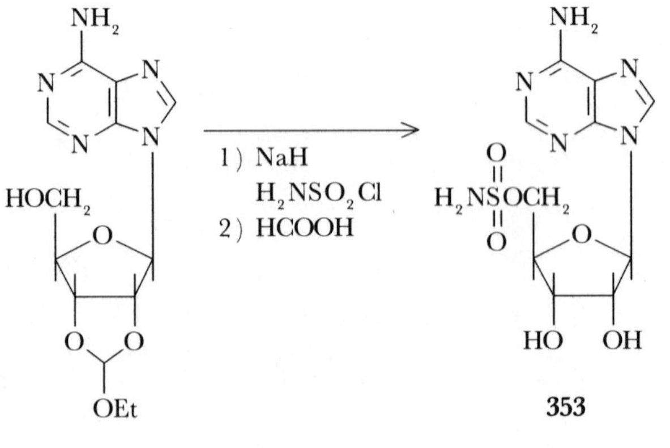

352   **353**

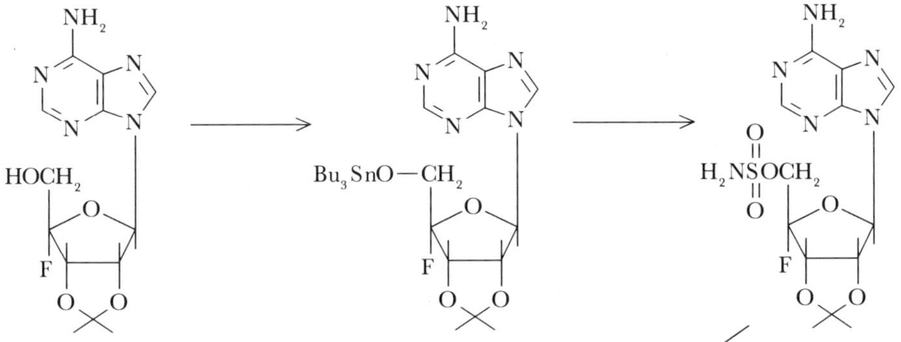

distannoxane, giving the intermediate 5′-*O*-tributylstannylene derivative (**355**). Compound (**355**) was treated directly with sulfamoyl chloride to furnish 4′-fluoro-2′,3′-*O*-isopropylidine-5′-*O*-sulfamoyladenosine (**356**) in 87% yield. Treatment of (**356**) with 90% trifluoroacetic acid then gave nucleocidin (**357**).[375]

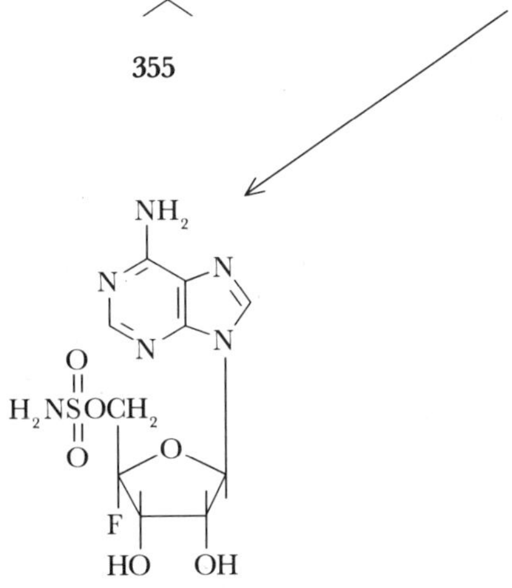

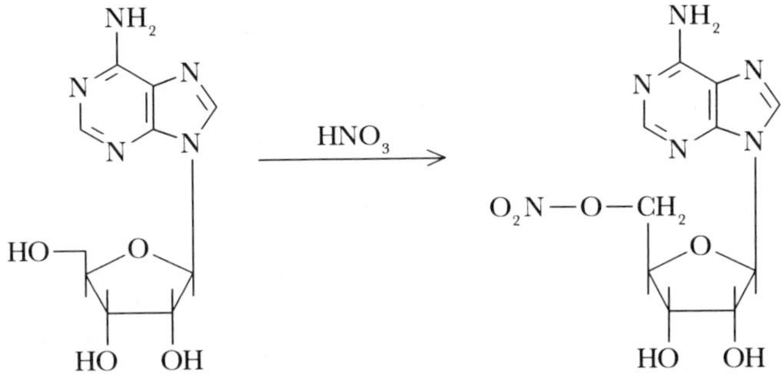

Duschinsky and co-workers successfully prepared adenosine 5′-nitrate (**358**) by adding adenosine (**128**) to cold fuming nitric acid, which had been pretreated with urea (to remove nitrous acid) and acetic anhydride (to scavange any water).[376] A series of di- and trinitrates of adenosine as well as nitrates of inosine, guanosine, and 3'- and 5'-AMP were prepared in a similar manner.[376] In a series of pyrimidine nucleoside nitrates, the ability of these nitrate esters to serve as leaving groups as well as hydroxyl blocking groups (removable by hydrogenolysis) was established.[377,378] The use of nitrate esters in purine nucleosides has not yet been exploited or studied.

The deoxynucleoside, in the absence of *cis* vicinal hydroxy groups, often react very differently from the nucleosides with *cis* vicinal hydroxy groups. As discussed before, Srivastava *et al.*[335] reported on the 5'-chlorination of 2',3'-O-isopropylidineadenosine by thionyl chloride (neat). The application of this reaction to uridine gave a 5',5'-bis(sulfite)[379] similar to the bis (3'-O-5-chloro-2,5-dideoxyadenosine)sulfoxide (**359**) formed by the thionyl chloride treatment of 2'-deoxyadenosine (**313**) in the presence of HMPA.[380] The bis(sulfite) group is hydrolyzed in methanol and concentrated ammonium hydroxide to give 5'-chloro-2',5'-dideoxyadenosine (**360**).[380]

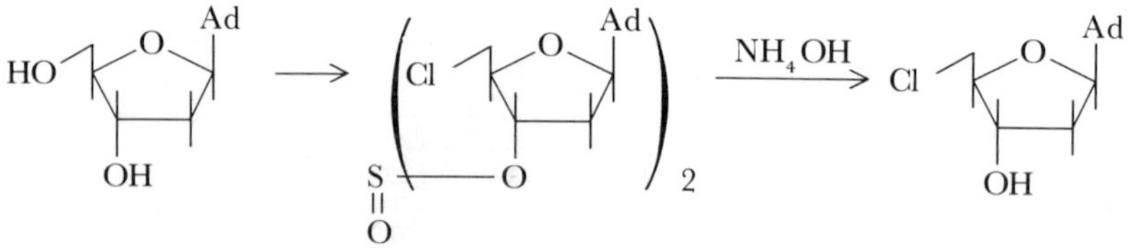

Some excellent new chemistry has been introduced by the reaction of 2'-deoxy-5'-O-trityladenosine (**361**) with triphenylphosphane/diethyl azodicarboxylate and NH_3. The reaction proceeds without neighboring-group participation and with inversion of configuration to yield 9-(3-azido-2,3-dideoxy-5-O-trityl-β-D-*threo*-pentofuranosyl)adenine (**362**).[381] The method could potentially be useful for the synthesis of other modified xylonucleosides.

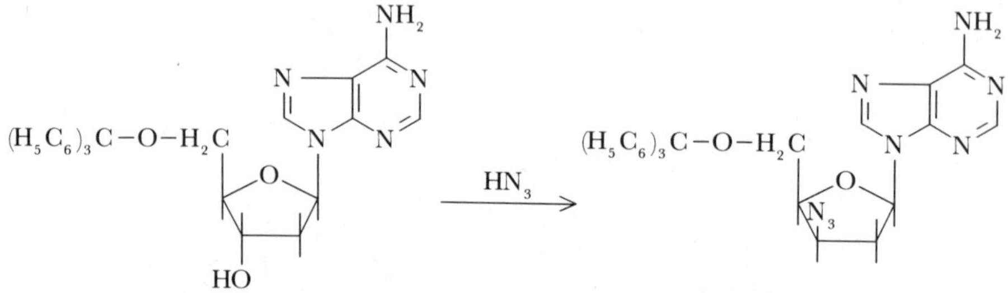

3.2. Reactions of the Purine Ring of Purine Nucleosides and Nucleotides

3.2.1. Electrophilic Substitution

Electrophilic substitution on the heterocyclic ring of purine nucleosides and nucleotides are considered in two types of processes: electrophilic substitution at carbon and electrophilic substitution at nitrogen (alkylation, acylation, oxidation).

3.2.1.1. Electrophilic Substitution at Carbon

Hydrogen Exchange

The simplest case of electrophilic substitution at a carbon atom of the purine ring system is hydrogen exchange. This reaction is usually measured by deuterium incorporation or detritiation, which occurs readily at the 8 position of nucleosides and nucleotides of guanine, adenine, and hypoxanthine and much more slowly at the 2 position of the latter two compounds.[382]

Whereas one study[383] has shown that the rate of exchange of adenosine was constant between pH 4 and 11, others have shown that the rate of exchange

Scheme V

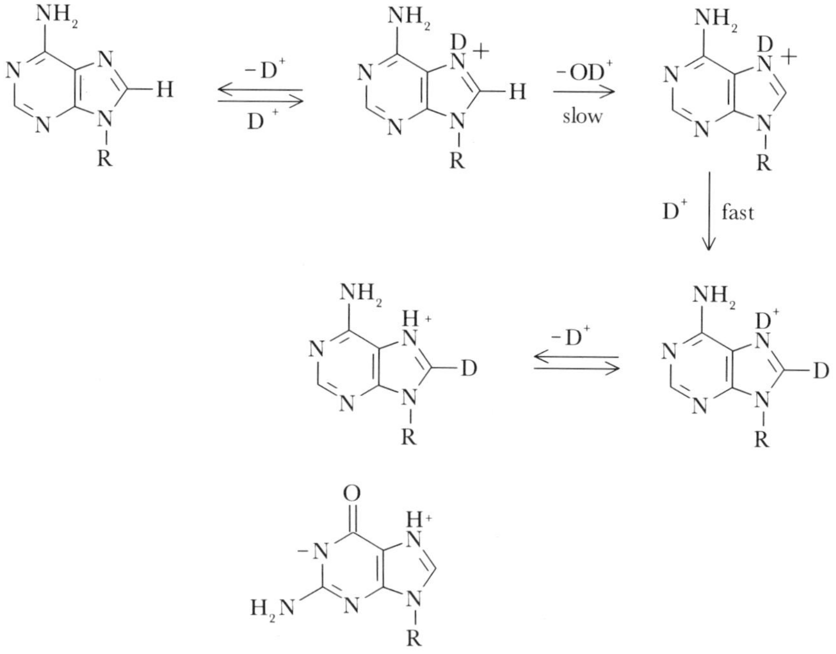

363

increased in the range of pH 2–4, was roughly constant up to pH 10[384] (or, in one case, pH 12),[385] and then abruptly increased with increasing pH. These data have been interpreted to suggest a mechanism involving an initial protonation at N-7, followed by proton abstraction from C-8 by hydroxide (or deuteroxide) (see Scheme V).

With guanosine and inosine, an ionization of the hydrogen on N-1 supplies an additional factor in the analysis. The rate of exchange for guanosine was pH independent up to pH 7, then rose sharply to a plateau some tenfold faster at pH 10.[383,386] Inosine was found to behave in a similar manner.[385,386] The high-pH rate boost has been attributed to an attack of hydroxide on the zwitterionic tautomer of guanosine.[383]

The rates of exchange for several purine nucleosides in neutral solution have been studied. Of that group, 9-(β-D-ribofuranosyl)purin-6(1H)-thione showed the fastest exchange.[385] The hydrogen at the 8 position of the purine moiety can also be exchanged with lithium as discussed in the latter part of this section.

Bromination

The introduction of bromine into the 8 position of the heterocyclic ring of various purine nucleosides and nucleotides has provided intermediates that have proved invaluable for the preparation of many 8-substituted derivatives. Mild conditions were necessary for this reaction in order to avoid a glycosidic cleavage. Holmes and Robins were successful in obtaining 8-bromo-2′,3′,5′-tri-O-acetylguanosine (**365**) in a 76% yield by treatment of 2′,3′,5′-tri-O-acetylguanosine (**364**) with Br_2 in sodium acetate–glacial acetic acid.[387] Likewise, 8-bromo-2′,3′,5′-tri-O-acetyladenosine (**367**) was obtained from 2′,3′,5′-tri-O-acetyladenosine (**366**) in 59% yield.[387]

These conditions were too acidic for the bromination of the more labile deoxynucleosides and prompted the study of more neutral apolar condtions. Treatment of 2′,3′,5′-tri-O-acetyladenosine or 2′,3′,5′-tri-O-acetylinosine with N-bromoacetamide in chloroform also gave the corresponding 8-bromo derivatives. This procedure was also adopted for the treatment of 3′,5′-di-O-acetyldeoxyadenosine to furnish the corresponding 8-bromo derivative.[387] All

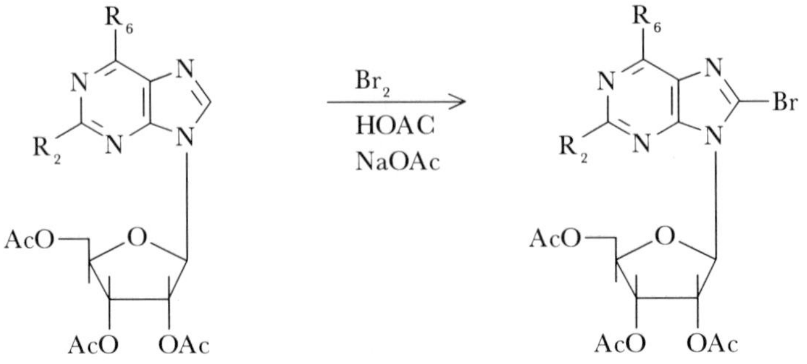

364, $R_2 = H_2N$, $R_6 = OH$
366, $R_2 = H$, $R_6 = NH_2$

365, $R_2 = H_2N$, $R_6 = OH$
367, $R_2 = H$, $R_6 = NH_2$

these 8-bromo derivatives gave, on treatment with methanolic ammonia, the corresponding deacetylated 8-bromopurine nucleoside or deoxynucleoside.[387] *N*-Bromosuccinimide has also been shown to brominate guanosine and adenosine, but not inosine, using DMF as the solvent.[388]

It is perhaps conceivable that the reaction of the purine nucleoside with *N*-bromoacetamide proceeds *via* a free radical and not by an electrophilic mechanism, but the exact nature of this reaction has not been elucidated. The fact that the reaction proceeds readily without the usual radical initiators (e.g., light, peroxide) would seem to support the heterolytic mechanism.

The disodium salts of purine nucleotides undergo bromination in water containing an additional 1 equivalent of sodium hydroxide and 1 equivalent of bromine.[389] In a similar fashion, it was possible to brominate adenosine, 2′-deoxyadenosine, and 2′,3′-*O*-isopropylidine adenosine, using dioxane and 10% aqueous Na_2HPO_4 in the latter case.[389]

The bromination of purine nucleosides and nucleotides in aqueous solution has been adopted as strandard practice. Treatment of the very labile 2′-deoxyguanosine with bromine water in an aqueous suspension[390] gave the 8-bromo derivative in 46% yield. Xanthosine gave 8-bromoxanthosine under similar conditions.[390] The appropriate use of buffers can minimize the possibility of a glycosidic cleavage. Thus, adenosine, AMP, ADP, ATP,[391] and cyclic AMP[391,392] gave the corresponding 8-bromo derivatives on treatment with bromine in a pH 4 sodium acetate buffer. Cyclic GMP likewise gave a good yield of 8-bromoguanosine cyclic 3′,5′-phosphate,[393] while GDP and GTP were brominated in formamide with bromine.[394]

The direct introduction of chlorine or iodine into the 8 position of a purine nucleoside or nucleotide is more difficult than bromination and therefore has received little attention. However, the direct chlorination of adenosine and its mono- and diphosphate derivatives has been achieved by using tetrabutyl-ammonium iodotetrachloride in DMF, although in low yield.[395] More recently, a reaction between *m*-chloroperbenzoic acid and purine nucleosides in dipolar aprotic solvents containing HCl has been described to result in a direct chlorination. The reaction has provided good yields of 8-chloroadenosine, 8-chloroguanosine, and 8-chloro-9-(β-D-arabinofuranosyl)adenine from the corresponding nucleoside substrates.[396]

The direct iodination of guanosine, giving 8-iodoguanosine in 70% yield, has been reported to occur in DMSO as the solvent in the presence of an excess of *N*-iodosuccinimide and a catalytic amount of dibutyl disulfide.[397] The reaction does not occur without the sulfide catalyst.[397] Deoxyguanosine and guanosine 2′(3′)-phosphate were also iodinated, but adenosine was found to react very slowly.[397] Little subsequent use has been made of this interesting reaction.

Diazo Coupling

A study on the coupling of various nucleic acids with 2,5-disulfobenzene-diazonium chloride (**368**) revealed that, at pH 9, 2′-deoxyguanylic acid (**369**) reacted 60 times faster than any other nucleotide studied.[398] This product was inferred to be 8-(2,5-disulfophenyldiazo)-2′-deoxyguanosine 5′-phosphate (**370**). This assumption was made on the basis of the report by Fischer[399] that guanine couples with phenyldiazonium salts in the 8 position. Kossel also found[400] that guanylic acid reacted much faster with diazotized sulfanilic acid than adenylic or cytidylic acids but assigned the structure of his products as being coupled at the N^2 position. Xanthylic acd was a hydrolytic product of this reaction. It was also found that guanosine (**371**) reacts with diazotized sulfanilic acid (**372**) at pH 9, with nitrogen then being lost to give the 8-arylated products, 8-(*p*-sulfophenyl) guanosine (**373**) and 8-(*p*-sulfophenyl)xanthine (**374**).[401] The reactions of ben- zenediazonium ions with guanosine and 5′-guanylic acid in basic solution (pH 8.5 or 10.5) have recently been reinvestigated.[402] Guanosine was found to react slowly to give 8-arylguanosines while 5′-guanylic acid reacted slowly, at ambient temperature, only with compounds bearing strong electron-withdrawing groups to yield N-2 triazenes. No 8-aryl- or 8-arylazo-5′-guanylic acids are formed at

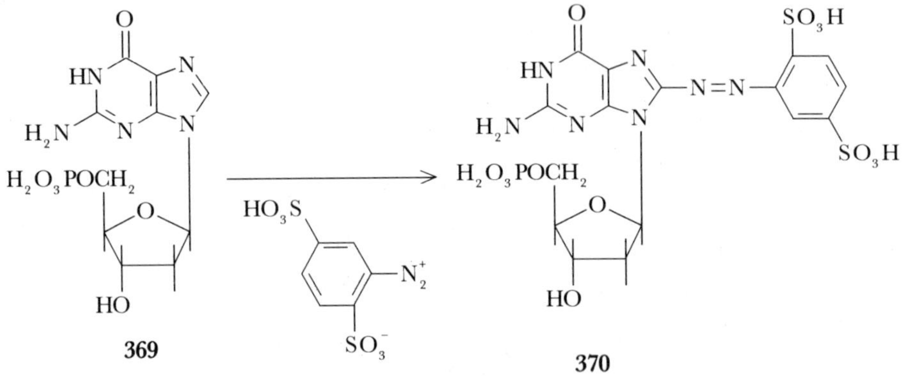

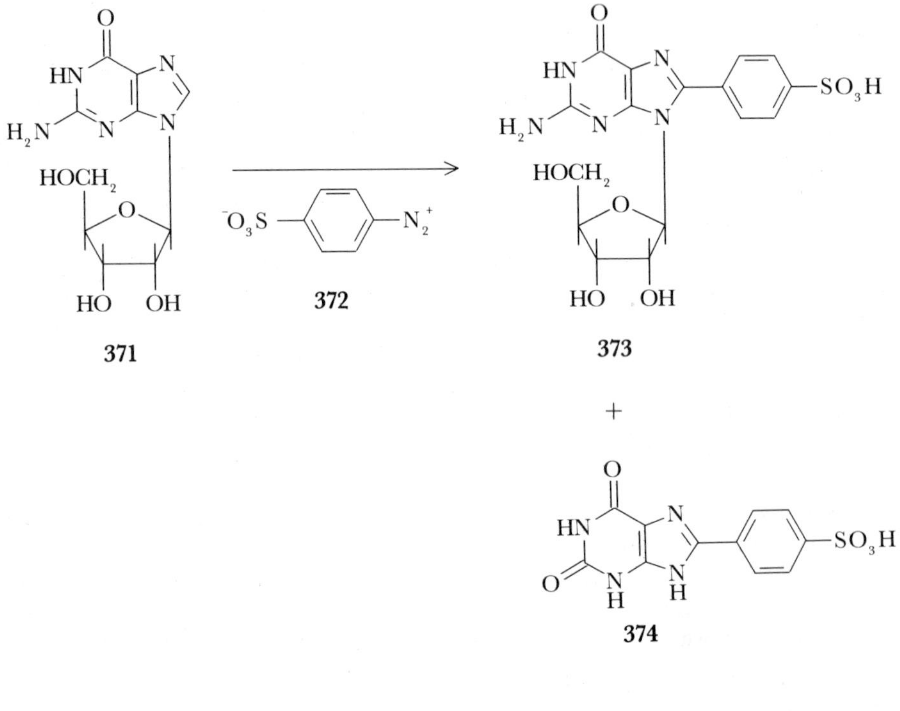

ambient temperature.[402] However, certain 4-substituted benzenediazonium ions do react with 5'-guanylic acid at a higher temperature to give 8-aryl-5'-guanylic acids in low yield. The structures of these compounds were established by a hydrolytic conversion to the corresponding 8-arylguanines.[402] The reaction of diazotized sulfanilic acid and adenosine 1-oxide at pH 10.5 was found to give N^6-(*p*-sulfophenylazo)adenosine 1-oxide.[403] The product was converted to adenosine 1-oxide at pH 3.

Amination

Several purine nucleotides and nucleosides have been observed to react with compounds containing electrophilic nitrogen. Many of these reactions may be related to the mechanism by which certain compounds display carcinogenesis. Treatment of guanosine **(371)** with hydroxylamine-*O*-sulfonic acid at pH values between 2 and 4 gave[404] 8-aminoguanosine **(375)** (approximately 25%). At pH 7, this reagent gave 1-aminoguanosine **(376)**, indicative of attack on the monoanion of guanosine.[405] The mechanism for the former reaction is not well understood and may be a mechanism other than a simple electrophilic attack at C-8. Adenosine did not react under the above-described conditions.

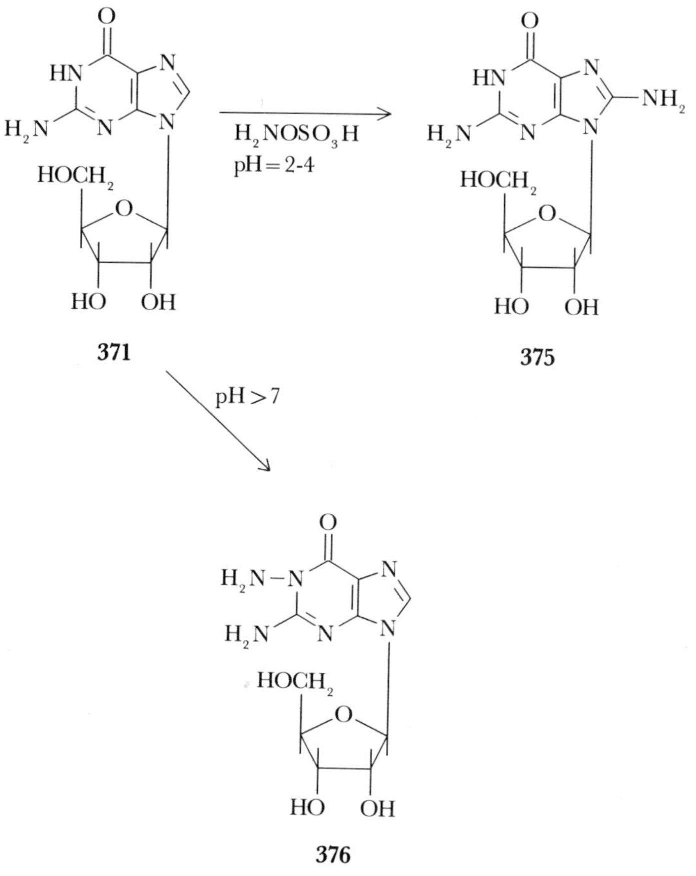

3.2.1.2. Electrophilic Substitution at Nitrogen

N-Oxidation

The first report on the synthesis of a purine N-oxide by a direct oxidation was the preparation of an adenine monooxide from the treatment of adenine with acetic acid/hydrogen peroxide.[406] The direct preparation of N-oxides of adenosine (**377**) and 2′,3′-O-isopropylidine adenosine was also reported.[406] The position of electrophilic attack was established as the 1 position.[407]

Brown's group then prepared the N-1-oxides of adenosine 2′-phosphate, 3′-phosphate, 5′-phosphate, and 5′-diphosphate, as well as 2′-deoxyadenosine 5′-phosphate.[408] The N-1-oxides of 2′-deoxyadenosine and 2′-deoxyadenylic acid were prpared from the corresponding nucleoside and nucleotide by oxidation with monoperphthalic acid in water, at pH 5. The yields were somewhat better than that obtained by the $AcOH/H_2O_2$ procedure.[409]

Attempts to prepare an oxide of inosine or its 5′-phosphate by this method failed, probably because of the lack of nucleophilicity of N-1 in the keto tautomer of inosine.[410] Inosine 1-oxide (**378**) was subsequently prepared *via* a nitrosyl chloride deamination of adenosine 1-oxide (**377**).

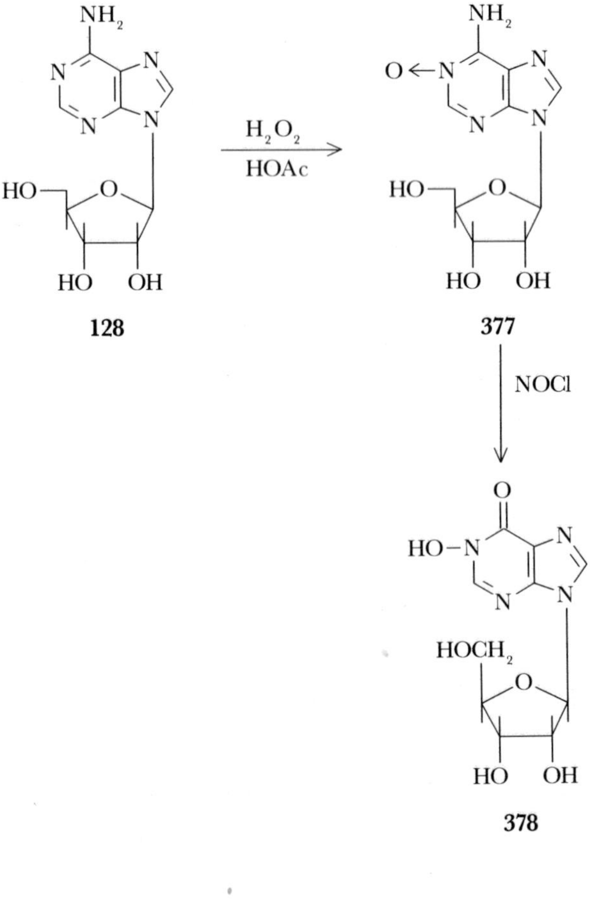

The synthesis of N-oxide nucleosides has been accomplished on a large scale, for example, adenosine cyclic 3′,5′-phosphate N-1-oxide has been prepared from cyclic AMP by treatment with *m*-chloroperbenzoic acid in a two-phase system. Acidification of the aqueous phase gave the crystalline product in a 90% yield.[411]

The major synthetic utility of the adenine N-1-oxide nucleotides and nucleosides lies in the labilization of the pyrimidine portion of the purine ring after the oxidation. This subject is covered in Section 3.2.4.

Treatment of 2′-deoxyadenosine 5′-phosphate with hydrogen peroxide, for 5 days at 37°C and pH 7.4, gave no detectable N-1-oxide. Instead, dAMP-7-oxide was the only adenine nucleotide oxide[412] found under these conditions. Substituted adenosine derivatives have also been found to undergo N-oxidation. 8-Bromoadenosine cyclic 3′,5′-phosphate gave the N-1-oxide on treatment with *m*-chloroperbenzoic acid.[413] 8,5′-Anhydro-8-mercaptoadenosine also gave the N-1-oxide in approximately 94% yield when treated with perphthalic acid with no S-oxides being formed.[414]

C- and N-Alkylation of the Purine Ring

Studies on the alkylation of nitrogen atoms in purine nucleosides and nucleotides has contributed to the understanding of the mechanism of action of some mutagenic alkylating agents and cancer chemotherapeutic agents. Additionally, many of the heterocycles in certain species of tRNA contain methyl groups that have been added after the synthesis of the tRNA strand.[415]

Early studies on the methylation of adenosine with dimethyl sulfate in aqueous sodium hydroxide gave an adenosine derivative that was permethylated on the sugar hydroxyls and with one methyl group on the purine ring. The position of the latter methyl group was not determined at this time.[416] Further studies[417-420] provided conflicting results regarding the position of this methylation. The products were subsequently characterized as adenines, after degradation of the nucleoside, and this work has been thoroughly reviewed.[421,422]

Selective methods for the direct C-alkylation of purine nucleosides are much fewer in number than N-alkylation. Various 8-alkylpurine nucleosides have been synthesized *via* a lithiation procedure. The feasibility of a carbon alkylation process was established[423] by a successful transformation of N^6,N^6-dimethyl-5′-O-methyl-2′,3′-O-isopropylidineadenosine (**379**) into the corresponding 8-methyl (**380**) and 8-ethyl (**381**) derivatives *via* lithiation and treatment with the appropriate alkyl iodide.[423] Butyl lithium in tetrahydrofuran at −78°C was used for the lithiation of nucleosides. A similar lithiation of 6-chloro-9-(2,3-O-isopropylidine-β-D-ribofuranosyl)purine (**382**) with butyl lithium gave a mixture from which 8-butyl-6-chloro-9-(2,3-O-isopropylidine-β-D-ribofuranosyl)purine (**383**) was isolated.[424] This indicated, in part, an attack of butyl lithium on the 6 position of purine to generate butyl chloride.[424] However, lithiation of (**382**), with lithium diisopropylamide, in THF below −70°C, proceeded at the 8 position in more than an 80% yield. Treatment of the lithiated intermediate with various electrophiles gave C-8-substituted nucleosides (**384**).[424] The C-substitution of nucleosides by the Eschemoser sulfide contraction is discussed in Section 3.2.2.

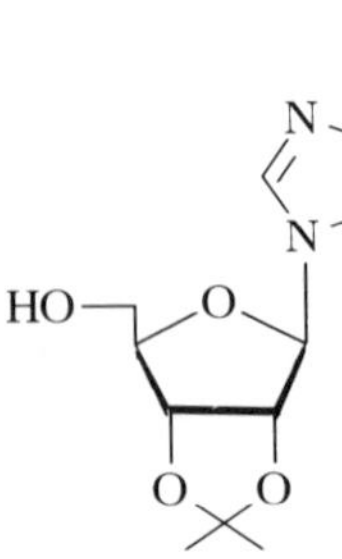

382

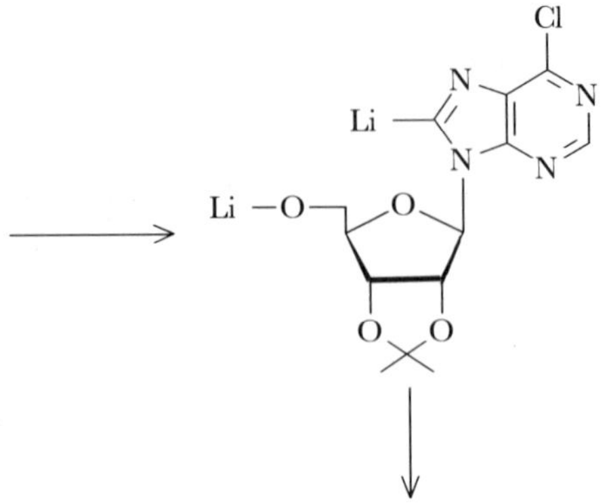

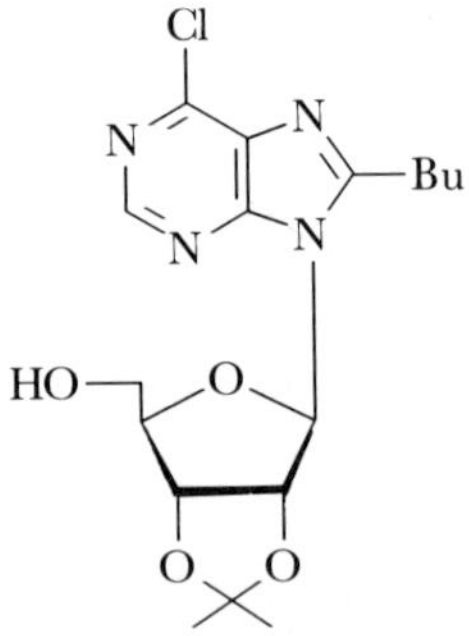

383

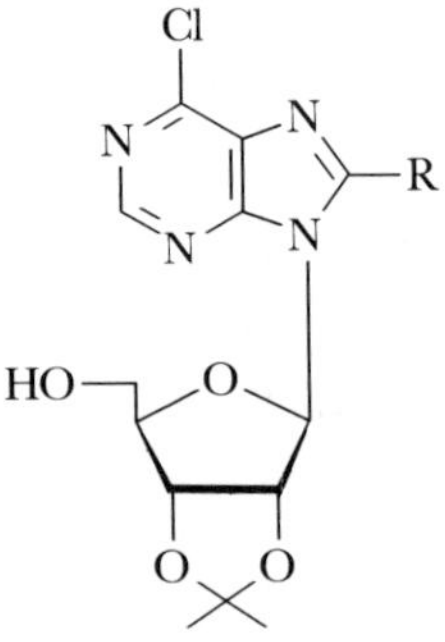

384

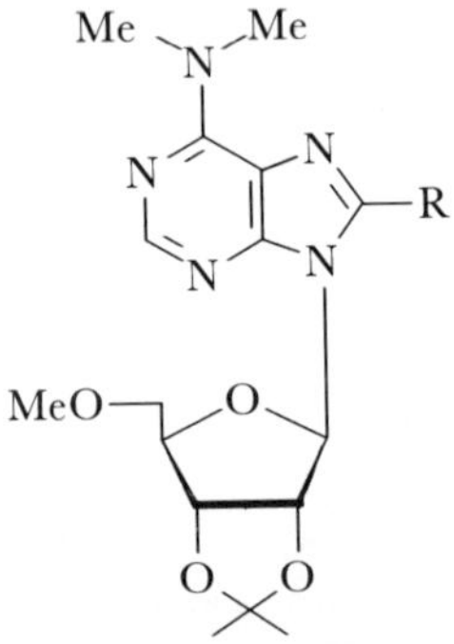

379, R = H
380, R = Me
381, R = Et

A careful investigation of the reaction of methyl *p*-toluenesulfonate (Ts-OCH$_3$) with adenosine (**128**) in *N*,*N*-dimethylacetamide at room temperature revealed that the *sole* product was 1-methyladenosine (**385**), isolated as the tosylate salt.[421] The structure proof for (**385**) was by a Dimroth rearrangement in NaOH (see Section 3.2.4) to give the known N^6-methyladenosine. 1-Methyl-adenosine was also obtained from a reaction of adenosine with excess methyl iodide in dimethylacetamide. The free base of this compound was then liberated by an adjustment of the pH to 8–9 with NH_4OH. It has also been found that ethylene oxide and adenosine formed 1-(2-hydroxyethyl)adenosine exclusively at room temperature.[425] Deoxyadenosine also yielded only the 1-methyl derivative from MeI/DMF.[421]

Early reports[418,426] of the preparation of 1-methylguanosine from guanosine and diazomethane, at approximately neutral pH, were later shown to be incorrect since the product obtained by an acid hydrolysis of this methylated nucleoside was 7-methylguanine.[421] In fact, guanosine (**371**) reacts readily with MeI at ambient temperature in dimethylacetamide to give the HI salt of 7-methyl-guanosine (**386**) in good yield.[421] It was also found that deoxyguanosine and inosine (**387**) give 7-methyl-2′-deoxyguanosine and 7-methylinosine (**388**), respectively, under similar conditions. In the presence of K_2CO_3, inosine was methylated in the 1 position to give 1-methylinosine (**389**).[421] It had previously been shown that inosine could be alkylated on the 1 position with benzyl chloride[427] or methyl chloromethyl ether[428] with basic catalysis. Furthermore, treatment of 2′,3′,5′-tri-O-acetylinosine in sodium hydride/DMF with *p*-toluenesulfonyl chloride gave 1-*p*-toluenesulfonylinosine triacetate.[429] These 1-substituents of inosine were found to induce a labilization of the purine ring (see Section 3.2.4).

This concept of alkylating the anion of a 6-oxopurine nucleoside was extended to guanosine (**371**) and deoxyguanosine, with their N-1-methyl derivatives [i.e., 1-methylguanosine; (**390**)] being obtained from their treatment with MeI in K_2CO_3/DMSO.[422] Furthermore, it was found that the free base of 7-methylguanosine (**386**) could be methylated at the N-1 position to give 1,7-dimethylguanosine iodide (**391**).[422] A later investigation found that guanosine reacts with methyl or ethyl iodide in K_2CO_3/DMF to give, in addition to the predominant N-1-alkyl guanosine, smaller amounts of 7-alkyl, 6-O-alkyl, 1,7-dialkyl, and other guanosine derivatives.[430]

Treatment of inosine with 2-dimethylamino-1,3-dioxolane (dimethylform-amide ethyleneacetal) at 110°C gave a 42% yield of 1-(2-hydroxyethyl) inosine.[431] This apparently represents one of the few cases where N-1-alkylation of inosine is accomplished without basic catalysis. This reagent did not alkylate on a ring nitrogen of guanosine or adenosine, but instead gave the exocyclic-*N*-dimethylaminomethylene derivatives of these compounds.[432] However,

128   385

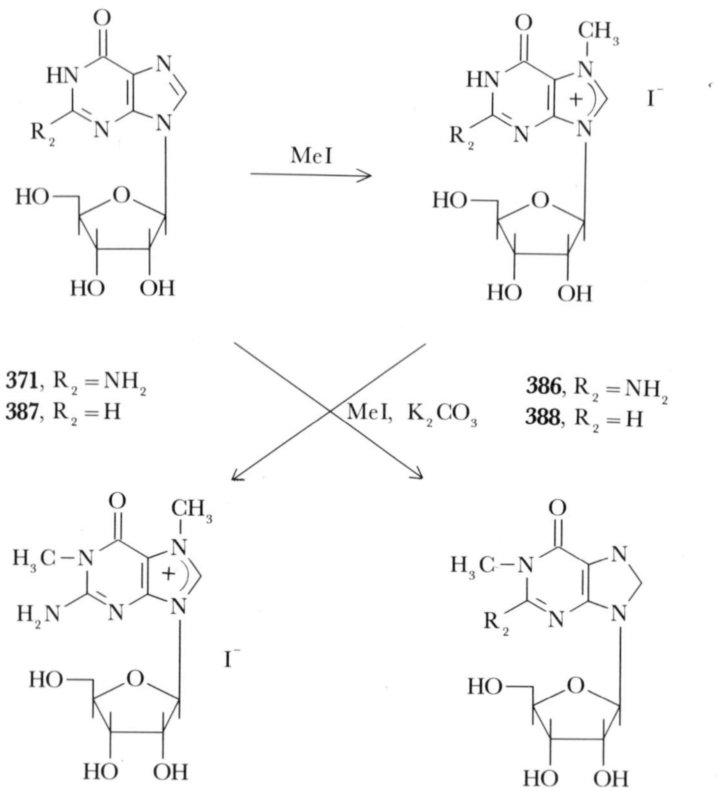

371, $R_2 = NH_2$
387, $R_2 = H$

386, $R_2 = NH_2$
388, $R_2 = H$

391

389, $R_2 = H$
390, $R_2 = NH_2$

guanosine did react with the dibenzyl acetal of dimethylformamide to give 1-benzyl-N^2-(dimethylaminomethylene)guanosine.[433] The benzyl group was readily removed by sodium naphthalene in dioxane.[433]

These reactions have been adapted for synthetic usage. The preparation of N^6-monoalkyladenine nucleosides and nucleotides is readily accomplished *via* a N-1-alkylation and subsequent rearrangement (see Section 3.2.4).

An N^6-methoxy group in adenosine was found to have a directive effect on the alkylation of the nucleoside. Treatment of N-methoxyadenosine (392) with methyl iodide gave the 7-methyl isomer (393) as the major product with the N^6-methyl isomer (394) being the minor product. Hydrogenolysis of (393) gave the first preparation of the very labile 7-methyladenosine (395).[434]

Ring alkylation of purines, especially derivatives of adenine and guanine, with bifunctional reagents of the type XCHRCOR′ or RCOCOR′, where X = halogen, results in a new heterocyclic ring being fused to the purine. One of the first and most useful examples of this was the reaction of chloroacetaldehyde with adenosine (128) to form N-1,N^6-ethenoadenosine [3-(β-D-ribofuranosyl) imidazo[2,1-i]purine, (396)].[435,436] This compound was found to be highly fluorescent and of use as a fluorescent probe into the structure of more complex

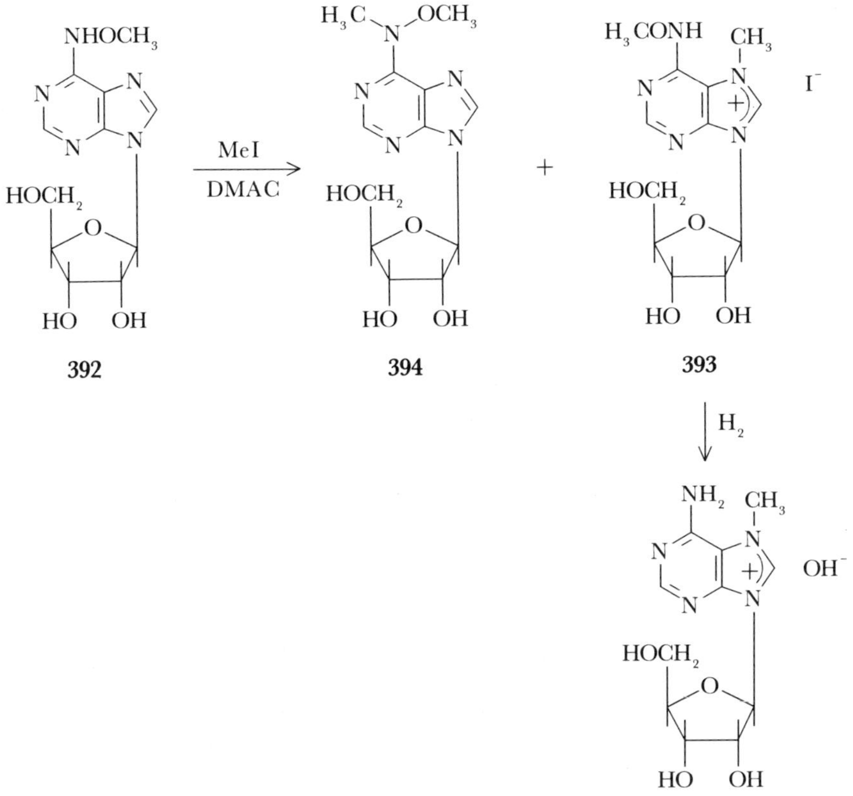

392   394   393

395

oligonucleotides.[436] Since chloroacetaldehyde was found to be unreactive with 1-methyladenosine, it was presumed that the reaction proceeds *via* an initial nucleophilic attack of N-1 on chloroacetaldehyde, displacing chloride, followed by ring closure of the aldehyde with the N^6 group followed by dehydration.[436]

This reaction, which takes place in an aqueous solution between the adenine derivative and excess chloroacetaldehyde at pH 4–5, has been extended to the preparation of N-1,N^6-etheno-cyclic AMP,[437] NAD,[438] and various other adenine-containing compounds.[439] Furthermore, it has been shown to occur between chloroacetaldehyde and a variety of 8-substituted adenine derivatives; for example, 8-bromo- and 8-alkylthioadenosine cyclic 3′,5′-phosphates yield the corresponding N-1,N^6-etheno-8-substituted cyclic AMP derivatives.[440] The reaction has also been expanded with cyclic AMP (**397**) to include the introduction of substituents on the etheno bridge by use of an alkylating agent such as $R-CO-CH_2-X$. In this case R appears at the 8 position (**398**) of the imidazo[2,1-*i*] purine, while with $R-CHX-CHO$ the R group resides at the 7 position (**399**) of the new ring system.[411,440]

An additional ring has also been fused onto the guanosine ring system by ring closure using electrophilic reagents. Thus, 1-aminoguanosine (**376**) and the Vilsmeyer reagent, $Me_2N=CClH^+PO_2Cl_2^-$ (from DMF and $POCl_3$), gave

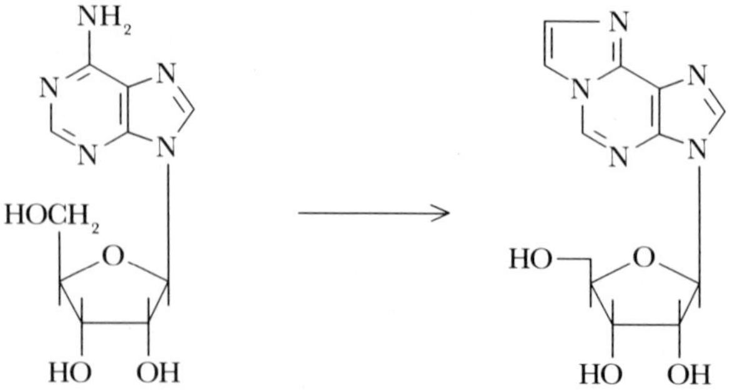

128   396

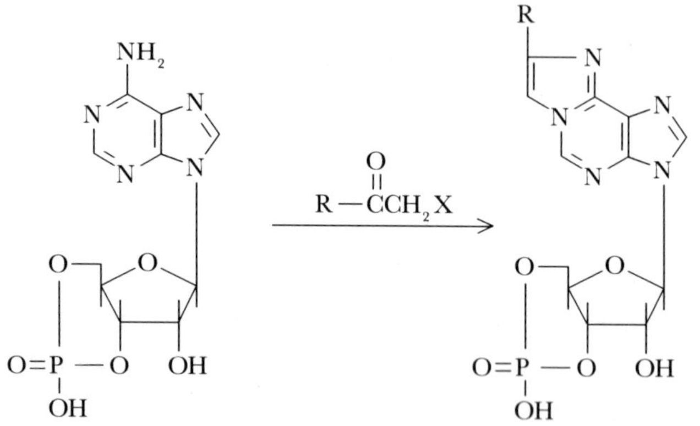

397   398

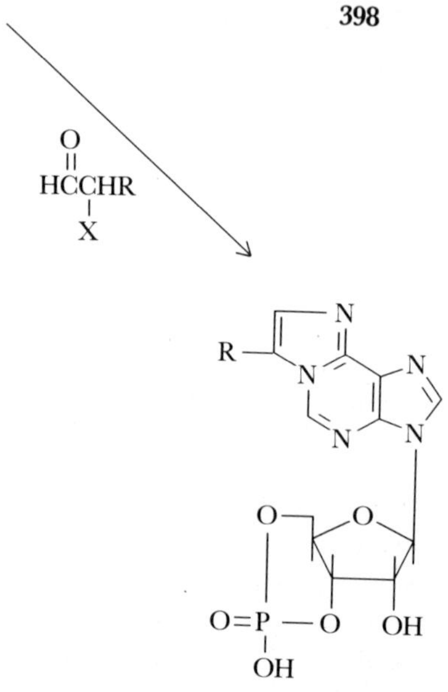

399

9-oxo-3-(β-D-ribofuranosyl)-1,2,4-triazolo[2,3-a]purine **(400)**.[441] Compound **(400)** was fluorescent, as was the 6,7-dimethyl-10-oxo-3-(β-D-ribofuranosyl)-1,2,4-triazino[2,3-a]purine **(401)** obtained from the reaction of 1-aminoguanosine **(376)** and biacetyl.[441] Substitution of glyoxal for biacetyl had been expected to give **(402)**; however, it was found that this compound existed largely as its covalent hydrate **(403)** in solution.[441] Several adducts between N-1 and N^2 of guanosine and bifunctional electrophilic reagents such as glyoxal,[442] ninhydrin,[443] and glyceraldehyde[443] are known.

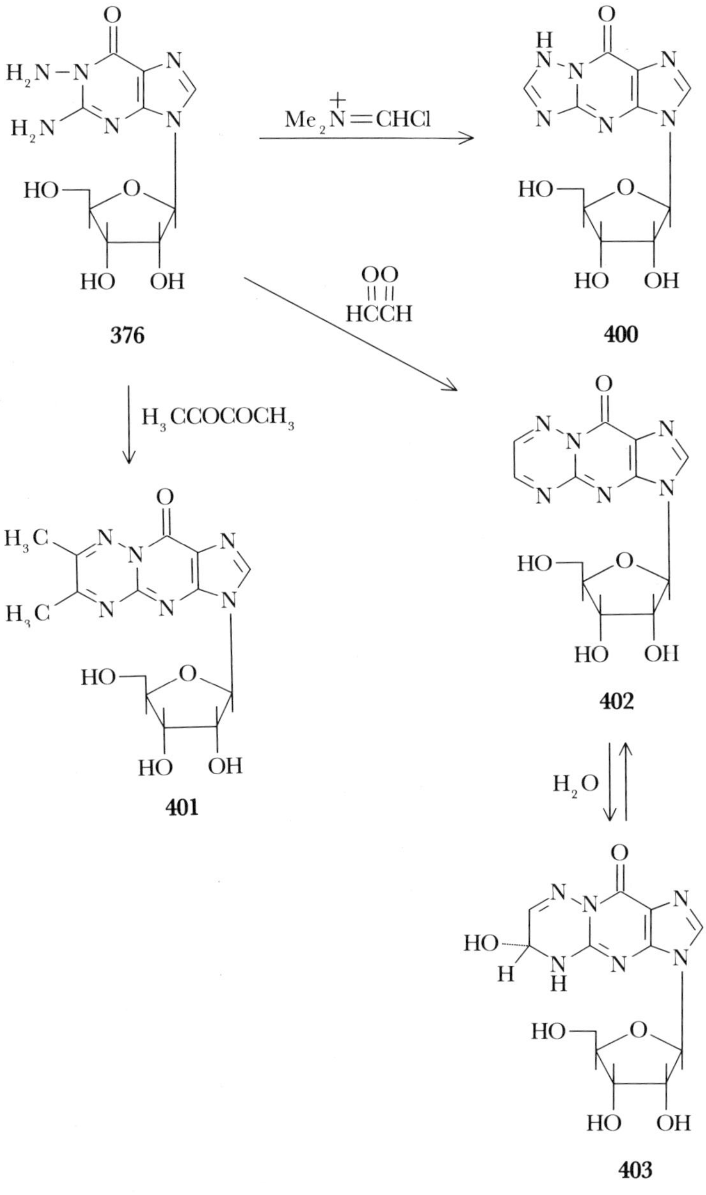

Other interesting nucleosides with a heterocyclic ring fused to the purine ring system have been prepared using β-acylvinylphosphonium salts as the electrophile. Adenosine (128) and β-acetylvinyltriphenylphosphonium bromide (404) in ethanol gave 8-methyl-3-(β-D-ribofuranosyl)imidazo[2,1-i]-purine-7-ylmethyl-triphenylphosphonium bromide (405). The structure was assigned by using the nuclear Overhauser effect observed between the 8-methyl group and H-5.[444,445] Reaction of the same phosphonium salt (404) and guanosine (371) gave 6-methyl -3-(β-D-ribofuranosyl)imidazo[1,2-a]purine-9-on-7-ylmethyltriphenylphosphonium bromide (406). The structure assignment of this compound was rationalized by postulating an attack of the electrophilic vinyl-β-carbon on N-1 of guanosine, followed by a ring closure.[446] These reactions also were found to occur in hot water and offer possible new means of nucleic acid modification and the preparation of compounds related to the Y-base[447] of tRNA.

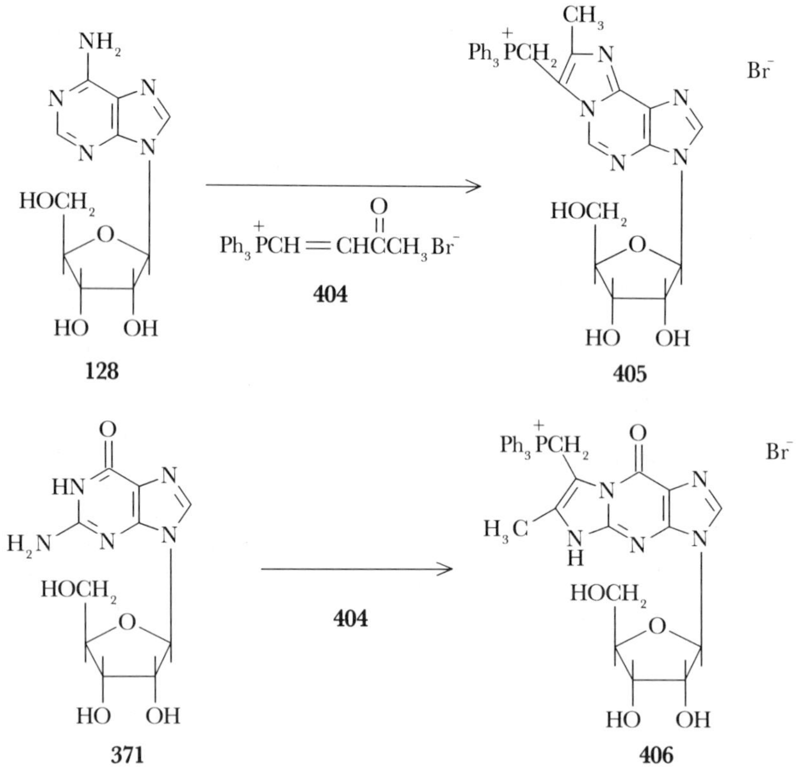

Another new tricyclic nucleotide system was obtained from the reaction of 5'-AMP (42) with cyanoacetylene. This reaction occurs *via* a Michael addition at N-1 of the adenine nucleus to the triple bond, followed by a ring closure of the N^6 group to the cyano function. The zwitterionic product, 9-amino-3-(β-D-ribofuranosyl)pyrimido[2,1-i]purin-6-ium 5'-phosphate (407), was formed in good yield in aqueous alcohol containing mercuric chloride at 0°C.[448] This suggests that reagent may also see some development as a nucleic acid modification reagent.

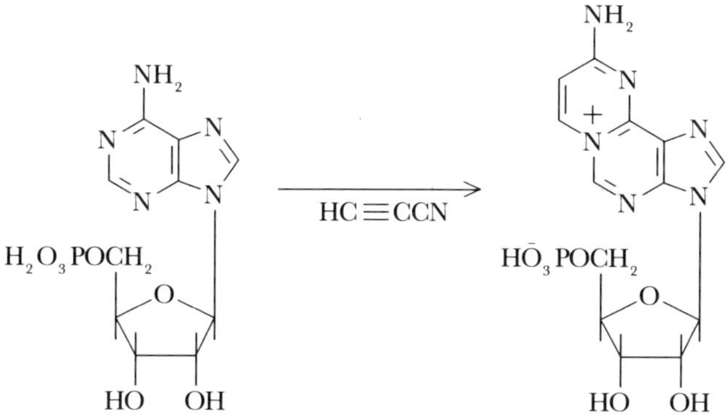

$$42 \xrightarrow{HC\equiv CCN} 407$$

N-Amination

Another possible mechanism of nucleic acid damage by nascent nitrogen cations is amination at a ring nitrogen. The study of N-amination might also be expected to provide new, easily removed blocking groups for the ionizable positions of guanosine and inosine. The first instance of direct purine amination by electrophilic attack at nitrogen was the synthesis of N-1-aminoguanosine (**376**). The corresponding deoxyguanosine and inosine analogues were also obtained by a treatment of the parent nucleoside with hydroxylamine-*O*-sulfonic acid in sodium hydroxide solution.[405] Furthermore, N-1-amination was observed with 8-benzyloxyguanosine under similar conditions. After 8-debenzylation, an amino group could then be introduced into the 7 position to give 1,7-diamino-8-oxoguanosine.[405] Also, 7-amino-8-oxoadenosine (**409**) was prepared from 8-oxoadenosine (**408**) by the same procedure.[405] As mentioned earlier, hydroxylamine-*O*-sulfonic acid has been shown to react with guanosine at pH 2–4 to give 8-aminoguanosine, presumably but not unequivocally by an electrophilic attack at C-8. However, under basic conditions, the N-1 position seems to be the most nucleophilic position on guanosine.

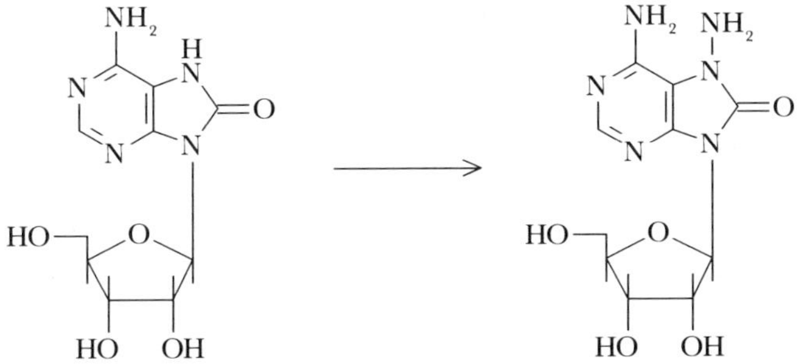

$$408 \longrightarrow 409$$

Other hydroxylamine derivatives were later found to be more potent aminating reagents. Adenosine (128) and cyclic AMP gave the N-1-amino derivatives [e.g., N-1-aminoadenosine, (410)] when treated with O-(2,4-dinitrophenyl)hydroxylamine in DMF.[449]

3.2.2. Nucleophilic Substitution

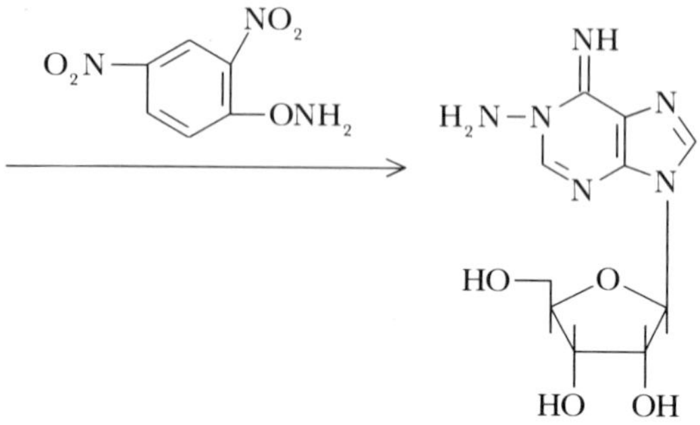

The displacement of a substituent on a purine nucleoside or nucleotide by a nucleophile has been by far the most convenient method of introducing new substituents into the purine ring. This versatile procedure involves (1) functionalization of the purine, if no readily displaceable groups are present, and (2) nucleophilic displacement by nitrogen, oxygen, sulfur, carbon, or other nucleophiles containing the desired substituents. Functionalization of the 8 position is normally accomplished by an electrophilic process that has already been discussed (e.g., bromination). Displacement functions are most commonly introduced at positions 2 and 6 by a functional group exchange, the essence of which is a nucleophilic reaction, using the naturally occurring nucleosides, or by the synthesis of a functionalized nucleoside.

3.2.2.1. Conversion of Hydroxypurines and Mercaptopurines to Halopurines. The preparation of a large number of 6-substituted purine nucleosides was made possible by the introduction of methods for directly converting the readily available 6-hydroxypurine nucleosides into 6-chloropurine nucleosides. The first method, and perhaps the most often used, was the treatment of 2',3',5'-tri-O-acetylinosine (411) or 2',3',5'-tri-O-acetylguanosine (364) with phosphoryl chloride and N,N-diethylaniline to give the corresponding 6-chloro-9-(2,3,5-tri-O-acetyl-β-D-ribofuranosyl)purine (412) and its 2-amino counterpart (413), respectively.[450] Omission of the tertiary amine from the reaction resulted in a glycosidic cleavage. The acetyl groups were readily removed with cold methanolic ammonia to give 6-chloro-9-β-D-ribofuranosylpurine (414) and the 2-amino analogue (415).[450]

Another versatile method of converting the blocked inosine to the blocked 6-chloropurine riboside was found[451–453] to be treatment with thionyl chloride and DMF in $CHCl_3$ or CH_2Cl_2. Because of the relatively mild conditions

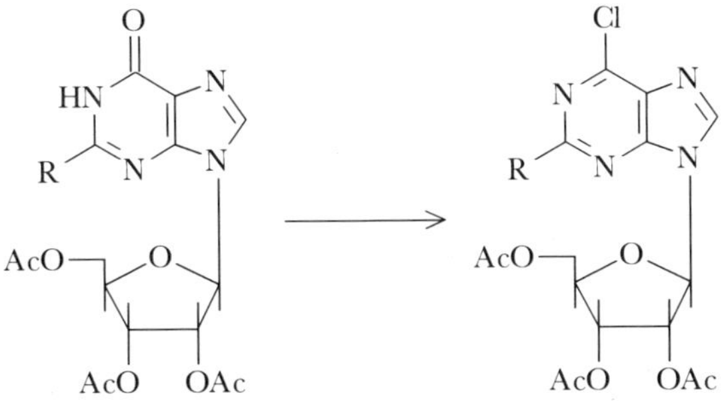

411, R = H
364, R = NH$_2$

412, R = H
413, R = NH$_2$

NH$_3$/MeOH

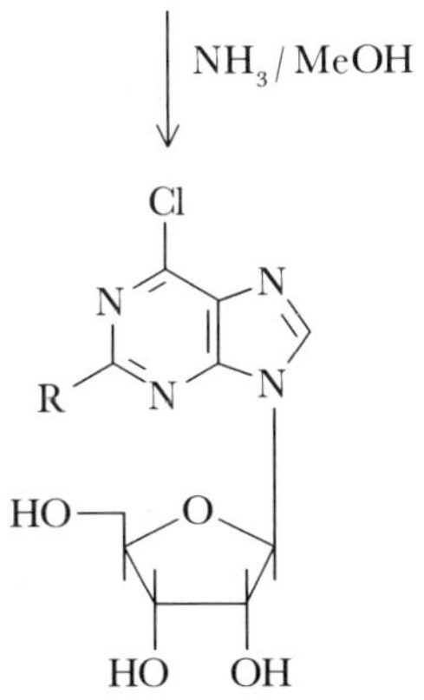

414, R = H
415, R = NH$_2$

employed (the 2′,3′-*O*-isopropylidine moiety was stable)[452] and the ease of workup, this procedure is used increasingly where solubility of the starting nucleoside is not a problem.

A reaction of the hydroxypurines with POCl$_3$ is analogous to the reaction of that reagent with alcohols. There is an initial phosphorylation of the oxygen (see Scheme VI) with the release of 1 mol of HCl; the chloride ion then displaces the dichlorophosphonate ion to give the 6-chloropurine. Since HCl is probably mostly undissociated in POCl$_3$ solution, addition of the tertiary base would be necessary to provide a free chloride ion. The SOCl$_2$/DMF procedure is analogous, with the intermediate chloromethylenedimethylammonium chloride reagent adding to the hydroxypurine, followed by an attack of the chloride ion that arises from the excess reagent.

The ring closure of a nucleoside 5′-phosphate, to a 3′,5′-cyclic phosphate, is a tedious procedure requiring high dilution (see Section 2.4.1). Therefore, the introduction of a displaceable chloride atom directly into the 6 position of a preformed cyclic nucleotide affords a more versatile route to a diverse number of compounds. Thus, 2′-*O*-acetylinosine cyclic 3′,5′-phosphate (**416**) was treated with POCl$_3$ to give, with or without a tertiary amine after deblocking, 6-chloro-9-

Scheme VI

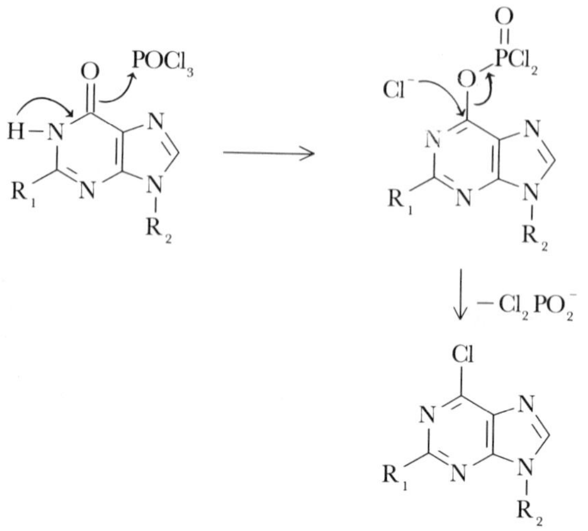

(β-D-ribofuranosyl)purine cyclic 3',5'-phosphate (**417**).[454,455] 2'-O-Acetyl-cyclic GMP (**418**) was also converted to 2-amino-6-chloro-9-(β-D-ribofuranosyl)purine cyclic 3',5'-phosphate (**419**) with $POCl_3$/diethylaniline.[456]

The generality of the $SOCl_2$/DMF reaction was exemplified by a conversion of the tri-O-acetate of 2-bromo-, 2-chloro-, and 2-butoxyinosine (**420**) to the corresponding 6-chloro analogues (**421**).[457] The conversion of a series of 2-alkyl- and 2-arylinosines[458] to their corresponding 6-chloro derivatives was also reported. The phosphoryl chloride method is equally adaptable. Treatment of 2',3',5'-tri-O-acetylinosine bearing 2-alkyl, 2-alkylthio and 2-alkylamino substituents gave good yields of the corresponding 6-chloro derivatives upon treatment with $POCl_3$.[459,460] Inosine derivatives with an 8-substituent have been converted into the 6-chloro derivative by a similar procedure. Thus,

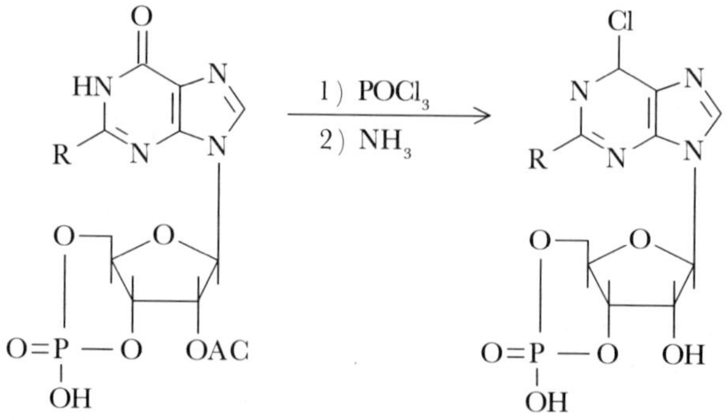

416, R = H
418, R = NH$_2$

417, R = H
419, R = NH$_2$

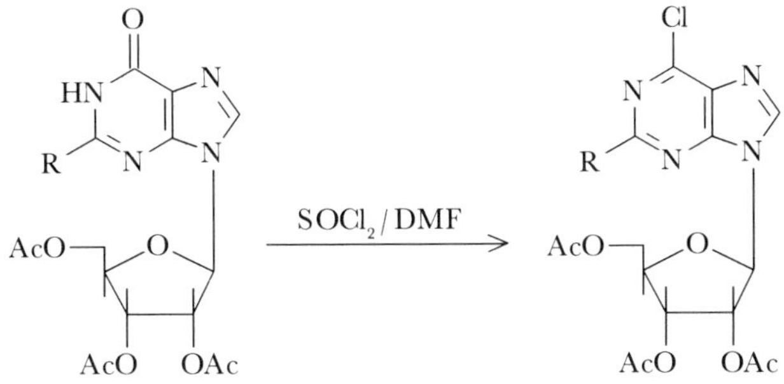

$$R = Cl, \ Br, \ O\text{-Alkyl,}$$
$$Alkyl, \ Aryl$$

8-benzylthio-, 8-methylthio-, and 8-*p*-chlorophenylthio-2'-*O*-acetylinosine cyclic 3',5'-phosphate, as their triethylammonium salts, were converted into the corresponding 6-chloro derivatives with $POCl_3$.[461] In this case, the triethylamine served as the acid acceptor in the reaction. Treatment of 8-acetamido-2',3',5'-tri-*O*-acetylinosine under similar conditions gave the 6-chloro derivative.[462]

The 2'-deoxynucleosides are much more prone to glycosidic cleavage under the above conditions. Studies on an alternative procedure furnished the interesting and useful approach that involved a trifluoroacetylation of the nucleoside substrate. Treatment of 2'-deoxy-3',5'-di-*O*-trifluoroacetylinosine with $SOCl_2$/DMF in methylene chloride at reflux resulted in an introduction of the 6-chloro function.[463] Subsequent chromatography on neutral alumina removed the trifluoroacetyl blocking groups, and 6-chloro-9-(2-deoxy-*β*-D-*erythro*-pentofuranosyl)purine was obtained in an 81% yield.[463] The efficiency of this conversion, on a very acid-labile nucleoside, was attributed to the electron-withdrawing nature of the trifluoroacetates. It was assumed that they destabilize glycosyl carbonium ion formation, a postulated step in nucleoside cleavage (see Section 3.4).

N-oxidation of the starting nucleosides provided another variant on this chlorination reaction. The blocked nucleoside 2',3',5'-tri-*O*-acetylinosine N-1-oxide (**422**) was treated with $POCl_3$ and tertiary amines to yield 2,6-dichloro-9-(2,3,5-tri-*O*-acetyl-*β*-D-ribofuranosyl)purine (**423**).[464] There were no chlorinated products obtained from blocked adenosine 1-oxide, which indicates that a mechanism such as that shown in Scheme VII must be invoked where the 6-chloro group is introduced first. This reaction did not proceed without a tertiary amine present. Tertiary alkyl amines, 2-picoline, and *N,N*-dimethylaniline were found to be the best catalysts.[464]

These reaction conditions were also used on 2'-*O*-acetylinosine cyclic 3',5'-phosphate 1-oxide. The prolonged reflux time (2 hr) gave the expected conversion of the purine, but the chlorination was accompanied by a cleavage of the cyclic phosphate.[456]

When 8-bromo-2',3',5'-tri-*O*-acetylguanosine (**365**) was treated with $POCl_3$ and *N,N*-diethylaniline, the 6-chloro group was introduced as expected. In

Scheme VII

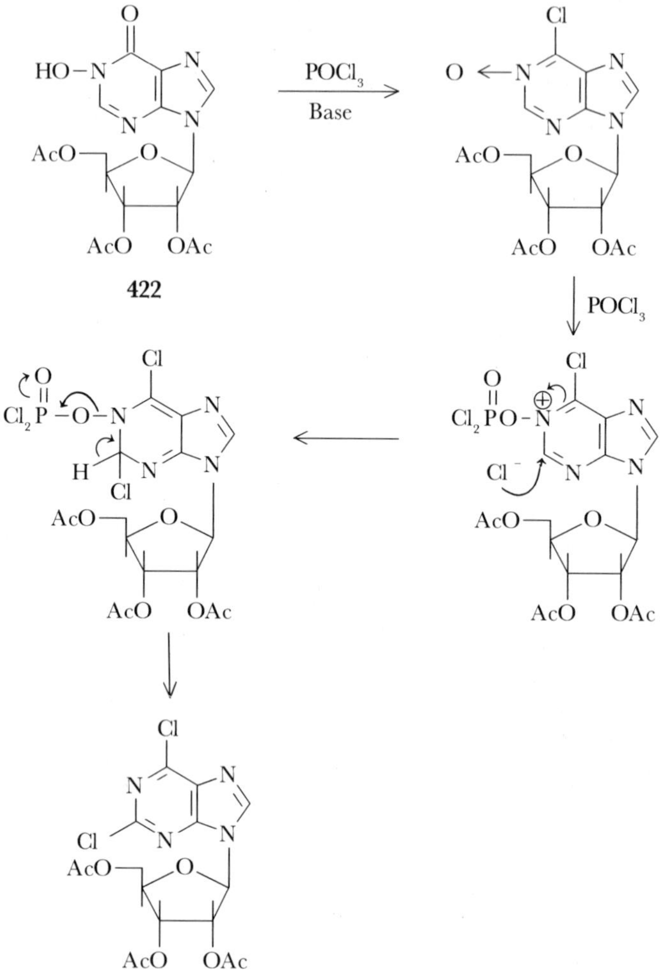

422

423

addition, the 8-bromo function was displaced by a chloride ion to give, after ammonia treatment, 2-amino-6,8-dichloro-9-(β-D-ribofuranosyl)purine **(424)** as the final product in 52% yield.[465] A similar reaction was also found to occur with 8-bromo-2′,3′,5′-tri-O-acetylinosine and gave 6,8-dichloro-9-(2,3,5-tri-O-acetyl-β-D-ribofuranosyl)purine in good yield, although it was not isolated.[462] The triethylammonium salt of 8-bromo-2′-O-butyrylinosine cyclic 3′,5′-phosphate **(425)** gave 6,8-dichloro-9-(2-O-butyryl-β-D-ribofuranosyl)purine cyclic 3′,5′-phosphate **(426)** in 56% yield.[461]

It thus appears that these reactions have wide synthetic utility for the introduction of a halogen into the purine nucleus in place of a hydroxyl function.

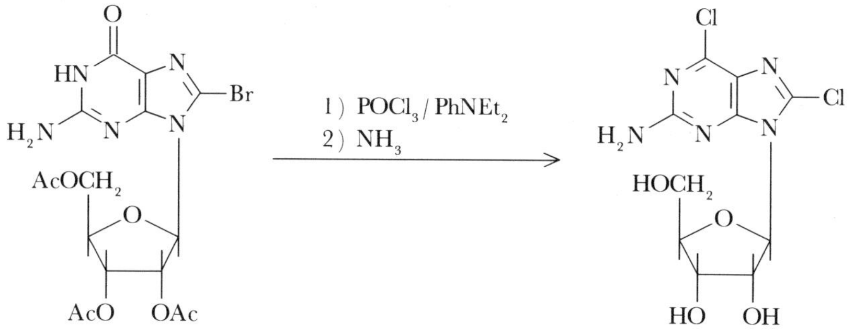

In particular, the $SOCl_2/DMF$ procedure appears to be the mildest and most convenient method available where solubility factors are not limiting.

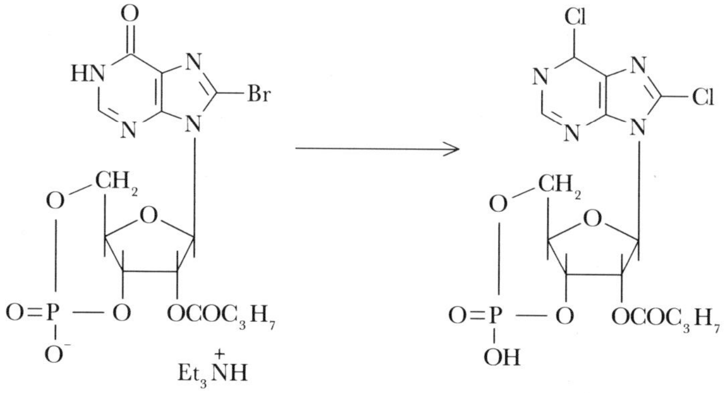

The feasibility of a replacement of another type of a functional group on a purine nucleoside by halogen was established by the synthesis of 6-chloro-9-(β-D-ribofuranosyl)purine (**414**), in 80% yield, from 6-mercapto-9-(β-D-ribofuranosyl) purine (**427**) by treatment with chlorine gas in cold methanol.[466,467] A similar treatment of 6-methylthio-9-(β-D-ribofuranosyl)purine gave a 90% yield of the same product.[466] While 2-amino-6-methylthio-9-(β-D-ribofuranosyl)purine (**428**) gave a 51% yield of 2-amino-6-chloro-9-(β-D-ribofuranosyl)purine (**415**). This procedure has already provided a variety of halopurine nucleosides and nucleotides[468–470] and is obviously applicable for the preparation of other diverse chloropurine nucleosides and nucleotides. 8-Chloroadenosine cyclic 3′,5′-phosphate was prepared from 8-mercapto cyclic AMP by a treatment with chlorine in methanolic HCl. The mechanism of this reaction has been rationalized[470] as an initial formation of the intermediate (**429**) followed by a nucleophilic displacement by chloride.

The preparation of iodopurine compounds, which are otherwise difficult to synthesize, has been accomplished by replacing chlorine with iodine in a modification of the preceding method. Thus, 2′-deoxy-8-mercaptoadenosine (**430**)

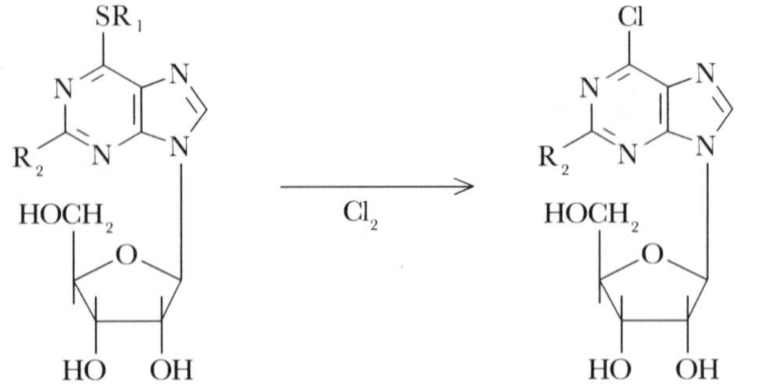

427, $R_1 = H$, $R_2 = H$
428, $R_1 = CH_3$, $R_2 = NH_2$

414, $R_2 = H$
415, $R_2 = NH_2$

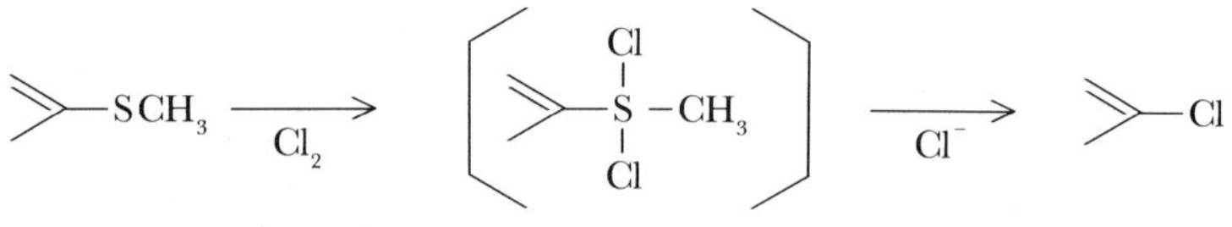

429

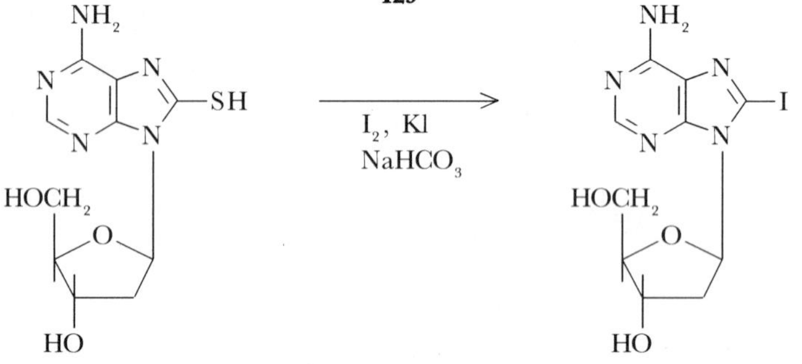

430

431

on treatment with iodine in aqueous sodium bicarbonate containing potassium iodide gave 2′-deoxy-8-iodoadenosine (431).[471] 8-Iodo-cyclic AMP was prepared in a similar manner from 8-mercapto-cyclic AMP.[472]

3.2.2.2. Nucleophilic Substitution on Aminopurine Nucleosides and Nucleotides Mediated by Nitrous Acid. The reaction of aminopurines with nitrous acid was investigated as early as 1861.[473] Since the early demonstration of Levene and Jacobs that nitrous acid can convert adenosine (128) to inosine (387) and guanosine (371) to xanthosine (432),[474] this reagent has found extensive use in nucleic acid chemistry. The reaction is very useful in a preparative sense and has found other uses as discussed below. The hydroxypurine products are useful intermediates, particularly in the synthesis of mercapto- and chloropurine nucleosides and nucleotides. The reaction is readily adaptable to a large scale, as exemplified by the preparation of cyclic IMP from cyclic AMP[454] in high yield. The reaction of nitrous acid with guanosine (371) produced a very interesting side product.

The structure for this side product was found to be 2-nitroinosine (**433**). This nucleoside (**433**) was found in yields of up to 5% if the reaction was run at 0°C with excess nitrite—conditions that might be expected to improve the efficacy of a Sandmeyer-type reaction. 8-Nitroxanthine (**434**) was also isolated from this reaction, but none of the corresponding nucleoside was obtained from the reaction.[475,476]

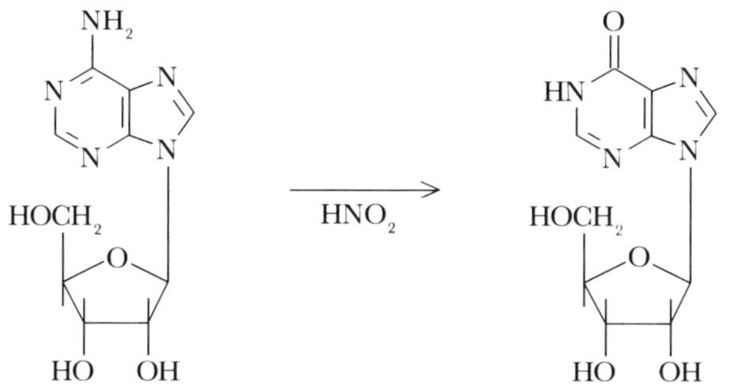

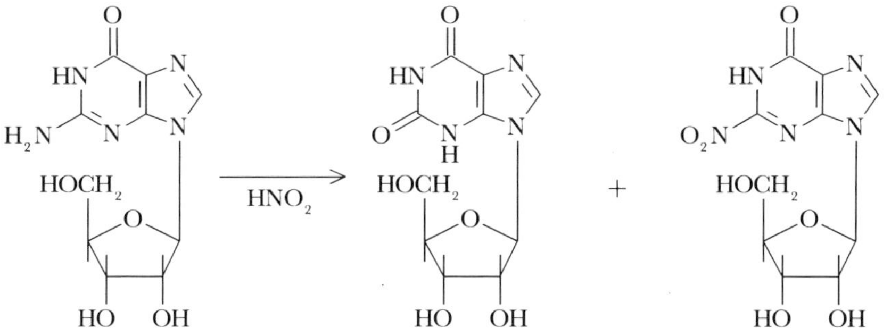

Nitrosyl chloride in DMF has been used to prepare inosine N-1-oxide and IMP-N-1-oxide from adenosine N-1-oxide and AMP-N-1-oxide,[410] respectively. The standard aqueous acetic acid/sodium nitrite method has also been used for the deamination of cyclic AMP-N-1-oxide[411] and adenosine N-1-oxide.[477]

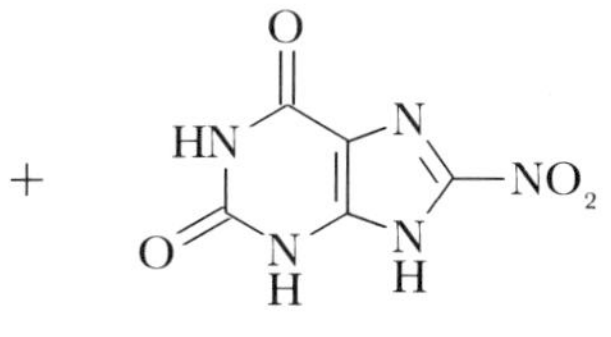

The nitrous-acid-mediated method for the conversion of adenosine to inosine derivatives has also been demonstrated in the cyclic nucleotide series by the conversion of a series of 8-substituted cyclic AMP derivatives to 8-substituted cyclic IMP derivatives.[478] 8-Bromoadenosine and 8-bromoguanosine were converted to 8-bromoinosine and 8-bromoxanthosine,[471,479] respectively.

Relative rates for the reaction of nitrous acid with the amine-containing nucleosides cytidine, adenosine, and guanosine have been found[476] to be $1:2.2:6.5$ at pH 5.4 and 21.5°C. The mechanism of the adenine reaction has been studied by Bunton and Wolfe.[480] They found that 9-propyladenine (435) was nitrosated to first yield the stable diazotate (436). Species (437) rapidly decomposes to give the adenine (435) at pH < 1 or the hypoxanthine (438) at any higher pH. Although the decomposition of (437) to (438) could conceivably simply proceed through a diazonium salt, very few products except the 6-hydroxypurine derivatives have ever been detected.

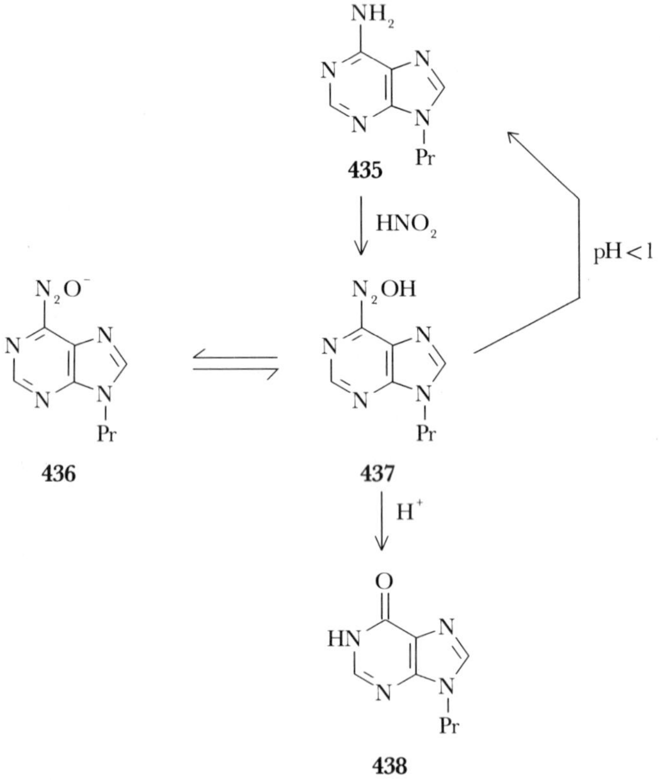

When there is a substituent present on the exocyclic nitrogen, formation of the diazohydroxide cannot take place. Thus, *N*-methyladenosine (439) gave a 39% yield of *N*-methyl-*N*-nitrosoadenosine (440) on treatment with nitrous acid, while *N*-hydroxyadenosine (441) gave[481] *N*-hydroxy-*N*-nitrosoadenosine (442) in an 11% yield.

Another reaction of fundamental importance in the interconversion of substituents of purine nucleosides and nucleotides is the conversion of a 2-amino group to a halogen. 2-Aminoadenosine (443) was diazotized with nitrous acid, in the cold and in the presence of HBF_4,[482,483] to furnish 2-fluoroadenosine (444). This reaction has been termed the modified Schiemann reaction, presumably because (but not necessarily, as shown above) a diazonium fluoroborate salt is

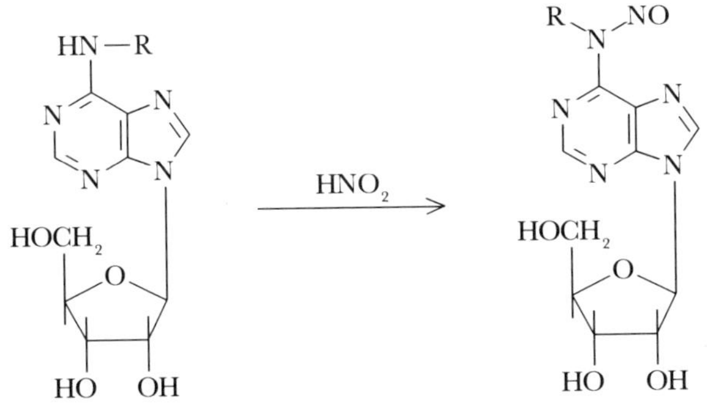

HN—R → R—N—NO

HNO$_2$

439, R = CH$_3$
441, R = OH

440, R = CH$_3$
442, R = OH

NaNO$_2$
HBF$_4$

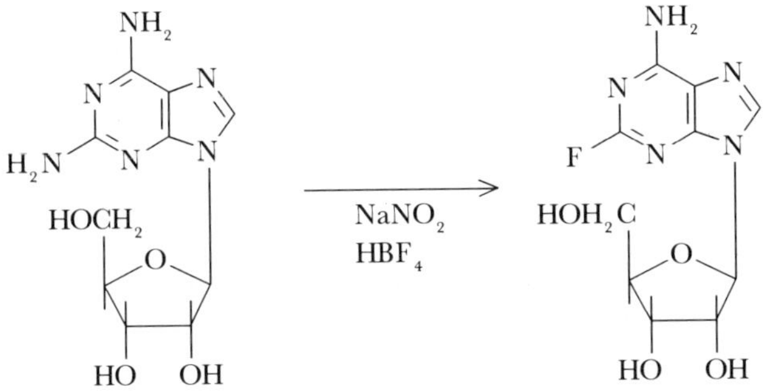

443

444

formed, which decomposes *in situ* to yield the 2-fluoro derivative. The same conditions led to decomposition in an attempt to prepare 2-fluoro-2'-deoxyadenosine from 2-amino-2'-deoxyadenosine.[484] However, the reaction was successful when applied to 2-amino-2'-deoxy-3',5'-di-*O*-acetyl adenosine. In a similar manner, the 5'-deoxy-2'3'-*O*-isopropylidine analogue of (**443**) gave 5'-deoxy-2-fluoroadenosine after a removal of the isopropylidine group in the acidic medium.[485]

The isolation of a small yield of 2,6-difluoro-(2,3,5-tri-*O*-acetyl-β-D-ribofuranosyl)purine, from the treatment of 2-amino-2',3',5'-tri-*O*-acetyl-adenosine,[484] was one of the few reports of the introduction of a halogen at position 6 of the purine ring by the diazotization method.

The synthesis of 2-fluoro-6-benzyloxy-9-(β-D-ribofuranosyl)purine (**446**) from the 2-amino nucleoside (**445**), by a diazotization procedure,[486] provided a facile synthesis of a series of *N*²-substituted guanosines. The fluoride group was displaced by various nucleophiles such as dimethylamine,[486] L-phenylalanine,[487] and monosodium L-glutamate,[487] respectively, to give 6-benzyloxy-2-substituted-9-(β-D-ribofuranosyl)purines (**447**). Catalytic hydrogenolysis then gave *N*,*N*-dimethylguanosine (**448a**),[486] *N*-(9-β-D-ribofuranosylpurin-6-on-2-yl)-L-phenyl-alanine (**448b**),[487] and *N*-(9-β-D-ribofuranosylpurin-6-on-2-yl)-L-glutamic acid (**448c**),[487] respectively. The fluoride group in the 6-benzylthio analogue (**449**) has also been replaced with various L-amino acid nucleophiles such as

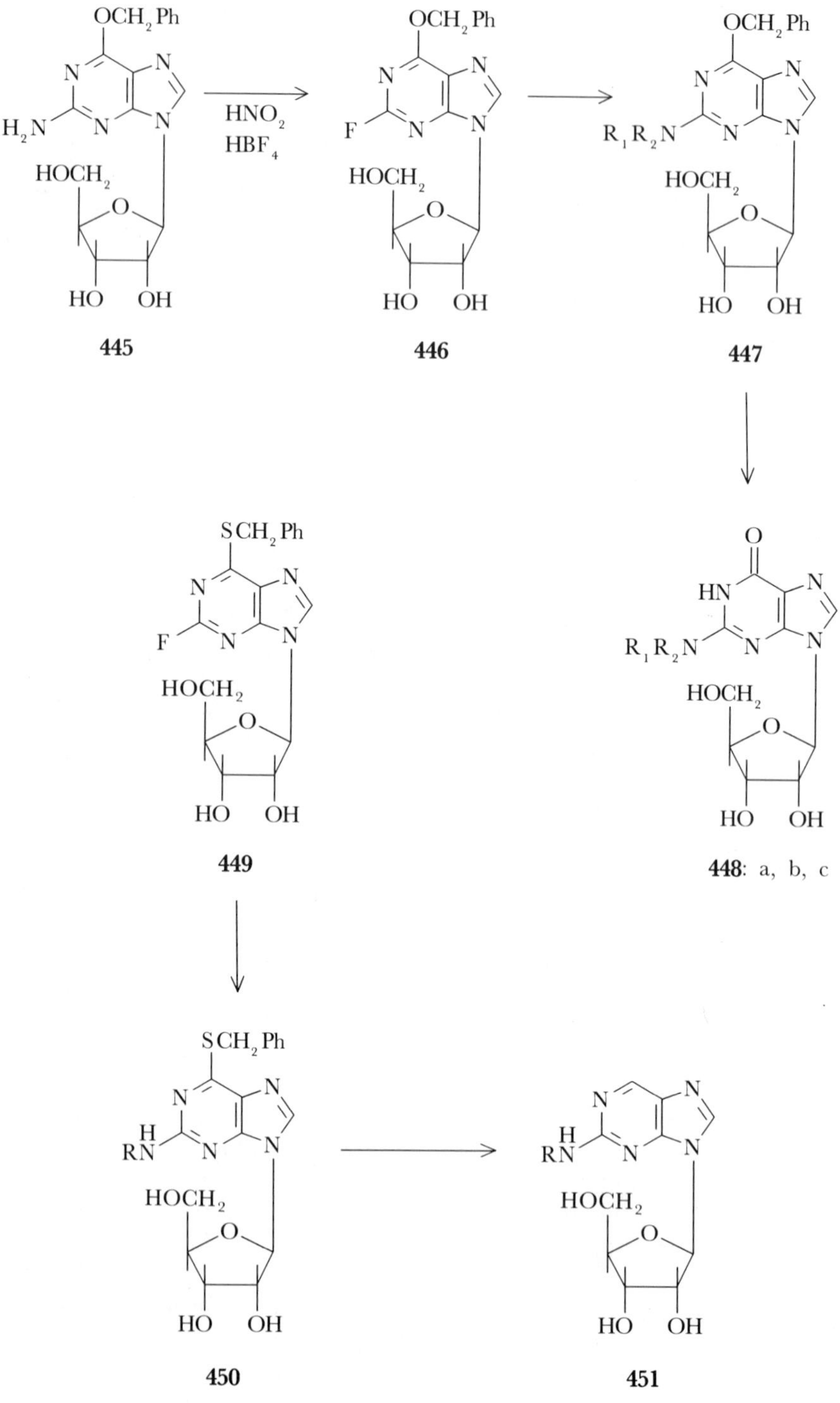

a: R, CH(CH$_2$C$_6$H$_5$)COOH **c**: R, CH$_2$COOH

b: R, CH(CH$_2$CH$_2$COOH)COOH **d**: R, CH(CH$_3$)COOH

phenylalanine, glutamic acid, glycine, and alanine to give (**450a–d**). The alanine analogue (**450d**), after debenzylsulfurization with Raney Ni, gave N-2-nebularinyl-L-alanine (**451d**).[487]

This reaction has been extended to the preparation of 2-chloropurine nucleosides and nucleotides from their 2-amino counterparts. 2-Amino-9-(β-D-ribofuranosyl)purin-6-thione (**452**) and concentrated HCl/NaNO$_2$ gave 2-chloro-9-(β-D-ribofuranosyl)purin-6-thione (**453**) in a 19% yield.[488] The same starting material (**452**) and HBF$_4$/NaNO$_2$ gave the 2-fluoro nucleoside

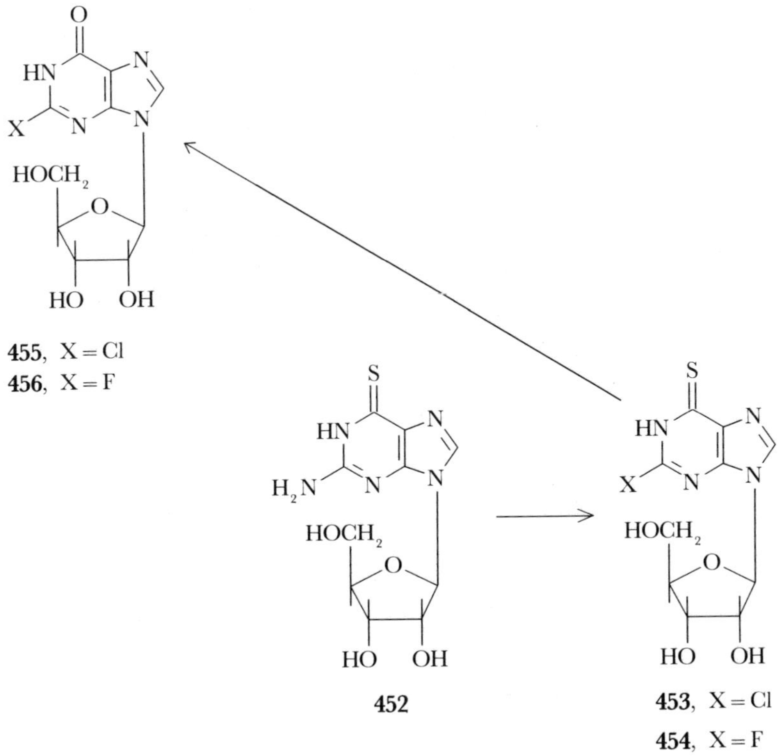

455, X = Cl
456, X = F

452

453, X = Cl
454, X = F

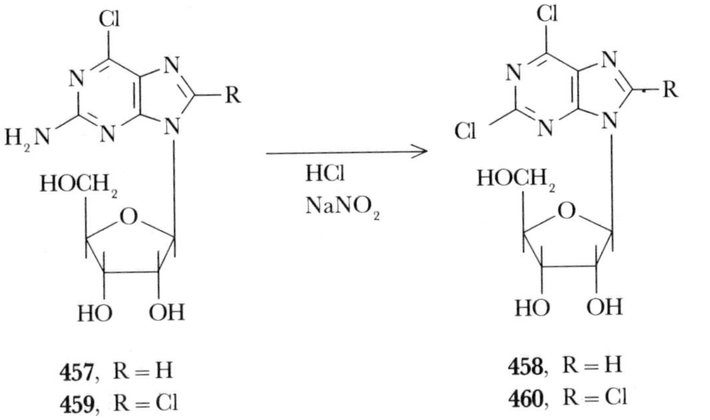

HCl
NaNO$_2$

457, R = H
459, R = Cl

458, R = H
460, R = Cl

(**454**).[489] These reactions do not appear to occur with guanosine. Treatment of (**453**) or (**454**) with alkaline hydrogen peroxide gave 2-chloro- (**455**) and 2-fluoroinosine (**456**),[488] respectively. The latter was also produced by a catalytic debenzylation of 6-benzyloxy-2-fluoro-9-(β-D-ribofuranosyl)purine (**446**).[488] 2-Amino-6-chloro-9-(β-D-ribofuranosyl)purine (**457**) and concentrated HCl/NaNO$_2$ gave 2,6-dichloro-9-β-D-ribofuranosylpurine (**458**). Treatment of the triacetate of (**457**) with HBF$_4$/NaNO$_2$ gave 6-chloro-2-fluoro-9-(2,3,5-tri-*O*-acetyl-β-D-ribofuranosyl)purine.[488] These dihalogenated compounds are of considerable synthetic utility. In the cyclic nucleotide series, 2,6-dichloro-9-(β-D-ribofuranosyl)purine cyclic 3',5'-phosphate was prepared, but not isolated, by an analogous route and converted directly into 2-chloro cyclic AMP.[456]

2-Amino-6,8-dichloro-9-(β-D-ribofuranosyl)purine (**459**) was converted into 2,6,8-trichloro-9-(β-D-ribofuranosyl)purine (**460**) by similar methods.[488] This compound is very difficult to prepare from a blocked nucleoside because of the high susceptibility of the halogens to the usual nucleophilic deblocking reagents.

A nonaqueous medium has been used for the diazotization/halodediazotization of aminonucleoside substrates (**461**). The diazotization mediums[490] such as *tert*-butyl nitrite (TBN) in 60% anhydrous hydrogen fluoride/pyridine, TBN/pyridine hydrochloride in dichloromethane or TBN/antimony trichloride in 1,2-dichloroethane, and TBN/antimony tribromide in dibromomethane[491] were used for the synthesis of a variety of fluoro-, chloro- and bromonucleosides[491] (**462**).

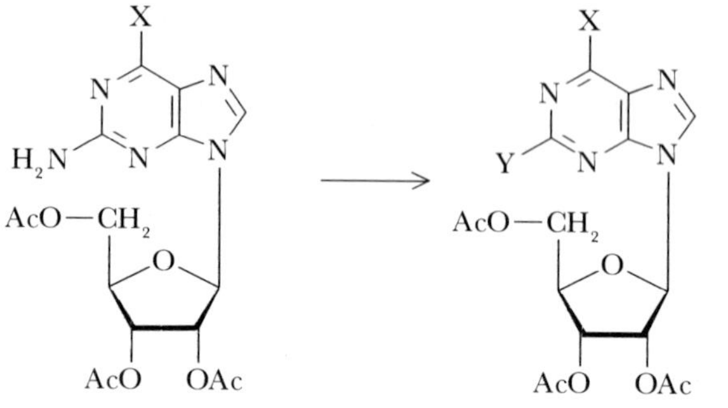

461, X = F, Cl, NH$_2$　　　　　**462**, Y = F, Cl, Br

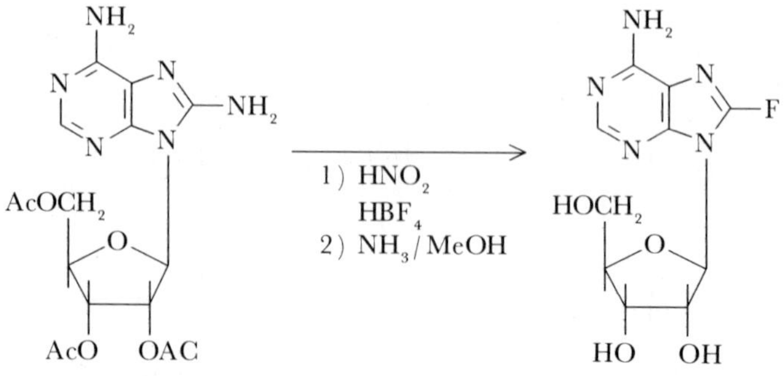

1) HNO$_2$
   HBF$_4$
2) NH$_3$/MeOH

463　　　　　　　　**464**

The above reactions indicate both the preference of a 2,6-diamino purine nucleoside to undergo attack of nitrous acid on N^2 preferentially and of the N-2 diazotate (or subsequently formed diazonium compound) to undergo a Sandmeyer (or Schiemann) type reaction with the introduction of a halogen (or nitro group). The same type of "aromatic" reactivity has been noted for an 8-amino function. 8-Aminoadenosine, when treated with $HBF_4/NaNO_2$, presumably gave a mixture of 8-hydroxyadenosine and 8-fluoroadenosine.[492a] The labilization of the glycosidic linkage was circumvented by the use of 8-amino-2′,3′,5′-tri-O-acetyladenosine (**463**) and the formation of triacetate of 8-fluoro-adenosine (**464**) was claimed in 29% yield.[492a] However, the synthesis of 8-fluoroadenosine has eluded recent attempts,[492b] which also contradict the above report.

3.2.2.3. General Metathetic Nucleophilic Reactions of Substituted Purine Nucleosides and Nucleotides. This process of nucleophilic displacement is the most general for the introduction of substituents into positions 2, 6, and 8 of the purine ring. Because of the overall electron-deficient nature of the purine ring, a variety of functional groups are susceptible to displacement by suitable nucleophiles. The halogen-substituted purine nucleosides discussed in Sections 3.2.2.1 and 3.2.2.2 serve as pivotal substrates for nucleophilic substitutions (e.g., by amines that are important for hydrogen bonding in certain biological reactions). A large body of literature has accumulated regarding the application of these reactions to special situations, and some pertinent aspects are considered here.

Direct Displacement of Oxygen, Sulfur, and Nitrogen Functions

The conversion of an exocyclic oxygen function on the purine ring to a halogen has already been discussed. One very similar reaction concerns what is termed the "thiation" reaction, or direct replacement of oxygen by sulfur. Treatment of 2′,3′,5′-tri-O-benzoylinosine (**465**) or 2′,3′,5′-tri-O-benzoylguanosine (**466**) with phosphorus pentasulfide in pyridine gave, after debenzoylation with sodium methoxide, very good yields of 6-mercapto-9-(β-D-ribofuranosyl)purine [6-thioinosine, (**467**)] and its 2-amino derivative [6-thioguanosine, (**468**)], respectively.[493] Phosphorus pentasulfide in dioxane/pyridine converts 5′-inosinic acid and 5′-guanylic acid into their 6-mercapto derivatives without the necessity of blocking the sugar hydroxyls.[494]

A convenient one-pot method, albeit indirect, for the replacement of an oxygen with amines *via* a trimethylsilylation has been reported by Vorbruggen. When inosine, guanosine, or xanthosine (**469**) were heated in a bomb at 160°C in the presence of hexamethyldisilazane, the substituted amine, and a Lewis acid catalyst (mercuric chloride was most efficient), the respective N-substituted adenosine (or N-substituted-2-hydroxy or aminoadenosine) (**470**) was obtained in excellent yield.[495-497] Furthermore, the 5′-phosphates of inosine and guanosine underwent the same one-pot, two-step reaction with amines in the presence of hexamethyldisilazane and either ammonium sulfate or p-toluene sulfonic acid catalyst to give the N-substituted 5′-AMP or 2-amino-5′-AMP or 2-amino-5′-AMP derivatives in good yields.[497]

Interestingly, the 2,6-bis(trimethylsilyloxy)purine nucleoside underwent dis-

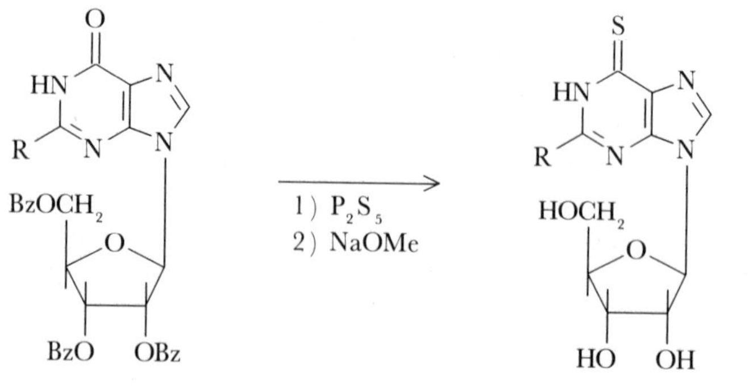

465, R = H
466, R = NH$_2$

467, R = H
468, R = NH$_2$

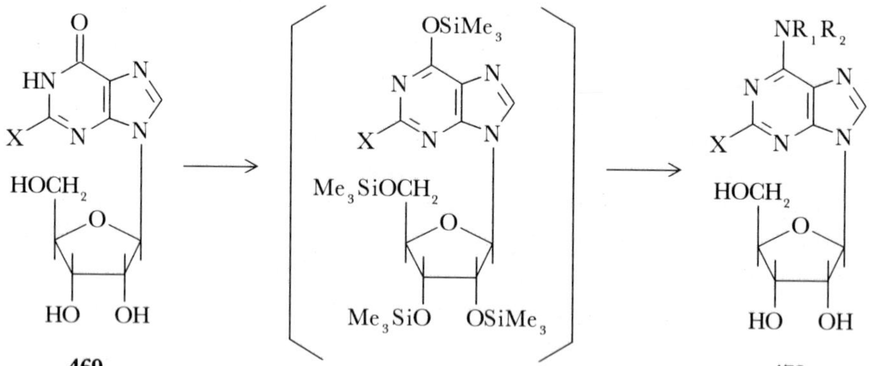

469

X = H
X = NH$_2$
X = OH

X = H
X = NHSiMe$_3$
X = OSiMe$_3$

470

X = H
X = NH$_2$
X = OH

placement by phenylethylamine only at C-6 to give 2-hydroxy-N-phenylethyl-adenosine [N-phenylethylisoguanosine (**470**), X = OH, R$_1$ = PhCH$_2$CH$_2$, R$_2$ = H].[497]

The displacement of an alkoxy group by nucleophiles is a well-known reaction that has preparative value in pyrimidine nucleoside chemistry since an alkoxy group results from the Hilbert–Johnson synthesis. Although this is a known reaction in purine nucleoside chemistry, the alkoxypurines are prepared *via* intermediates (i.e., chloropurine nucleosides), which are themselves more advantageously used for nucleophilic displacements.[491,498]

A displacement of sulfur from the purine ring by nitrogen nucleophiles is an alternative method for the synthesis of aminopurine nucleosides and nucleotides, especially where the chloropurine derivative may be difficult to obtain. Displacement of the alkylthio group is particularly well documented; for example, 6-methylthio-9-(β-D-ribofuranosyl)purine and ammonia in a bomb readily gave adenosine.[493]

The S-alkyl derivatives of purine nucleosides are also good substrates for the

Eschenmoser sulfide contraction[499] and react with triphenylphosphine in the presence of a strong base to give the appropriate *C*-alkylated nucleosides. Thus, 2′,3′,5′-tri-*O*-benzoyl-6-phenacylthioinosine[500] [(471) or the corresponding tri-*O*-acetyl derivative[501]] and 6-acetonylthioinosine (472)[500] were converted into 6-phenacyl-9-(β-D-ribofuranosyl)purine (473) and 6-acetonyl-9-(β-D-ribo-furanosyl)purine (474), respectively. The 2-amino analogue of (473) was also prepared by heating 2-trimethylsilylamino-2′,3′,5′-tri-*O*-trimethylsilyl-6-phenacyl-thioinosine with triphenylphosphine and potassium *t*-butoxide in xylene and then in methanol.[501] Compounds (473) and (474) gave an intense color with a ferric chloride solution, which indicated that the enol is the preferred tautomeric form for these compounds.[500] The extrusion of sulfur from alkylthionucleosides appeared to be driven by the electron-withdrawing character of the carbonyl moiety attached to the thiomethylene group.[500] Such compounds are potentially useful for the synthesis of *C*-alkylpurine nucleosides after a reduction of the keto group.

The direct placement of an unalkylated mercapto group is generally very difficult to accomplish. However, the treatment of 6-mercapto-9-(β-D-ribofuranosyl) purine with hydrazine in methoxyethanol gave *N*-aminoadenosine in an excellent yield.[502]

The oxidized sulfur substituents undergo a very facile nucleophilic displacement. Displacement by halide *in situ* during the oxidation process has already been discussed. 8-Methylsulfonylguanosine and sodium methoxide gave 8-methoxyguanosine under the same conditions (140–150°C) as when this transformation was performed on 8-bromoguanosine.[503] Treatment of 2-methylsulfonylinosine 5′-phosphate (475) with methylamine at 140°C gave *N*-methylguanosine 5′-phosphate (476), albeit in low yield.[504] Oxidation of 2′,3′-*O*-isopropylidine-2-mercaptoinosine (477) with hydrogen peroxide gave 2′,3′-*O*-isopropylidineinosine-2-sulfonic acid (478), which when treated, without isolation, with alkylamines at 120°C gave good yields of *N*-alkyl-2′,3′-*O*-isopropyl-idineguanosines (479).[505] The displacement of 6-methyl (or *para*-toluene) sulfonyl groups of purine nucleosides with carbanions offers an alternative to the Eschen-

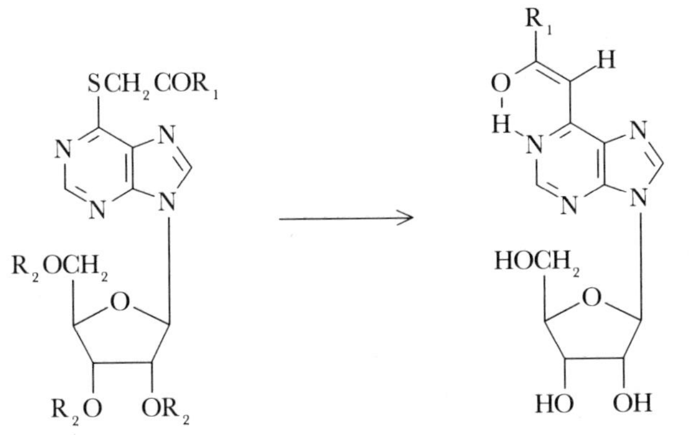

471, $R_1 = C_6H_5$, $R_2 = Bz$
472, $R_1 = CH_3$, $R_2 = H$

473, $R_1 = C_6H_5$
474, $R_1 = CH_3$

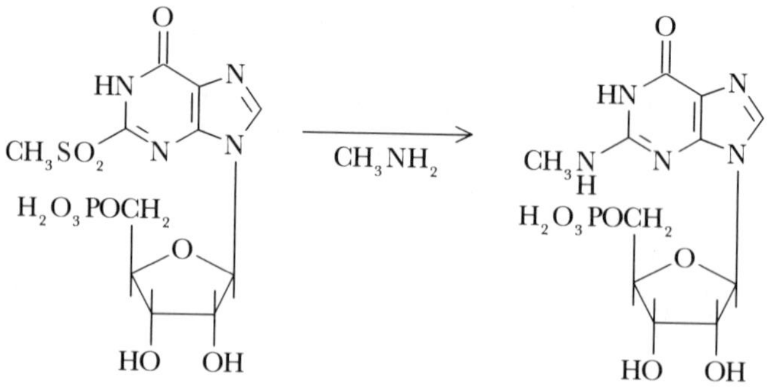

475   476

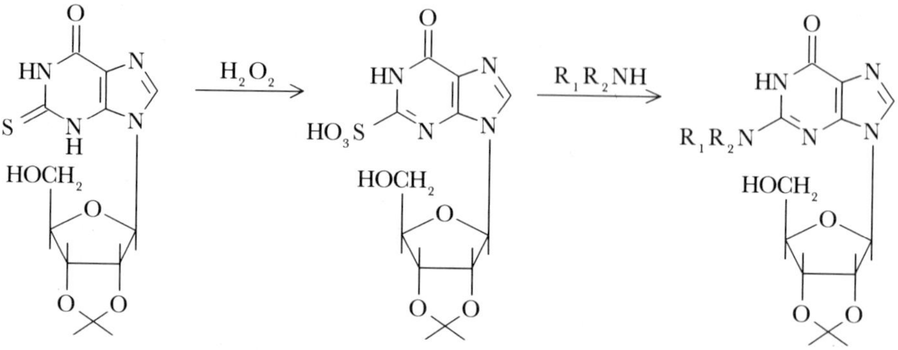

477   478   479

moser contraction reaction for the synthesis of 6-*C*-substituted purine nucleosides. Treatment of 6-methylsulfonyl-9-(2,3,5-tri-*O*-benzoyl-β-D-ribofuranosyl)purine with a carbanion derived from ethyl acetoacetate, diethyl malonate, ethyl cyanoacetate, malononitrile, nitromethane, and sodium cyanide gave the corresponding 6-*C*-substituted purine nucleosides.[506] The 6-ethoxycarbonyl-methyl analogues of purine[506] and 2-aminopurine[507] nucleosides, on decar-boxylation, facilitated the preparation of the hitherto almost inaccessible *C*-alkylpurine nucleosides.

    Desulfurization of purine nucleosides with Raney nickel, on activated catalyst with hydrogen preadsorbed to the surface, has been used to remove substituents with the introduction of hydrogen. This reagent has been used, for example, for the synthesis of nebularine (9-β-D-ribofuranosylpurine) from 6-methylthio-9-(β-D-ribofuranosyl)purine[508] or from 6-mercapto-9-(β-D-ribofuranosyl)purine[493] and for the synthesis of inosine from 2-mercaptoinosine.[505] One particularly useful application developed by Ikehara and co-workers has been the preparation of certain deoxynucleosides *via* purine 8-cyclonucleosides (see Section 3.3).

    The reductive desulfurization reaction has also been used in the presence of other reactive groups. For instance, 6-benzylthio-2-fluoro-9-(β-D-ribofuranosyl) purine, when treated with Raney nickel, gave 2-fluoronebularine.[509]

    There have been a few recorded examples of a direct nucleophilic dis-placement of a 6-amino or 6-alkoxyamino function from purine nucleosides and

Scheme VIII

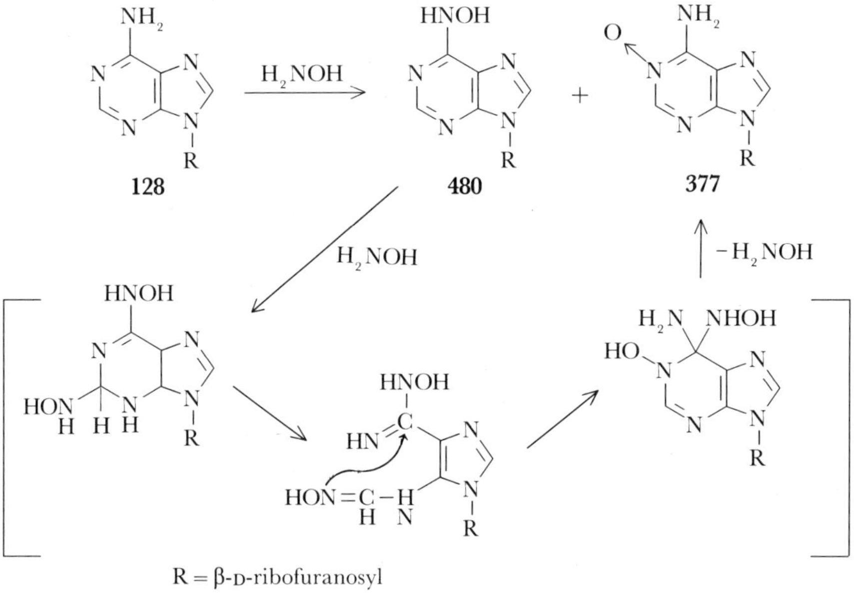

R = β-D-ribofuranosyl

nucleotides. Treatment of adenosine (**128**) with hydroxylamine acetate at 100°C gave *N*-hydroxyadenosine (**480**).[510] This was probably formed by a direct displacement of the amino function, possibly under the influence of acid catalysis. An additional product was found to be adenosine 1-oxide (**377**). Since the latter product was also formed from the reaction of hydroxylamine with *N*-hydroxyadenosine it was postulated that a rearrangement occurred after the initial attack of hydroxylamine on the 2 position of *N*-hydroxyadenosine as shown in Scheme VIII.[510]

Ueda and co-workers have developed a useful synthetic route to 6-mercaptopurine nucleotides and nucleosides, which involves the reaction of *N*-methoxyadenosine (**481**) with H_2S in pyridine. A good yield of 6-mercapto-9-(β-D-ribofuranosyl)purine (**467**) was obtained[511,512] using this approach. The same reaction gave a good yield of 6-mercapto-9-(β-D-ribofuranosyl)purine cyclic 3′,5′-phosphate from *N*-methoxy-cyclic AMP.[411] The reaction was also found to be successful in replacing the amino group of 1-methyladenosine with sulfur, but a direct treatment of adenosine under the same conditions only gave extensive decomposition.[511,512] In a similar manner, 2-amino-*N*-methoxyadenosine (**482**) gave 2-amino-6-mercapto-9-(β-D-ribofuranosyl)purine (**468**) with liquid H_2S/pyridine.[513]

An analogous reaction has also been observed to occur when adenosine (**128**), 2-aminoadenosine, 9-(β-D-arabinofuranosyl)adenine, 1-methyl-AMP, cyclic AMP, and 1-methyl-cyclic AMP were treated with an excess of H_2Se in aqueous pyridine. In all cases the corresponding 6-seleno derivative [i.e., (**483**)] was

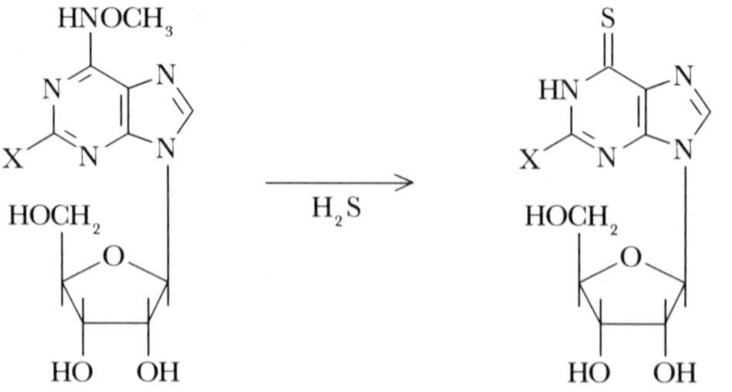

481, X=H
482, X=NH$_2$

467, X=H
468, X=NH$_2$

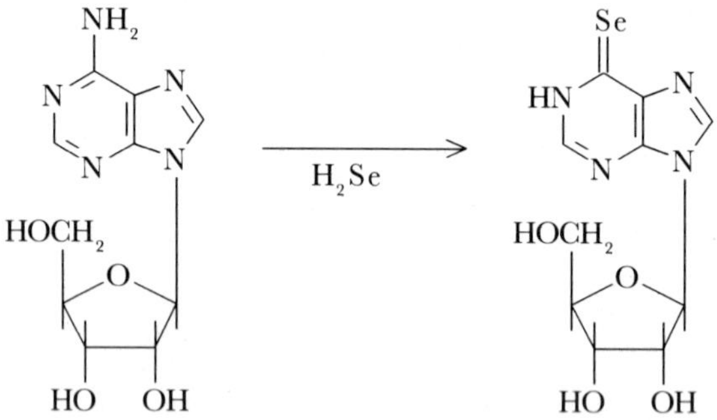

128

483

produced in 21–75% yield.[514] Furthermore, the H$_2$Se reagent gives a much more facile displacement of the amino function than H$_2$S.

Townsend and co-workers have found that Raney nickel very rapidly deselenized 2-amino-6-seleno-9-(β-D-ribofuranosyl)purine (6-selenoguanosine) to give 2-amino-9-(β-D-ribofuranosyl)purine in 97% yield.[515] This reaction seemed to be much more facile than the analogous desulfurization reaction. The alkylseleno group has proved to be highly reactive to nucleophilic displacement. 2-Amino-6-benzylseleno-9-(β-D-ribofuranosyl)purine (484) gave 2-amino-6-methoxy-9-(β-D-ribofuranosyl)purine (485) upon treatment with sodium methoxide in methanol much more rapidly than the corresponding 6-benzylthio congener (486), under the same conditions.[515] Furthermore, 6-methylseleno-9-(β-D-ribofuranosyl)purine cyclic 3′,5′-phosphate and sodium hydrosulfide gave 6-mercapto-9-(β-D-ribofuranosyl)purine cyclic 3′,5′-phosphate.[516] A variety of purine nucleoside and nucleotide transformations via the corresponding selenium intermediates have recently been reviewed.[517]

A convenient preparation of the otherwise difficult-to-obtain 6-fluoro-9-(β-D-ribofuranosyl)purine (487) was accomplished by treatment of 9-(β-D-ribofuranosyl)purin-6-yltrimethylammonium chloride (488) with potassium fluoride in *N*,*N*-dimethylformamide in 31% total yield.[518,519] 2-Amino-6-fluoro-

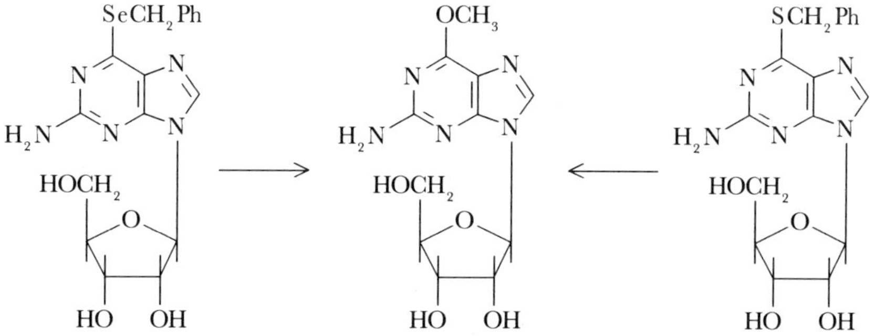

484 → 485 ← 486

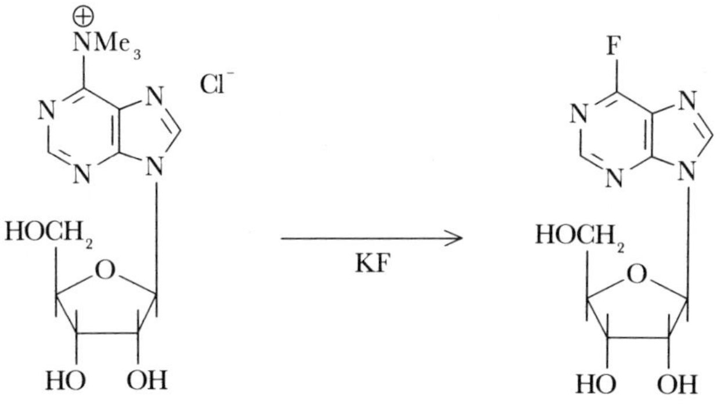

488 —KF→ 487

9-(β-D-ribofuranosyl)purine was obtained in a similar manner from its 6-trimethylammonium counterpart in 47% yield.[519]

Trifluoromethyl-copper[520] in HMPA solution reacts with the halopurine nucleosides (**489**) at temperatures greater than 100°C to give the corresponding trifluoromethylated nucleosides (**490**). Free copper (insoluble in HMPA) under the reaction conditions acts as a reducing agent, causing hydrodehalogenation, for example, the formation of 2',3',5'-tri-*O*-acetyladenosine from the corresponding 8-iodo analogue.[520]

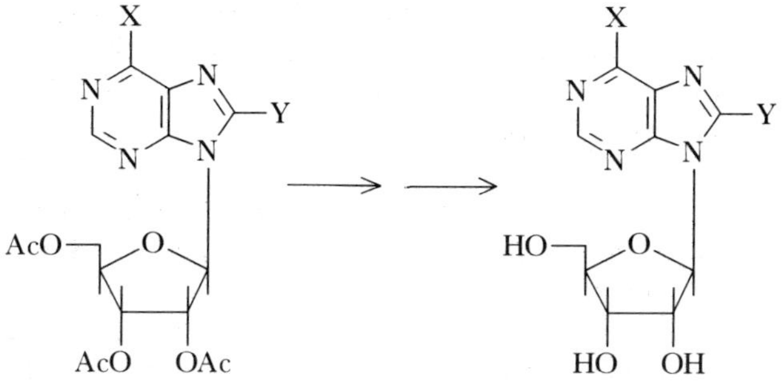

489

X=NH$_2$, Y=I
X=OH, Y=Br
X=Cl, Y=H

490

X=NH$_2$, Y=CF
X=OH, Y=CF$_3$
X=CF$_3$, Y=H

Scheme IX

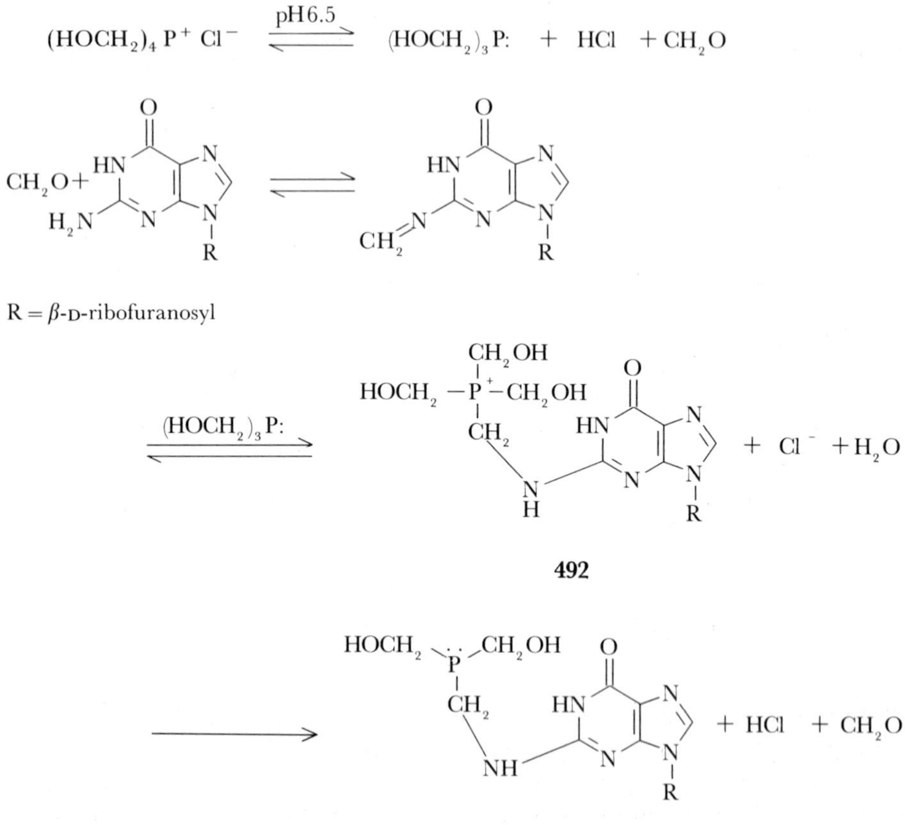

$$(HOCH_2)_4 P^+ Cl^- \underset{}{\overset{pH6.5}{\rightleftharpoons}} (HOCH_2)_3 P: \quad + \quad HCl \quad + CH_2O$$

$R = \beta\text{-}\text{D-ribofuranosyl}$

492

491

3.2.2.4. Reactions of an Amine and Keto/Enol Group on the Purine Moiety. The amine and keto/enol groups of purine nucleosides react with a variety of reagents to give intermediates of biological and synthetic utility. Most of this work has been oriented toward blocking these groups in order to prevent interference with reactions on the sugar hydroxyls. Some special reactions displayed by these groups are described here.

Guanosine reacts with tetrakis(hydroxymethyl)phosphonium chloride (THPC) to give the 2-aminosubstituted phosphine product (**491**) (Scheme IX). Initially, tris(hydroxymethyl)phosphine, formaldehyde, and hydrochloric acid are formed from THPC through a preliminary equilibrium reaction. Formaldehyde is known to react with purine amines[521,522] to form the Schiff base, which then reacts with phosphine to form the phosphonium compound (**492**). The phosphonium compound (**492**) is obviously converted to the phosphine (**491**) in a manner analogous to the first step.[523]

2′-Deoxyadenosine (**313**), after silylation with trimethylchlorosilane, reacts with phthaloyl chloride to give N^6-phthaloyldeoxyadenosine (**493**).[524] The phthaloyl group could potentially be useful for stabilizing the glycosyl bond in deoxynucleosides or for retardation of depurination.[525]

The *p*-nitrophenylethyl group has been introduced into the O^6-position of 2′-deoxyguanosine and guanosine[526] *via* the Mitsunobu reaction.[527]

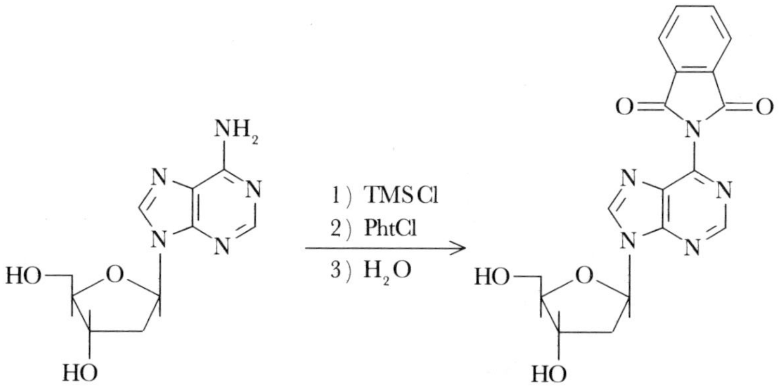

313   493

3.2.3. Free Radical and Photochemical Additions to the Purine Base

3.2.3.1. Photochemically Induced Additions. Light or γ-rays catalyze the addition of alcohols to the purine ring system. Isopropanol added to the 7,8 double bond of adenosine, guanosine, and deoxyguanosine under the influence of light at $\lambda > 260$ nm to give, initially, the 7,8-dihydro-8-(2-hydroxy-2-propyl)nucleosides, which were oxidized in air to give the aromatic 8-substituted purine nucleosides;[528] for example, adenosine (**128**) gave a 50% yield of 8-(2-hydroxyl-2-propyl)adenosine (**494**).[528]

Guanosine can be substituted at C-8 by the free radical produced on the ultraviolet photolysis of 8-bromoadenosine to form 8-(8-adenosyl)guanosine

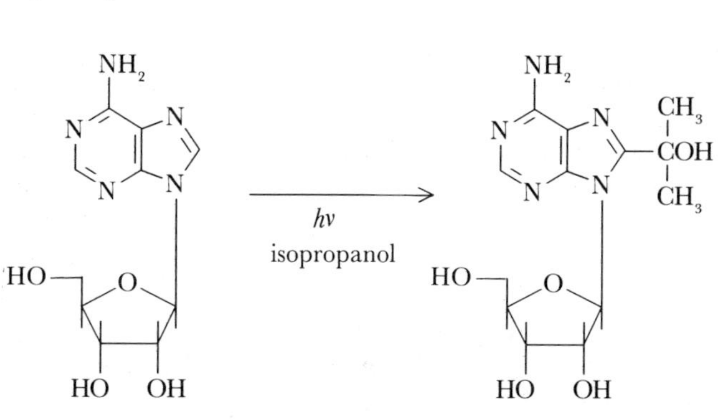

128   494

(**495**).[529] Other common purine ribonunucleosides such as adenosine, inosine, and xanthosine are also susceptible to substitution by a purinyl free radical produced photochemically from 8-bromoadenosine, 8-bromoguanosine, 8-bromo-inosine, or 8-bromoxanthosine to form the (8 → 8) coupled diribonucleosides,[530] for example, 8-(8-adenosyl)adenosine, 8-(8-adenosyl)xanthosine, 8-(8-guanosyl)xanthosine, 8-(8-inosyl)adenosine, 8-(8-inosyl)xanthosine, and 8-(8-xanthosyl)xanthosine. Inosine is preferentially substituted at C-2 to give the (8 → 2) coupled diribonucleosides, for example, 2-(8-xanthosyl)inosine (**496**).[530] The 1,6

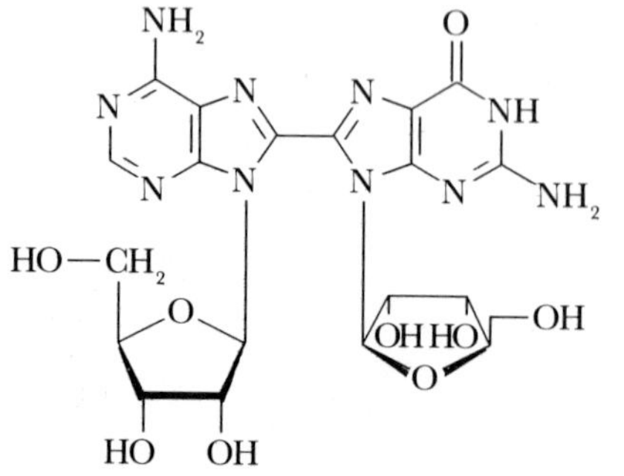

495

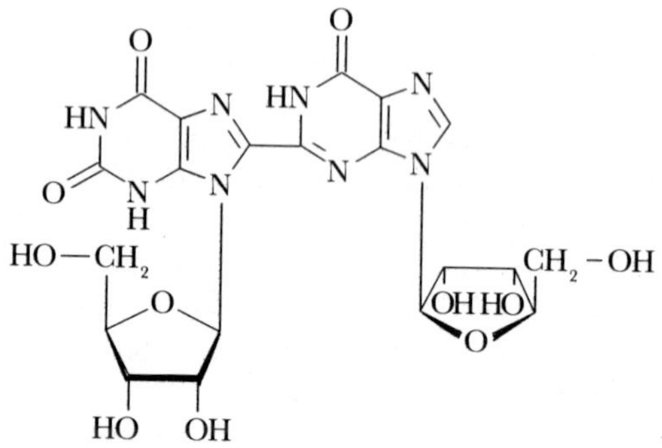

496

double bond of unsubstituted purine has been found to be even more reactive than the 7,8 double bond.[531] Treatment of 2',3',5'-tri-*O*-acetylnebularine (**75**) with methanol and ultraviolet irradiation gave a 96% yield of 1,6-dihydro-(*R*)-6-hydroxymethyl-9-(2,3,5-tri-*O*-acetyl-β-D-ribofuranosyl)purine (**497**).[532] A mixture of isomers was obtained under different reaction conditions.[533] This intermediate was used in an innovative synthesis of the naturally occurring adenosine deaminase inhibitor, coformycin (**500**). Treatment of (**497**) with methanesulfonyl chloride gave the mesylate (**498**), which was cyclized to the aziridinopurine (**499**). Treatment of (**499**) with aqueous base gave coformycin (**500**).[532,534] The methanol addition step was assumed to have been totally stereospecific, as only the *R*-isomer of (**500**) was obtained.

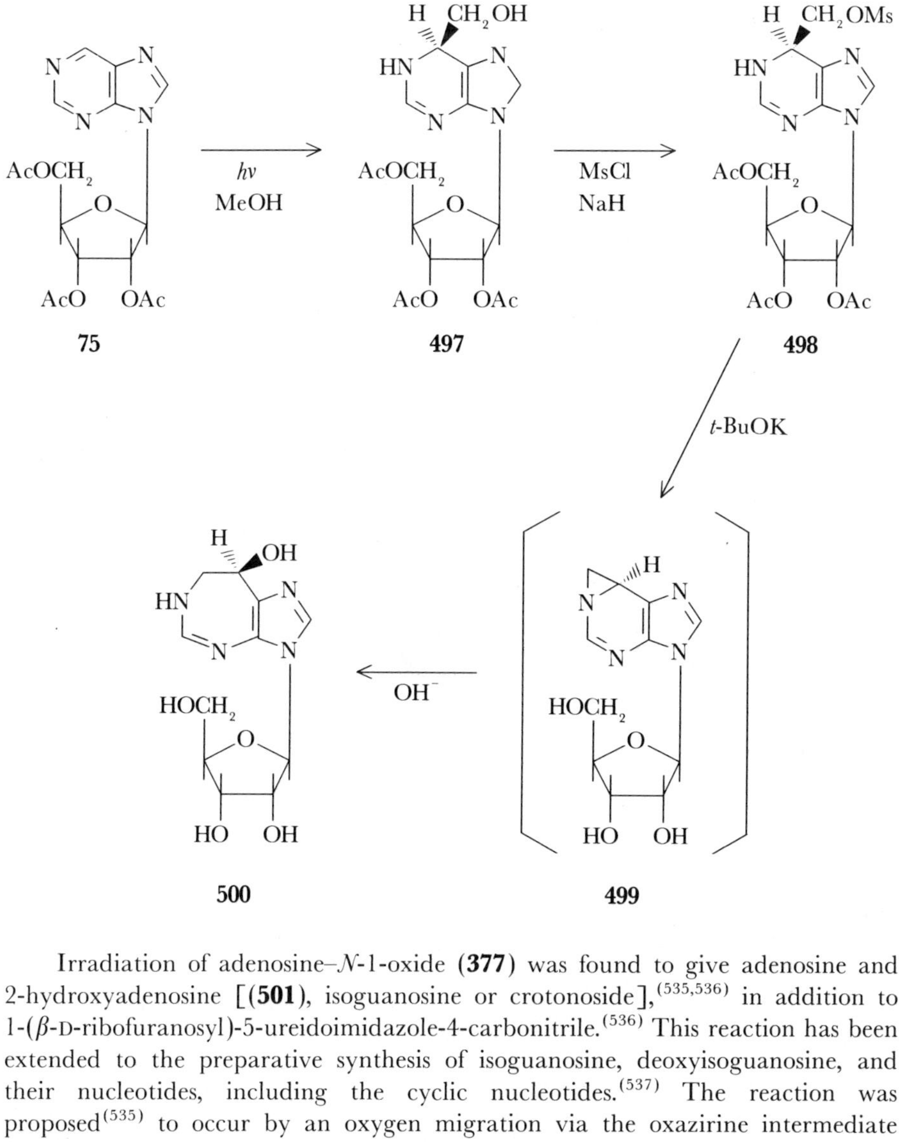

Irradiation of adenosine–*N*-1-oxide (**377**) was found to give adenosine and 2-hydroxyadenosine [(**501**), isoguanosine or crotonoside],[535,536] in addition to 1-(β-D-ribofuranosyl)-5-ureidoimidazole-4-carbonitrile.[536] This reaction has been extended to the preparative synthesis of isoguanosine, deoxyisoguanosine, and their nucleotides, including the cyclic nucleotides.[537] The reaction was proposed[535] to occur by an oxygen migration via the oxazirine intermediate (**502**).

Photolysis of 2,6-dihalogenated purine nucleosides produced, through cleavage of the carbon–halogen bond, both purin-2-yl and purin-6-yl radicals, which in an aromatic (benzene) or heteroaromatic (e.g., *N*-methylpyrrole, 2-methylfuran, thiophene, or pyridine) system gave the photostable mono- and diarylated or heteroarylated nucleosides.[538]

Photochemical cyclization of 2′,3′-*O*-isopropylidine-8-phenylthioadenosine in the presence of peroxides afforded, after deacetonation, the 8,5′(*S*)- and 8,5′(*R*)-cycloadenosines [(**503**), *S* and *R*].[539] Irradiation of adenosine 5′-monophosphate (AMP) produces only the C(5′) epimer of 8,5′-cycloadenosine 5′-monophosphate [(**504**), *S*]. This nucleoside (**504**) is inactive as a substrate for various AMP-utilizing enzymes.[540] The x-ray study of the cyclonucleoside, corresponding to the inactive irradiation nucleotide product, has established the absolute configuration to be C(5′)-*S*.[541]

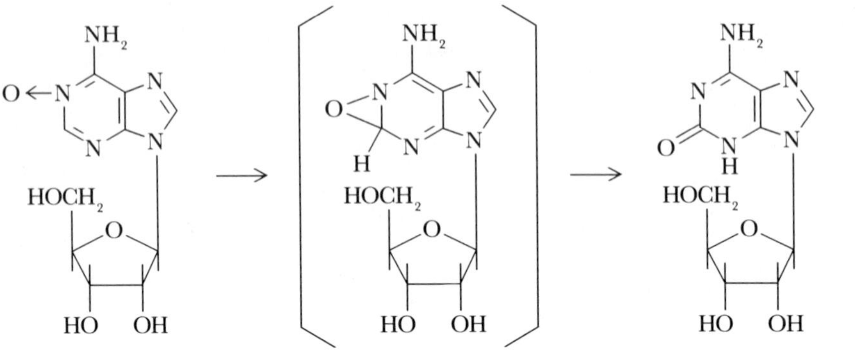

377   502   501

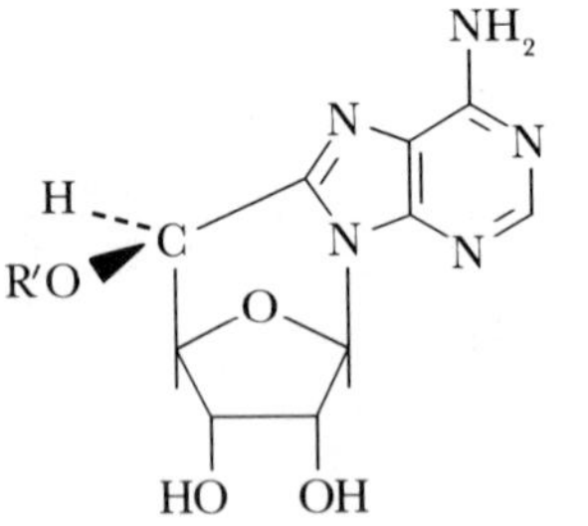

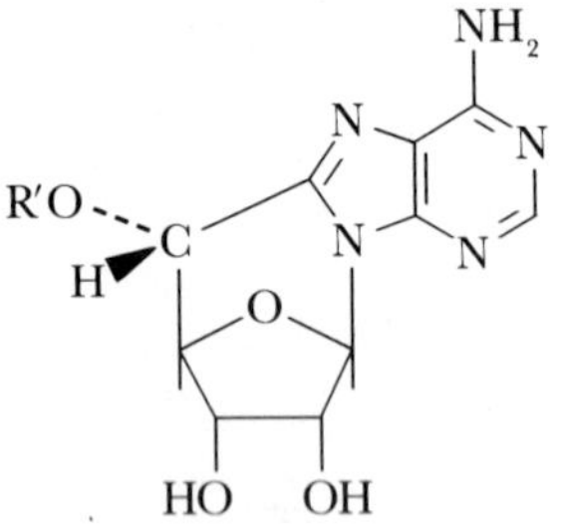

503,   S, R′ = H
504,   S, R′ = H$_2$PO$_3$ 503,   R, R′ = H

3.2.3.2. Free Radical Reactions.   Very few cases involving the addition of free radicals to purine nucleosides are known, even though this type of reaction may have important implications in understanding genetic lesions caused by these agents. Kawazoe and co-workers found that treatment of guanosine and 5′-guanylic acid with *tert*-butyl hydroperoxide and ferrous sulfate in 1 N H$_2$SO$_4$[542,543] gave good yields of their 8-methyl derivatives. The acid conditions were important in this reaction since it had been previously shown by Minisci that heterocyclic systems exhibited unusual regiospecificity in their reactions with free radicals when they were protonated.[544] Furthermore, inosine was *C*-methylated predominantly in the 8 position although some methylation (2%) occurred in the 2 position. Under similar conditions, adenosine gave small amounts of the 2- and 8-monomethyl and 2,8-dimethyl analogues.[543] The syntheses of 2-halo-6-chloropurine nucleosides have been described from the corresponding 2-amino-6-chloropurine nucleosides *via* a neutral purin-2-yl radical formed in a halogenated methane solvent in the presence of *n*-pentyl nitrite.[545]

     The scope of the addition of free radicals to cyclic GMP was studied and it was found that many other groups, besides methyl, would add to the 8 position.[546] The 8-benzyl- and 8-neopentyl-cyclic GMP derivatives were prepared by a free-radical addition. Acyl radicals can be generated by the action of ammonium persulfate and ferrous sulfate in aldehydes.[547] These radicals have been added to cyclic GMP to give a series of 8-acyl-cyclic GMP derivatives. For instance, treatment of cyclic GMP with acetaldehyde, ferrous sulfate, and ammonium persulfate in dilute acid gave a 43% yield of 8-acetylguanosine cyclic 3',5'-phosphate.[546] The acyl radicals were also added to the 8 position of cyclic IMP, in lower yield, while no reaction was found to occur with cyclic AMP.[546]

3.2.4. Reactions Involving the Opening of the Purine Ring: Degradations and Rearrangements

Depending on the other nuclear substituents present, the carbon atoms at position 2 and 8 of the purine ring may be susceptible to attack by nucleophilic reagents. The imidazoles or pyrimidines thus produced may undergo rearrangements to afford new purines or new ring systems, or they may have intrinsic utility as synthetic intermediates.

3.2.4.1. By Nucleophilic Attack at C-2. The first example of the purine ring being labilized by *N*-alkylation was provided by Baker and Joseph.[548] They found that alkaline treatment of 3'-amino-3,5'-anhydro-3'-deoxy-*N*,*N*-dimethyl-adenosine cyclic 2',3'-carbamate methanesulfonate (505) gave an unusual imidazole cyclonucleoside, N^5,5'-anhydro-1-(3-amino-3-deoxy-β-D-ribofuranosyl)-5-formamidoimidazole-4(*N*,*N*-dimethyl)-carboxamidine cyclic 2',3'-carbamate (506). The *N*-formyl group was readily removed by further alkaline treatment.

Ring opening reactions of purine nucleosides and nucleotides have since become very valuable synthetic tools. One general reaction of purine nucleosides and nucleotides, with no substituents on C-2 but with a substituent on N-1, is an opening of the pyrimidine ring to give a derivative of 5-aminoimidazole. Treatment of 1-benzylinosine [(507), R = PhCH$_2$—] with ethanol containing aqueous NaOH at reflux gave a 70% yield of 5-amino-*N*-benzyl-1-[β-D-ribofuranosyl) imidazole-4-carboxamide [(508), R = PhCH$_2$—].[427] This compound was deben-

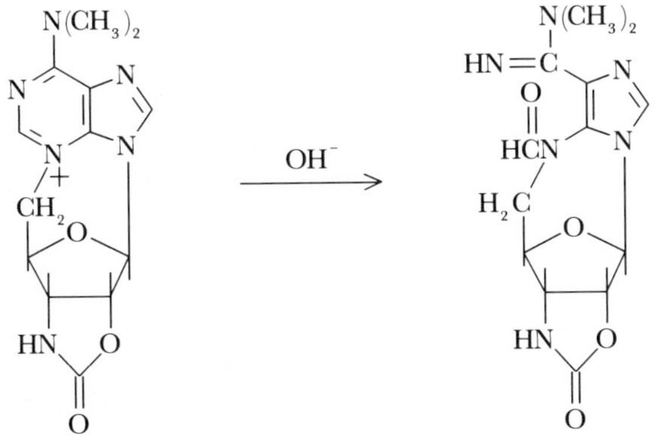

505   506

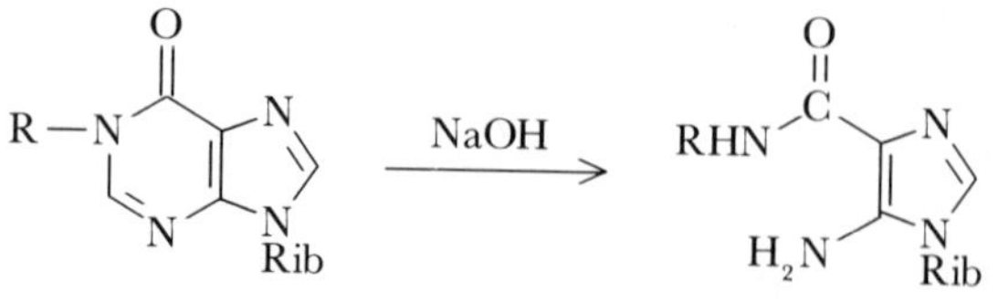

Rib : β-D-Ribofuranosyl

zylated with sodium in liquid ammonia to give 5-amino-1-(β-D-ribofuranosyl)-imidazole-4-carboxamide (AICA-riboside).[427] The 5′-phosphate derivative of AICA-riboside (AICAR) is an important intermediate in the *de novo* purine biosynthesis. It was subsequently found that the purine ring of inosine could be labilized by more readily removable blocking groups such as the 1-methoxymethyl[428] (removed by mild acid) and the 1-amino-group[549] (removed by catalytic hydrogenation). This ring opening reaction has also been observed with 1-methoxymethylinosine cyclic 3′,5′-phosphate.

Scheme X

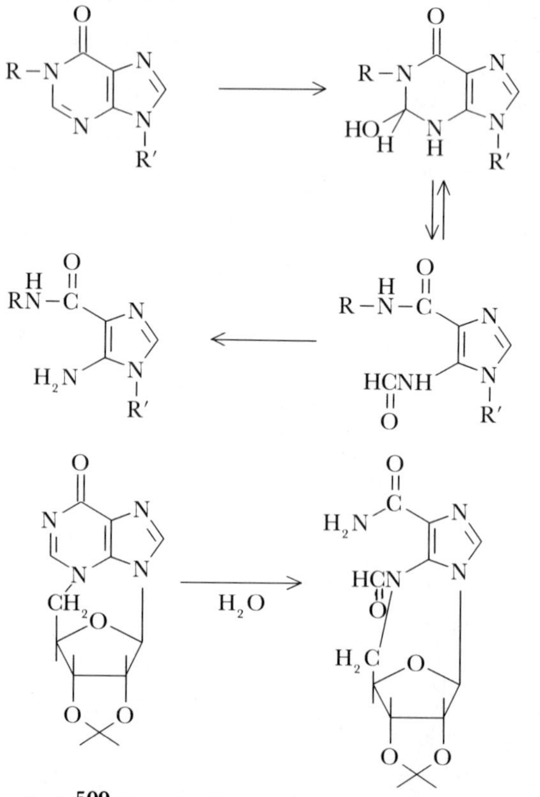

This reaction most likely occurs by an initial hydroxide attack at C-2. This gives the intermediate 5-formamidoimidazole-4-carboxamide and the formyl group is rapidly removed by a base-catalyzed hydrolysis as shown in Scheme X.

Inosine *per se* undergoes cleavage by attack of hydroxide at C-2 in water at pH 9–10. The stability of inosine toward base at higher pH values may be rationalized by an electrostatic repulsion of hydroxide attack at C-2 by the anionic form of inosine at these pH values. At pH 10 there must be enough hydroxide available to attack, but still enough inosine in the un-ionized form to be available for cleavage. Thus, AICA-riboside was obtained in approximately 30% yield and another product, 9-(β-D-ribopyranosyl)hypoxanthine, was obtained in a 9% yield.[550] Ring opeing also occurs readily when inosine is alkylated at the N-3 position. The purine cyclonucleoside 3,5′-anhydro-2′,3′-O-isopropylideneinosine (**509**) was readily opened in water when attempts were made to convert the cyclonucleoside from a salt to the free base, although this fact was not initially appreciated.[551] It was only later that the free base thus obtained[552–554] was assigned the structure 5′,N^5-anhydro-1-(2,3-O-isopropyl-idine-β-D-ribofuranosyl)-5-formamidoimidazole-4-carboxamide (**510**).[555] The desired cyclonucleoside was obtained by a very careful neutralization in the cold.[555] A similar ring opening of 2′,3′-O-isopropylidine-5′-O-mesyl-8-bromo-guanosine with hydrazine *via* an N^3,5′-cyclization has been reported.[556]

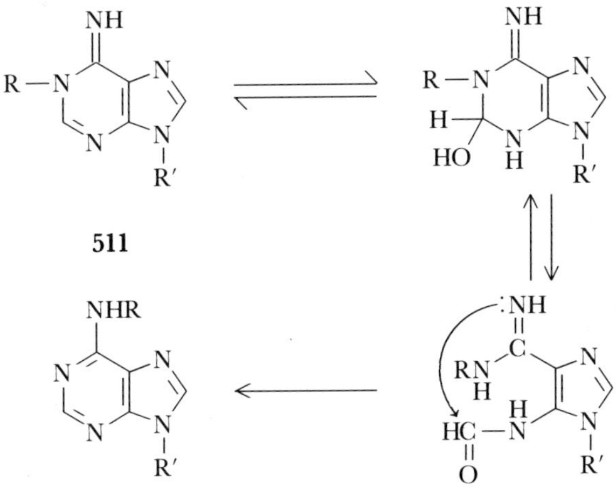

The synthesis of N^6-monoalkyl derivatives of adenosine by the Dimroth rearrangement of the corresponding 1-alkyl derivatives has seen widespread use since its original application to adenine.[557–559] This rearrangement proceeds *via* the initial hydroxide attack on the unprotonated species of a N-1-alkyl adenosine derivative (**511**) (which is a strong base, $pK_a = 8.25$ for the 1-methyladenosine[560]). This is followed by a ring opening to afford the intermediate N-alkyl-5-formamidoimidazole-4-carboxamidine (**512**). Whereas the formyl group was readily hydrolyzed during the alkaline degradation of 1-alkyl inosines, the carboxamidine generated in this case rotates and recyclizes before saponification can occur. This gives the N^6-alkyladenosine analogue (**513**) in good yield. The kinetics of this reaction have been studied by Macon and Wolfenden.[560] Windmueller and Kaplan used this procedure to convert the 1-(2-hydroxyethyl) derivatives of adenosine and NAD to the corresponding N^6-(2-hydroxyethyl) derivatives.[425] Proof that the conversion of 1-methyladenosine proceeds *via* the ring opening and not by methyl migration has been obtained recently by double-labeling experiments.[561] The synthetic utility of this method has been abundantly demonstrated. Grimm and Leonard prepared N-(3-methyl-2-butenyl)adenosine and the 5'-AMP derivative by alkylation of adenosine and adenosine 5'-β-cyanoethyl phosphate with 3-methyl-2-butenyl bromide, followed by alkali-induced rearrangement.[562] Robins and Tripp introduced 3-methyl-2-butenyl and a benzyl group into 2'-deoxyadenosine, 3'-deoxyadenosine, and 9-β-D-arabinofuranosyladenine by a similar procedure.[563] The procedure has also been found applicable for the synthesis of N^6,8-di-substituted derivatives of cyclic AMP. Thus, alkylation of either 8-bromo- or 8-methylthio-cyclic AMP with benzyl bromide, followed by a rearrangement in aqueous Na_2CO_3, gave the corresponding N^6-benzyl derivatives.[461]

A rearrangement of the N-1-alkyladenosine derivatives cannot be interrupted at the intermediate ring-opened step (512) to give an aminoimidazole carboxamidine or its N^5-formylated derivative. However, adenine-N-1-oxides and their O-alkylated counterparts readily give such imidazole nucleosides and nucleotides. G. B. Brown and co-workers initially showed that adenosine-N-1-oxide (377) was converted by hot dilute alkali to 5-amino-1-(β-D-ribofuranosyl)imidazole-4-carboxamidoxime (514), which in fact constituted the structure proof for the monooxide of adenosine.[408]

The same reaction occurred with the 5'- and 3'(2') phosphate of adenosine N-1-oxide.[408] Later, it was shown that the Dimroth rearrangement apparently does occur to a limited extent, since N^6-hydroxy-cyclic AMP (517) was isolated as a minor product (9%) in the alkaline degradation of cyclic AMP-N-1-oxide (516) to afford 5-amino-1-(β-D-ribofuranosyl)imidazole-4-carboxamidoxime cyclic 3',5'-phosphate (515).[411]

377, R = β-D-ribofuranosyl

516, R = β-D-ribofuranosyl
       cyclic 3',5'-phosphate

514, R = β-D-ribofuranosyl

515, R = β-D-ribofuranosyl
       cyclic 3',5'-phosphate

+

517, R = β-D-ribofuranosyl cyclic 3',5'-phosphate

Scheme XI

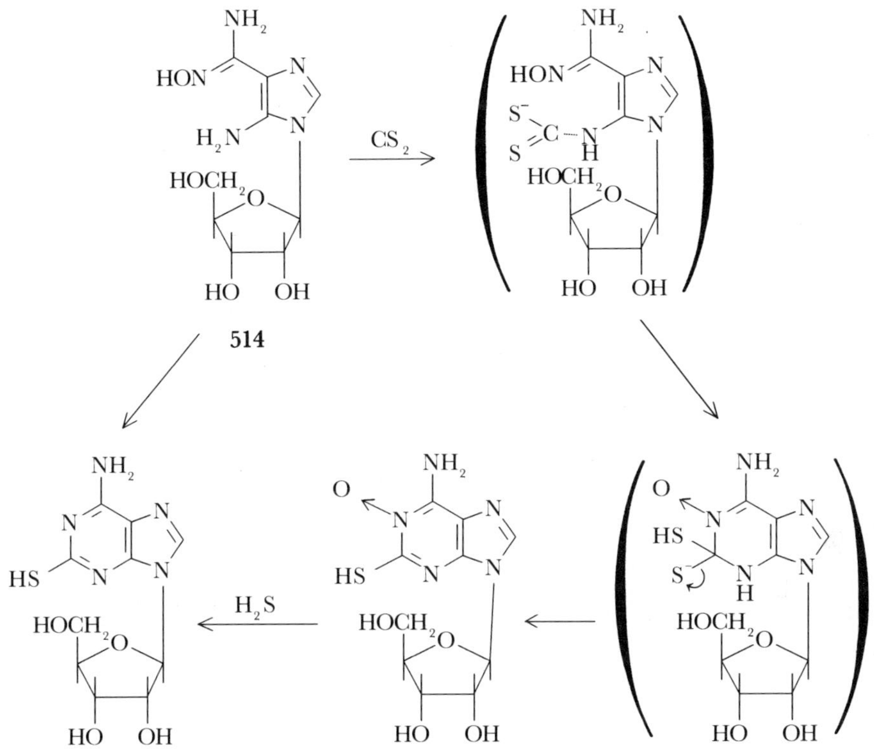

Kikugawa *et al.*[564] described a facile synthesis of 2-thioadenosine (**518**) from adenosine-N-1-oxide (**377**). The reaction of (**514**), obtained from (**377**), with $CS_2 - MeOH - H_2O$ at 120°C in an autoclave gave (**518**) in an overall yield of 60%. The formation of (**518**) was assumed to occur *via* cyclization and reduction of the cyclized 2-thio-N-1-oxide intermediate (Scheme XI).[564]

An elegant method for the conversion of adenine nucleosides into guanine nucleosides was described by Ueda *et al.*[565] Adenosine-N-1-oxide (**377**) reacts with cyanogen bromide to give 2-imino-6-(β-D-ribofuranosyl)-[1,2,4]-oxadiazolo [3,2-*f*]purine hydrobromide (**519**), which exists in a pH-dependent equilibrium with N^6-cyanoadenosine-N-1-oxide (**520**). Compound (**520**), on methylation and alkaline treatment, rearranges to form 2-amino-N^6-methoxyadenosine (**521a**). Compound (**521a**) converted into the 2,6-diamino- (**522a**) and 2-amino-6-thiopurine (**523a**) ribosides on catalytic hydrogenation and sulfhydrolysis, respectively.[565] The N-1-oxides of 2'-deoxyadenosine, adenosine 5'-phosphate, 2'-deoxyadenosine 5'-phosphate, and 9-β-D-arabinofuranosyladenine 5'-phosphate were also found to be suitable substrates for such transformations and gave the corresponding 2-amino-N^6-methoxyadenine analogues. The N-6-methoxy

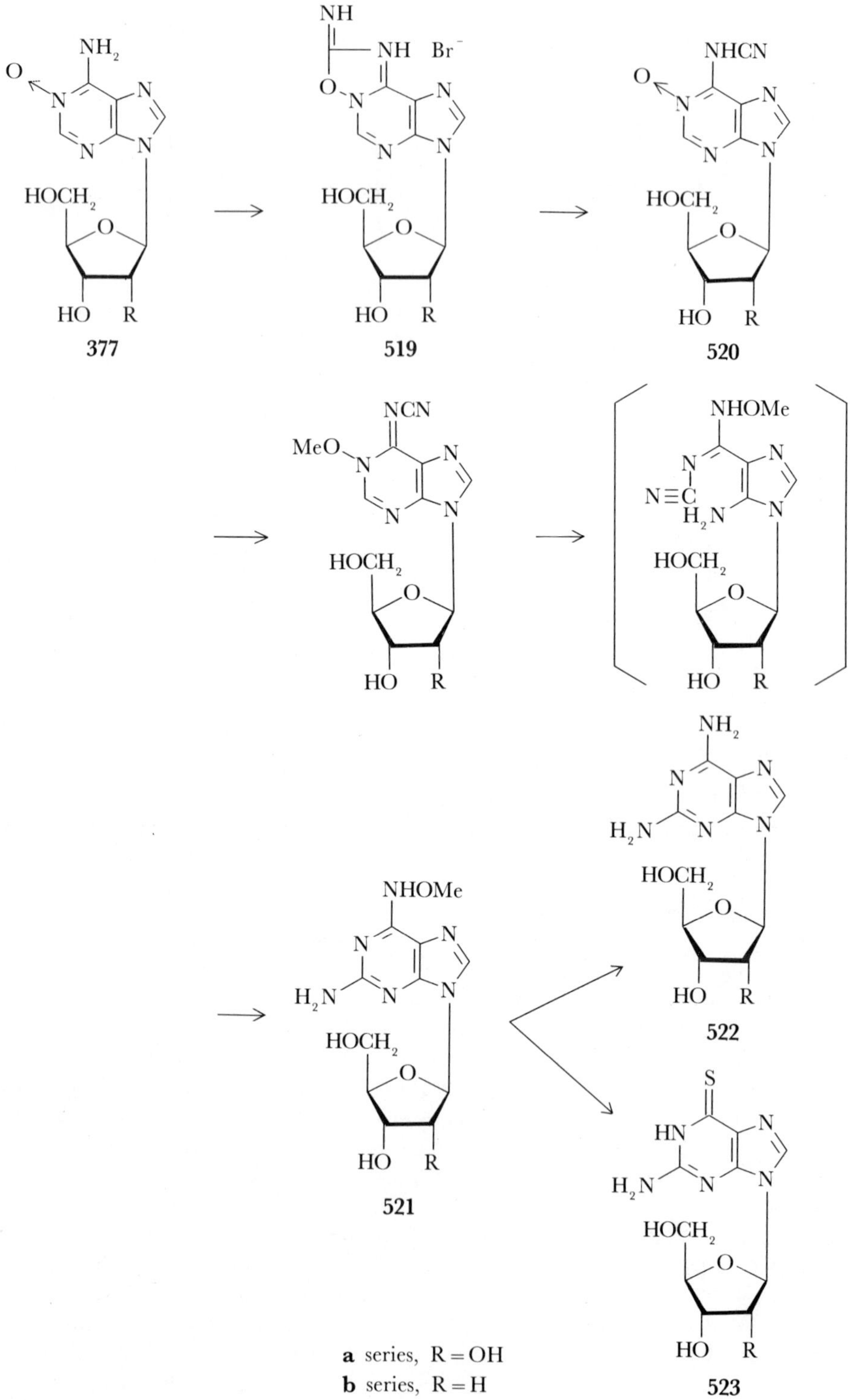

a series, R = OH
b series, R = H

nucleotides, however, were subjected only to sulfhydrolysis to yield the corresponding 6-thioguanine nucleotides.[565]

Perhaps more versatile than the adenine-N-1-oxide nucleosides and nucleotides are their *O*-alkylated counterparts. Through the extensive work of Fujii, Itaya, and their colleagues, it has been demonstrated that the 1-alkoxy-9-alkyladenine compounds (**524**) undergo a Dimroth rearrangement to give N^6-alkoxyadenines (**526**), when boiled with water.[566,567] These compounds were conveniently prepared by *O*-alkylation of the 9-alkyladenine-N-1-oxides (**525**).[568,569] Alternatively, when the free base of these 1-alkoxy-9-alkyladenines (**524**) was treated with cold water, a 1-alkyl-*N*-alkoxy-5-formamidoimidazole-4-carboxamidine derivative (**527**) was isolated.[566,567] Treatment of (**524**) with strong base gave[567] the 5-aminoimidazole (**528**). These reactions were efficiently applied to the appropriate adenosine[567,568] derivatives.

This set of interconversions has found synthetic utility for the preparation of 5-aminoimidazole-4-carboxamidine nucleosides (**529**) and nucleotides, which can easily be prepared from the corresponding amidoximes or *N*-alkoxyamidines by a catalytic hydrogenation. Thus, alkaline cleavage of 1-benzyloxyadenosine gave 5-amino-*N*-benzyloxy-1-(β-D-ribofuranosyl)imidazole-4-carboxamidoxime. A catalytic reduction of this compound furnished 5-amino-1-(β-D-ribofuranosyl)-

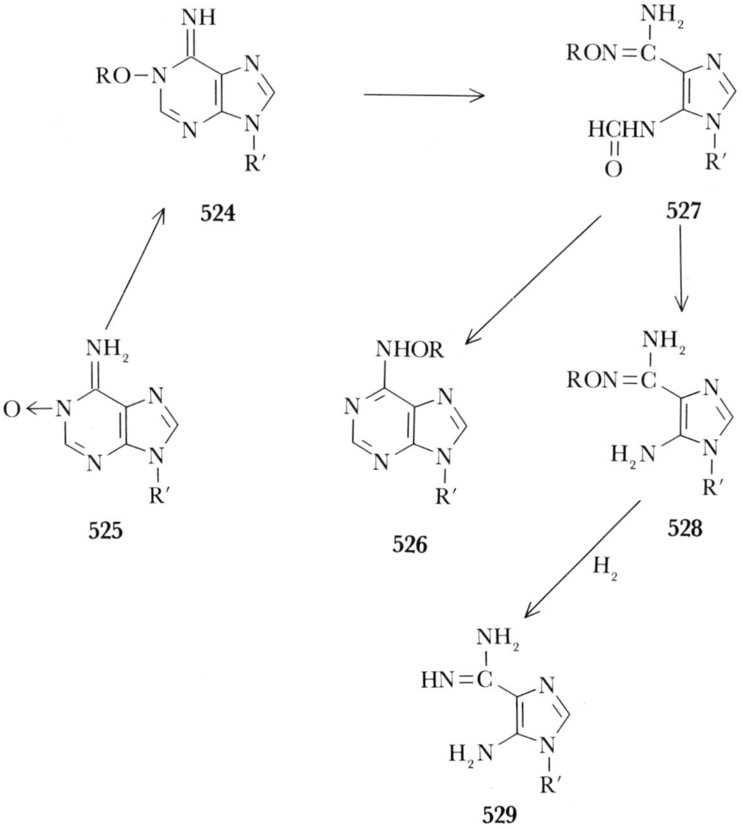

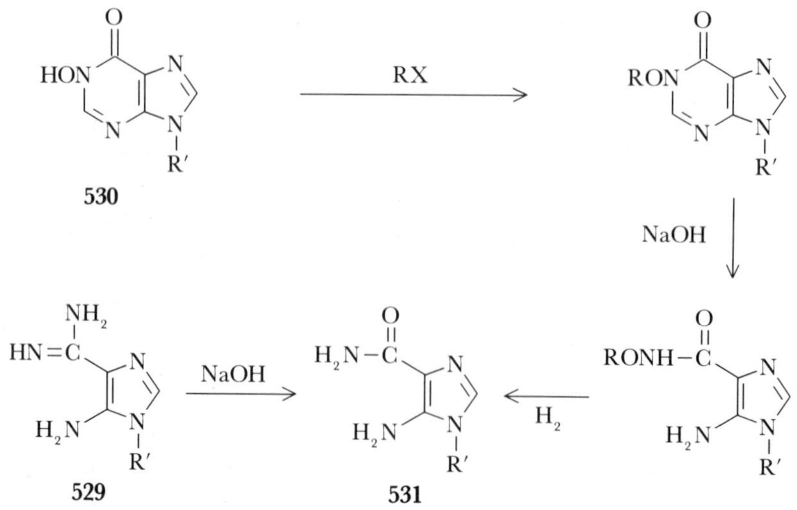

imidazole-4-carboxamidine, which was then cyclized with nitrous acid to afford 2-azaadenosine.[570] The same series of conversions have also been performed on the 9-cyclopentyl-, 9-(2-deoxy-β-D-ribofuranosyl)-, 9-(β-D-arabinofuranosyl)-, 9-(β-D-xylofuranosyl)-,[571] and 9-(β-D-ribofuranosyl)-cyclic 3',5'-phosphate[411] derivatives of adenine. Additionally, the N^6-alkoxy derivatives of adenosine have been prepared by heating the 1-alkoxy derivatives, as the free base, in water.[571]

For a series of 1-alkoxy-cyclic AMP derivatives, it was found that a NaHCO$_3$ solution at reflux temperature gave 53% yield of the rearrangement product (526), while a 2 N NaOH solution at reflux gave a 65% yield of the ring opened aminoimidazole derivatives (528).[411]

The cyclic 3',5'-phosphate derivative of AICA-riboside could be obtained by alkaline hydrolysis of the corresponding carboxamidine (529).[411] Another preparation for the nucleosides of 5-aminoimidazole-4-carboxamide from purine precursors, involved a route using 1-N-oxides of hypoxanthine nucleosides. Alkylation of inosine, deoxyinosine, and 9-β-D-arabinosylhypoxanthine-1-oxide (530) (prepared by nitrous acid deamination of the adenine 1-oxide derivative) with benzyl bromide, followed by NaOH hydrolysis and catalytic reduction, gave the respective glycosyl derivatives of 5-aminoimidazole-4-carboxamide (531).[572]

Coward and co-workers found an unusual case of 5'-hydroxyl participation in the opening of the purine ring at position 2. When N-(dimethylamino-methylene)-2,3-O-isopropylidineadenosine (532) was treated with *tert*-butyl-β-iodobutyrate and sodium hydride, initial alkylation at N-1 occurred (533). This alkylation was followed by an attack of the anion of the 5'-hydroxyl at C-2 and ring opening to give an amidine intermediate (534). This intermediate was then converted *in situ* to the novel imidazole cyclonucleoside (535).[573]

3.2.4.2. By Nucleophilic Attack at C-8. The susceptibility of purine nucleosides, which do not contain electron-donating substituents in the 6 position, to hydroxide attack at the 8 position is well known. Nebularine (9-β-D-ribofuranosyl-purine) and 9-(β-D-ribofuranosyl)-6-methylpurine [(536), R = H or CH$_3$]

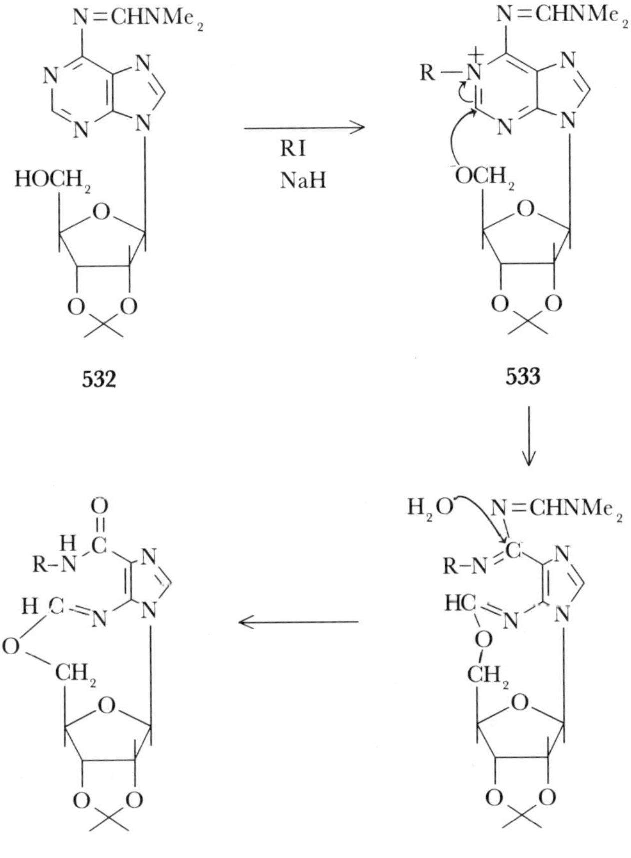

undergo hydroxide-induced ring opening in the imidazole portion of the purine ring to give a 4-ribofurano(or pyrano)sylamino-5-formamidopyridimidine [(537) or (538), respectively]. The formyl group is rapidly cleaved under the reaction conditions in the former case.[574] 6-Choro-9-(2,3-O-isopropylidine-β-D-ribofuranosyl)purine underwent alkaline ring opening at the 8 position to give a 50% yield of 4-chloro-5-formamido-6-(2,3-O-isopropylidine-β-D-ribofuranosyl) aminopyrimidine.[575,576] This ring opening reaction had been observed earlier with 6-chloropurine nucleosides.[574,577]

The electron density of the imidazole moiety of the purine ring is decreased by *N*-alkylation, which labilizes the imidazole ring. Thus, methylation of guanosine under neutral conditions gives 7-methylguanosine (386),[421,578] which under basic conditions opens to give 2-amino-4-hydroxy-5-methylamino-6-(β-D-ribofuranosylamino)pyrimidine (539).[579] Alkali-catalyzed cleavage of 7-methylxanthosine has also been noted.[579] When adenosine (128) was treated with diethyl pyrocarbonate in aqueous solution at pH 5, followed by methanolic ammonia, two major products were obtained: 4,5-di(ethoxycarbonylamino)-6-(β-D-ribofuranosylamino)pyrimidine (540) and its 4-amino-5-ethoxycarbonyl-

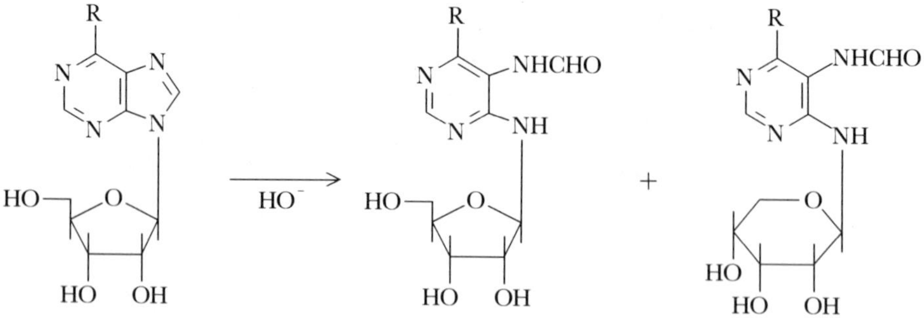

R = H or CH$_3$

536   537   538

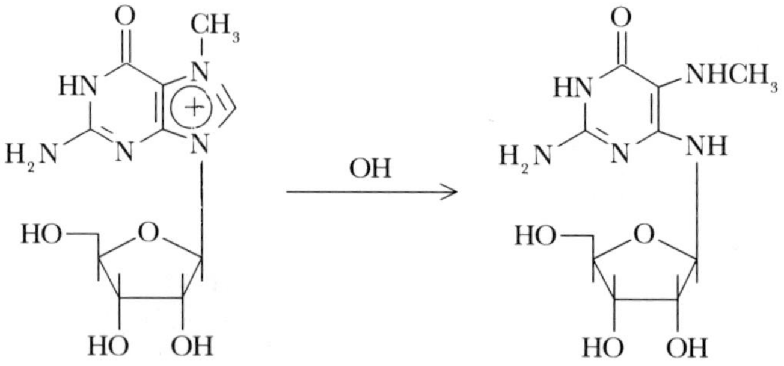

386   539

amino counterpart (541). Formation of these products was rationalized on the basis of N-7 ethoxycarbonylation followed by hydrolytic ring opening.[580] Diethyl pyrocarbonate has also been shown to cause a similar degradation of guanosine (371) at pH 6. The N^2 of guanosine was not acylated under these conditions, and the only product (40%) was 2-amino-5-ethoxycarbonylamino-4-hydroxy-6-(β-D-ribofuranosylamino)pyrimidine (542).[581]

A most interesting rearrangement was observed when 6-cyano-9-(2,3,5-tri-*O*-acetyl-β-D-ribofuranosyl)purine (543) was treated with methanolic ammonia in an attempt to form purine-6-carboxamidine nucleoside.[582,583] The product was identified by x-ray diffraction studies[583] as 4-amino-8-(β-D-ribofuranosylamino) pyrimido[5,4-*d*]pyrimidine (544). Since electron-deficient purines are apparently labile to attack at ring carbons, it was proposed that attack of ammonia at C-8, followed by cyclization to the cyano group, must have occurred.[583]

As a part of their investigations into pteridines, Eistetter and Pfleiderer[584] have demonstrated a direct transformation of a purine nucleoside into a pteridine nucleoside. When 5'-*O*-acetyl-7-cyanomethyl-2',3'-*O*-isopropylidineguanosinium bromide (545) was allowed to stand in water at pH 8.8, water presumably attacked the 8 position to give the pyrimidine nucleoside (546). After treatment with sodium methoxide, then water, 8-(2,3-*O*-isopropylidine-β-D-ribofuranosyl) isoxanthopteridine (547) was obtained, albeit in quite low yield. A total glycosidic cleavage was observed without the presence of the isopropylidine group.

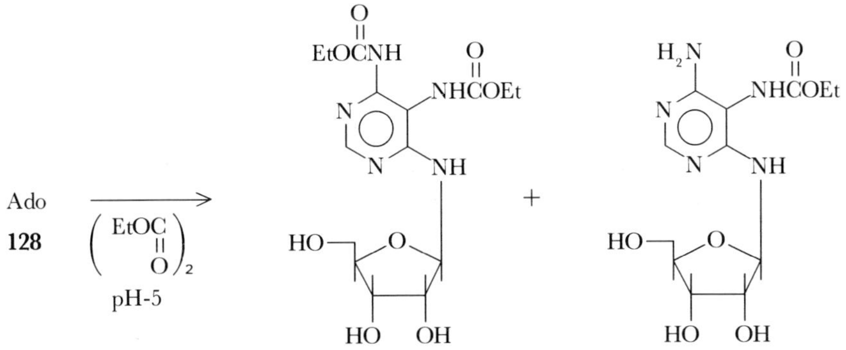

540   **541**

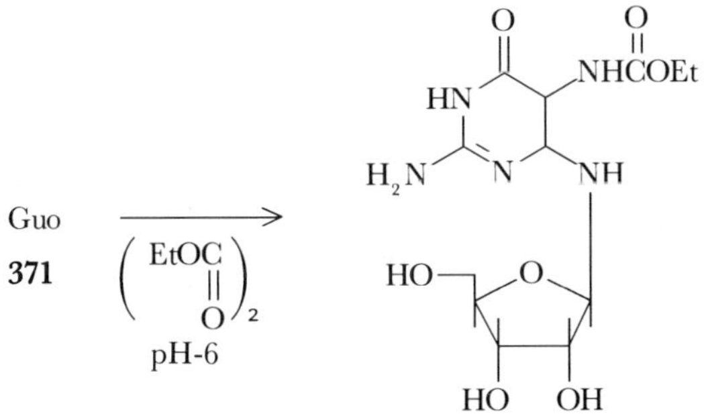

542

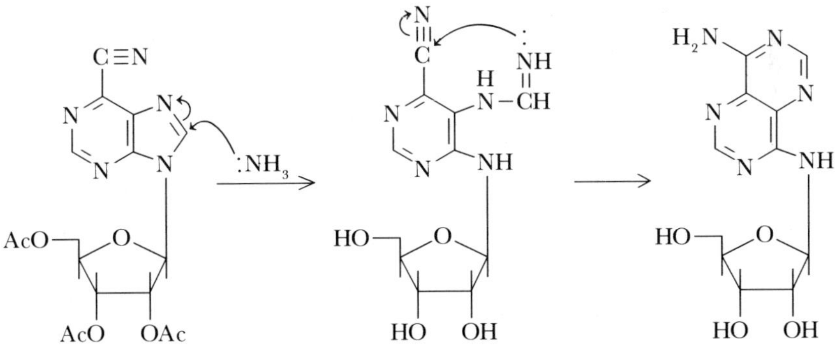

543   **544**

3.2.4.3. Reactions in Aqueous Solution. The reaction of guanosine **(371)** with *p*-methylbenzyl chloride in neutral aqueous solution at 40°C furnished[585] an unanticipated product, 4-(*p*-methylbenzyl)-5-guanidine-1-β-D-ribofuranosyl-imidazole **(548)** in addition to the expected *N*²-, *O*⁶-, and 7-(*p*-methylbenzyl) guanosine derivatives.[586] Dissociation of *O*⁶-(*p*-methoxybenzyl)guanosine in

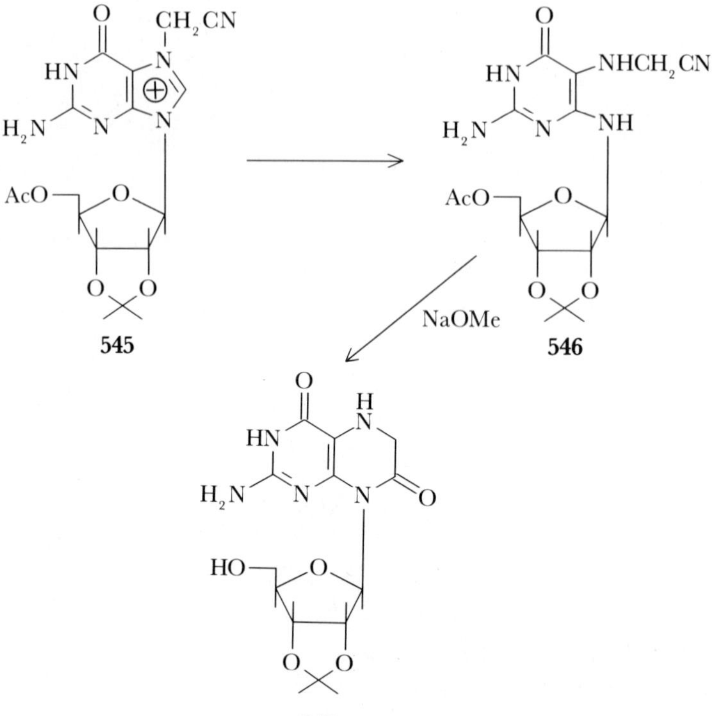

Scheme XII

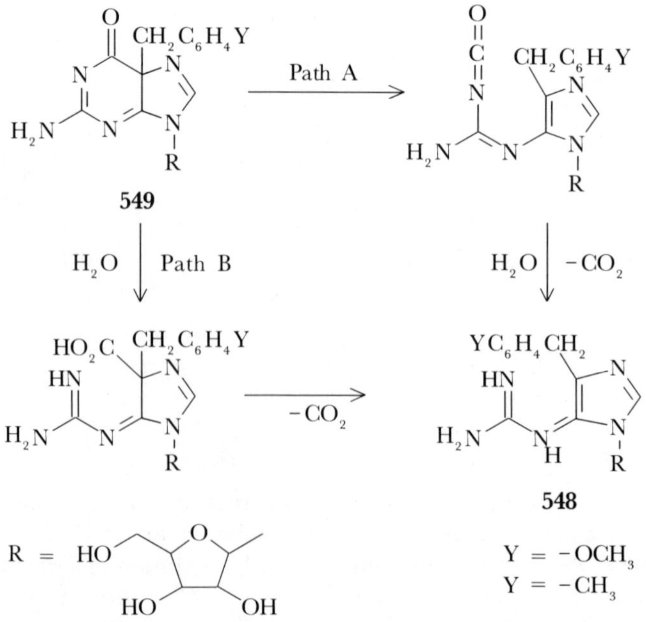

aqueous solution similarly yielded the *p*-methoxybenzyl analogue of (**548**).[587] A plausible mechanistic route was proposed (Scheme XII)[587] and involves the intermediate formation of a 5-substituted guanosine (**549**), which by either an electrocyclic rearrangement to an isocyanate derivative and subsequent hydrolysis and loss of CO_2 (path A, Scheme XII) or by a hydrolytic cleavage of the pyrimidine ring followed by decarboxylation (path B, Scheme XII) could yield (**548**).

3.3. Interaction of the Sugar and the Purine Ring: Purine Cyclonucleosides

A nucleoside with a covalent bond between the heterocyclic base and the sugar, in addition to the glycoside linkage, is defined as a cyclonucleoside and this group of compounds has established a special niche in nucleic acid chemistry. The vast majority of this class of compounds involve a removal of the elements of water from a carbon on the sugar containing a hydroxyl group and either an atom in the purine ring or a group ($-OH$, $-SH$, $-NHR$) attached to it; hence, the term "anhydro" enters the nomenclature to describe the attachment points of the linkage where the elements of water have been removed. Without this trivial system, the nomenclature of these compounds would be a very difficult task.

Although purine cyclonucleosides *per se* have not proved to be very useful as therapeutic agents, their uses in chemical transformations of nucleosides and their phosphorylated derivatives and in studying the physical properties of nucleosides of rigid conformation have been invaluable. Only the more pertinent transformations are considered here. There is an excellent review[588] by Ikehara on the purine 8-cyclonucleosides.

3.3.1. N-3 Cyclonucleosides

The first cyclonucleoside, 3,5′-anhydro-2′,3′-*O*-isopropylidine adenosine (**551**), was formed spontaneously by an intramolecular nucleophilic displacement of the tosylate group from 2′,3′-*O*-isopropylidine-5′-*O*-*p*-toluenesulfonyladenosine (**550**).[319] The corresponding 3,5′-anhydroadenosine has been prepared from

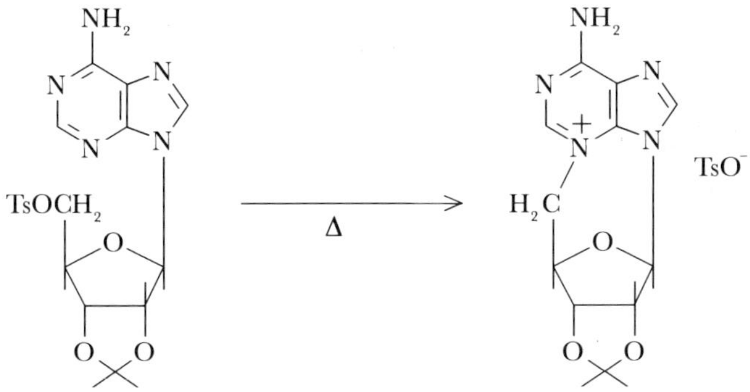

550   551

Ts = *p*-toluenesulfonyl

5'-*O*-*p*-toluenesulfonyladenosine.[320] Formation of the 3,5'-cyclonucleoside has also been observed to occur with 2',3'-*O*-isopropylidine-5'-*O*-*p*-toluenesulfonyl-guanosine[589] and 2',3'-*O*-isopropylidine-5'-*O*-*p*-toluenesulfonylinosine.[551] The instability of these compounds, which involves a hydrolytic cleavage of the purine ring,[548,555] is noted in Section 3.2.4. The formation of a cyclonucleoside by this method has been used as a chemical verification of the β-D-configuration assignment of the ribofuranosides.[590]

The acylation of N^6 of adenosine derivatives with a leaving group on the 5'-carbon has been shown to decrease 3,5'-cyclonucleoside formation.[321] Alternatively, acylation of the N^6 position, after cyclonucleoside formation, facilitates a nucleophilic displacement of the purine from the 5' position. Thus, *N*-acetyl-3,5'-anhydro-2',3'-*O*-isopropylidineadenosine gave, on treatment with dibenzyl phosphate, dibenzyl 2',3'-*O*-isopropylidineadenosine 5'-phosphate.[591]

Cyclonucleosides have also been formed between N-3 and other positions of the sugar moiety. When the epoxy nucleoside, 2',3'-anhydroadenosine (**321**) was heated, a cyclonucleoside, which was ultimately identified as the aminoimidazole-carboxamidine nucleoside (**553**), was obtained. This nucleoside must obviously arise from the degradation of 3,3'-anhydroadenosine (**552**).[316,592] A 3,4'-cyclo-nucleoside was formed by the treatment of 9-(2,3-*O*-isopropylidine-5-deoxy-β-D-*erythro*-pent-4-enofuranosyl)adenine (**554**) with bromine in chloroform. The product, 3,4'-cyclo-9-(5-bromo-5-deoxy-2,3-*O*-isopropylidine-β-L-lyxofuranosyl)

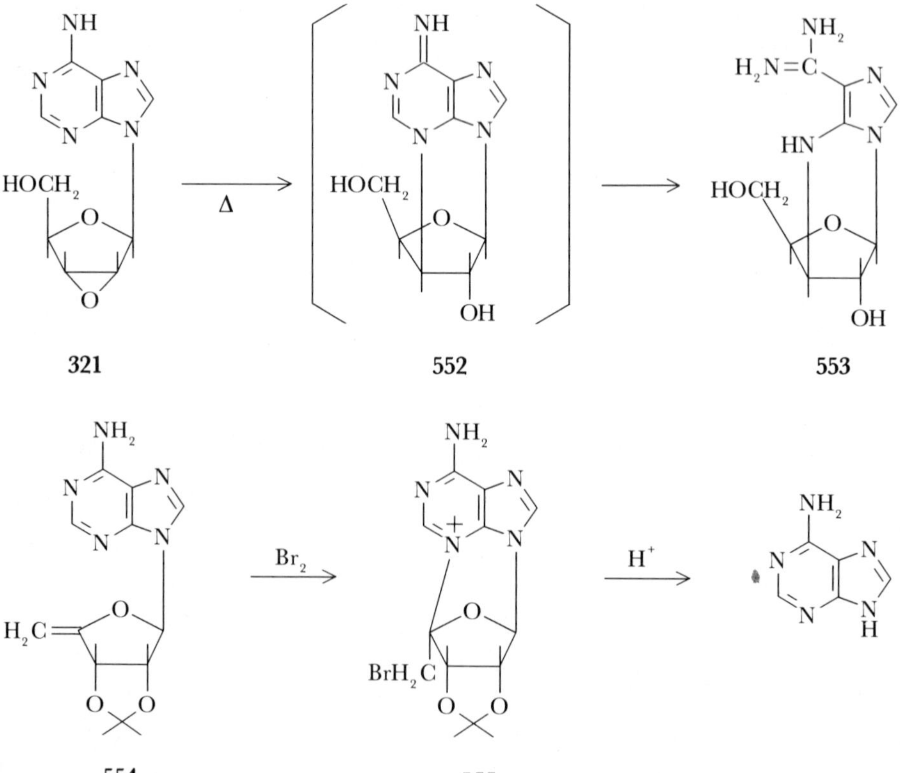

adenine (555), precipitated immediately and arose presumably by an attack of N-3 on the carbonium ion formed at 4′ *in situ*, after the addition of bromonium ion to the double bond.[593]

3.3.2. S-Cyclonucleosides

The thiocyclonucleosides are formed very easily owing to the nucleophilicity of sulfur. Furthermore, a facile removal of the sulfur can give a deoxynucleoside directly from a ribonucleoside. Thus, treatment of 5′-O-acetyl-8-bromo-2′ (or 3′)-O-p-toluenesulfonyladenosine (556) with thiourea gave a mixture of 8,2′- (557) and 8,3′-thiocyclonucleosides (558). The yield was improved by the use of 2,4,6-triisopropylbenzenesulfonyl chloride in NaH/DMF as the sulfonylating agent, followed by NaSH treatment.[595] 2′-deoxyadenosine (313) and 3′-deoxyadenosine [cordycepin, (14)] were obtained by Raney nickel desulfurization[594] of the appropriate thiocyclonucleosides. 2′- and 3′-deoxyinosine[596] and 2′- and 3′-deoxyguanosine[597] were also prepared by an analogous route. 2,6-Diamino-8-bromo-9-(2,3-O-dimethylstannylene-β-D-ribofuranosyl)purine has been tosylated selectively at the 2′ position. The intermediate steps involving thiourea ring closure and Raney nickel reduction gave 2,6-diamino-9-(2-deoxy-β-D-ribofuranosyl)purine.[598]

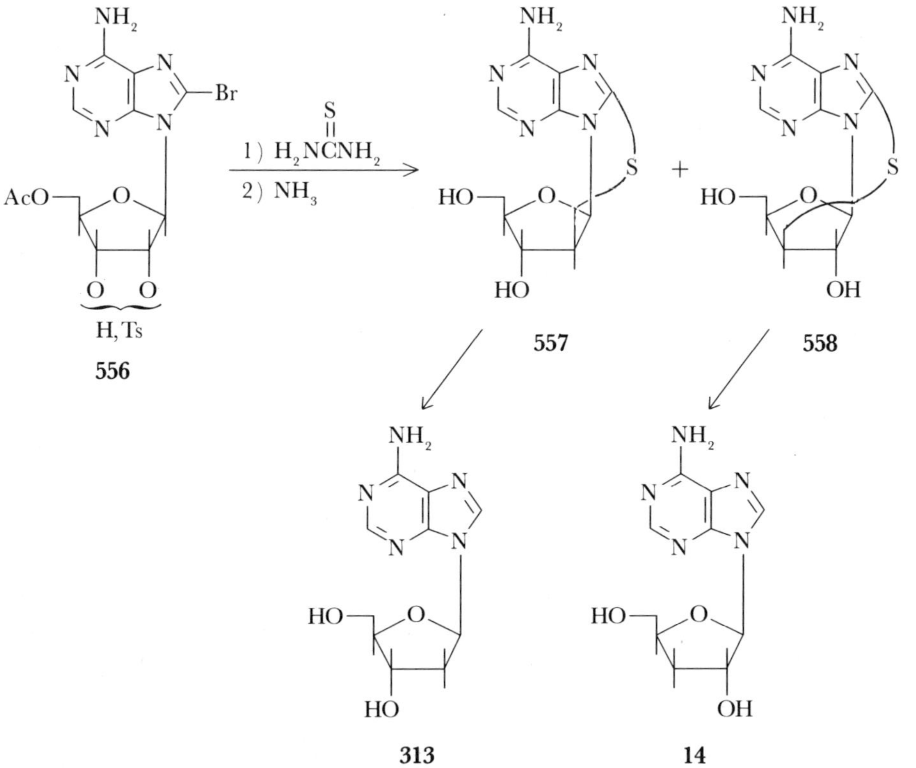

The propensity of N-3 to displace leaving groups at 5′ has been well documented. Thus, the preparation of the 8,5′-thioanhydronucleoside was approached with some caution. However, 8,5′-anhydronucleoside formation was successful with both guanosine[599–601] and adenosine.[602,604] In the latter case, a low-temperature tosylation of 8-bromo-2′,3′-O-isopropylidine adenosine (**559**) was followed by a low-temperature treatment with H_2S/pyridine to give the cyclonucleoside (**560**). Removal of the isopropylidine group from the 8,5′-anhydro-2′-3′-O-isopropylidine-8-mercaptoadenosine (**560**) gave very low yields of the corresponding cyclonucleoside.[603] The more labile 2′,3′-O-ethoxymethylidine group[605] proved more suitable for blocking the *cis*-hydroxyl groups. Raney nickel treatment of 8,5′-anhydro-8-mercaptoadenosine gave 5′-deoxyadenosine (**561**).[603,606]

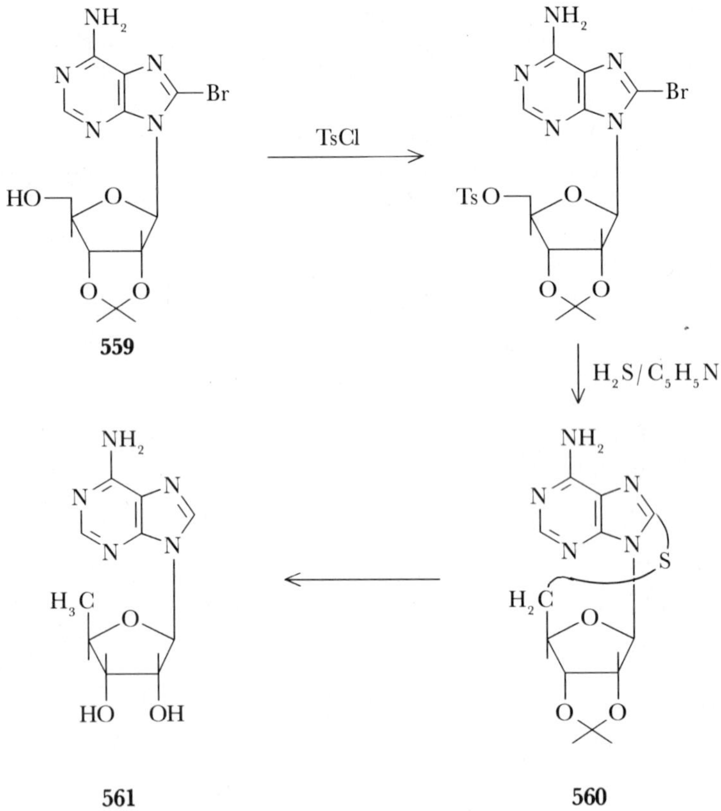

These procedures have also been used in the direct synthesis of 2′-deoxy-cyclic AMP[607] and 2′-deoxy-cyclic GMP.[393] Thus, treatment of 8-bromo-cyclic AMP (or GMP) (**562**) with *p*-toluenesulfonyl chloride in NaOH solution gave the 2′-O-tosyl derivative (**563**). The reaction of (**563**) with NaSH gave the 8,2′-anhydro-8-mercapto-cyclic nucleotide (**564**), which by treatment with Raney nickel gave the 2′-deoxy derivative (**565**). Similarly, 2′-deoxyadenylic acid was prepared directly from 5′-AMP.[608]

The problem of separating isomers obtained in the tosylation of the 2′,3′-vicinal hydroxyls, followed by cyclonucleoside formation, was largely circumvented by the treatment of the 2′,3′-cyclic carbonate of 8-bromoadenosine (**566**) with thiourea in *n*-butanol[609] at reflux. A 66% yield of 8,2′-anhydro-9-β-D-arabinofuranosyl-8-mercaptoadenosine (**557**) was obtained using this approach. Subsequently, it was shown that a direct treatment of 8-mercapto-adenosine[610] and 8-mercaptoguanosine[611] with diphenyl carbonate at elevated temperatures gave the corresponding 8,2′-thioanhydronucleosides.

TsCl
Ni
$R_6 = NH_2$; $R_2 = H$
$R_6 = OH$; $R_2 = NH_2$
562
563
564
565
566
557

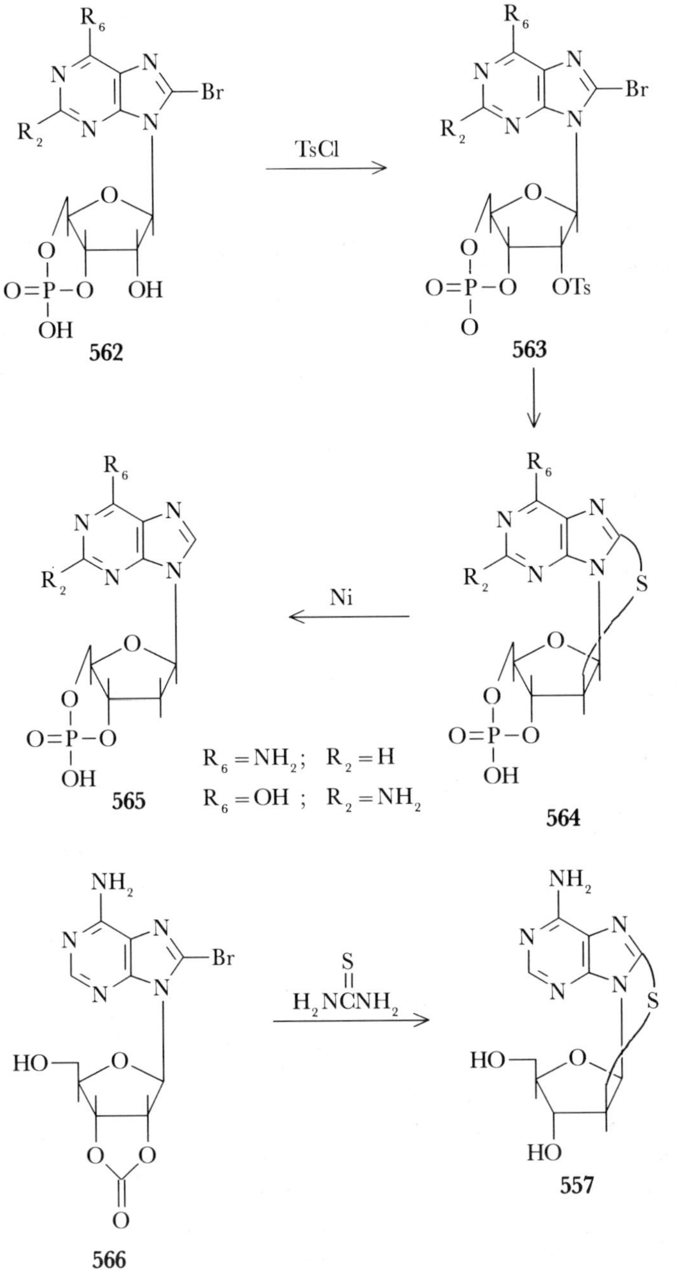

A convenient cyclonucleoside synthesis was provided by the treatment of 8-bromoadenosine (**567**) with thionyl chloride to give 8-bromo-2',3'-O-sulfonyl-adenosine (**568**). Treatment of (**568**) with thiourea then gave a 70% yield of 8,2'-anhydro-9-β-D-arabinofuranosyl-8-mercaptoadenine (**557**).[612] In a reaction very similar to the formation of 2,2'-cyclocytidine 3'-phosphate from cytidine cyclic 2',3'-phosphate, 8-mercaptoadenosine cyclic 2',3'-phosphate (**569**) was heated with trimethylsilyl chloride to give a 75% yield of 8,2'-anhydro-9-β-D-arabinofuranosyl-8-mercaptoadenine 3'-phosphate (**570**).[613]

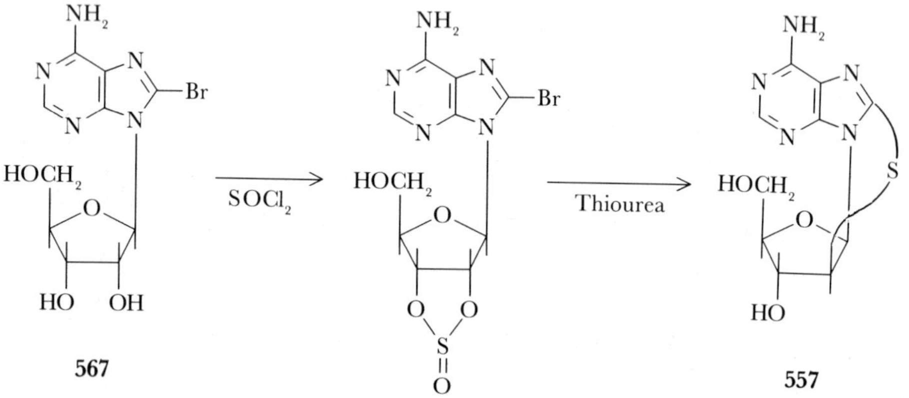

8,5'-Anhydro-2',3'-O-isopropylidine-8-mercaptoadenosine has been shown to undergo the expected reactions of a blocked adenosine derivative, including N-1 oxidation, N-1-methylation and Dimroth rearrangement, deamination to the inosine derivative with nitrous acid and conversion of that compound to the 6-chloropurine derivative with POCl$_3$, and displacement of the 6-chloro group with various nucleophiles.[614] Additionally, the 8,2'-, 8,3'-, and 8,5'-anhydro-8-mercaptoinosines (prepared from the corresponding adenosine derivative by deamination) have been converted with phosphorus pentasulfide to the corresponding cyclonucleosides of 6-mercaptopurine.[615]

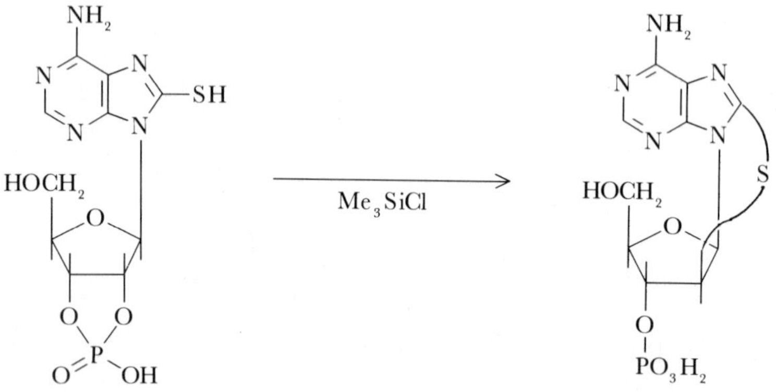

Oxidation of 8,2'-, 8,3'-, and 8,5'-anhydro-8-mercaptoadenosine with *tert*-butyl hypochlorite in methanol gave the respective sulfoxide.[616-618] More extensive oxidation of thioanhydronucleosides, for example, 8,5'-anhydro-8-mercapto-9-β-D-xylofuranosyladenine **(571)** with chlorine in methanolic hydrogen chloride gave 8-chloro-9-(5-deoxy-5-sulfo-β-D-xylofuranosyl)adenine **(572)** in 30% yield.[617] Treatment of the same starting material with hydrogen peroxide in acetic acid gave a 81% yield of 9-β-D-xylofuranosyladenin-8-sulfonic acid **(573)**.[617]

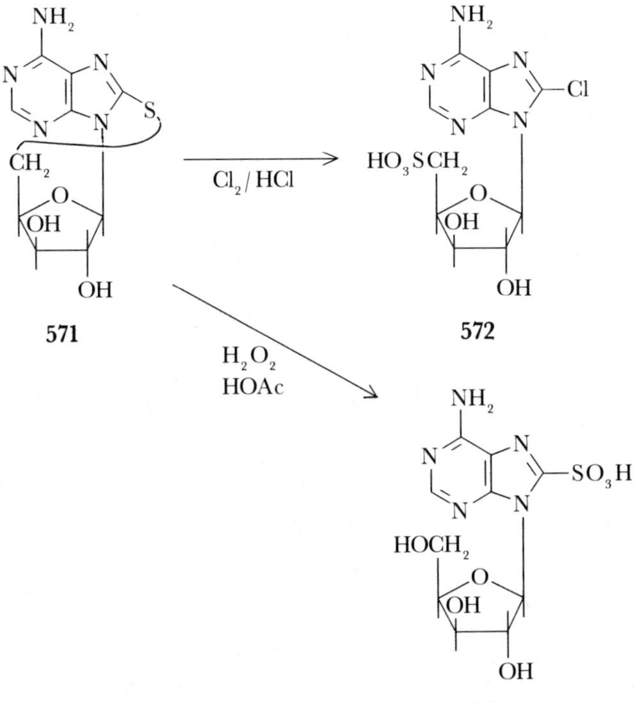

One particularly innovative method of nucleoside synthesis involves the alkylation of 8-mercaptoadenine **(574)** with 5-deoxy-5-iodo-1,2-O-isopropylidine-xylofuranose **(575)**, followed by treatment with acetic acid/acetic anhydride/sulfuric acid and then ammonia to give 8,5'-anhydro-8-mercapto-9-β-D-xylofuranosyladenine **(576)** in 44% yield. Oxidation of **(576)** with aqueous *N*-bromosuccinimide gave the sulfoxide **(577)**. When **(577)** was treated with benzoic anhydride, the blocked thiocyclonucleoside **(578)**, a product of the Pummerer rearrangement, was obtained. Desulfurization of **(578)** with Raney nickel gave 9-β-D-xylofuranosyladenine **(579)**.[619]

3.3.3. O-Cyclonucleosides

Treatment of 5'-O-acetyl-8-bromo-2'-O-*p*-toluenesulfonyladenosine **(580)** with sodium acetate in acetic anhydride gave the intermediate 8-hydroxypurine

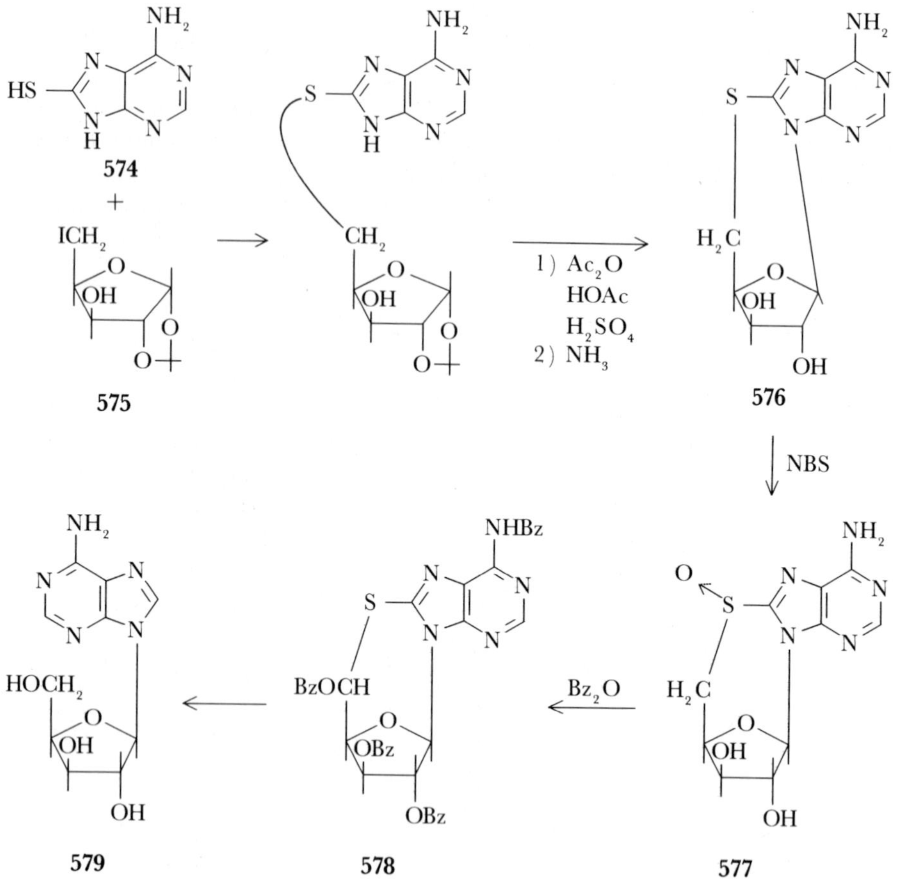

HS — **574**

+

ICH$_2$ ... **575**

CH$_2$ ...

1) Ac$_2$O
HOAc
H$_2$SO$_4$
2) NH$_3$

H$_2$C ... **576**

NBS

NH$_2$ ... HOCH$_2$ ... **579**

NHBz ... BzOCH ... **578**

Bz$_2$O

H$_2$C ... **577**

nucleoside (**581**), which was then cyclized to afford 8,2′-anhydro-9-β-D-arabino-furanosyl-8-oxyadenine (**582**).[620] Other variants of this procedure involved potassium *tert*-butoxide cyclization of the 8-hydroxynucleoside by displacement of a 3′-O-(2,4,6-triisopropylbenzenesulfonyloxy) group to give the 8,3′-anhydro derivative, and ammonia-catalyzed displacement of the same leaving group at the 2′ position to give an 8,2′-cyclonucleoside.[621–623] 8,5′-Anhydro-8-oxyguanosine and 8,5′-anhydro-8-oxyadenosine (**583**) have been prepared by a base-catalyzed displacement of a bromo group at the 8 position by the 5′-hydroxyl of 8-bromo-guanosine[624] and 8-bromoadenosine[625–627] (**584**), respectively. The cyclization of 8-bromo-9-(2,3-O-isopropylidine-β-D-ribofuranosyl)purines with sodium hydride in DMF also furnished the corresponding purine cyclonucleosides.[628] The successful deacetonation of these cyclonucleoside intermediates proved to be rather difficult.[628]

The major obstacle in the preparation of the 8,2′-cyclonucleosides from ribonucleosides (particularly guanosine) has been a facile conversion of the 2′-hydroxyl group to a leaving group. This problem has been resolved by the finding that 2′,3′-O-dibutylstannylene nucleosides react with *p*-toluenesulfonyl chloride to give selective 2′-O-tosyl nucleosides.[629] Using this procedure,

8-bromo-2'-O-p-toluenesulfonylguanosine (**585**) was prepared from the stannylene derivative (**586**) and then converted to 8,2'-anhydro-9-β-D-arabinofuranosyl-8-oxyguanine (**587**) by treatment of (**585**) with acetic anhydride/acetic acid, followed by ammonia.[630] 9-β-D-Arabinofuranosyl-8-chloroadenine undergoes a silica-gel-catalyzed cyclization to provide 8,2'-anhydro-β-D-arabinofuranosyl-8-oxyadenine (**582**).[396]

The synthetic utility of 8,2'-anhydro-8-oxypurine nucleosides has proved substantial. Mild acid treatment of 8,2'-anhydro-8-oxyadenosine (**582**) cleaved

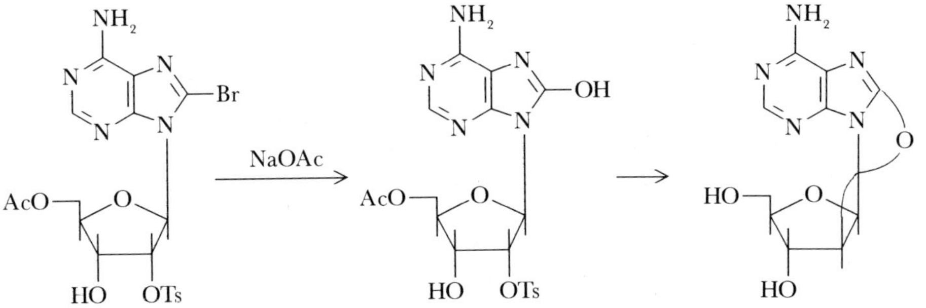

$$R_6 = OH; \quad R_2 = NH_2$$
$$R_6 = NH_2; \quad R_2 = H$$

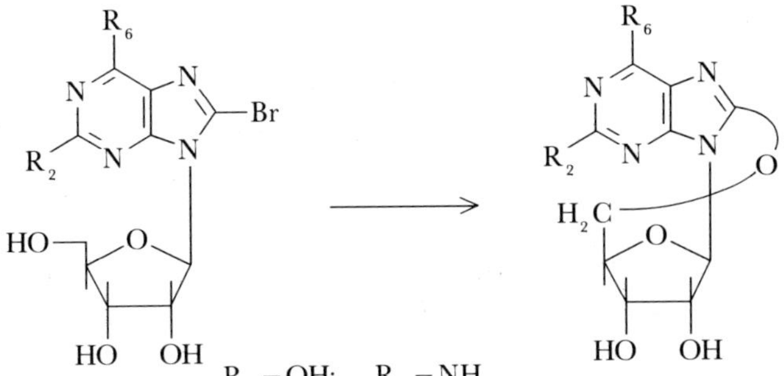

580   581   582

584   583

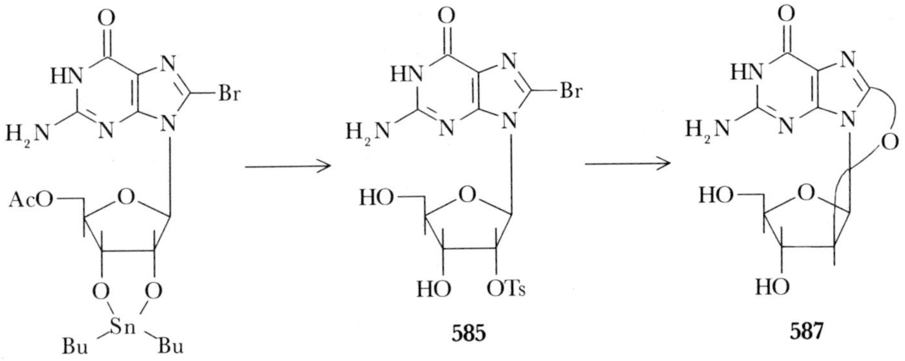

586   585   587

the cyclonucleoside linkage to give 9-β-D-arabinofuranosyl-8-oxyadenosine
(**588**),[621,622] while treatment of the same compound (**582**) with sodium benzoate in DMF gave a reinversion of configuration at the 2′ position to yield
2′-*O*-benzoyl-8-hydroxyadenosine (**589**).[622] Treatment of (**582**) with sodium
hydrogen sulfide gave the intermediate 9-β-D-arabinofuranosyl-8-mercapto-
adenine (**590**). Subsequent Raney nickel desulfurization provided a convenient
route for the preparation of arabinonucleosides from ribonucleosides. This was
first demonstrated by the conversion of adenosine to 9-β-D-arabinofuranosyl-
adenine (**254**),[631,632] a compound of interest in antiviral and antineoplastic
chemotherapies. A similar route was also used for a synthesis of the 9-β-D-arabino-
furanosyladenine[633] and 9-β-D-arabinofuranosylguanine[393] cyclic 3′,5′-
phosphates from cAMP and cGMP. The hydrolytic cleavage of 8,2′-anhydro-8-
oxypurine nucleosides requires some caution since (**582**) has been reported to
undergo a rearrangement to afford 2′,3′-anhydro-8-oxyadenosine under mild
basic conditions.[634] The corresponding 5′-deoxy analogue also underwent similar
transformations.[635] The synthesis of araAMP has been carried out from AMP *via*
8,2′-anhydronucleotide formation followed by a ring opening of the anhydro ring
with H$_2$S followed by desulfurization.[636] It has also been demonstrated that
8,5′-anhydro-8-oxyguanosine (**591**) will undergo an attack at the C-5′ position by
a variety of nucleophilic reagents to give the corresponding 5′-substituted-
8-hydroxyguanosines (**592**).[637] A similar reaction, with H$_2$S/AcOH, was also

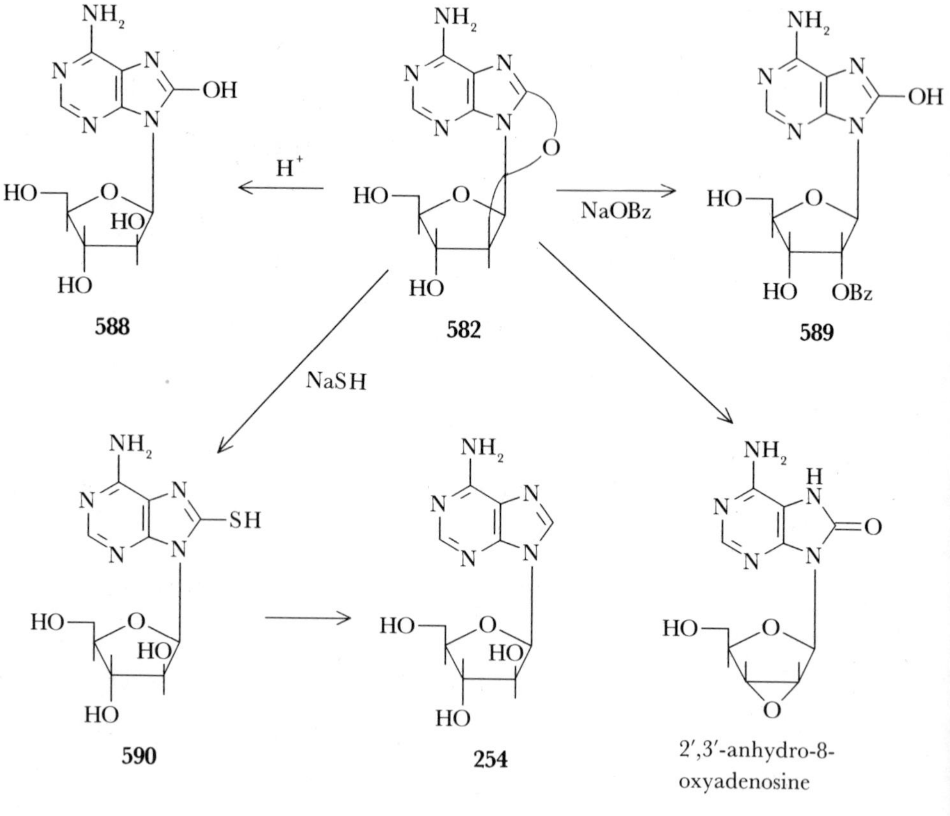

588

582

589

590

254

2′,3′-anhydro-8-
oxyadenosine

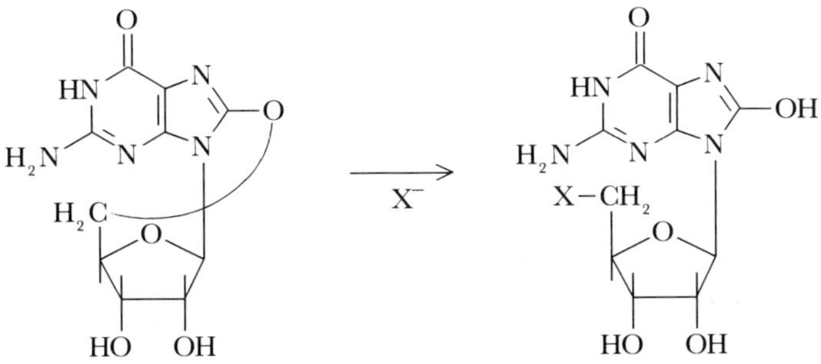

591   592

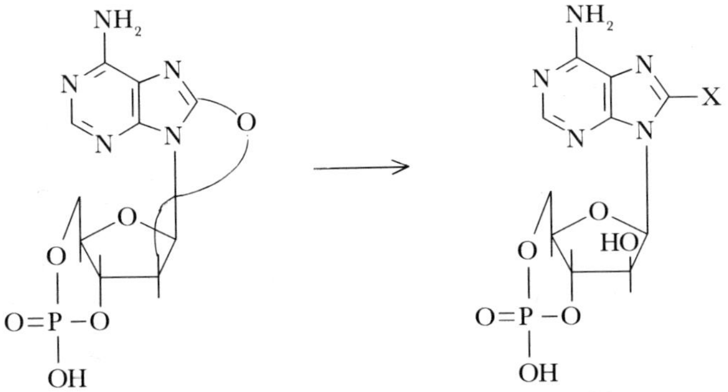

593   594

X = NH₂, OCH₃

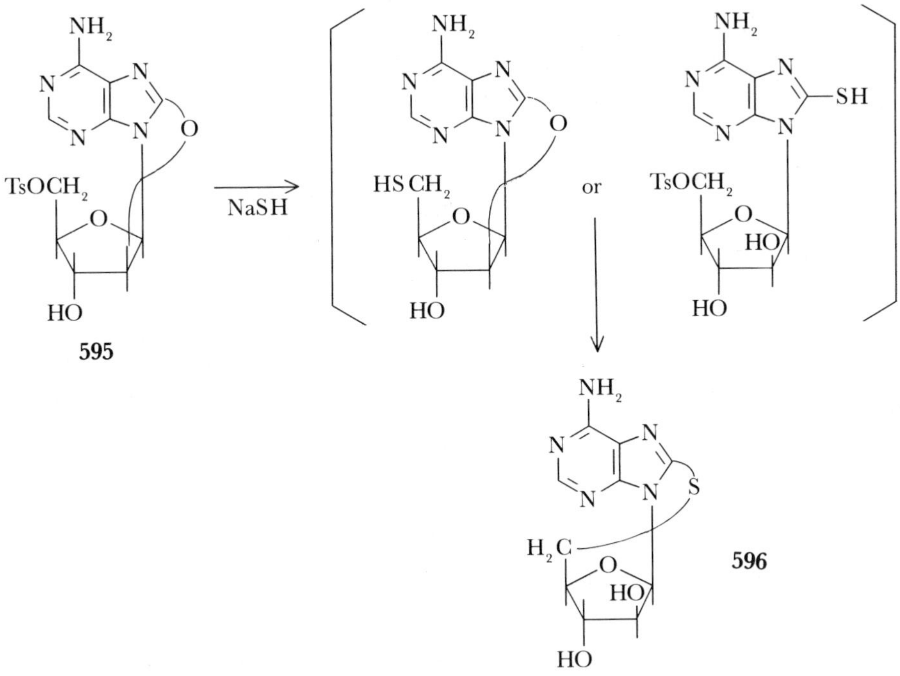

595   596

accomplished in the adenosine series.[626] Additionally, treatment of 8,2'-anhydro-9-β-D-arabinofuranosyladenine cyclic 3',5'-phosphate (**593**) with ammonia or sodium methoxide gave the 8-amino and 8-methoxy derivatives of 9-β-D-arabino-furanosyladenine cyclic 3',5'-phosphate (**594**),[638] respectively.

Some interesting rearrangements have been observed for the cyclonucleo-sides. Tosylation of 8,2'-anhydro-9-β-D-arabinofuranosyladenine gave the 5'-O-toluenesulfonyl derivative (**595**). Treatment of this compound (**595**) with NaSH gave 8,5'-anhydro-9-β-D-arabinofuranosyl-8-mercaptoadenine (**596**) in 46% yield. This compound (**596**) was presumably formed by an attack of hydrosulfide at either the 5' or the 8 position, followed by recyclization.[639] A similar reaction was found to occur with the 8,3'-cyclonucleoside. Ring opening of (**582**) with sodium azide occurs at the 2' position to give 2'-azido-8-oxoadenosine(ribo configuration).[640]

When 8,3'-anhydro-9-(5-O-p-toluenesulfonyl-β-D-xylofuranosyl)-8-mercapto-adenine (**597**) was treated with NaSH, a mixture of 9-(3,5-dideoxy-3,5-dimercapto-β-D-xylofuranosyl)-8-mercaptoadenine (**598**) and the corresponding 8,3'-anhydro derivative (**599**) were obtained. This result was rationalized by invoking an initial formation of the sulfonium intermediate (**600**).[641]

Another interesting rearrangement occurred as a result of chloride attack on the 5' position of 8,5'-anhydro-2',3'-O-isopropylidineadenosine (**601**). It was presumed that the intermediate 5'-chloro-5'-deoxynucleoside (**602**), formed by

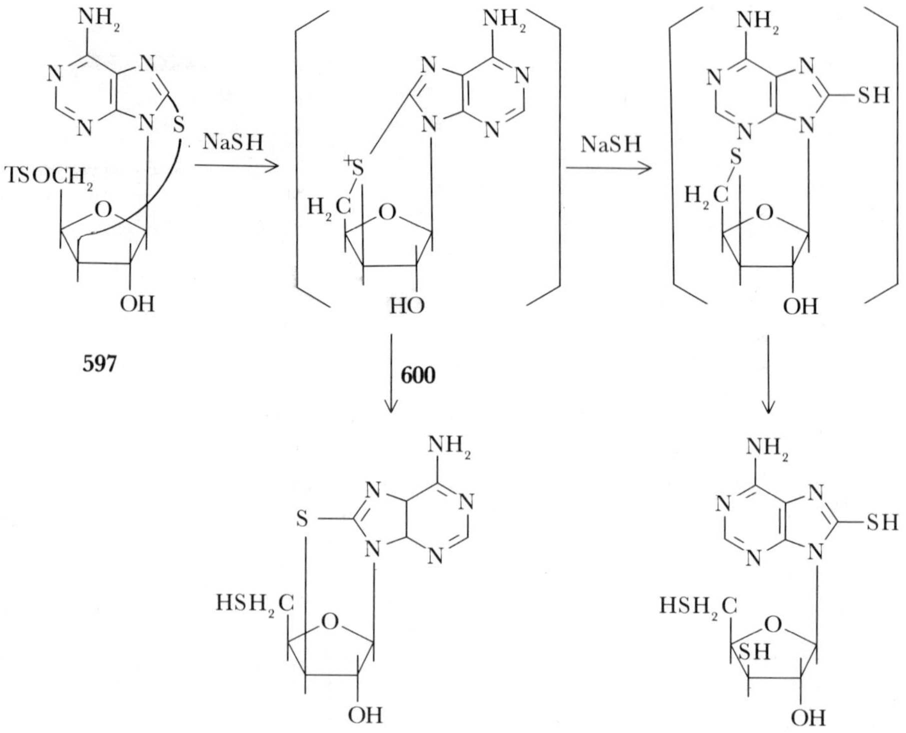

597   600   598

599

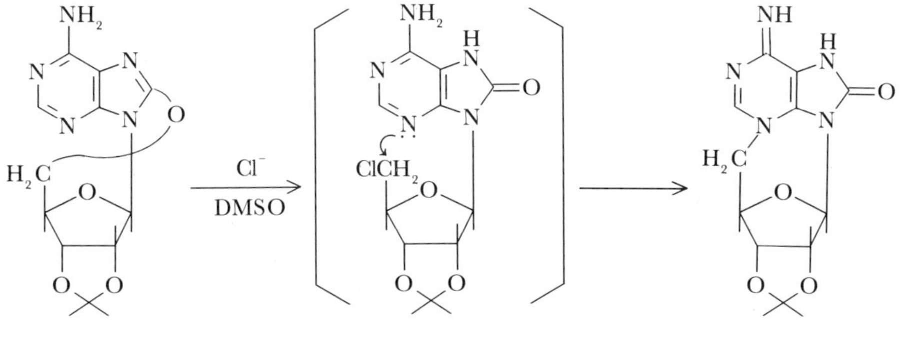

601 **602** **603**

treatment with sodium chloride in dimethylsulfoxide at 130–140°C, immediately recyclized to give 3,5′-anhydro-8-hydroxy-2′3′-O-isopropylidineadenosine (**603**).[642]

3.3.4. N-Cyclonucleosides

The first examples reported for this class of compounds were 8,2′-anhydro-8-amino(and methylamino)-9-β-D-arabinofuranosyladenine (**605**). These compounds were prepared by treatment of 8-bromo-2′-O-triisopropylbenzenesulfonyladenosine (**604**) with ammonia (or methyl amine), followed by cyclization in NaOAc/DMF at reflux.[643] 8-Bromo-2′-O-p-toluenesulfonyladenosine cyclic 3′,5′-phosphate gave 8,2′-anhydro-8-amino(or methylamino)-9-β-D-arabinofuranosyladenine cyclic 3′,5′-phosphate upon treatment with ammonia in methanol at 80°C or with methylamine followed by NaOAc/DMF.[638]

The reaction of 2′,3′-O-isopropylidine-5′-O-tosyl-8-bromoadenosine (**606**) in a 44-fold excess of hydrazine has been reported to yield 2′,3′-O-isopropylidine-8,5′cyclo-8-aminoiminoadenosine (**607**). Compound (**607**) was converted to 8,5′-cyclo-8-aminoiminoadenosine (**608a**), 8,5′-cyclo-8-aminoadenosine (**608b**), and the corresponding inosine analogues (**608**).[644] For a similar 8,5′-cyclization in the guanosine series, it was found necessary to proceed from a N^1-methoxy-

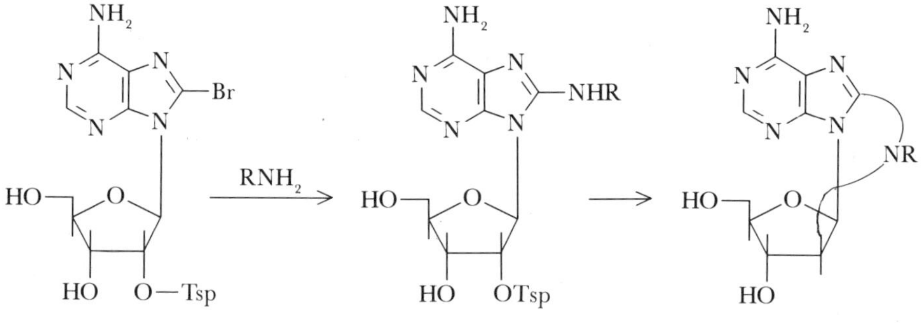

604 **605**

R = H, CH$_3$

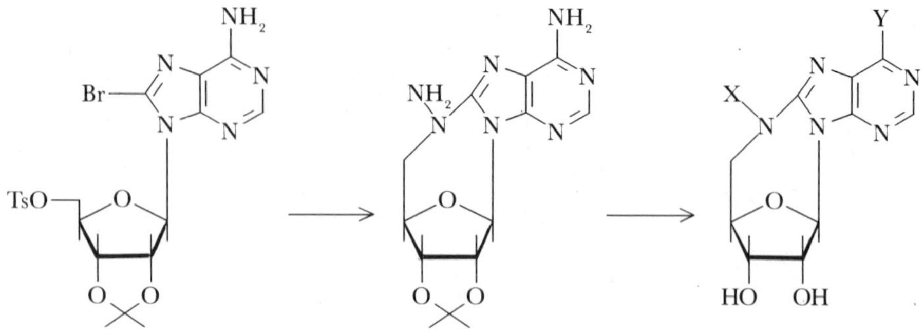

606   **607**   **608**

a: X = Y = NH$_2$
b: X = H, Y = NH$_2$
c: X = H, Y = OH
d: X = NH$_2$, Y = OH

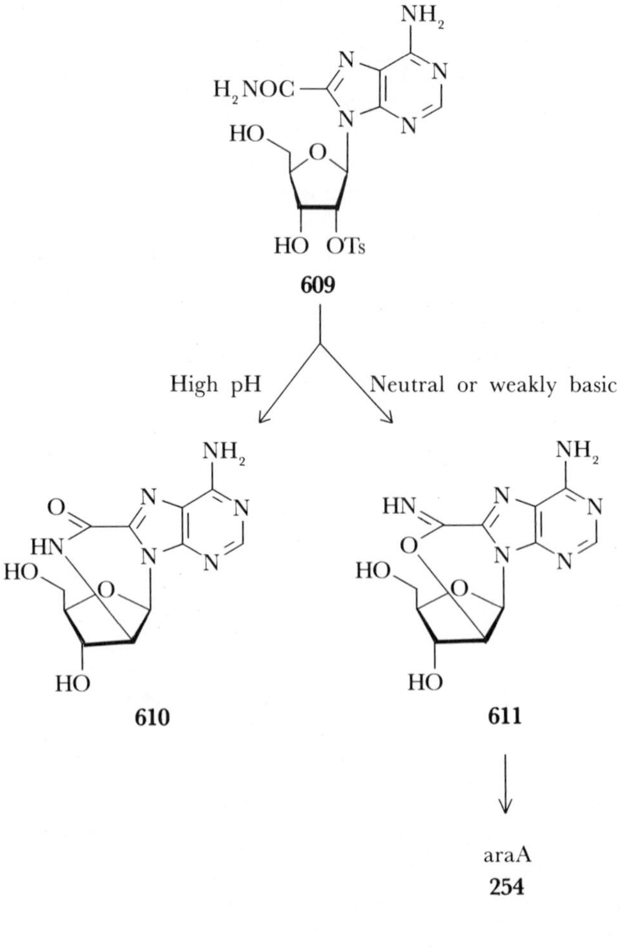

609

High pH   Neutral or weakly basic

610   **611**

araA
254

methylene-N^2-dimethylaminomethylidine guanosine derivative to avoid a facile N^3,5'-cyclization.[556]

The cyclization of 8-carboxamido-2'-O-tosyladenosine (609)[645] with N^1,N^1, N^3,N^3-tetramethylguanidine in a dioxane–water mixture gave 2'-amino-2'-deoxy-8,2'-cyclo-8-carbamoyl-9-β-D-arabinofuranosyladenine (610). The carboxamide moiety in this case behaves like a nitrogen nucleophile. However, in neutral or weakly basic solution (609) gave araA (254) in almost quantitative yield. The formation of (254) could undoubtedly be visualized to occur *via* a cyclonucleoside intermediate (611) arising from the nucleophilic behavior of the carboxamide oxygen. The cyclonucleoside (611) must then hydrolyze and decarboxylate[646] readily, since none of the intermediates, including cyclonucleoside (611), were detected in the experiment.[645]

3.3.5. Cyclonucleosides Containing Carbon–Carbon Bonds

This novel type of cyclonucleoside was first reported by Hampton and co-workers. Treatment of 1-(adenin-9-yl)-2,3-O-isopropylidine-β-D-ribofurano-uronic acid (612) with methyl lithium gave, *via* the 8-lithio derivative (613), the unusual cyclonucleoside (614) with a carbon–carbon bond between the 8 and 5'

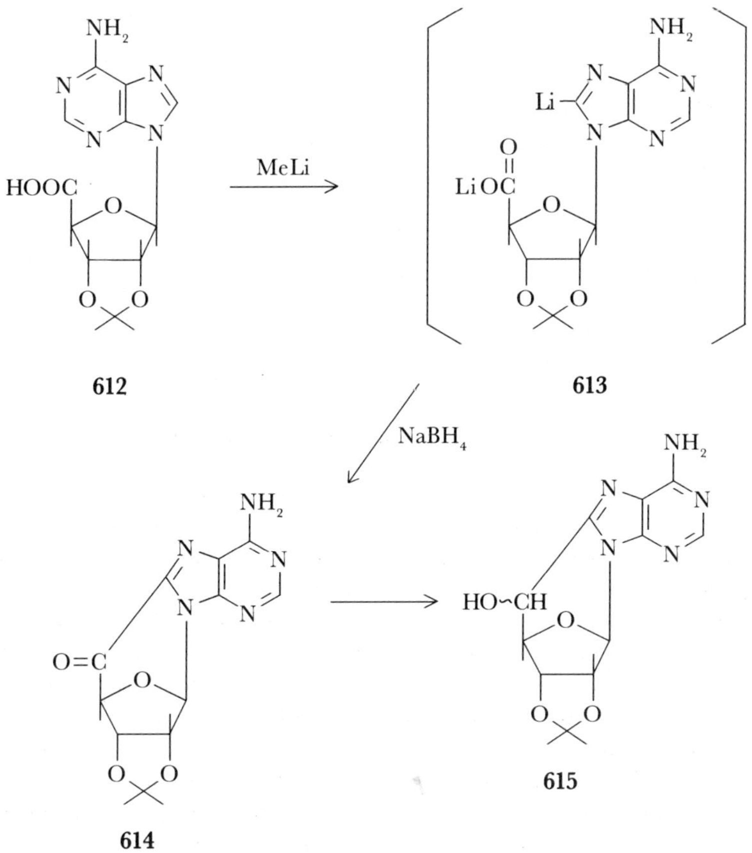

positions in 15% yield.[647] The systematic name of this compound, $3\alpha(R)$-$(3\alpha,4\beta,12\beta,12\alpha)$-9-amino-$3\alpha,4,12,12\alpha$-tetrahydro-2,2-dimethyl-11-H-1,3-dioxolo[5,6]azepino[1,2-d]purin-11-one, defies ready identification and the more convenient 2′,3′-O-isopropylidine-5′-keto-8,5′-cycloadenosine is used. The reduction of (614) with sodium borohydride gave 2′,3′-O-isopropylidine-8,5′-cyclo-adenosine (615). The 5′-phosphate analogue, a close structural analogue of 5′-AMP, was prepared[648] by the phosphoryl chloride/trimethyl phosphate procedure.[649]

The homologue of this cyclonucleoside was prepared by the treatment of 1-(adenin-9-yl)-2,3-O-isopropylidine-α-L-talofuranouronic acid (616) with methyl lithium. The product mixture, isolated in 5% yield and not further fractionated, contained the expected product (617) and the isomer (618) resulting from a base-catalyzed isomerization.[650]

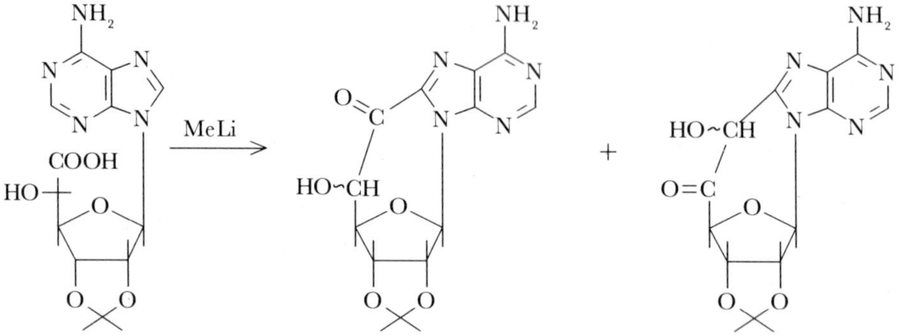

| 616 | 617 | 618 |

Various methods (including photochemical, Section 3.2.3.1) for the synthesis of carbon-linked 8,5′-cyclopurine nucleosides have been reported,[651,652] as has the stereochemistry at the 5′-carbon studied by NMR[653] and single-crystal x-ray analysis.[541]

3.4. Hydrolysis of the Glycosyl Bond: Stabilities and Mechanism

Hydrolysis of the glycosidic bond in nucleosides and nucleotides has important chemical and biochemical applications. The naturally occurring purine nucleosides and nucleotides undergo biochemical degradation by certain enzymes to yield the corresponding purine bases. Various structurally modified nucleosides are also the substrates for these enzymes and thus mimic their natural counterparts. Chemically, both acidic and basic hydrolyses have been observed in model systems. However, the acidic hydrolysis has been investigated in a more thorough manner. The acidic hydrolysis of nucleosides has been suggested to operate by two different mechanisms—Schiff base or A-1 (Scheme XIII). Early reports[654–656] favored the Schiff-base mechanism in which an initial protonation of the glycosyl ring oxygen leads to an opening of the sugar ring giving a Schiff-base intermediate. The attack of water on this intermediate can then liberate the "aglycon."

Scheme XIII

Schiff base

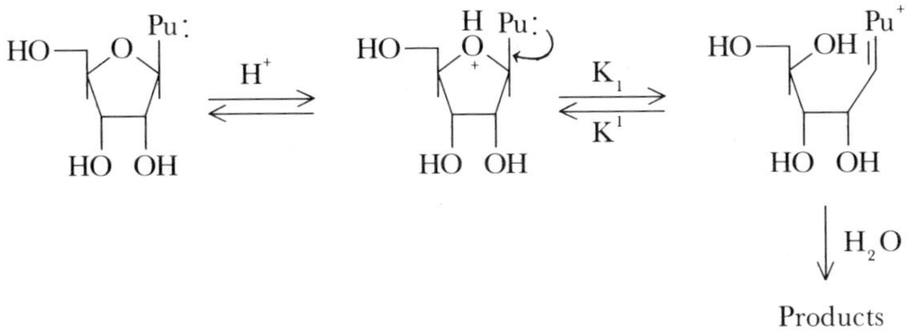

Products

A − 1

Ribose

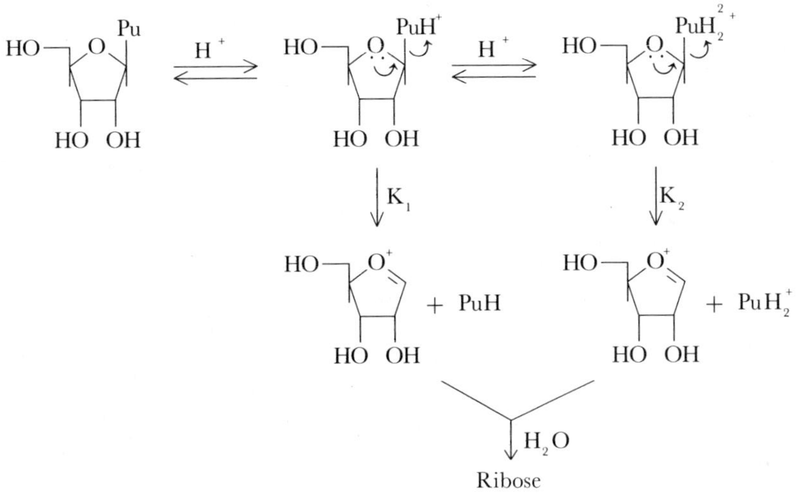

Pu = Purine

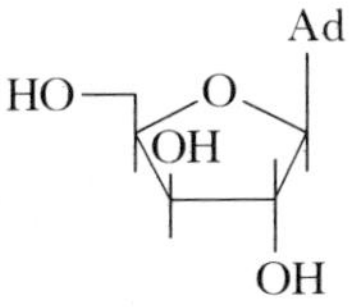

579 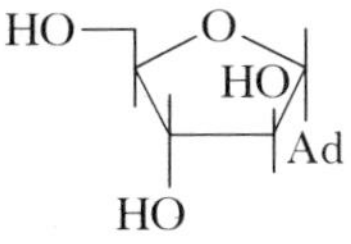   **619**

A hydrolysis that occurs concurrent with anomerization—not commonly observed in purine nucleosides—would tend to support this mechanism.[657] More recent experimental evidence supports the A-1 mechanism,[658-664] in which a rapid initial protonation of the base moiety is followed by a rate-limiting dissociation of the protonated nucleoside to a glycosyl oxocarbenium ion and the free purine. The oxocarbenium ion then reacts with water to form ribose.

It has been suggested[658,659] that the hydrolysis of purine nucleosides by an A-1 mechanism could take place *via* both mono- and diprotonated substrates. It has been shown that the rate of hydrolysis at low pH increased with increasing hydronium ion concentration[658] and leveled off in very strong acid, supposedly where diprotonation of the aglycon was completed. This observation was further substantiated by the reported[665] log k_{obs}–pH profiles of 8-aminoguanosines, which showed that both monobase and dibase protonated nucleosides undergo hydrolysis.

The pH rate profiles for the acidic hydrolysis of 9-β-D-ribofuranosylpurine and the corresponding 6-substituted purine nucleosides have been determined and the products analyzed.[666] 9-β-D-Ribofuranosylpurine in strongly acidic medium yields purine *via* the dication formation. The dication in this case is hydrolyzed to yield a protonated purine and ribosyl oxocarbenium ion analogous to the 6-substituted purine nucleosides, which yield 6-substituted purines on hydrolysis. At low acidic concentrations, the hydrolysis of 9-β-D-ribofuranosylpurine proceeds *via* the formation of a monocation to yield 4-amino-5-formamidopyrimidine by ring opening of the imidazole ring. Hydrolysis of the monocation appears to proceed in a manner similar to basic hydrolysis,[667] in which case a nucleophilic attack at C-8 of the monocation would be expected to provide the pyrimidine analogue after a C-8, N-9 bond cleavage and sugar detachment.[666]

3.4.1. Effect of Substitution on Base and Sugar on Acid Hydrolysis

In general, a decrease in electron density at N-9 facilitates the deglycosylation of purine nucleosides. Consequently, substitution of an electron-donating group on the purine ring should stabilize the C−N glycosidic linkage while an electron-withdrawing group, on the contrary, would make the bond more labile toward acidic hydrolysis. For example, 8-aminoguanosine and 8-methylaminoguanosine were hydrolyzed slower than the parent guanosine,[665] while 8-bromoguanosine was hydrolyzed faster. Similar inductive effects on the acid hydrolysis of 8-bromo- and 8-methoxyadenosine have been reported.[661] Theoretical calculations indicate that amino and chloro substituents at the 8 position increase and decrease, respectively, the electron density at N-9 as compared to a hydrogen at C-8 in guanines.[668] These reports also simultaneously support the A-1 mechanism for hydrolysis, since for the Schiff-base mechanism the opposite electronic requirement for the substituents on purine would be expected. Kinetic data for the hydrolysis of a number of 7-β-D- and 9-β-D-ribofuranosyl-purines indicate that steric crowding near the glycosyl bond accelerates the hydrolysis.

The acid hydrolyses of β-ribo-, deoxy-, dideoxy-, xylo-, arabino-, and psicofuranosides of adenine have been reported and the effect of a hydroxyl group in the sugar on hydrolysis has been evaluated.[661,669–671] In the adenosine series, the rate of hydrolysis increases in the order of 4, 15, 1200, and 4×10^4 times with the removal of 5'-, 3'-, 2'-hydroxyls, and both 2' and 3'-hydroxyls, respectively. The hydroxyl groups induce an electron-withdrawing effect to destabilize the glycosyl carbonium ion formation and thus make the ribonucleosides more stable than the corresponding deoxynucleosides.[658,672] An important application of this observation has been the introduction of an electron-withdrawing trifluoroacetyl group in the deoxynucleoside sugar to make the substrate more stable in an acidic medium[463] (see Section 3.2.2.1). The configuration at the anomeric carbon also plays an important role in the stability of the purine nucleosides. For example, β-xyloadenosine (**579**) hydrolyzes less rapidly than α-arabinoadenosine (**619**), although in both nucleosides the orientation of $2' + 3'$ sugar hydroxyl groups with respect to the adenine moiety is the same. In the β-anomer, however, adenine is occupying a quasiequatorial position and can be protonated to develop a dipole opposite to the dipole of the ring oxygen, resulting in an increased stability of the β-anomer relative to the α-anomer.[671] Furthermore, the *cis* orientation of the hydroxyl group (or groups) with respect to the aglycon weakens the stability of the glycosidic bond by imposing a steric strain and increases the rate of hydrolysis. Consequently, the lyxoadenosine is probably the most labile of the adenine pentosides.[661]

Acknowledgment

Research presented in this chapter was supported by the Office of Health and Environmental Research, U.S. Department of Energy, under contract DE-AC05-84OR21400 with Martin Marietta Energy Systems, Inc.

4. References

1. E. Fischer and B. Helferich, *Chem. Ber.* **47**, 210 (1914).
2. J. M. Gulland and L. F. Story, *J. Chem. Soc.*, 259 (1938).
3. J. Davoll, B. Lythgoe, and A. R. Todd, *J. Chem. Soc.*, 967, 1685 (1948).
4. J. Davoll and B. A. Lowy, *J. Am. Chem. Soc.* **73**, 1650 (1951).
5. R. A. Earl and L. B. Townsend, *J. Carbohydr. Nucleosides Nucleotides* **1**, 177 (1974).
6. H. M. Kissman, C. Pidacks, and B. R. Balser, *J. Am. Chem. Soc.* **77**, 18 (1955).
7. A. M. Michelson, *Chemistry of Nucleosides and Nucleotides*, Academic Press, New York, 1963.
8. J. A. Montgomery and H. J. Thomas, *Adv. Carbohydr. Chem.* **11**, 301 (1962).
9. P. K. Gupta and D. S. Bhakuni, *Indian J. Chem.* **18B**, 248 (1979).
10. N. K. Saxena and D. S. Bhakuni, *Indian J. Chem.* **18B**, 348 (1979).
11. P. K. Gupta and D. S. Bhakuni, *Indian J. Chem.* **20B**, 534 (1981).
12. E. Lazzari, A. Vigerani, and F. Avcamone, *Carbohydr. Res.* **56**, 35 (1977).
13. B. R. Baker, *Ciba Found. Symp. Chem. Biol. Purines*, 120 (1957).
14. B. Shimizu and M. Miyaki, *Chem. Pharm. Bull. (Tokyo)* **15**, 1066 (1967).
15. B. Shimizu and M. Miyaki, *Chem. Pharm. Bull. (Tokyo)* **18**, 510 (1970).
16. M. Miyaki and B. Shimizu, *Chem. Pharm. Bull. (Tokyo)* **18**, 732 (1970).
17. M. Miyaki and B. Shimizu, *Chem. Pharm. Bull. (Tokyo)* **18**, 1446 (1970).
18. S. R. Jenkins, F. W. Holly, and E. Walton, *J. Org. Chem.* **30**, 285 (1965).
19. A. J. Freestone, L. Hough, and A. C. Richardson, *Carbohydr. Res.* **28**, 378 (1973).
20. R. K. Ness and H. G. Fletcher, Jr., *J. Am. Chem. Soc.* **82**, 3434 (1960).
21. L. Vargha and J. Kuszmann, *Ann. der Cheme* **684**, 231 (1965).
22. J. Kuszmann and L. Vargha, *Chem. Ber.* **96**, 2327 (1963).
23. E. Walton, R. F. Nutt, S. R. Jenkins, and F. W. Holly, *J. Am. Chem. Soc.* **86**, 2952 (1964).
24. E. Walton, F. W. Holly, G. E. Boxer, R. F. Nutt, and S. R. Jenkins, *J. Med. Chem.* **8**, 659 (1965).
25. B. R. Baker, R. E. Schaub, and H. M. Kissman, *J. Am. Chem. Soc.* **77**, 5911 (1955).
26. D. H. Murray and J. Prokop, *J. Pharm. Sci.* **54**, 1468 (1965).
27. J. Prokop and D. H. Murray, *J. Pharm. Sci.* **54**, 359 (1965).

28. J. Prokop and D. H. Murray, *J. Pharm. Sci.* **57**, 1697 (1968).
29. G. L. Tong, W. W. Lee, and L. Goodman, *J. Org. Chem.* **32**, 1984 (1967).
30. K. J. Ryan, E. M. Acton, and L. Goodman, *J. Org. Chem.* **33**, 1783 (1968).
31. R. F. Nutt, M. J. Dickinson, F. W. Holly, and E. Walton, *J. Org. Chem.* **33**, 1789 (1968).
32. H. P. Albrecht and J. G. Moffatt, *Tetrahedron Lett.* **13**, 1063 (1970).
33. H. Yanagisawa, M. Kinoshita, S. Nakada, and S. Umezawa, *Bull. Chem. Soc. Jpn.* **43**, 246 (1970).
34. D. H. Murray and J. Prokop, *J. Pharm. Sci.* **56**, 865 (1967).
35. J. M. J. Tronchet and J. Tronchet, *Carbohydr. Res.* **34**, 263 (1974).
36. J. M. J. Tronchet and J. Tronchet, *Carbohydr. Res.* **59**, 594 (1977).
37. S. R. Jenkins and E. Walton, *Carbohydr. Res.* **26**, 71 (1973).
38. J. M. J. Tronchet, R. Graf, and J. Tronchet, *Helv. Chim. Acta* **58**, 1497 (1975).
39. J. M. J. Tronchet and J. F. Tronchet, *Helv. Chim. Acta* **62**, 689 (1979).
40. A. J. Grant and L. M. Lerner, *J. Med. Chem.* **22**, 1016 (1979).
41. L. M. Lerner, *J. Org. Chem.* **40**, 2400 (1975).
42. W. A. Szarek, R. G. S. Ritchie, and D. M. Vyas, *Carbohydr. Res.* **62**, 89 (1978).
43. L. M. Lerner, *J. Org. Chem.* **44**, 4359 (1979).
44. V. K. Srivastava and L. M. Lerner, *Tetrahedron* **34**, 2627 (1978).
45. W. W. Lee, A. P. Martinez, R. W. Blackford, V. J. Bartuska, E. J. Reist, and L. Goodman, *J. Med. Chem.* **14**, 819 (1971).
46. G. L. Tong, K. J. Ryan, W. W. Lee, E. M. Acton, and L. Goodman, *J. Org. Chem.* **32**, 859 (1967).
47. T. Ukita and H. Hayatsu, *J. Am. Chem. Soc.* **84**, 1879 (1962).
48. R. W. Wright, G. M. Tener, and H. G. Khorana, *J. Am. Chem. Soc.* **80**, 2004 (1958).
49. M. Ikehara, E. Ohtsuka, E. Honda, and A. Nomura, *J. Org. Chem.* **30**, 1077 (1965).
50. L. M. Lerner and P. Kohn, *J. Org. Chem.* **31**, 339 (1966).
51. L. M. Lerner, B. D. Kohn, and P. Kohn, *J. Org. Chem.* **33**, 1780 (1968).
52. P. Kohn, R. H. Samaritano, and L. M. Lerner, *J. Org. Chem.* **31**, 1503 (1966).
53. P. Kohn, L. M. Lerner, and B. D. Kohn, *J. Org. Chem.* **32**, 4076 (1967).
54. M. L. Wolfrom, H. G. Garg, and D. Horton, *Chem. Ind. (London)*, 930 (1964).
55. M. L. Wolfrom, H. G. Garg, and D. Horton, *J. Org. Chem.* **30**, 1556 (1965).
56. E. J. Reist, D. E. Gueffroy, and L. Goodman, *Chem. Ind. (London)*, 1364 (1964).
57. G. S. Ritchie and W. A. Szarek, *J. Chem. Soc. Chem. Commun.*, 686 (1973).
58. E. J. Reist, D. E. Gueffroy, R. W. Blackford, and L. Goodman, *J. Org. Chem.* **31**, 4025 (1966).
59. J. A. Montgomery and H. J. Thomas, *J. Am. Chem. Soc.* **87**, 5442 (1965).
60. J. A. Montgomery and H. J. Thomas, *J. Org. Chem.* **31**, 1411 (1966).
61. J. A. Montgomery and H. J. Thomas, *J. Org. Chem.* **28**, 2304 (1963).
62. D. E. Cowley, C. C. Duke, A. J. Liepa, J. K. MacLeod, and D. S. Letham, *Aust. J. Chem.* **31**, 1095 (1978).
63. Y. Mizuno, W. Limn, K. Tsuchida, and K. Ikeda, *J. Org. Chem.* **37**, 39 (1972).
64. N. Yamaoka, K. Aso, and K. Matsuda, *J. Org. Chem.* **30**, 149 (1965).
65. J. A. Montgomery and K. Hewson, *J. Med. Chem.* **12**, 498 (1969).
66. D. L. Leland and M. P. Kotick, *Carbohydr. Res.* **38**, C9 (1974).
67. M. Hoffer, *Chem. Ber.* **93**, 2777 (1960).
68. C. C. Bhat, in *Synthetic Procedures in Nucleic Acid Chemistry* (W. W. Zorbach and R. S. Tipson, eds.), p. 521, Wiley, New York, 1968.
69. W. W. Lee, A. P. Martinez, L. Goodman, and D. W. Henry, *J. Org. Chem.* **37**, 2923 (1972).
70. J. J. K. Novak and F. Sorm, *Coll. Czech. Chem. Commun.* **38**, 1173 (1973).
71. H. Hrebabecky and J. Farkas, *Coll. Czech. Chem. Commun.* **39**, 2115 (1974).
72. T. Sato, T. Shimidate, and Y. Ishido, *Nippon Kagaku Zasshi* **81**, 1440 (1960).
73. Y. Ishido and T. Sato, *Bull. Chem. Soc. Jpn.* **34**, 1347 (1961).
74. T. Shimidate, Y. Ishido, and T. Sato, *Nippon Kagaku Zasshi* **82**, 938 (1961).
75. M. J. Robins and R. K. Robins, *J. Am. Chem. Soc.* **37**, 4934 (1965).
76. M. J. Robins and R. K. Robins, *J. Org. Chem.* **34**, 2160 (1969).
77. L. F. Christensen, A. D. Broom, M. J. Robins, and A. Bloch, *J. Med. Chem.* **15**, 735 (1972).
78. Y. Ishido, A. Hosono, K. Fujii, Y. Kikuchi, and T. Sato, *Nippon Kagaku Zasshi* **87**, 752 (1966); *Chem. Abstr.* **65**, 10734 (1966).
79. K. Onodera, S. Hirano, and H. Fukumi, *Agric. Biol. Chem. (Tokyo)* **28**, 173 (1964).

80. T. Hashizume and H. Iwamura, *Tetrahedron Lett.*, 3095 (1965).
81. K. Imai, A. Nohara, and M. Honjo, *Chem. Pharm. Bull. (Tokyo)* **14**, 1377 (1966).
82. T. Sugiyama and T. Hashizume, *Agric. Biol. Chem.* **42**, 1791 (1978).
83. Y. Ishido, T. Matsuba, A. Hosono, K. Fujii, T. Sato, S. Isome, A. Maruyama, and Y. Kikuchi, *Bull. Chem. Soc. Jpn.* **40**, 1007 (1967).
84. J. A. Montgomery and K. Hewson, *J. Heterocycl. Chem.* **1**, 213 (1964).
85. J. A. Montgomery and K. Hewson, *J. Med. Chem.* **11**, 48 (1968).
86. H. Iwamura and T. Hashizume, *J. Org. Chem.* **33**, 1976 (1968).
87. Y. H. Pan, R. K. Robins, and L. B. Townsend, *J. Heterocycl. Chem.* **4**, 246 (1967).
88. N. Nakazaki, T. Takeda, T. Yoshino, M. Sekiya, and Y. Ishido, *Carbohydr. Res.* **44**, 215 (1975).
89. N. Pravdic and I. Franzie-Mihalic, *Carbohydr. Res.* **62**, 301 (1978).
90. S. Shimizu, T. Nishimura, and M. Ikehara, *Agric. Biol. Chem. (Tokyo)* **31**, 637 (1967).
91. B. Shimizu, M. Asai, and T. Nishimura, *Chem. Pharm. Bull. (Tokyo)* **13**, 230 (1965).
92. B. Shimuzu, M. Asai, and T. Nishimura, *Chem. Pharm. Bull. (Tokyo)* **15**, 1847 (1967).
93. W. A. Bowles and R. K. Robins, *J. Am. Chem. Soc.* **36**, 1252 (1964).
94. E. E. Leutzinger, W. A. Bowles, R. K. Robins, and L. B. Townsend, *J. Am. Chem. Soc.* **90**, 127 (1968).
95. M. Fuertes, G. Garcia-Munoz, R. Madronero, and M. Stud, *Tetrahedron* **26**, 4823 (1970).
96. J. J. Fox and K. A. Watanabe, *Tetrahedron Lett.*, 896 (1966).
97. E. E. Leutzinger, R. K. Robins, and L. B. Townsend, *Tetrahedron Lett.* **43**, 3751 (1970).
98. R. J. Ferrier and M. M. Ponpipom, *J. Chem. Soc. C*, 553 (1971).
99. R. J. Ferrier and M. M. Ponpipom, *J. Chem. Soc. C*, 560 (1971).
100. E. E. Leutzinger, T. Meguro, L. B. Townsend, D. A. Shuman, M. P. Schweizer, C. M. Stewart, and R. K. Robins, *J. Org. Chem.* **37**, 3695 (1972).
101. T. Kondo and T. Goto, *Agric. Biol. Chem. (Tokyo)* **35**, 912 (1971).
102. H. Iwamura, M. Miyakado, and T. Hasizume, *Carbohydr. Res.* **27**, 149 (1973).
103. S. Susaki, A. Yamazaki, A. Kamimura, K. Mitsugi, and I. Jumashiro, *Chem. Pharm. Bull. (Tokyo)* **18**, 172 (1970).
104. K. J. Ryan, E. M. Acton, and L. Goodman, *J. Org. Chem.* **36**, 2646 (1971).
105. W. W. Lee, G. L. Tong, R. W. Blackford, and L. Goodman, *J. Org. Chem.* **35**, 3808 (1970).
106. J. A. Montgomery, K. Hewson, A. G. Laseter, and M. C. Thorpe, *J. Am. Chem. Soc.* **94**, 7176 (1972).
107. R. Vince and R. G. Almquist, *Carbohydr. Res.* **36**, 214 (1974).
108. J. A. Montgomery, K. Hewson, and A. G. Laseter, *J. Med. Chem.* **18**, 571 (1975).
109. D. A. Baker, R. A. Harder, J. Tolman, and R. L. Tolman, *J. Chem. Soc. Chem. Commun.*, 167 (1974).
110. Y. Furukawa and M. Honjo, *Chem. Pharm. Bull. (Tokyo)* **16**, 1076 (1968).
111. W. W. Lee, A. P. Martinez, and L. Goodman, *J. Org. Chem.* **36**, 842 (1971).
112. G. E. Hilbert and T. B. Johnson, *J. Am. Chem. Soc.* **52**, 4489 (1930).
113. F. W. Lichtenthaler, P. Voss, and A. Heerd, *Tetrahedron Lett.* **24**, 2141 (1974).
114. F. W. Lichtenthaler, P. Voss, and G. Bamback, *Bull. Chem. Soc. Jpn.* **47**, 2297 (1974).
115. C. L. Schmidt and L. B. Townsend, *J. Chem. Soc. Perkin Trans. I*, 1257 (1975).
116. C. L. Schmidt and L. B. Townsend, *J. Org. Chem.* **37**, 2300 (1972).
117. V. Nelson, H. S. El Khadem, B. K. Whitten, and D. Sesselman, *J. Med. Chem.* **26**, 1071 (1983).
118. V. Nelson and H. S. El Khadem, *J. Med. Chem.* **26**, 1527 (1983).
119. J. D. Stoecler, C. Cambor, and R. E. Parks, Jr., *Biochemistry* **19**, 102 (1980).
120. I. S. Kazmers, B. S. Mitchell, P. E. Dadonna, L. L. Wotring, L. B. Townsend, and W. N. Kelley, *Science* **214**, 1137 (1981).
121. H. Vorbruggen, K. Krolikiewicz, and B. Bennua, *Chem. Ber.* **114**, 1234 (1981).
122. H. Verbruggen and G. Hofle, *Chem. Ber.* **114**, 1256 (1981).
123. T. Itoh and Y. Mizuno, *Heterocycles* **5**, 285 (1976).
124. G. E. Wright and L. W. Dudyez, *J. Med. Chem.* **27**, 175 (1984).
125. H. Vorbruggen and B. Bennua, *Chem. Ber.* **114**, 1279 (1981).
126. A. V. Azhayev and J. Smrt, *Coll. Czech. Chem. Commun.* **43**, 1520 (1978).
127. F. Leclercq, J. Jumelet-Bach, and K. Antonakis, *Carbohydr. Res.* **62**, 73 (1978).
128. M. Imazawa and F. Eckstein, *J. Org. Chem.* **43**, 3044 (1978).

129. M. Miyaki, A. Saito, and B. Shimuzu, *Chem. Pharm. Bull. (Tokyo)* **18**, 2459 (1970).
130. F. W. Lichtenthaler and K. Kitahara, *Angew. Chem. Int. Ed. Engl.* **14**, 815 (1975).
131. M. Imazawa and F. Eckstein, *J. Org. Chem.* **44**, 2039 (1979).
132. T. Azuma, K. Isono, P. F. Crain, and J. A. McCloskey, *Tetrahedron Lett.*, 1687 (1976).
133. T. Utagawa, H. Morisawa, T. Miyoshi, F. Yoshinaga, A. Yamazaki, and K. Mitsugi, *FEBS Lett.* **109**, 261 (1980).
134. T. Utagawa, H. Morisawa, T. Nakamatsu, A. Yamazaki, and S. Tamanaka, *FEBS Lett.* **119**, 101 (1980).
135. H. Morisawa, T. Utagawa, S. Tamanaka, and A. Tamazaki, *Chem. Pharm. Bull. (Tokyo)* **29**, 3191 (1981).
136. A. L. Schwartz and L. M. Lerner, *J. Org. Chem.* **40**, 24 (1975).
137. W. P. Fuller, R. A. Sanchez, and L. E. Orgel, *J. Mol. Evol.* **1**, 249 (1972).
138. G. Schramm, H. Groetsch, and W. Pollmann, *Angew. Chem. Int. Ed.* **1**, 1 (1962).
139. J. A. Carbon, *J. Am. Chem. Soc.* **86**, 720 (1964).
140. T. Nakagawa, T. Sakakibara, and S. Kumazawa, *Tetrahedron Lett.* **19**, 1645 (1970).
141. A. Holy and F. Sorm, *Coll. Czech. Chem. Commun.* **34**, 3383 (1969).
142. J. A. Montgomery, S. D. Clayton, and H. J. Thomas, *J. Org. Chem.* **40**, 1923 (1975).
143. N. J. Leonard and R. A. Laursen, *Biochemistry* **4**, 354 (1965).
144. M. Asai, M. Kiyaki, and B. Shimuzu, *Agric. Biol. Chem. (Tokyo)* **31**, 319 (1967).
145. C. L. Schmidt and L. B. Townsend, *J. Heterocycl. Chem.* **10**, 687 (1973).
146. Z. Kazimierczuk, H. B. Cottom, G. R. Revankar, and R. K. Robins, *J. Am. Chem. Soc.* **106**, 6379 (1984).
147. L. B. Townsend, *Chem. Rev.* **67**, 533 (1967).
148. A. Yamazaki and M. Okutsu, *J. Heterocycl. Chem.* **15**, 353 (1978).
149. R. J. Rousseau, L. B. Townsend,ηand R. K. Robins, *Chem. Commun.*, 265 (1966).
150. R. J. Rousseau, R. K. Robins, and L. B. Townsend, *J. Am. Chem. Soc.* **90**, 2661 (1968).
151. R. J. Rousseau, R. P. Panzica, S. M. Reddick, R. K. Robins, and L. B. Townsend, *J. Org. Chem.* **35**, 631 (1970).
152. G. Shaw and D. V. Wilson, *J. Chem. Soc.*, 2937 (1962).
153. A. Yamazaki, I. Kumashiro, and T. Takenishi, *J. Org. Chem.* **32**, 3258 (1967).
154. M. Okutsu and A. Yamazaki, *Nucleic Acids Res.* **3**, 231 (1976).
155. A. Yamazaki, I. Kumashiro, and T. Takenishi, *J. Org. Chem.* **32**, 3032 (1967).
156. K. Imai, R. Marumoto, K. Kobayashi, Y. Yoshioka, J. Toda, and M. Honjo, *Chem. Pharm. Bull. (Tokyo)* **19**, 576 (1971).
157. K. Omura, R. Marumoto, and Y. Furukawa, *Chem. Pharm. Bull. (Tokyo)* **29**, 1870 (1981).
158. C. G. Wong and R. B. Meyer, Jr., *J. Med. Chem.* **27**, 429 (1984).
159. A. Yamazaki, I. Kumashiro, and T. Takenishi, *J. Org. Chem.* **32**, 1825 (1967).
160. A. Yamazaki, M. Okutsu, and Y. Yamada, *Nucleic Acids Res.* **3**, 251 (1976).
161. M. Ikutsu and A. Yamazaki, *Nucleic Acids Res.* **3**, 237 (1976).
162. R. J. Quinn, R. P. Gregson, A. F. Cook, and R. T. Bartlett, *Tetrahedron Lett.* **21**, 567 (1980).
163. F. A. Fuhrman, G. J. Fuhrman, Y. H. Kim, L. A.ηPavelka, and H. S. Mosher, *Science* **207** 193 (1980).
164. Y. H. Kim, R. J. Nachman, L. Pavelka, H. S. Mosher, F. A. Fuhrman, and G. J. Fuhrman, *J. Nat. Prod.* **44**, 206 (1981).
165. T. Itaya and T. Harada, *Tetrahedron Lett.* **23**, 2203 (1982).
166. A. F. Cook, R. T. Bartlett, R. P. Gregson, and R. J. Quinn, *J. Org. Chem.* **45**, 4020 (1980).
167. R. T. Bartlett, A. F. Cook, M. J. Holman, W. W. McComas, E. F. Nowoswait, M. S. Poonian, J. A. Baird-Lambert, B. A. Baldo, and J. F. Marwood, *J. Med. Chem.* **24**, 947 (1981).
168. R. Marumoto, Y. Yoshioka, O. Miyashita, S. Shima, K. Imai, K. Kawazoe, and M. Honjo, *Chem. Pharm. Bull. (Tokyo)* **23**, 759 (1975).
169. R. J. Rousseau and L. B. Townsend, *J. Org. Chem.* **33**, 2828 (1968).
170. M. Hatton, K. Ienaga, and W. Pfleiderer, *Liebigs Ann. Chem.*, 1796 (1978).
171. J. A. Montgomery and H. J. Thomas, *J. Med. Chem.* **15**, 1334 (1972).
172. R. B. Meyer, Jr., D. A. Shuman, and R. K. Robins, *J. Med. Chem.* **16**, 1319 (1973).
173. R. B. Meyer, Jr., D. A. Shuman, and R. K. Robins, *J. Am. Chem. Soc.* **96**, 4962 (1974).

174. R. B. Meyer, Jr., H. Uno, R. K. Robins, L. N. Simon, and J. P. Miller, *Biochemistry* **14**, 3315 (1975).
175. K. Kikugawa and H. Suehiro, *J. Carbohydr. Nucleosides Nucleotides* **2**, 159 (1975).
176. R. M. Creswell and G. B. Brown, *J. Org. Chem.* **28**, 2560 (1963).
177. (a) P. C. Srivastava, G. A. Ivanovics, R. J. Rousseau, and R. K. Robins, *J. Org. Chem.* **40**, 2920 (1975); (b) R. B. Meyer, Jr., unpublished results.
178. G. W. Kenner, C. W. Taylor, and A. R. Todd, *J. Chem. Soc.*, 1620 (1949).
179. M. E. Wolff and A. Burger, *J. Am. Pharm. Assoc. (Sci. Ed.)* **48**, 56 (1959).
180. D. Autenrieth, H. Schmid, K. Harzer, M. Ott, and W. Pfleiderer, *Angew. Chem. Int. Ed.* **10**, 927 (1971).
181. H. Rokos and W. Pfleiderer, *Chem. Ber.* **104**, 748 (1971).
182. R. Vince, J. Brownell, and S. Daluge, *J. Med. Chem.* **27**, 1358 (1984).
183. Y. F. Shealy and J. D. Clayton, *J. Pharm. Sci.* **62**, 1432 (1973).
184. Y. F. Shealy and J. D. Clayton, *J. Am. Chem. Soc.* **88**, 3885 (1966).
185. Y. F. Shealy and C. A. O'Dell, *Tetrahedron Lett.*, 2231 (1969).
186. Y. F. Shealy and J. D. Clayton, *J. Pharm. Sci.* **62**, 1252 (1973).
187. S. Daluge and R. Vince, *J. Med. Chem.* **15**, 171 (1972).
188. R. Vince and S. Daluge, *J. Med. Chem.* **17**, 578 (1974).
189. S. Pestka, R. Vince, S. Daluge, and R. Harris, *Antimicrob. Agents Chemother.* **4**, 37 (1973).
190. R. Ranganathan, *Tetrahedron Lett.*, 1185 (1975).
191. G. Weiman and H. G. Khorana, *J. Am. Chem. Soc.* **84**, 4324 (1962).
192. F. R. Atherton, H. T. Openshaw, and A. R. Todd, *J. Chem. Soc.*, 382 (1945).
193. R. W. Chambers, J. G. Moffatt, and H. G. Khorana, *J. Am. Chem. Soc.* **79**, 3747 (1957).
194. G. M. Tener, *J. Am. Chem. Soc.* **83**, 159 (1961).
195. P. T. G. Iham and G. M. Tener, *Chem. Ind. (London)*, 542 (1959).
196. J. A. Montgomery and H. J. Thomas, *J. Org. Chem.* **26**, 1926 (1961).
197. T. A. Khwaja and C. B. Reese, *Tetrahedron* **27**, 6189 (1971).
198. M. Yoshikawa, T. Kato, and T. Takenishi, *Tetrahedron Lett.*, 5065 (1967).
199. M. Yoshikawa, T. Kato, and T. Takenishi, *Bull. Chem. Soc. Jpn.* **42**, 3505 (1969).
200. D. E. Macfarlane, D. C. B. Mills, and P. C. Srivastava, *Biochemistry* **21**, 544 (1982).
201. D. E. Macfarlane, P. C. Srivastava, and D. C. B. Mills, *J. Clin. Invest.* **71**, 420 (1983).
202. B. G. Hughes, P. C. Srivastava, D. D. Muse, and R. K. Robins, *Biochemistry* **22**, 2116 (1983).
203. K. Imai, S. Fujii, K. Takanohasi, Y. Furukawa, T. Matsuda, and M. Honjo, *J. Org. Chem.* **34**, 1547 (1969).
204. T. Sowa and S. Ouchi, *Bull. Chem. Soc. Jpn.* **48**, 2084 (1975).
205. M. Morr and M. R. Kula, *Tetrahedron Lett.*, 23 (1974).
206. A. W. Murray and M. R. Atkinson, *Biochemistry* **7**, 4023 (1968).
207. E. W. Sutherland and T. W. Rall, *J. Am. Chem. Soc.* **79**, 3608 (1957).
208. L. N. Simon, D. A. Shuman, and R. K. Robins, in *Advances in Cyclic Nucleotide Research*, Vol. 3, p. 225 (P. Greengard and G. A. Robison, eds.), Raven Press, New York, 1973.
209. D. Lipkin, W. H. Cook, and R. Markham, *J. Am. Chem. Soc.* **81**, 6198 (1959).
210. D. Lipkin, R. Markham, and W. H. Cook, *J. Am. Chem. Soc.* **81**, 6075 (1959).
211. M. Smith, G. I. Drummond, and H. G. Khorana, *J. Am. Chem. Soc.* **83**, 698 (1961).
212. G. I. Drummond, M. W. Gilgan, E. J. Reiner, and M. Smith, *J. Am. Chem. Soc.* **86**, 1626 (1964).
213. R. K. Bordon and M. Smith, *J. Org. Chem.* **31**, 3247 (1966).
214. A. Murayama, B. Jastorff, and H. Hettler, *Angew. Chem.* **82**, 666 (1970).
215. A. Murayama, B. Jastorff, F. Cramer, and H. Hettler, *J. Org. Chem.* **36**, 3029 (1971).
216. A. Murayama, B. Jastorff, H. Hettler, and F. Cramer, *Chem. Ber.* **106**, 3127 (1973).
217. B. Jastorff and T. Krebs, *Chem. Ber.* **105**, 3192 (1972).
218. F. Eckstein, L. P. Simonson, and H.-P. Baer, *Biochemistry* **13**, 3806 (1974).
219. D. A. Shuman, J. P. Miller, M. B. Scholten, L. N. Simon, and R. K. Robins, *Biochemistry* **12**, 2781 (1973).
220. G. H. Jones, H. P. Albrecht, N. P. Damodaran, and J. G. Moffatt, *J. Am. Chem. Soc.* **92**, 3510 (1970).
221. M. Morr, M.-R. Kula, and L. Ernst, *Tetrahedron* **31**, 1619 (1975).

222. M. Ikehara and J. Yano, *Nucleic Acids Res.* **1**, 1783 (1974).
223. R. Marumoto, T. Nishimura, and M. Honjo, *Chem. Pharm. Bull. (Tokyo)* **23**, 2295 (1975).
224. J. P. Berry, J. R. Arnold, and A. F. Isbell, *J. Org. Chem.* **33**, 1664 (1968).
225. M. Morr, L. Ernst, and R. Mengel, *Liebigs Ann. Chem.*, 651 (1982).
226. G. R. Revankar and R. K. Robins, in *Cyclic Nucleotides, Vol. I, Handbook of Experimental Pharmacology*, pp. 17–151 (J. A. Nathanson and J. W. Kebabian, eds.), Springer-Verlag, Berlin, 1982.
227. A. Holy and J. Smrt, *Coll. Czech. Chem. Commun.* **31**, 1528 (1966).
228. T. Ueda and I. Kawai, *Chem. Pharm. Bull. (Tokyo)* **18**, 2303 (1970).
229. J. D. Smith and D. B. Dunn, *Biochim. Biophys. Acta* **31**, 573 (1959).
230. A. D. Broom and R. K. Robins, *J. Am. Chem. Soc.* **87**, 1145 (1965).
231. J. A. Haines, C. B. Reese, and A. R. Todd, *J. Chem. Soc.*, 5281 (1962).
232. J. W. Jones and R. K. Robins, *J. Am. Chem. Soc.* **85**, 193 (1963).
233. T. A. Khwaja and R. K. Robins, *J. Am. Chem. Soc.* **88**, 3641 (1966).
234. J. Yano, L. S. Kan, and P. O. P. Ts'O, *Biochim. Biophys. Acta* **629**, 178 (1980).
235. M. J. Robins and S. R. Naik, *Biochim. Biophys. Acta* **246**, 341 (1971).
236. M. J. Robins, S. R. Naik, and A. S. K. Lee, *J. Org. Chem.* **39**, 1891 (1974).
237. M. J. Robins and E. M. Trip, *Biochemistry* **12**, 2179 (1973).
238. L. F. Christensen and A. D. Broom, *J. Org. Chem.* **37**, 3399 (1972).
239. D. G. Bartholomew and A. D. Broom, *J. Chem. Soc. Chem. Commun.*, 38 (1975).
240. Y. Mizuno, T. Endo, and K. Ikeda, *J. Org. Chem.* **40**, 1385 (1975).
241. A. Myles and W. Pfleiderer, *Chem. Ber.* **105**, 3327 (1972).
242. I. Tazawa, S. Tazawa, J. L. Alderfer, and P. O. P. Ts'O, *Biochemistry* **11**, 4931 (1972).
243. K. K. Ogilvie, S. L. Beaucage, A. L. Schifman, N. Y. Theriault, and K. L. Sadana, *Can. J. Chem.* **56**, 2768 (1978).
244. K. K. Ogilivie and N. Y. Theriault, *Tetrahedron Lett.*, 2111 (1979).
245. R. Charubala and W. Pfleiderer, *Tetrahedron Lett.*, 1933 (1980).
246. J. A. McCloskey, A. M. Lawson, K. Tsuboyama, P. M. Krueger, and R. N. Stillwell, *J. Am. Chem. Soc.* **90**, 4182 (1968).
247. J. A. McCloskey, in *Basic Principles in Nucleic Acid Chemistry*, Vol. I, p. 209 (P. O. P. Ts'O, ed.), Academic Press, New York, 1974.
248. L. F. Christensen, P. D. Cook, R. K. Robins, and R. B. Meyer, Jr., *J. Carbohydr. Nucleosides Nucleotides* **4**, 175 (1977).
249. K. K. Ogilvie, S. L. Beaucage, D. W. Entwistle, E. A. Thompson, M. A. Quilliam, and J. B. Wetmore, *J. Carbohydr. Nucleosides Nucleotides* **3**, 197 (1976).
250. E. J. Corey and A. Venkateswarlo, *J. Am. Chem. Soc.* **94**, 6190 (1972).
251. K. K. Ogilvie, Krishan L. Sadana, E. A. Thompson, M. A. Quilliam, and J. B. Westmore, *Tetrahedron Lett.*, 2861 (1974).
252. K. K. Ogilvie, *Can. J. Chem.* **51**, 3799 (1973).
253. S. S. Jones and C. B. Reese, *J. Chem. Soc. Perkin Trans. I*, 2762 (1979).
254. W. T. Markiewicz and M. Wiewiorowski, *Nucleic Acids Res. Spec. Publ.*, S158 (1978).
255. M. J. Robins, J. S. Wilson, and F. Hansske, *J. Am. Chem. Soc.* **105**, 4059 (1983).
256. (a) C. Gioeli, M. Kwiatkowski, B. Oberg, and J. B. Chattopadhyaya, *Tetrahedron Lett.* **22**, 1741 (1981). (b) K. W. Pankiewicz, K. A. Watanabe, H. Takayanagi, T. Itoh, and H. Ogura, *J. Heterocycl. Chem.* **22**, 1703 (1985); (c) K. W. Pankiewicz and K. A. Watanabe, *Nucleic Acids Res. Symp. Ser.* **11**, 9 (1982).
257. A. Hampton and A. W. Nichol, *Biochemistry* **5**, 2076 (1966).
258. J. P. H. Verheyden, I. D. Jenkins, G. R. Owen, S. D. Dimitrijevich, C. M. Richards, P. C. Srivastava, N. Hong, and J. G. Moffatt, *Ann. N.Y. Acad. Sci.* **255**, 151 (1974).
259. P. Drasar, L. Hein, and J. Beranek, *J. Nucleic Acids Res. Publ.* **1**, 561 (1975).
260. A. Hampton, *J. Am. Chem. Soc.* **83**, 3640 (1961).
261. S. Chladek and J. Smrt, *Coll. Czech. Chem.* **28**, 1301 (1963).
262. A. Hampton, J. C. Fratantoni, P. M. Carroll, and S. Wong, *J. Am. Chem. Soc.* **87**, 5481 (1965).
263. M. Smith, D. H. Rammler, I. H. Goldberg, and H. G. Khorana, *J. Am. Chem. Soc.* **84**, 430 (1964).
264. F. Cramer, W. Saenger, K. Scheit, and J. Tennigkeit, *Ann. der Chemie* **679**, 156 (1964).
265. J. Zemlicka, *Chem. Ind. (London)*, 581 (1964).

266. M. Jarman and C. B. Reese, *Chem. Ind. (London)*, 1493 (1964).
267. C. B. Reese and J. E. Sulston, *Proc. Chem. Soc.*, 214 (1964).
268. H. P. M. Fromageot, B. E. Griffin, C. B. Reese, and J. E. Sulston, *Tetrahedron* **23**, 2315 (1967).
269. B. E. Griffin, M. Jarman, C. B. Reese, J. E. Sulston, and D. R. Trentham, *Tetrahedron* **23**, 2301 (1967).
270. C. B. Reese and D. R. Trentham, *Tetrahedron Lett.*, 2459 (1965).
271. C. B. Reese ad D. R. Trentham, *Tetrahedron Lett.*, 2467 (1965).
272. M. Smith, D. H. Rammler, I. H. Goldberg, and H. G. Khorama, *J. Am. Chem. Soc.* **84**, 430.
273. B. E. Griffin, M. Jarman, and C. B. Reese, *Tetrahedron Lett.* **24**, 639 (1968).
274. C. B. Reese, R. Saffhill, and J. E. Sulston, *J. Am. Chem. Soc.* **89**, 3366 (1967).
275. B. E. Griffin and C. B. Reese, *Tetrahedron Lett.* **25**, 4057 (1969).
276. D. P. L. Green, T. Ravindranathan, C. B. Reese, and R. Saffhill, *Tetrahedron Lett.* **26**, 1031 (1970).
277. M. J. Robins and J. S. Wilson, *J. Am. Chem. Soc.* **103**, 932 (1981).
278. D. C. Baker, T. H. Haskell, and S. R. Putt, *J. Med. Chem.* **21**, 1218 (1978).
279. D. C. Baker, S. D. Kumar, W. J. Waites, G. Arnett, W. M. Shanon, W. I. Higuchi, and W. J. Lambert, *J. Med. Chem.* **27**, 270 (1984).
280. Y. Ishido, N. Nakazaki, and N. Sakairi, *J. Chem. Soc. Perkin Trans. I*, 2088 (1979).
281. H. Takaku, T. Nomoto, and K. Kimura, *Chem. Lett. (Chem. Soc. Jpn.)*, 1221 (1981).
282. K. E. Pfitzner and J. G. Moffatt, *J. Am. Chem. Soc.* **85**, 3027 (1963).
283. K. E. Pfitzner and J. G. Moffatt, *J. Am. Chem. Soc.* **87**, 5661 (1965).
284. K. E. Pfitzner and J. G. Moffatt, *J. Am. Chem. Soc.* **87**, 5670 (1965).
285. R. S. Ranganathan, G. H. Jones, and J. G. Moffatt, *J. Org. Chem.* **39**, 290 (1974).
286. J. D. Albright and L. Goldman, *J. Am. Chem. Soc.* **87**, 4214 (1965).
287. J. D. Albright and L. Goldman, *J. Am. Chem. Soc.* **89**, 2410 (1967).
288. W. Sowa and G. H. S. Thomas, *Can. J. Chem.* **44**, 836 (1966).
289. K. Antonakeis and F. Leclercq, *C.R. Acad. Sci. Ser. C* **271**, 1197 (1970).
290. K. Antonakeis and F. Leclercq, *Bull. Soc. Chim. Fr.*, 4309 (1971).
291. J. Herscovici and K. Antonateis, *J. Chem. Soc. Perkin Trans. I*, 979 (1974).
292. D. C. Baker and D. Horton, *Carbohydr. Res.* **21**, 393 (1972).
293. G. H. Jones and J. G. Moffatt, *J. Am. Chem. Soc.* **90**, 5337 (1968).
294. J. A. Montgomery, A. G. Laseter, and K. Hewson, *J. Heterocycl. Chem.* **11**, 211 (1974).
295. T. E. Walker, H. Follmann, and H. P. C. Hogenkamp, *Carbohydr. Res.* **27**, 225 (1973).
296. H. Follmann, *Angew. Chem.* **86**, 41 (1974).
297. D. C. Baker and T. H. Haskel, *J. Med. Chem.* **18**, 1041 (1975).
298. D. C. Baker, R. P. Crews, T. H. Haskell, S. R. Putt, G. Arnett, W. M. Shanon, C. M. Reinke, N. B. Katlama, and J. C. Drach, *J. Med. Chem.* **26**, 1530 (1983).
299. A. F. Cook and J. G. Moffatt, *J. Am. Chem. Soc.* **89**, 2697 (1967).
300. A. Brodbech and J. G. Moffatt, *J. Org. Chem.* **35**, 3552 (1970).
301. A. Rosenthal, M. Sprinzl, and D. A. Baker, *Tetrahedron Lett.*, 4233 (1970).
302. K. Antonakis and M. J. Arvor-Egron, *Carbohydr. Res.* **27**, 468 (1973).
303. A. S. Jones and A. R. Williamson, *Chem. Ind. (London)*, 1624 (1960).
304. J. P. Moss, C. B. Reese, K. Schofield, R. Shapiro, and A. Todd, *J. Chem. Soc.*, 1149 (1963).
305. R. R. Schmidt, U. Schloz, and D. Schwille, *Chem. Ber.* **101**, 590 (1968).
306. P. J. Harper and A. Hampton, *J. Org. Chem.* **35**, 1688 (1970).
307. R. E. Harmon, C. V. Zenarosa, and S. K. Gupta, *Chem. Ind. (London)*, 1141 (1969).
308. R. R. Schmidt and H.-J. Fritz, *Chem. Ber.* **103**, 1867 (1970).
309. H.-J. Fritz, R. Machat, and R. R. Schmidt, *Chem. Ber.* **105**, 642 (1972).
310. H. H. Stein, *J. Med. Chem.* **16**, 1306 (1973).
311. J. J. Baker, Am M. Mian, and J. R. Tittensor, *Tetrahedron* **30**, 2939 (1974).
312. F. W. Lichtenthaler and H. P. Albrecht, *Chem. Ber.* **102**, 964 (1969).
313. F. W. Lichtenthaler, P. Emig, and D. Bommer, *Chem. Ber.* **102**, 971 (1969).
314. H. H. Baer and M. Bayer, *Can. J. Chem.* **49**, 568 (1971).
315. K. Miyai, R. K. Robins, and R. L. Tolman, *J. Med. Chem.* **15**, 1092 (1972).
316. M. J. Robins, Y. Fouron, and R. Mengel, *J. Org. Chem.* **39**, 1564 (1974).
317. P. A. Levene and R. S. Tipson, *J. Biol. Chem.* **109**, 623 (1935).
318. A. Todd and T. L. V. Ulbricht, *J. Chem. Soc.*, 3275 (1960).

319. M. Clark, A. Todd, and J. Zussman, *J. Chem. Soc.*, 2952 (1951).

320. S. H. Mudd, G. A. Jameson, and G. L. Cantoni, *Biochim. Biophys. Acta* **38**, 164 (1960).

321. W. Jahn, *Chem. Ber.* **98**, 1705 (1965).

322. R. Kuhn and W. Jahn, *Chem. Ber.* **98**, 1699 (1965).

323. S. R. Landaver and H. N. Rydon, *J. Chem. Soc.*, 2224 (1953).

324. J. P. H. Verheyden and J. G. Moffatt, *J. Org. Chem.* **35**, 2319 (1970).

325. E. J. Prisbe, J. Smejkal, J. P. H. Verheyden, and J. G. Moffatt, *J. Org. Chem.* **41**, 1836 (1976).

326. S. D. Dimitrijevich, J. P. H. Verheyden, and J. G. Moffatt, *J. Org. Chem.* **44**, 400 (1979).

327. P. C. Srivastava and R. K. Robins, *J. Carbohydr. Nucleosides Nucleotides* **4**, 93 (1977).

328. K. Haga, M. Yoshikawa, and T. Kato, *Bull. Chem. Soc. Jpn.* **43**, 3922 (1970).

329. J. P. H. Verheyden and J. G. Moffatt, *J. Org. Chem.* **37**, 2289 (1972).

330. J. B. Lee and I. M. Downie, *Tetrahedron* **23**, 359 (1967).

331. S. Hanessian, M. M. Ponpipom, and P. Lavallee, *Carbohydr. Res.* **24**, 45 (1972).

332. K. K. Kugawa and M. Ichino, *Tetrahedron Lett.*, 87 (1971).

333. I. Yamamoto, M. Sekine, and T. Hata, *J. Chem. Soc. Perkin Trans. I*, 306 (1980).

334. T. Adachi, Y. Arai, I. Inoue, and M. Saneyoshi, *Carbohydr. Res.* **78**, 67 (1980).

335. P. C. Srivastava, K. L. Nagpal, and M. M. Dhar, *Experientia* **25**, 356 (1969).

336. H. P. C. Hogenkamp, *Biochemistry* **13**, 2736 (1974).

337. S. Greenberg and J. G. Moffatt, *J. Am. Chem. Soc.* **95**, 4016 (1973).

338. A. F. Russell, S. Greenberg, and J. G. Moffatt, *J. Am. Chem. Soc.* **95**, 4025 (1973).

339. M. J. Robins, F. Hansske, N. H. Low, and J. I. Park, *Tetrahedron Lett.* **25**, 367 (1984).

340. T. C. Jain, I. D. Jenkins, A. F. Russell, J. P. H. Verheyden, and J. G. Moffatt, *J. Org. Chem.* **39**, 30 (1974).

341. M. J. Robins, R. Mengel, and R. A. Jones, *J. Am. Chem. Soc.* **95**, 4074 (1973).

342. R. Mengel and W. Muhs, *Liebigs Ann. Chem.*, 1585 (1977).

343. R. Mengel and W. Muhs, *Chem. Ber.* **112**, 625 (1979).

344. K. Kondo, T. Adachi, and I. Inoue, *J. Org. Chem.* **42**, 3967 (1977).

345. A. J. Freestone, L. Hough, and A. C. Richardson, *Carbohydr. Res.* **43**, 239 (1975).

346. M. Ikehara and J. Imura, *Chem. Pharm. Bull. (Tokyo)* **29**, 3281 (1981).

347. M. J. Robins, R. Mengel, R. A. Jones, and Y. Fouran, *J. Am. Chem. Soc.* **98**, 8204 (1976).

348. Y. Wang, H. P. C. Hogenkamp, R. A. Long, G. R. Revankar, and R. K. Robins, *Carbohydr. Res.* **59**, 449 (1977).

349. K. Ramalingam and R. W. Woodard, *J. Org. Chem.* **49**, 1291 (1984).

350. N. Zylber and J. Zylber, *J. Chem. Soc. Chem. Commun.* **24**, 1084 (1978).

351. P. H. C. Mundill, R. W. Ries, C. Woenckhaus, and B. V. Plapp, *J. Med. Chem.* **24**, 474 (1981).

352. C. D. Anderson, L. Goodman, and B. R. Baker, *J. Am. Chem. Soc.* **81**, 3967 (1959).

353. W. W. Lee, A. Benitez, C. D. Anderson, L. Goodman, and B. R. Baker, *J. Am. Chem. Soc.* **83**, 1906 (1961).

354. G. L. Tong, W. W. Lee, and L. Goodman, *J. Org. Chem.* **30**, 2854 (1965).

355. J. Baddiley and G. A. Jamieson, *J. Chem. Soc.*, 4280 (1954).

356. P. A. Levene and R. S. Tipson, *J. Biol. Chem.* **111**, 313 (1935).

357. J. K. Coward and E. P. Slisz, *J. Med. Chem.* **16**, 460 (1973).

358. R. T. Borchart, J. A. Huber, and Y. S. Wu, *J. Med. Chem.* **17**, 868 (1974).

359. J. K. Coward, M. D'urso-Scott, and W. D. Sweet, *Biochem. Pharmacol.* **21**, 1200 (1972).

360. R. T. Borchart and Y. S. Wu, *J. Med. Chem.* **17**, 862 (1974).

361. M. Ikehara and M. Kanebo, *Tetrahedron* **26**, 4251 (1970).

362. M. Ikehara, H. Tada, and Kei Muneyama, *Chem. Pharm. Bull. (Tokyo)* **13**, 639 (1965).

363. M. Ikehara and S. Uesugi, *Tetrahedron* **28**, 3687·(1972).

364. E. J. Reist, V. J. Bartuska, D. F. Calkins, and L. Goodman, *J. Org. Chem.* **30**, 3401 (1965).

365. W. W. Lee, A. Benitez, L. Goodman, and B. R. Baker, *J. Am. Chem. Soc.* **82**, 2648 (1960).

366. E. J. Reist, A. Benitez, L. Goodman, B. R. Baker, and W. W. Lee, *J. Org. Chem.* **27**, 3274 (1962).

367. E. J. Reist and L. Goodman, *Biochemistry* **3**, 15 (1964).

368. R. Mengel and M. Bartke, *Angew. Chem. Int. Ed. Engl.* **17**, 679 (1978).

369. B. R. Baker and R. E. Schaub, *J. Am. Chem. Soc.* **77**, 5900 (1955).

370. E. J. Reist, D. F. Calkins, and L. Goodman, *J. Org. Chem.* **32**, 169 (1967).

371. M. Ikehara, T. Maruyama, and H. Miki, *Tetrahedron* **34**, 1133 (1978).

372. E. J. Backus, H. D. Tresner, and T. H. Campbell, *Antibiot. Chemother.* **7**, 532 (1957).
373. G. O. Morton, J. E. Lancaster, G. E. Van Lear, W. Fulmori, and W. E. Meer, *J. Am. Chem. Soc.* **91**, 1535 (1969).
374. D. A. Shuman, M. J. Robins, and R. K. Tobins, *J. Am. Chem. Soc.* **92**, 3434 (1970).
375. I. D. Jenkins, J. P. H. Verheyden, and J. G. Moffatt, *J. Am. Chem. Soc.* **98**, 3346 (1976).
376. R. Duschinsky, H. Walker, W. Wojnarowski, W. Arnold, C. Englert, and B. Pellmont, *J. Carbohydr. Nucleosides Nucleotides* **1**, 411 (1974).
377. R. Duschinsky and U. Eppenberger, *Tetrahedron Lett.*, 5103 (1967).
378. F. W. Lichtenthaler and H. J. Miller, *Angew. Chem.* **85**, 765 (1973).
379. P. C. Srivastava and R. J. Rosseau, *Carbohydr. Res.* **27**, 455 (1973).
380. L. M. Beacham, *J. Org. Chem.* **44**, 3100 (1979).
381. H. Loibner and E. Zbiral, *Liebigs Ann. Chem.*, 78 (1978).
382. J. A. Elvidge, J. R. Jones, C. O'Brien, and E. A. Evans, *Chem. Commun.*, 394 (1971).
383. M. Tomaz, J. Olsen, and C. M. Mercado, *Biochemistry* **11**, 1235 (1972).
384. J. A. Elvidge, J. R. Jones, C. O'Brien, E. A. Evans, and H. C. Sheppard, *J. Chem. Soc. Perkin Trans. II*, 2138 (1973).
385. M. Maeda, M. Saneyoshi, and Y. Kawazoe, *Chem. Pharm. Bull. (Tokyo)* **19**, 1641 (1971).
386. J. A. Elvidge, J. R. Jones, C. O'Brien, E. A. Evans, and H. C. Sheppard, *J. Chem. Soc. Perkin Trans. II*, 174 (1974).
387. R. E. Holmes and R. K. Robins, *J. Am. Chem. Soc.* **86**, 1242 (1964).
388. P. C. Srivastava and K. L. Nagpal, *Experientia* **26**, 220 (1970).
389. M. Ikehara, S. Uesugi, and M. Kuneko, *Chem. Commun.*, 17 (1967).
390. R. A. Long, R. K. Robins, and L. B. Townsend, *J. Org. Chem.* **32**, 2751 (1967).
391. M. Ikehara and S. Uesugi, *Chem. Pharm. Bull. (Tokyo)* **17**, 348 (1969).
392. K. Muneyama, R. J. Bauer, D. A. Shuman, R. K. Robins, and L. N. Simon, *Biochemistry* **10**, 2390 (1971).
393. A. M. Mian, R. Harris, R. W. Sidwell, R. K. Robins, and T. A. Khwaja, *J. Med. Chem.* **17**, 259 (1974).
394. A. M. Kapuler and E. Reich, *Biochemistry* **10**, 4050 (1971).
395. H. J. Brentnall and D. W. Hutchinson, *Tetrahedron Lett.*, 2595 (1972).
396. E. K. Ryu and M. MacCoss, *J. Org. Chem.* **46**, 2819 (1981).
397. D. Lipkin, F. B. Howard, D. Nowothy, and M. Sano, *J. Biol. Chem.* **238**, 2249 (1963).
398. E. N. Moudrianakis and M. Beer, *Biochim. Biophys. Acta* **95**, 23 (1965).
399. E. Fischer, *Z. Physiol. Chem.* **69**, 69 (1909).
400. H. Kossel, *Z. Physiol. Chem.* **340**, 210 (1965).
401. H. D. Hoffman and W. Muller, *Biochim. Biophys. Acta* **123**, 421 (1966).
402. M. Hung and L. M. Stock, *J. Org. Chem.* **47**, 448 (1982).
403. H. Kossell and S. Doehring, *Biochim. Biophys. Acta* **95**, 663 (1965).
404. Y. Kawazoe and G.-F. Huang, *Chem. Pharm. Bull. (Tokyo)* **20**, 2073 (1972).
405. A. D. Broom and R. K. Robins, *J. Org. Chem.* **34**, 1025 (1969).
406. M. A. Stevens, D. I. Magrath, H. W. Smith, and G. B. Brown, *J. Am. Chem. Soc.* **80**, 2755 (1958).
407. M. A. Stevens and G. B. Brown, *J. Am. Chem. Soc.* **80**, 2759 (1958).
408. M. A. Stevens, H. W. Smith, and G. B. Brown, *J. Am. Chem. Soc.* **81**, 1734 (1959).
409. H. Klenow and S. Frederiksen, *Biochim. Biophys. Acta* **52**, 384 (1961).
410. H. Sigel and H. Brintzinger, *Helv. Chim. Acta* **48**, 433 (1965).
411. R. B. Meyer, Jr., D. A. Shuman, R. K. Robins, J. P. Miller, and L. N. Simon, *J. Med. Chem.* **16**, 1319 (1973).
412. H.-J. Rhaese, *Biochim. Biophys. Acta* **166**, 311 (1968).
413. H. Uno, R. B. Meyer, Jr., D. A. Shuman, R. K. Robins, L. N. Simon, and J. P. Miller, *J. Med. Chem.* **19**, 419 (1976).
414. M. Ikehara and Y. Ogiso, *Chem. Pharm. Bull. (Tokyo)* **23**, 2534 (1975).
415. R. H. Hall, *The Modified Nucleosides in Nucleic Acids*, Columbia University Press, New York, 1971.
416. P. A. Levene and R. S. Tipson, *J. Biol. Chem.* **94**, 809 (1932).
417. H. Bredereck, G. Muller, and E. Berger, *Chem. Ber.* **73**, 1059 (1940).
418. H. Bredereck, H. Haas, and A. Martini, *Chem. Ber.* **81**, 307 (1948).
419. A. Wacker and M. Ebert, *Z. Naturforsch.* **14b**, 709 (1959).

420. P. Brookes and P. D. Lawley, *J. Chem. Soc.*, 539 (1960).
421. J. W. Jones and R. K. Robins, *J. Am. Chem. Soc.* **85**, 193 (1963).
422. A. D. Broom, L. B. Townsend, J. W. Jones, and R. K. Robins, *Biochemistry* **3**, 494 (1964).
423. D. H. R. Barton, C. J. R. Hedgecock, E. Lederer, and W. B. Matherwall, *Tetrahedron Lett.*, 279 (1979).
424. H. Tanaka, Y. Uchida, M. Shinozaki, H. Hayakawa, A. Matsuda, and T. Miyasaka, *Chem. Pharm. Bull. (Tokyo)* **31**, 787 (1983).
425. H. G. Windmueller and N. O. Kaplan, *J. Biol. Chem.* **236**, 2616 (1961).
426. H. Bredereck and A. Martini, *Chem. Ber.* **80**, 401 (1948).
427. E. Shaw, *J. Am. Chem. Soc.* **80**, 3899 (1958).
428. E. Shaw, *J. Am. Chem. Soc.* **83**, 4770 (1961).
429. E. Shaw, *J. Am. Chem. Soc.* **81**, 6021 (1959).
430. B. Singer, *Biochemistry* **11**, 3939 (1972).
431. A. Holy, R. W. Bald, and N. D. Hong, *Coll. Czech. Chem. Commun.* **36**, 2658 (1971).
432. J. Zemlicka and A. Holý, *Coll. Czech. Chem. Commun.* **32**, 3159 (1967).
433. K. D. Philips and J. P. Horwitz, *J. Org. Chem.* **40**, 1856 (1975).
434. T. Fujii, F. Tanaka, K. Mohri, T. Itaya, and T. Suito, *Tetrahedron Lett.*, 4873 (1973).
435. N. D. Kochetkov, V. W. Shibaev, and A. A. Kost, *Tetrahedron Lett.*, 1993 (1971).
436. J. R. Barrio, J. A. Secrist III, and N. J. Leonard, *Biochem. Biophys. Res. Commun.* **46**, 597 (1972).
437. J. A. Secrist III, J. R. Barrio, N. J. Leonard, C. Villar-Palasi, and A. G. Gilman, *Science* **177**, 279 (1972).
438. J. A. Secrist III, J. R. Barrio, N. J. Leonard, and G. Weber, *Biochemistry* **11**, 3499 (1972).
439. J. E. Roberts, Y. Aizonzo, M. Sonenberg, and N. I. Swislocki, *Bioorg. Chem.* **4**, 181 (1975).
440. G. H. Jones, D. V. K. Murthy, D. Tegg, R. Golling, and J. G. Moffat, *Biochem. Biophys. Res. Commun.* **53**, 1338 (1973).
441. G. L. Anderson, B. H. Rizkalla, and A. D. Broom, *J. Org. Chem.* **39**, 937 (1974).
442. R. Shapiro and J. Hachmann, *Biochemistry* **9**, 2799 (1966).
443. B. M. Goldschmidt, T. P. Blazej, and B. L. Ban Dunren, *Tetrahedron Lett.*, 1583 (1968).
444. E. Zbrial and E. Hugh, *Tetrahedron Lett.*, 439 (1972).
445. E. Hugl, G. Schulz, and E. Zbiral, *Liebigs Ann. Chem.*, 278 (1973).
446. C. Ivanovics and E. Zbiral, *Monatsch. Chem.* **106**, 417 (1975).
447. K. Nakanishi, N. Furutachi, M. Funamizu, D. Grunberger, and I. B. Weinstein, *J. Am. Chem. Soc.* **92**, 7617 (1970).
448. Y. Furukawa, O. Miyashita, and M. Honjo, *Chem. Pharm. Bull. (Tokyo)* **22**, 2552 (1974).
449. G.-F. Huang, T. Okamoto, M. Maeda, and Y. Kawazve, *Tetrahedron Lett.*, 4541 (1973).
450. J. F. Gerster, J. W. Jones, and R. K. Robins, *J. Org. Chem.* **28**, 945 (1963).
451. M. Ikehara, H. Uno, and F. Ishikawa, *Chem. Pharm. Bull. (Tokyo)* **12**, 267 (1964).
452. M. Ikehara and H. Uno, *Chem. Pharm. Bull. (Tokyo)* **13**, 221 (1965).
453. J. Zemlicka and F. Sorm, *Coll. Czech. Chem. Commun.* **30**, 1880 (1965).
454. R. B. Meyer, Jr., D. A. Shuman, R. K. Robins, R. J. Bauer, M. K. Dimmitt, and L. N. Simon, *Biochemistry* **11**, 2704 (1972).
455. G. Michal, K. Muhlegger, M. Nelboeck, C. Thiessen, and G. Weimann, *Pharmacol. Res. Commun.* **6**, 203 (1974).
456. R. B. Meyer, Jr., H. Uno, D. A. Shuman, R. K. Robins, L. N. Simon, and J. P. Miller, *J. Cyclic Nucleotide Res.* **1**, 159 (1975).
457. R. Marumoto, Y. Yoshioka, O. Miyashita, S. Shima, K. Imai, K. Kawazoe, and M. Honjo, *Chem. Pharm. Bull. (Tokyo)* **23**, 759 (1975).
458. R. B. Meyer, Jr., P. D. Cook, and R. K. Robins, *Nucleic Acid Chemistry: Improved and New Synthetic Procedures, Methods and Techniques*, Part 2, p. 607 (L. B. Townsend and R. S. Tipson, eds.), Wiley, New York, 1978.
459. A. Yamazaki, I. Kumashiro, and T. Takenishi, *J. Org. Chem.* **33**, 2583 (1968).
460. A. Yamazaki, T. Furukawa, M. Akiyama, M. Okutsu, I. Kumashiro, and M. Ikehara, *Chem. Pharm. Bull. (Tokyo)* **21**, 692 (1973).
461. K. H. Boswell, L. F. Christensen, D. A. Shuman, and R. K. Robins, *J. Heterocycl. Chem.* **12**, 1 (1975).
462. G. L. Szekeres, R. K. Robins, K. H. Boswell, and R. A. Long, *J. Heterocycl. Chem.* **12**, 15 (1975).

463. M. J. Robins and G. L. Basom, *Can. J. Chem.* **51**, 3161 (1973).

464. H. Kawashima and I. Kumashiro, *Bull. Chem. Soc. Jpn.* **40**, 639 (1967).

465. J. F. Gerster, B. C. Hinshaw, R. K. Robins, and L. B. Townsend, *J. Org. Chem.* **33**, 1070 (1968).

466. R. K. Robins, *J. Am. Chem. Soc.* **82**, 2654 (1960).

467. R. K. Robins, *Biochem. Prep.* **10**, 145 (1963).

468. R. K. Robins, *J. Org. Chem.* **26**, 447 (1961).

469. A. G. Beaman and R. K. Robins, *J. Appl. Chem.* **12**, 432 (1962).

470. C. W. Noell and R. K. Robins, *J. Am. Chem. Soc.* **81**, 5997 (1959).

471. R. E. Holmes and R. K. Robins, *J. Am. Chem. Soc.* **86**, 1242 (1964).

472. K. M. Umeyama, D. A. Shuman, K. H. Boswell, R. K. Robins, L. N. Simon, and J. P. Miller, *J. Carbohydr. Nucleosides Nucleotides* **1**, 1 (1974).

473. A. Strecker, *Ann. der Chemie*, 118 (1861).

474. P. A. Levene and W. A. Jacobs, *Chem. Ber.* **43**, 3150 (1910).

475. R. Shapiro, *J. Am. Chem. Soc.* **86**, 2948 (1964).

476. R. Shapiro and S. H. Pohl, *Biochemistry* **7**, 448 (1968).

477. J. C. Parham, J. Fissekis, and G. B. Brown, *J. Org. Chem.* **31**, 966 (1966).

478. J. P. Miller, K. H. Boswell, K. Muneyama, L. N. Simon, R. K. Robins, and D. A. Shuman, *Biochemistry* **12**, 5310 (1973).

479. P. C. Srivastava, K. L. Nagpal, and M. M. Dhar, *Indian J. Chem.* **7**, 1 (1969).

480. C. A. Bunton and B. B. Wolfe, *J. Am. Chem. Soc.* **96**, 7747 (1974).

481. A. Giner-Sorolla, J. Longley-Cook, M. McCravey, G. B. Brown, and J. H. Burchenal, *J. Med. Chem.* **16**, 365 (1973).

482. J. A. Montgomery and K. Hewson, *J. Am. Chem. Soc.* **79**, 4559 (1957).

483. J. A. Montgomery and K. Hewson, *J. Org. Chem.* **33**, 432 (1968).

484. J. A. Montgomery and K. Hewson, *J. Med. Chem.* **12**, 498 (1969).

485. P. C. Srivastava and R. K. Robins, *J. Carbohydr. Nucleosides Nucleotides* **4**, 93 (1977).

486. J. F. Gerster and R. K. Robins, *J. Am. Chem. Soc.* **87**, 3752 (1965).

487. M. Kawana, P. C. Srivastava, and R. K. Robins, *J. Carbohydr. Nucleosides Nucleotides* **8**, 131 (1981).

488. J. F. Gerster and R. K. Robins, *J. Am. Chem. Soc.* **87**, 3752 (1965).

489. J. A. Montgomery and K. Hewson, *J. Am. Chem. Soc.* **82**, 463 (1960).

490. M. J. Robins and B. Uznanski, *Can. J. Chem.* **59**, 2608 (1981).

491. M. Huang, T. L. Avery, R. L. Blakley, J. A. Secrist, and J. A. Montgomery, *J. Med. Chem.* **27**, 800 (1984).

492. (a) M. Ikehara and S. Yamada, *Chem. Pharm. Bull. (Tokyo)* **19**, 104 (1971); (b) J. A. Secrist, L. L. Bennett, Jr., P. W. Allen, L. M. Rose, C.-H. Chang, and J. A. Montgomery, *J. Med. Chem.* **29**, 2069 (1986).

493. J. J. Fox, I. Wempen, A. Hampton, and J. L. Doerr, *J. Am. Chem. Soc.* **80**, 1669 (1958).

494. M. Saneoshi, *Chem. Pharm. Bull. (Tokyo)* **19**, 493 (1971).

495. H. Vorbruggen, *Angew. Chem. Int. Ed.* **11**, 304 (1972).

496. H. Vorbruggen, *Ann. N.Y. Acad. Sci.* **255**, 82 (1975).

497. H. Vorbruggen and K. Krolikiewicz, *Liebigs Ann. Chem.*, 745 (1976).

498. M. J. Robins and B. Uznanski, *Can. J. Chem.* **59**, 2601 (1981).

499. M. Roth, P. Dubs, E. Gotschi, and A. Eschenmoser, *Helv. Chim. Acta* **54**, 710 (1971).

500. A. Yamane, H. Inoue, and T. Ueda, *Chem. Pharm. Bull. (Tokyo)* **28**, 157 (1980).

501. H. Vorbruggen and K. Krolikiewicz, *Angew. Chem. Int. Ed. Engl.* **15**, 869 (1976).

502. M. Saneyoshi and K. Terashima, *Chem. Pharm. Bull. (Tokyo)* **17**, 2373 (1969).

503. M. Ikehara and K. Muneyama, *Chem. Pharm. Bull. (Tokyo)* **14**, 46 (1966).

504. A. Yamazaki, I. Kumashiro, and T. Takenishi, *Chem. Pharm. Bull. (Tokyo)* **16**, 338 (1968).

505. A. Yamazaki, I. Kumashiro, and T. Takenishi, *J. Org. Chem.* **32**, 3032 (1967).

506. A. Yamane, A. Matsuda, and T. Ueda, *Chem. Pharm. Bull. (Tokyo)* **28**, 150 (1980).

507. S. Sakata, S. Yonei, and H. Yoshino, *Chem. Pharm. Bull. (Tokyo)* **30**, 2583 (1982).

508. A. Hampton, J. J. Biesele, A. E. Moore, and G. B. Brown, *J. Am. Chem. Soc.* **78**, 5695 (1956).

509. M. Kawana, R. J. Rousseau, and R. K. Robins, *J. Med. Chem.* **15**, 214 (1972).

510. D. M. Brown and M. R. Osborne, *Biochim. Biophys. Acta* **247**, 514 (1971).

511. T. Ueda, M. Imazawa, K. Miura, R. Iwata, and K. Odajima, *Tetrahedron Lett.*, 2507 (1971).

512. K. Miura and T. Ueda, *Chem. Pharm. Bull. (Tokyo)* **23**, 2064 (1975).

513. K. Miura, T. Kasai, and T. Ueda, *Chem. Pharm. Bull. (Tokyo)* **23**, 464 (1975).
514. C.-Y. Shive and S.-H. Chu, *J. Chem. Soc. Chem. Commun.*, 319 (1975).
515. G. H. Milne and L. B. Townsend, *J. Heterocycl. Chem.* **8**, 379 (1971).
516. C.-Y. Shive and S.-H. Chu, *J. Heterocycl. Chem.* **12**, 493 (1975).
517. Z. J. Witczak, *Nucleosides Nucleotides* **2**, 295 (1983).
518. J. Kiburis and J. H. Lister, *Chem. Commun.*, 381 (1969).
519. J. Kiburis and J. H. Lister, *J. Chem. Soc. C*, 3942 (1971).
520. Y. Kobayashi, K. Yamamoto, T. Asai, M. Nakano, and I. Kumadaki, *J. Chem. Soc. Perkin Trans. I*, 2755 (1980).
521. H. Fraenkel-Conrat, *Biochim. Biophys. Acta* **15**, 307 (1954).
522. J. Alexander, *J. Org. Chem.* **49**, 1453 (1984).
523. G. Loewengart and B. L. Van Duuren, *Tetrahedron Lett.*, 3473 (1976).
524. A. Kume, M. Sekine, and T. Hata, *Tetrahedron Lett.* **23**, 4365 (1982).
525. K. Miyoshi, T. Huang, and I. Itakura, *Nucleic Acids Res.* **8**, 5491 (1980).
526. T. Trichtinger, R. Charubala, and W. Pfleiderer, *Tetrahedron Lett.* **24**, 711 (1983).
527. O. Mitsunobu, *Synthesis*, 1 (1981).
528. H. Steinmaus, I. Rosenthal, and D. Elad, *J. Org. Chem.* **36**, 3594 (191).
529. S. N. Bose, R. J. H. Davies, D. W. Anderson, J. C. Van Niekerk, L. R. Nassimbeni, and R. D. Macfarlane, *Nature* **271** 783 (1978).
530. P. C. Joshi and R. J. H. Davies, *J. Chem. Res.* **5**, 227 (1981).
531. H. Linschity and J. S. Connolly, *J. Am. Chem. Soc.* **90**, 2980 (1968).
532. M. Ohno, N. Yagisawa, S. Shibahara, S. Kondo, K. Maeda, and H. Umezawa, *J. Am. Chem. Soc.* **96**, 4326 (1974).
533. B. Evans and R. Wolfenden, *J. Am. Chem. Soc.* **92**, 4751 (1970).
534. H. Nakamura, G. Koyama, Y. Iitaka, M. Ohna, N. Yagisawa, S. Kondo, K. Maeda, and H. Umezawa, *J. Am. Chem. Soc.* **96**, 4327 (1974).
535. G. B. Brown, G. Levin, and S. Murphy, *Biochemistry* **7**, 880 (1964).
536. F. Cramer and G. Schlingloff, *Tetrahedron Lett.*, 3201 (1964).
537. Z. Kazimierczuk and D. Sugar, *Acta Biochim. Pol.* **20**, 395 (1973).
538. V. Nair and D. A. Young, *J. Org. Chem.* **49**, 4340 (1984).
539. A. Matsuda, M. Tezuka, and T. Ueda, *Tetrahedron* **34**, 2449 (1978).
540. J. A. Raleigh and B. J. Blackburn, *Biochem. Biophys. Res. Commun.* **83**, 1061 (1978).
541. T. P. Haromy, J. Raleigh, and M. Sundaralingam, *Biochemistry* **19**, 1718 (1980).
542. Y. Kawazoe, M. Maeda, and K. Nushi, *Chem. Pharm. Bull. (Tokyo)* **20**, 1341 (1972).
543. M. Maeda, K. Nushi, and Y. Kawazoe, *Tetrahedron* **30**, 2677 (1974).
544. F. Minisci, R. Mondelli, G. P. Cardini, and O. Porta, *Tetrahedron* **28**, 2403 (1972).
545. V. Nair and G. Richardson, *Synthesis*, 670 (1982).
546. L. F. Christensen, R. B. Meyer, Jr., J. P. Miller, L. N. Simon, and R. K. Robins, *Biochemistry* **14**, 1490 (1975).
547. F. Minisci and O. Porta, *Adv. Heterocycl. Chem.* **16**, 123 (1974).
548. B. R. Baker and J. P. Joseph, *J. Am. Chem. Soc.* **77**, 15 (1955).
549. J. A. Montgomery and H. J. Thomas, *J. Med. Chem.* **15**, 1334 (1972).
550. Y. Suzuki, *Bull. Chem. Soc. Jpn.* **47**, 898 (1974).
551. R. E. Holmes and R. K. Robins, *J. Org. Chem.* **28**, 3483 (1963).
552. K. Kusashio and M. Yoshikawa, *Bull. Chem. Soc. Jpn.* **41**, 142 (1968).
553. J. P. H. Verheyden and J. G. Moffatt, *J. Org. Chem.* **35**, 2319 (1970).
554. A. Hampton, M. Bayer, V. S. Gupta, and S. Y. Chu, *J. Med. Chem.* **11**, 1229 (1968).
555. J. T. Witkowski, G. P. Kreishman, M. P. Schweizer, and R. K. Robins, *J. Org. Chem.* **38**, 180 (1973).
556. T. Sasaki, K. Minamoto, and H. Itoh, *Tetrahedron* **36**, 3509 (1980).
557. G. B. Elion, in *Ciba Foundation Symposium on the Chemistry and Biology of Purines* (G. E. W. Wolstenholme and C. M. O'Connor, eds.), p. 39, Little, Brown and Co., Boston, 1957.
558. A. R. Todd, in *Ciba Foundation Symposium on the Chemistry and Biology of Purines* (G. E. W. Wolstenholme and C. M. O'Connor, eds.), p. 57, Little, Brown and Co., Boston, 1957.
559. P. Brooks and P. D. Lawley, *J. Chem. Soc.*, 539 (1960).
560. J. B. Macon and R. Wolfenden, *Biochemistry* **7**, 3453 (1968).

561. M. H. Wilson and J. A. McCloskey, *J. Org. Chem.* **38**, 2247 (1973).
562. W. A. H. Grimm and N. J. Leonard, *Biochemistry* **6**, 3625 (1967).
563. M. J. Robins and E. M. Tripp, *Bochemistry* **12**, 2179 (1973).
564. K. Kikugawa, H. Suehiro, R. Yanase, and A. Aoki, *Chem. Pharm. Bull. (Tokyo)* **25**, 1959 (1977).
565. T. Ueda, K. Miura, and T. Kasai, *Chem. Pharm. Bull. (Tokyo)* **26**, 2122 (1978).
566. T. Fujii, C. C. Wu, T. Itaya, and S. Yamada, *Chem. Ind. (London)*, 1967 (1966).
567. T. Fujii, T. Itaya, C. C. Wu, and F. Tanaka, *Tetrahedron* **27**, 2415 (1971).
568. T. Fujii, T. Sato, and T. Itaya, *Chem. Pharm. Bull. (Tokyo)* **19**, 1731 (1971).
569. T. Fujii, C. C. Wu, T. Itaya, and S. Yamada, *Chem. Ind. (London)*, 1598 (1966).
570. J. A. Montgomery and H. J. Thomas, *Chem. Commun.*, 458 (1969).
571. J. A. Montgomery and H. J. Thomas, *J. Med. Chem.* **15**, 182 (1972).
572. J. A. Montgomery and H. J. Thomas, *J. Med. Chem.* **15**, 1334 (1972).
573. C.-D. Chang and J. K. Coward, *J. Med. Chem.* **19**, 684 (1976).
574. M. P. Gordon, V. S. Weliky, and G. B. Brown, *J. Am. Chem. Soc.* **79**, 3245 (1957).
575. J. A. Montgomery and H. J. Thomas, *Chem. Commun.*, 265 (1970).
576. J. A. Montgomery and H. J. Thomas, *J. Org. Chem.* **36**, 1962 (1971).
577. B. R. Baker and K. Hewson, *J. Org. Chem.* **22**, 959 (1957).
578. A. H. Haines, C. B. Reese, and A. R. Todd, *J. Chem. Soc.*, 5281 (1962).
579. L. B. Townsend and R. K. Robins, *J. Am. Chem. Soc.* **85**, 242 (1963).
580. N. J. Leonard, J. J. McDonald, R. E. L. Henderson, and M. E. Reichmann, *Biochemistry* **10**, 3335 (1971).
581. A. Vincze, R. E. L. Henderson, J. J. McDonald, and N. J. Leonard, *J. Am. Chem. Soc.* **95**, 2677 (1973).
582. Y. Ishido, T. Matsuba, A. Hosono, K. Fujii, T. Sato, S. Isome, A. Maruyama, and Y. Ikiuchi, *Bull. Chem. Soc. Jpn.* **40**, 1007 (1967).
583. H. M. Berman, R. J. Rousseau, R. W. Mancuso, G. P. Kreishman, and R. K. Robins, *Tetrahedron Lett.*, 3099 (1973).
584. K. Eistetter and W. Pfleiderer, *Chem. Ber.* **107**, 575 (1974).
585. R. C. Moschel, W. R. Hudgins, and A. Dipple, *Tetrahedron Lett.* **22**, 2427 (1981).
586. R. C. Moschel, W. R. Hudgins, and A. Dipple, *J. Org. Chem.* **44**, 3324 (1979).
587. R. C. Moschel, W. R. Hudgins, and A. Dipple, *J. Am. Chem. Soc.* **103**, 5489 (1981).
588. M. Ikehara, *Accounts Chem. Res.* **2**, 47 (1969).
589. R. W. Chambers, J. G. Moffatt, and H. G. Khorama, *J. Am. Chem. Soc.* **79**, 3747 (1957).
590. E. Reist, P. A. Hart, L. Goodman, and B. R. Baker, *J. Org. Chem.* **26**, 1557 (1961).
591. Y. Mizuno and T. Sasaki, *J. Am. Chem. Soc.* **88**, 863 (1966).
592. E. J. Reist, D. F. Calkins, and L. Goodman, *J. Org. Chem.* **32**, 2538 (1967).
593. J. R. McCarthy, Jr., R. K. Robins, and M. J. Robins, *J. Am. Chem. Soc.* **90**, 4993 (1968).
594. M. Ikehara and H. Tada, *Chem. Pharm. Bull. (Tokyo)* **15**, 94 (1967).
595. M. Ikehara and M. Kaneko, *Tetrahedron* **26**, 4251 (1970).
596. A. Yamazaki, M. Akiyama, I. Kumashiro, and M. Ikehara, *Chem. Pharm. Bull. (Tokyo)* **21**, 1143 (1973).
597. M. Kaneko, M. Kimura, and B. Shimizu, *Chem. Pharm. Bull. (Tokyo)* **20**, 635 (1972).
598. M. Muraoka, *Chem. Pharm. Bull. (Tokyo)* **29**, 3449 (1981).
599. M. Ikehara, H. Tada, and K. Muneyama, *Chem. Pharm. Bull. (Tokyo)* **13**, 639 (1965).
600. M. Ikehara and K. Muneyama, *J. Org. Chem.* **34**, 3039 (1967).
601. M. Ikehara and K. Muneyama, *J. Org. Chem.* **34**, 3042 (1967).
602. M. Ikehara, M. Kaneko, and M. Sagai, *Chem. Pharm. Bull. (Tokyo)* **16**, 1151 (1968).
603. M. Ikehara, M. Kaneko, and M. Sagai, *Tetrahedron* **26**, 5757 (1970).
604. K. L. Nagpal, P. C. Srivastava, and M. M. Dhar, *Indian J. Chem.* **9**, 183 (1970).
605. J. Pitha, S. Chladek, and J. Smrt, *Coll. Czech. Chem. Commun.* **31**, 1794 (1966).
606. H. M. Kissman and B. R. Baker, *J. Am. Chem. Soc.* **79**, 5534 (1957).
607. J. P. Miller, K. H. Boswell, A. M. Mian, R. B. Meyer, Jr., R. K. Robins, and T. A. Khwaja, *Biochemistry* **15**, 217 (1976).
608. M. Ikehara and S. Uesugi, *Tetrahedron* **28**, 3687 (1972).
609. K. K. Ogilvie and L. Slotin, *Chem. Commun.*, 890 (1971).
610. M. Ikehara and S. Tezuka, *Tetrahedron Lett.*, 1169 (1972).

611. M. Blandin and J. C. Catlin, *C.R. Acad. Sci. Ser. D* **275**, 1703 (1972).
612. T. Sowa and K. Tsunoda, *Bull. Chem. Soc. Jpn.* **48**, 3243 (1975).
613. M. Ikehara and T. Tezuka, *J. Carbohydr. Nucleosides Nucleotides* **1**, 67 (1974).
614. M. Ikehara and Y. Matsuda, *Chem. Pharm. Bull. (Tokyo)* **22**, 1313 (1974).
615. M. Ikehara and M. Muraoka, *Chem. Pharm. Bull. (Tokyo)* **20**, 550 (1972).
616. M. Ikehara, Y. Ogiso, and T. Morri, *Tetrahedron Lett.*, 2965 (1971).
617. Y. Mizuno, C. Kaneko, and Y. Oikawa, *J. Org. Chem.* **39**, 1440 (1974).
618. M. Ikehara, Y. Ogiso, and T. Morij, *Tetrahedron* **32**, 43 (1976).
619. Y. Mizuno, C. Knaeko, Y. Oikawa, T. Ikeda, and T. Itoh, *J. Am. Chem. Soc.* **94**, 4737 (1972).
620. M. Ikehara and M. Kaneko, *Chem. Pharm. Bull. (Tokyo)* **15**, 1261 (1967).
621. M. Ikehara, H. Tada, K. Muneyama, and M. Kaneko, *J. Am. Chem. Soc.* **88**, 3165 (1966).
622. M. Ikehara, H. Tada, and M. Kaneko, *Tetrahedron* **24**, 3489 (1968).
623. M. Ikehara and M. Kaneko, *Chem. Pharm. Bull. (Tokyo)* **18**, 2401 (1970).
624. M. Ikehara and K. Muneyama, *Chem. Pharm. Bull. (Tokyo)* **18**, 1196 (1970).
625. M. Ikehara and M. Kaneko, *J. Am. Chem. Soc.* **90**, 497 (1968).
626. M. Ikehara, M. Kaneko, and R. Okano, *Tetrahedron* **26**, 5675 (1970).
627. K. L. Nagpal and M. M. Dhar, *Tetrahedron Lett.*, 47 (1968).
628. P. C. Srivastava, K. L. Nagpal, and M. M. Dhar, *Indian J. Chem.* **7**, 1 (1969).
629. D. Wagner, J. P. H. Verheyden, and J. O. Moffatt, *J. Org. Chem.* **39**, 24 (1974).
630. M. Ikehara and T. Maruyama, *Tetrahedron* **31**, 1369 (1975).
631. M. Ikehara, M. Kaneko, and Y. Ogiso, *Tetrahedron Lett.*, 4637 (1970).
632. M. Ikehara and Y. Ogiso, *Tetrahedron* **28**, 3695 (1972).
633. T. A. Khwaja, R. Harris, and R. K. Robins, *Tetrahedron Lett.*, 4681 (1972).
634. J. B. Chattopadhyay and C. B. Reese, *J. Chem. Soc. Chem. Commun.*, 860 (1976).
635. M. Ikehara, H. Miki, and A. Hasegawa, *Chem. Pharm. Bull. (Tokyo)* **27**, 2647 (1979).
636. M. Kaneko, M. Kimura, B. Shimuzu, J. Yano, and M. Ikehara, *Chem. Pharm. Bull. (Tokyo)* **25**, 1982 (1977).
637. M. Ikehara and K. Muneyama, *Chem. Pharm. Bull. (Tokyo)* **18**, 1196 (1970).
638. T. A. Khwaja, K. H. Boswell, R. K. Robins, and J. P. Miller, *Biochemistry* **14**, 4238 (1975).
639. M. Ikehara and Y. Ogiso, *J. Carbohydr. Nucleosides Nucleotides* **2**, 121 (1975).
640. M. Ikehara, T. Maruyama, H. Miki, and Y. Takatsuka, *Chem. Pharm. Bull. (Tokyo)* **25**, 754 (1977).
641. M. Ikehara and Y. Ogiso, *J. Carbohydr. Nucleosides Nucleotides* **2**, 121 (1975).
642. M. Ikehara and S. Tanaka, *Tetrahedron Lett.*, 497 (1974).
643. M. Kaneko, B. Shimizu, and M. Ikehara, *Tetrahedron Lett.*, 3113 (1971).
644. T. Sasaki, K. Minamoto, and H. Itoh, *J. Org. Chem.* **43**, 2320 (1978).
645. K. J. Divakar and C. B. Reese, *J. Chem. Soc. Chem. Commun.*, 1191 (1980).
646. A. Matsuda, Y. Nomoto, and T. Ueda, *Chem. Pharm. Bull. (Tokyo)* **27**, 189 (1979).
647. P. J. Harper and A. Hampton, *J. Org. Chem.* **37**, 795 (1972).
648. A. Hampton, P. J. Harper, and T. Sasaki, *Biochemistry* **11**, 4965 (1972).
649. M. Yoshikawa, P. T. Kato, and T. Takenishi, *Bull. Chem. Soc. Jpn.* **42**, 3505 (1969).
650. A. Hampton and R. R. Chawla, *J. Carbohydr. Nucleotides Nucleosides* **2**, 281 (1975).
651. A. Matsuda, M. Tezuka, K. Nizuma, E. Sugiyama, and T. Ueda, *Tetrahedron* **34**, 2633 (1978).
652. J. Zylber, R. Pontikis, A. Merrien, C. Merrien, M. Baran-Marszak, and A. Gaudemer, *Tetrahedron* **36**, 1584 (1979).
653. R. Stolarski, L. Dudycz, and D. Shugar, *Eur. J. Biochem.* **108**, 111 (1980).
654. G. W. Kenner, in *The Chemistry and Biology of Purines* (G. E. W. Wolstenhome and C. M. O'Conner, eds.), p. 312, Little, Brown and Co., Boston, 1957.
655. C. A. Dekker, *Annu. Rev. Biochem.* **29**, 453 (1960).
656. F. Micheel and A. Heesing, *Chem. Ber.* **94**, 1814 (1961).
657. J. Cadet and R. Teoule, *J. Am. Chem. Soc.* **96**, 6517 (1974).
658. J. A. Zoltewicz, D. F. Clark, T. W. Sharpless, and G. Grahe, *J. Am. Chem. Soc.* **92**, 1741 (1970).
659. J. A. Zoltewicz and D. F. Clark, *J. Org. Chem.* **37**, 1193 (1972).
660. R. P. Panzica, R. J. Rousseau, R. K. Robins, and L. B. Townsend, *J. Am. Chem. Soc.* **94**, 4708 (1972).
661. E. R. Garrett and P. J. Mehta, *J. Am. Chem. Soc.* **94**, 8532 (1972).

662. L. Hevesi, E. Wolfson-Davidson, J. B. Nagy, O. B. Nagy, and A. Bruylants, *J. Am. Chem. Soc.* **94**, 4715 (1972).
663. R. Romero, R. Stein, H. G. Bull, and E. H. Cordes, *J. Am. Chem. Soc.* **100**, 7620 (1978).
664. R. Shapiro and S. Kang, *Biochemistry* **8**, 1806 (1969).
665. F. Jordon and H. Niv, *Nucleic Acids Res.* **4**, 697 (1977).
666. H. Lonnberg and P. Lehikoinen, *Nucleic Acids Res.* **10**, 4339 (1982).
667. M. P. Gordon, V. S. Welky, and B. Brown, *J. Am. Chem. Soc.* **79**, 3245 (1957).
668. F. Jordan, *J. Am. Chem. Soc.* **96**, 5911 (1974).
669. E. R. Garrett, *J. Am. Chem. Soc.* **82**, 827 (1960).
670. H. Venner, *Z. Physiol. Chem.* **339**, 14 (1964).
671. J. L. York, *J. Org. Chem.* **46**, 2171 (1981).
672. A. Streitweiser, Jr., *Solvolytic Displacement Reactions*, p. 112, Wiley, New York, 1962.

Chapter 3

Synthesis and Properties of Oligonucleotides

Morio Ikehara, Eiko Ohtsuka, Seiichi Uesugi, and Toshiki Tanaka

1. Introduction

Developments in the chemical synthesis of polynucleotides have been very rapid. A prediction made in a review article[1] that "advances in the techniques of chemical synthesis of polynucleotides would make nucleic acids more accessible than synthetic polypeptides" seems to have been realized by improvements in the solid-phase synthesis of DNA fragments. A short review, written in 1982, only covers early developments in the syntheses of genes.[2] DNA sequence analyses have been accelerated by recombinant DNA technology and now provide the primary structure of genes in many organisms. These new techniques and information have changed the way biochemists and chemists deal with biological subjects.

In the 1960s, synthetic oligonucleotides were used for deciphering the genetic code.[3] The chemical syntheses of gene fragments of yeast alanine tRNA[4] and *E. coli* suppressor tyrosine tRNA[5] were accomplished in the early 1960s and these DNA fragments were joined with the aid of DNA ligase to form the gene. This synthetic methodology can be described as the "complementally protruding fragments method" and has been used in the synthesis of genes for peptides.[6] Expression of these synthetic genes for the formation of human peptide hormones or interferons in *E. coli* lends considerable support for many theories and hypotheses espoused in modern molecular biology. The application of chemically synthesized oligonucleotides to biological problems has increased due to the recent advances in recombinant DNA technology. Synthetic DNA fragments are indispensable material not only as structural genes but also as probes for the sequence analysis or primers for site-directed mutageneses.

Morio Ikehara, Eiko Ohtsuka, Seeichi Uesugi, and ***Toshiki Tanaka*** • Faculty of Pharmaceutical Sciences, Osaka University, Osaka, Japan 565.

In this chapter, we feel that more recent advances in the synthesis of polynucleotides should receive the major emphasis. The reader is referred to a number of monographs[7] and reviews[1,2,8] on this and other related areas for a more comprehensive coverage. The phosphodiester approach[3] was established in the 1960s by Khorana and co-workers and led to the synthesis of tRNA genes.[4,5] The most commonly accepted concepts in the synthesis of polynucleotides were developed at that time. Some of these protecting groups for the functional groups are still being used in the phosphotriester method, which has been modified[8d] from the early work of Michelson and Todd.[9] This is now a major approach to gene synthesis.[6a,c–f,j] The phosphite triester method, which was first used for the synthesis of deoxyribo-oligonucleotides by Letsinger *et al.*,[10] has recently been improved[11] by a modification of the activation method and applying this to the solid-phase synthesis. By this modified phosphite method (the phosphoramidite method), larger oligo(deoxyribonucleotides) with a chain length of ~50 were synthesized mechanically.[12] Oligo(deoxyribonucleotides) with a chain length of 20 have already been obtained by using the phosphodiester method[13] and similar or slightly larger fragments have been obtained by the phosphotriester method in quantity,[14] both in solution phase. Synthesis by these techniques takes a much longer period of time when compared to the time required by the solid-phase synthesis. The rapid synthesis of gene fragments of α_1-interferon has been achieved by the solid-phase phosphotriester method.[6e] This demonstrates that genes for fairly large peptides are accessible by using chemical and enzymatic methods. A gene with 584 base pairs, for the human growth hormone, has been synthesized and expressed in *E. coli* very efficiently.[6j] This method has been applied to the synthesis of peptides or enzymes with ~200 amino acid residues. In these approaches, site-directed mutagenesis of proteins can be achieved by a replacement of certain deoxyoligonucleotide fragments in the gene. Site-directed *in vitro* mutagenesis of cloned genes can also be performed by using synthetic oligonucleotides.[15] The method should be extremely useful in studies on structure–function relationships of proteins whose active sites have been structurally located.

Synthesis of polyribonucleotide fragments requires more synthetic strategies than those required in the deoxyribo series. The efficient synthesis of ribo-oligonucleotides by the solid-phase method has yet to be developed. A joining enzyme (T4 RNA ligase) for single-stranded ribo-oligonucleotides has been found[16] and used for the total syntheses of tRNAs.[17,18] However, the usage of this technique has been rather limited, presumably because of the lack of a system for the amplification of RNA. A recently reported replication vehicle and Qβ replicase may extend the possibility of obtaining RNA with defined sequences.[19] A half molecule of the tRNA$_1^{Gly}$ with a chain length of 33 has been synthesized, and so far has been the largest synthetic RNA fragment.[20]

This chapter presents the synthetic methodologies that are currently used in the field of DNA and RNA. Physicochemical properties of polynucleotides, with defined sequences, and their sequence-dependent reactions are also discussed.

2. Protecting Groups and Phosphorylation Methods

Nucleic acids consist of polymers of phosphodiesters between the 3′- and 5′-hydroxyl groups of nucleosides. For the formation of this internucleotide linkage, phosphomonoesters (in the diester method) or phosphodiesters (in the triester method) must be activated to phosphorylate the hydroxyl group of an adjacent nucleoside as shown in Fig. 1. Functional groups, except those involved in the reaction, must be protected with suitable protecting groups during the reaction and subsequent purification. A selective removal of these protecting groups is one of the major problems in the synthesis of polynucleotides.

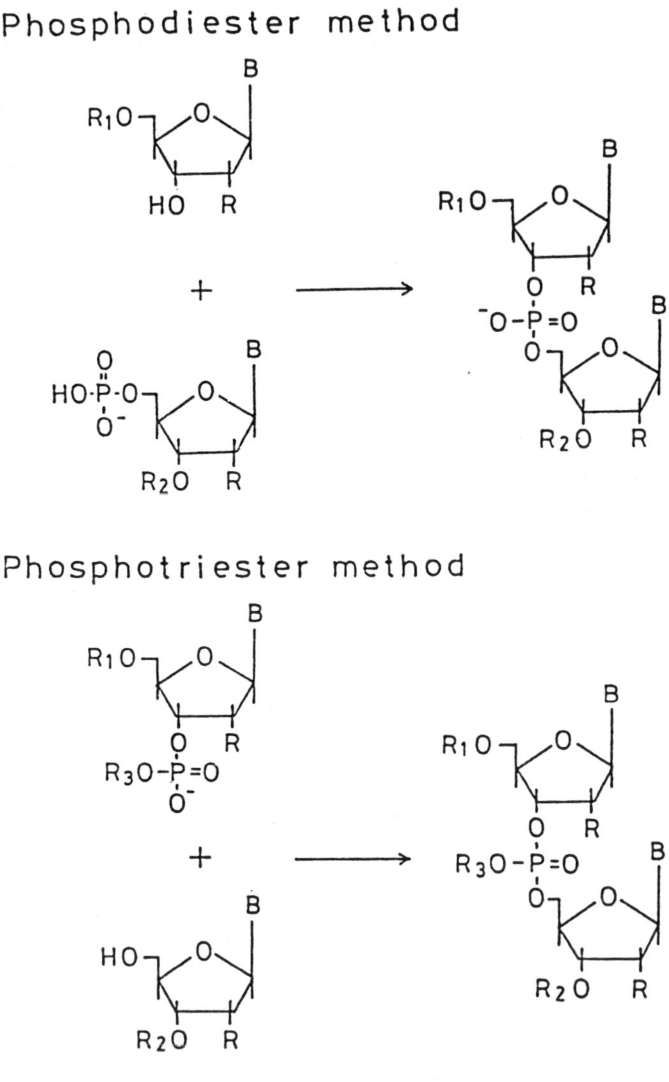

Phosphodiester method

Phosphotriester method

Figure 1. Formation of internucleotide linkages.

2.1. Protecting Groups for Bases

The exocyclic amino groups in adenine and guanine are rather poor nucleophiles when compared to the hydroxyl groups of sugars. However, the exocyclic amino group of cytosine was found to be more reactive than the sugar hydroxyl groups. A phosphoramidate between deoxycytidine and an activated thymidine 5′-phosphate has been obtained[21] using dicyclohexylcarbodiimide (DCC). Therefore, protection of the exocyclic amino group of cytidine and deoxycytidine is required. The exocyclic amino groups of other heterocyclic compounds are usually protected with acyl groups or trityl derivatives, although in most cases it is not imperative. These protecting groups increase the lipophilicity of nucleotides and their solubility in organic solvents.

Some acyl groups that are used for the protection of the exocyclic amino groups of nucleosides include the following: benzoyl or *p*-methoxybenzoyl (anisoyl) groups for adenine[22] and cytosine.[23] The more easily removed isobutyryl group has been used extensively for the amino group on guanine.[24] Investigations involving the search for other protecting groups have been reported.[25] Some examples of groups used for the protection of deoxynucleosides are shown in Fig. 2. Deoxyadenosine is perbenzoylated and the benzoyl groups on the sugar hydroxyl are then removed by treatment with a strong alkaline solution to yield N^6-benzoyldeoxyadenosine. One-pot syntheses for *N*-acyldeoxy-

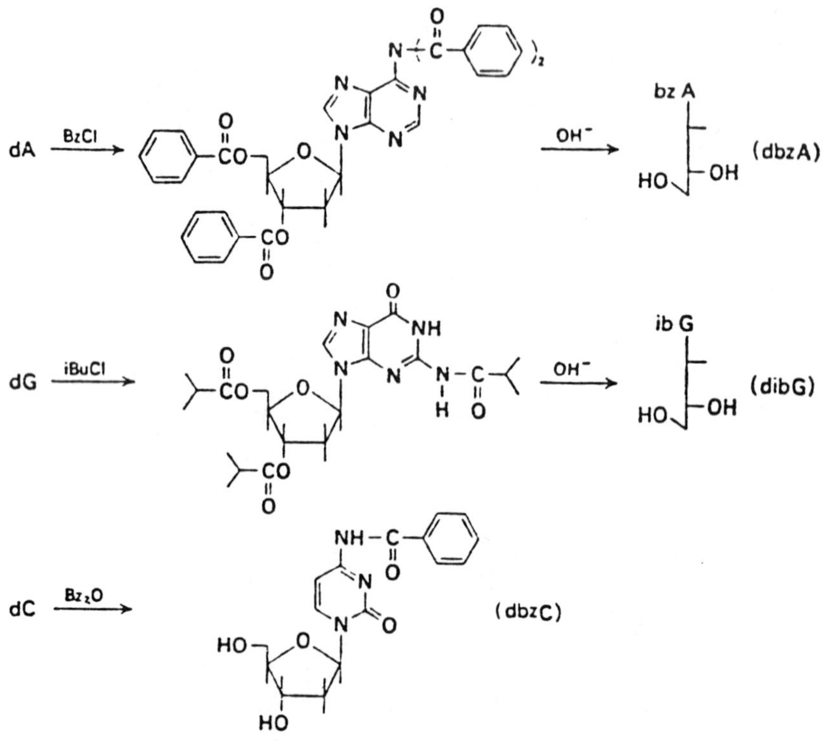

Figure 2. *N*-Acylation of deoxynucleosides.

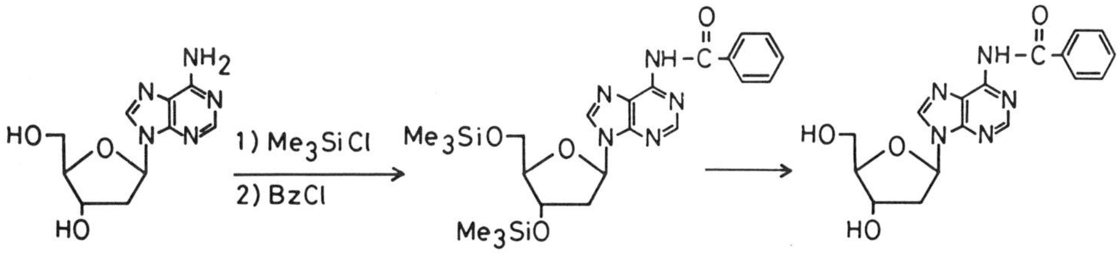

Figure 3. One-pot *N*-benzoylation of deoxyadenosine.

nucleosides, *via* trimethylsilyl intermediates, have been reported[26] (Fig. 3). In the ribo series, N^6-benzoyladenosine[27a] and N^2-isobutyrylguanosine[27b] were prepared by peracylation while N^4-acetyl-[27c] or N^4-benzoylcytidine[27d] could be prepared by a direct *N*-acylation using acid anhydride in the absence of basic conditions. These protecting groups are usually retained until the end of the complete synthesis and then removed by treatment with ammonia.

Trityl derivatives such as monomethoxy- and dimethoxytrityl groups,[28] more commonly used for the protection of a primary hydroxyl group, are occasionally used for the protection of exocyclic amino groups in combination with *other* alkaline labile protecting groups for the sugar hydroxyl moieties.[29]

The dimethylaminomethylene group is useful as a temporary protecting group for amino functions on nucleosides.[30] It is labile in acid or alkaline conditions and can be difficult to retain during reactions that consist of several steps.

Under certain phosphorylation conditions using arenesulfonyl azolides, the modification of guanine, uracil, and thymine has been observed.[31] Modified guanine and uracil derivatives were synthesized and identified as these azolide derivatives.[32] The O^4-phenyl derivatives of thymidine and O^6-phenyl derivatives of deoxyguanosine have been used as protecting groups in nucleosides.[25] Other protecting groups for the O-4 of uracil and thymine and the O-6 of guanine have been used.[33] Acylation of the N-3 position has also been reported.[34]

New protecting groups for the N-6 of deoxyadenosine, which stabilize the glycosidic linkage, have been reported.[35] These groups should increase the lipophilicity of oligonucleotides and decrease side reactions. However, a complete removal of the protecting groups must be feasible at the final stage of synthesis.

2.2. Protecting Groups for Sugar Hydroxyl Groups

2.2.1. Protecting Groups for Primary Hydroxyl Groups

Selectivity in conditions for removing a primary hydroxyl group protection is one of the important problems in the synthesis of polynucleotides. The acid labile triphenylmethyl (trityl) group has been used extensively in carbohydrate chemistry. This bulky group reacts with primary hydroxyl groups in preference to secondary hydroxyl groups. However, these trityl ether linkages are rather stable and occasionally remain intact under conditions in which other linkages (e.g., glycosidic, internucleotidic) become labile. Khorana and co-workers introduced *p*-methoxy group(s) to the phenyl ring to facilitate the removal of trityl derivatives

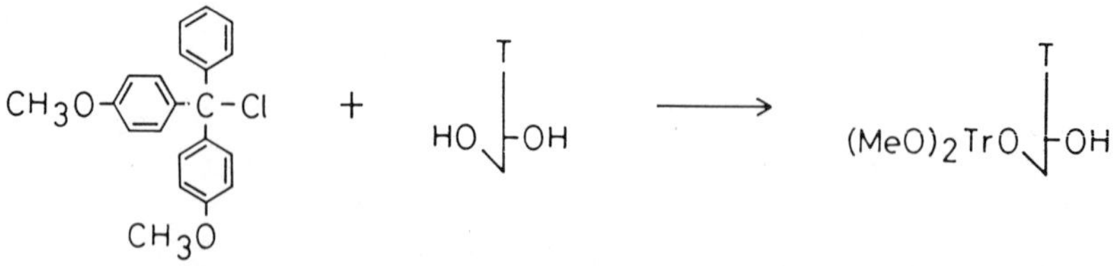

Figure 4. 5′-*O*-Dimethoxytritylation of thymidine.

with acid.[28] Figure 4 shows a route for the preparation of 5′-*O*-dimethoxytrityl-thymidine. Monomethoxytrityl groups are used extensively in the ribo series, since the glycosidic linkages of ribonucleosides are more stable under acidic conditions than those of the corresponding deoxynucleosides. The 5′-*O*-monomethoxytrityl groups of *N*-acylated ribonucleosides can be removed without observing a cleavage of the glycosidic bond.[36] *N*⁶-Benzoyl-2′-deoxyadenosine furnishes *N*⁶-benzoyladenine on treatment with 80% acetic acid but is more stable when incorporated into 3′-phosphorylated polymers.[37] Various acidic treatments have been devised for the removal of the dimethoxytrityl group in N-protected deoxypolynucleotides, for example, benzenesulfonic acid at low temperature,[37] dilute trichloroacetic acid,[38] or dichloroacetic acid.[12] Some of the conditions used for the complete removal of trityl derivatives are summarized in Table I. New trityl derivatives with different color have been reported as convenient reagents in the solid-phase synthesis.[39] Other acid labile protecting groups with selectivity for the primary alcohol have also been described.[40]

Table I. Conditions for Complete Hydrolysis of Trityl Derivatives

	Temperature	Time	Reference
(Tr)U	r.t.	48 hr[a]	28
(MeOTr)U	r.t.	2 hr[a]	28
(MeO)$_2$Tr U	r.t.	15 min[a]	28
(MeO)$_3$Tr U	r.t.	1 min[a]	28
d (MeO)$_2$Tr bzAp(CE, C$_6$H$_4$Cl)	0°C	2.5 min[b]	
d (MeO)$_2$Tr bzAp(CE, C$_6$H$_4$Cl)	r.t.	45 sec[c]	
d (MeO)$_2$Tr bzA	r.t.	3 min[d]	38
d (MeO)$_2$Tr bzApR	r.t.	2 min[e]	12
d (MeO)$_2$Tr bzA	r.t.	1 min[f]	113
d (MeO)$_2$Tr bzC	r.t.	7 min[f]	113
d (MeO)$_2$Tr ibG	r.t.	1 min[f]	113
d (MeO)$_2$Tr T	r.t.	3 min[f]	113

[a] 80% Acetic acid.
[b] 2% Benzenesulfonic acid in chloroform/methanol (7 : 3).
[c] 1% Benzenesulfonic acid in chloroform/methanol (7 : 3).
[d] 3% Trichloroacetic acid in dichloromethane.
[e] 0.2 M Dichloroacetic acid in dichloromethane.
[f] 1 M Zinc bromide in chloroform/isopropanol (85 : 15).

Table II. Alkaline Stability of 5'-O-Acylnucleosides

	$t_{1/2}$ 1 N Nacl[41a]	$t_{1/2}$ NH$_4$OH (pH 10.7)[42]
(Piv)T	3 min	
(Ac)T	30 sec	191 min
(Mesitoyl)T	80 hr	
(Methoxyacetyl)U		10.4 min
(Phenoxyacetyl)U		3.9 min
(Formyl)U		0.4 min
(Chloroacetyl)U		0.28 min
(Trityloxyacetyl)U2'(Thp)		20 min

Alkaline labile acyl groups are used as protecting groups for the 5'-hydroxyl functions when the other functional groups are protected with acid labile groups. Trimethylacetyl (pivaloyl) can be used and is a slightly more stable acyl group than acetic acid. Its steric bulk provides some preference for the acylation of primary hydroxyls.[41] Methoxy- and phenoxyacetyl groups are used when more alkaline labile protecting groups[42] are required. The trityloxyacetyl (trac) group was found to give a selective 5'-acylation and has been used in the synthesis of ribo-oligonucleotides.[43] The rate of alkaline hydrolysis of the 5'-O-trac group seems to be affected by other groups residing on the nucleosides.[44] A comparison of half times for the removal of 5'-O-acyl derivatives is shown in Table II.

In the presence of protecting groups on the 2'-hydroxyl function, protecting groups that can be removed specifically under conditions other than acidic or alkaline conditions are desirable for the protection of the 5'-hydroxyl group of ribonucleosides. Levulinyl-[45] and o-dibromomethylbenzoyl[46] groups on the 5'-O-hydroxyl functions are in this class and can be removed by treatment with dilute hydrazine and silver perchlorate, respectively.

2.2.2. Protecting Groups for Secondary Hydroxyl Groups

When the 5'-hydroxyl group is protected with an acid labile group such as mono- and dimethoxytrityl, the 3'-hydroxyl group of deoxynucleosides can be protected with acyl groups.[21–24] The benzoyl group is often used as a terminal protecting group of polynucleotides in the phosphotriester approach.[6c,14,47] The 3'-O-succinyl group of nucleosides is now commonly used as a protecting group as well as a linker between nucleosides and amino functions of polymers involved in a solid-phase synthesis. A typical reaction sequence for the preparation of a 3'-O-succinyldeoxynucleoside is shown in Fig. 5.[48]

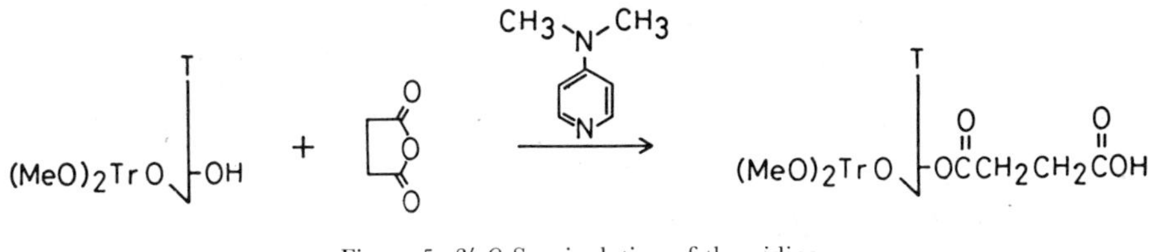

Figure 5. 3'-O-Succinylation of thymidine.

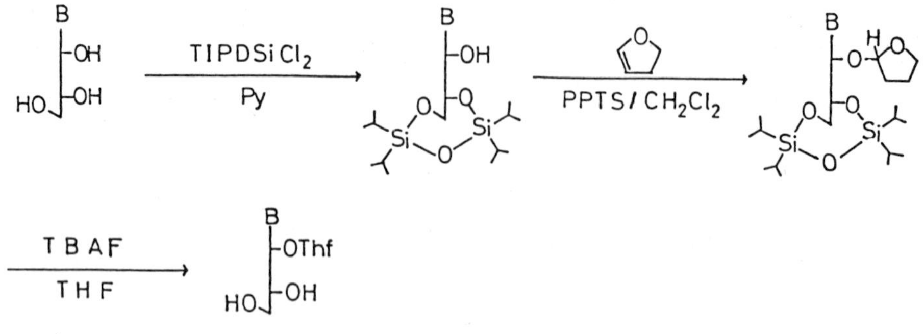

Figure 6. Preparation of 2'-*O*-tetrahydrofuranyl ribonucleosides.

For the synthesis of ribo-oligonucleotides with 3'–5' linkages, the 2'-hydroxyl group must be protected with a stable group that is usually removed at the last step of the synthesis. In the phosphodiester synthesis, acyl groups can be used as a protecting group for the 2'-hydroxyl function. However, in conjunction with the alkaline labile protecting groups for internucleotidic phosphates, stable alkaline protection is required in the phosphotriester method.

The tetrahydropyranyl group was introduced in the early 1960s by Khorana and his co-workers for the synthesis of dinucleoside phosphates.[28] Later the more acid labile tetrahydro-4-methoxypyranyl group was developed to avoid the formation of a diastereoisomeric mixture. This group was used in combination with alkaline labile 5'-protecting groups for the synthesis of ribo-oligonucleotides by the phosphotriester approach.[49] Stability of the internucleotidic linkage during the removal of this protecting group has been tested and a migration of the internucleotidic linkage has not been detected by HPLC.[50] The tetrahydrofuranyl group has been found to possess similar stability and 2'-*O*-tetrahydrofuranyl nucleosides have been prepared from the readily available dihydrofuran.[51a] The introduction of a bifunctional silyl protecting group, using 1,3-dichloro-1,1,3,3-tetraisopropyldisiloxane,[52] has facilitated the synthesis of 2'-*O*-substituted nucleosides.[51b] A scheme for the preparation of 2'-*O*-tetrahydrofuranylnucleosides is shown in Fig. 6. By using 2'-*O*-tetrahydrofuranyl nucleosides, a 33 mer, which has the sequence of the 5' half molecule of *E. coli* tRNA$_1^{Gly}$, has been synthesized.[20]

The *t*-butyldimethylsilyl function was introduced as a protecting group for nucleosides and has been used extensively for protection of the 2'-hydroxyl groups.[53] It was found that the silyl ether migrated during the isolation processes in alkaline media.[54] However, rapid phosphorylation of the vicinal hydroxyl group could prevent actual contamination of side products.[55] A scheme for the preparation of 2'- or 3'-*t*-butyldimethylsilyl nucleosides is shown in Fig. 7.

As a stable protecting group both in acid and alkaline conditions, the photolabile *o*-nitrobenzyl group was applied to the synthesis of ribo-oligonucleotides. *o*-Nitrobenzyl bromide[56] or *o*-nitrobenzyldiazomethane[57] was used to prepare 2'-*O*-(*o*-nitrobenzyl)nucleosides. Removal of the 2'-*O*-(*o*-nitrobenzyl) group in oligonucleotides was accomplished with irradiation by ultraviolet light at wavelengths longer than 280 nm to afford the product in ∼95%

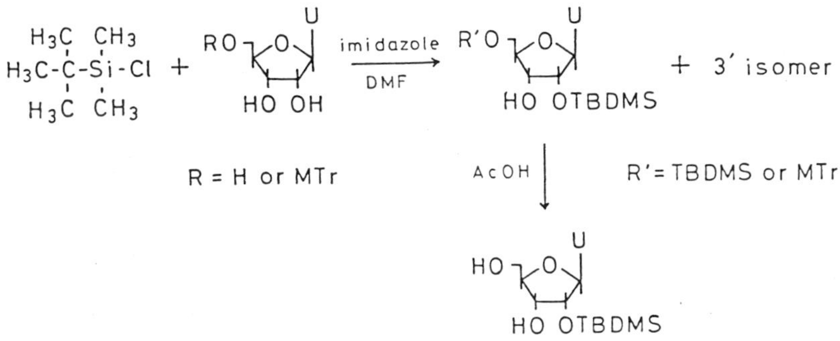

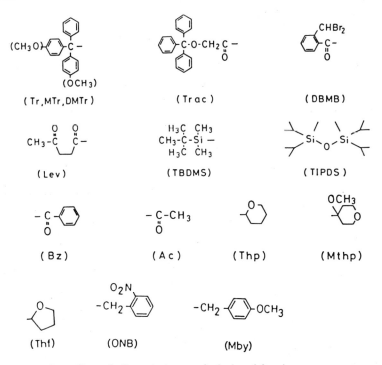

Figure 7. Preparation of 2'-*O*-(*t*-butyldimethylsilyl)ribonucleosides.

yield. Oligonucleotides, up to an eicosamer, containing sequences in various parts of tRNA were synthesized by using this method.[58] The 3'-*O*-(*o*-nitrobenzyl) nucleosides, which are obtained as side products, have been used for the preparation of 5'-triphosphorylated 2'–5' linked oligoadenylates (2–5A) or their core. Other protecting groups used for the 2'-hydroxyl group, in the synthesis of 3'–5' linked oligonucleotides, have been applied to the 3' protection of adenosine.[60] Another benzyl derivative, *p*-methoxybenzyl ether was also used for protection of the 2'-hydroxyl group.[61]

Figure 8 shows the structures of groups used for protecting the hydroxyl functions.

Figure 8. Protecting groups for hydroxyl functions.

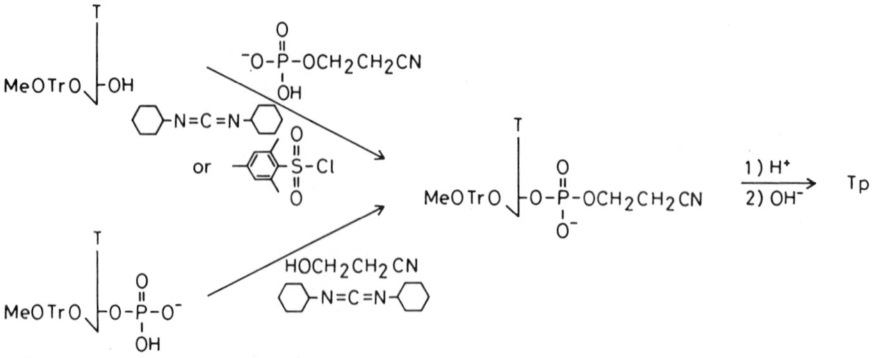

Figure 9. Phosphorylation of thymidine.

2.3. Protecting Groups for Phosphates and Phosphorylating Reagents

In the phosphodiester approach, only the terminal phosphate group needs protection. The 2-cyanoethyl group[62] has been used primarily because it is very easily removed. This group can also be incorporated into a phosphorylating reagent. Figure 9 shows the phosphorylation of a nucleoside with 2-cyanoethyl phosphate. The phenyl group has been used as a protecting group for carbohydrates where hydrogenolysis can be used. However, alkaline hydrolysis of the phenyl group on a nucleoside phosphate is much too slow unless a vicinal hydroxyl group participates in the hydrolysis. As a protecting group for internucleotidic phosphates, the 2-cyanoethyl group was found to be too labile and the alkaline lability of substituted phenyl derivatives was then investigated using the dimer shown in Fig. 10[63] (Table III) The *o*-chlorophenyl or *p*-chlorophenyl group is in common use for internucleotidic phosphates.

In the phosphotriester synthesis, internucleotidic phosphates must be blocked throughout the synthesis and terminal phosphates must have another temporary protecting group. Protecting groups that can be removed under conditions other than alkaline have been used as temporary protecting groups. Figure 11 shows several examples for the selective removal of temporary protecting groups from fully substituted phosphates. 2,2,2-Trichloroethyl can be removed by a reduction with zinc.[64] Aromatic phosphoroamidates are converted to the corresponding phosphates by treatment with isoamyl nitrate.[47,65] Dinitrobenzyl phosphate was

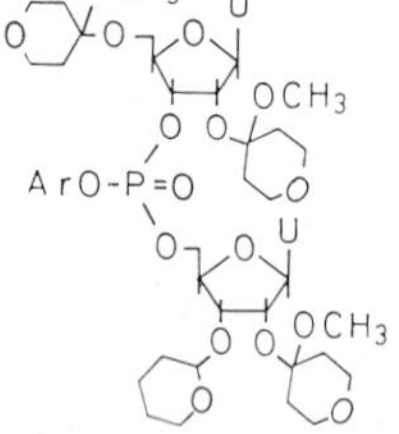

Figure 10. Arylesters of diuridine phosphate.

Table III. Alkaline Hydrolysis of Phenyl Esters

Ar	pK_a (ArOH)	$t_{1/2}{}^a$ (min)	By-product[b] (%)
C_6H_5-	9.92	40.0	4.2
2-FC_6H_4-	8.72	7.5	2.4
2-ClC_6H_4-	8.40	6.5	1.8
3,5-$Cl_2C_6H_3$-	8.19	3.0	1.2
2,4-$Cl_2C_6H_3$-	7.85	2.5	0.8
2,5-$Cl_2C_6H_3$-	7.51	1.9	0.7

[a] 0.1 M NaOH in dioxane/water (1 : 4, v/v) at 20°C.
[b] Uridine derivatives.

found to be stable in pyridine and could be used as a temporary protecting group, which was then dealkylated with toluene-p-thiol.[66] A stable protecting group, p-nitrophenylethyl, can be removed in aprotic solvents by a β-elimination mechanism.[67] S,S-Diphenyl phosphate derivatives of ribonucleotides were converted to phosphodiesters by treatment with phosphinic acid–triethylamine.[68a] 5-Chloro-8-quinolyl phosphate was found to be stable under alkaline conditions but could be removed by chelation with zinc chloride.[68b]

These protecting groups can be introduced easily by using the appropriate phosphorylating reagents. Figure 12 gives an example for the phosphorylation of nucleosides.

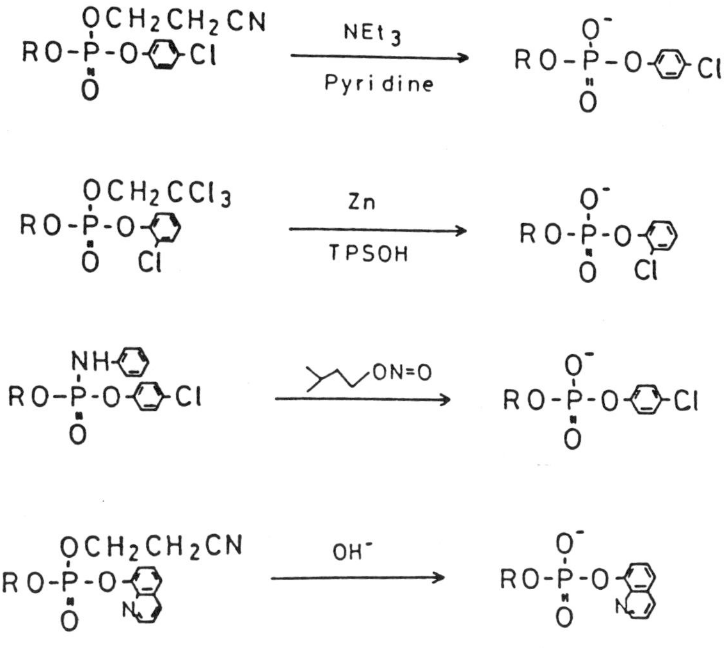

Figure 11. Selective removal of protecting groups for terminal phosphate.

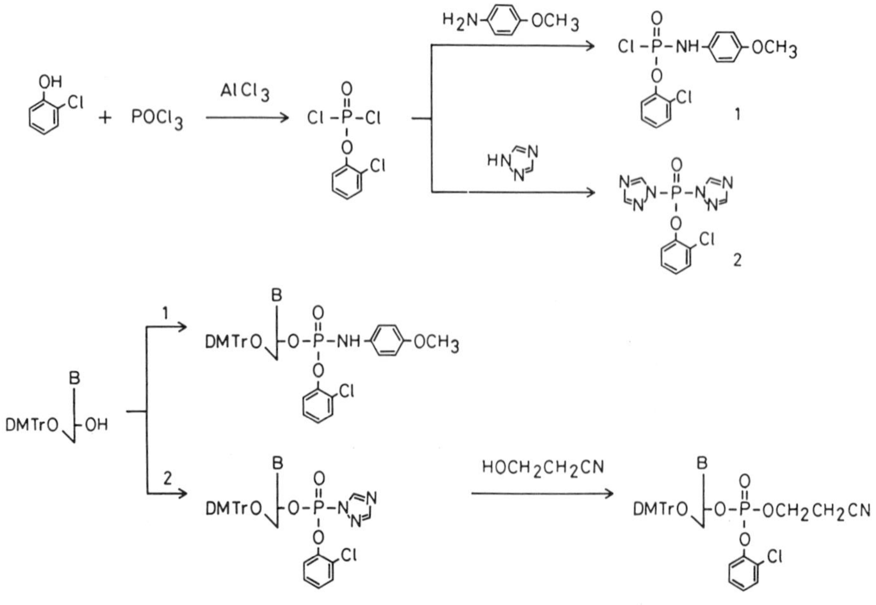

Figure 12. Preparation of fully protected nucleotides.

2.4. Condensing Reagents

The formation of internucleotide linkages requires either an activation of the hydroxyl group of nucleosides or that of the phosphate function of the adjacent molecule as illustrated in Fig. 13. Very few investigations on the former approach for the synthesis of oligonucleotides have been reported.[7a,69] Activation of phosphomonoesters is usually performed by the phosphodiester method using condensing reagents that can polarize the phosphorus. Dicyclohexylcarbodiimide (DCC) was used to activate nucleoside phosphates.[3] Arenesulfonyl chlorides were also used to condense nucleotide fragments in the phosphodiester method.[70] A mechanism was proposed for the activation of nucleotides with DCC (Fig. 14), and this mechanism may explain the lack of activation of bulky nucleotides with DCC. The polarization of phosphorus is not easily compared with that of a carbonyl carbon because of the $p\pi$–$d\pi$ back-bonding to empty phosphorus orbitals.[71]

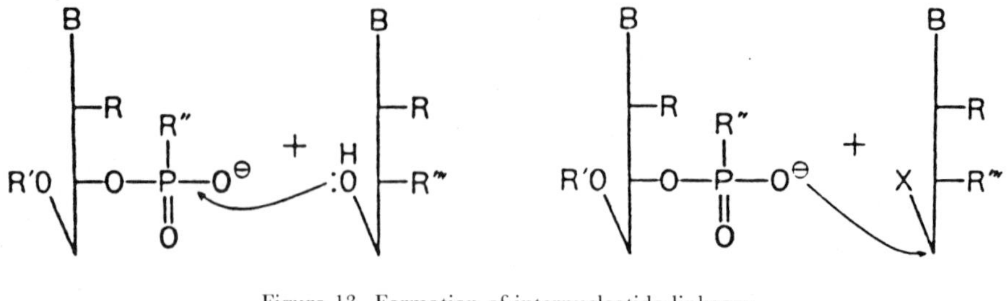

Figure 13. Formation of internucleotide linkages.

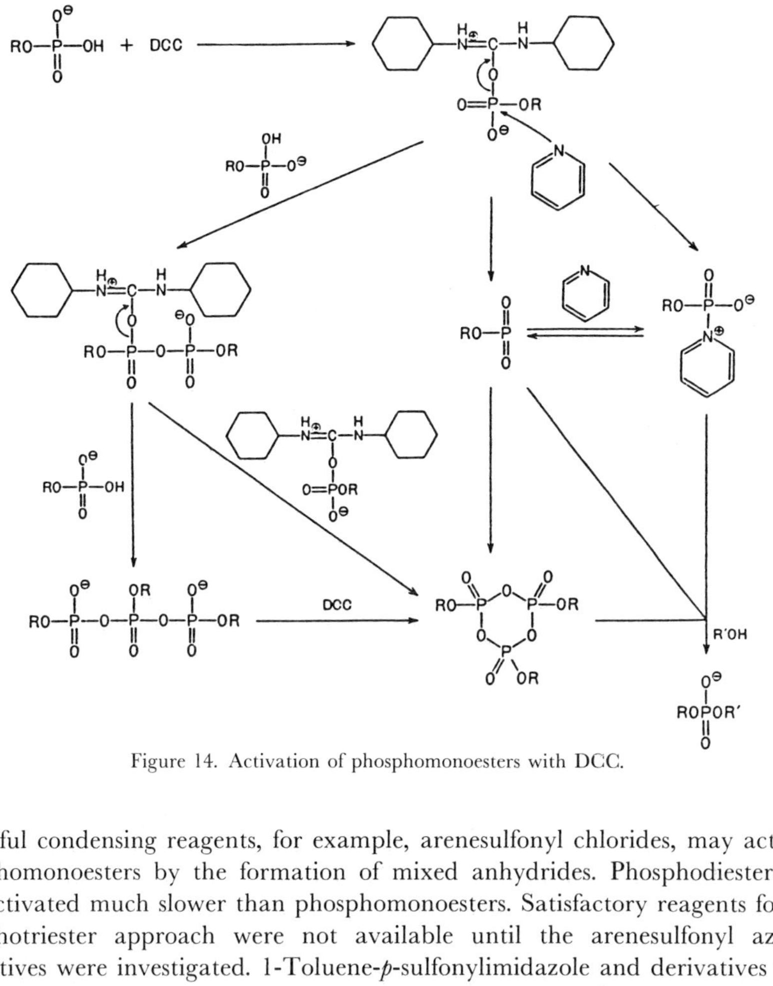

Figure 14. Activation of phosphomonoesters with DCC.

Powerful condensing reagents, for example, arenesulfonyl chlorides, may activate phosphomonoesters by the formation of mixed anhydrides. Phosphodiesters are also activated much slower than phosphomonoesters. Satisfactory reagents for the phosphotriester approach were not available until the arenesulfonyl azolide derivatives were investigated. 1-Toluene-*p*-sulfonylimidazole and derivatives were reported to act as activating reagents for phosphodiesters and were found to give less side products.[72] However, the reactions with these imidazolides were rather slow, especially with the 2,4,6-triisopropylbenzenesulfonyl (TPS) derivative. Later, several groups prepared arenesulfonyl azolides with an increasing number of nitrogens in the ring.[8d] TPS 1,2,3,4-tetrazole (TPSTe)[37] and 1-(mesitylenesulfonyl)-3-nito-1,2,4-triazole (MSNT)[31] are now the most commonly used agents to activate phosphodiesters. Structures of some condensing reagents are shown in Fig. 15. The mechanism for the activation of phosphodiesters with arenesulfonyl azolides has been proposed by several investigators.[73] The most probable route involves the formation of a mixed anhydride of a phosphodiester and arenesulfonic acid at the first step followed by the successive replacement with an azole and a hydroxyl group of a nucleoside (Fig. 16). Substituted TPSTe derivatives, 1-(mesitylenesulfonyl)-5-(pyridin-2-yl)tetrazole (MSPy), and 1-(2,4,6-benzenesulfonyl)-5-(pyridin-2-yl)tetrazole (TPSPy) afford one of the diastereomers of a phosphotriester when a *N*,5′-*O*-protected deoxynucleoside 3′-*O*-(*o*-chlorophenyl) phosphate was condensed with a *N*,3′-*O*-protected nucleoside.[74]

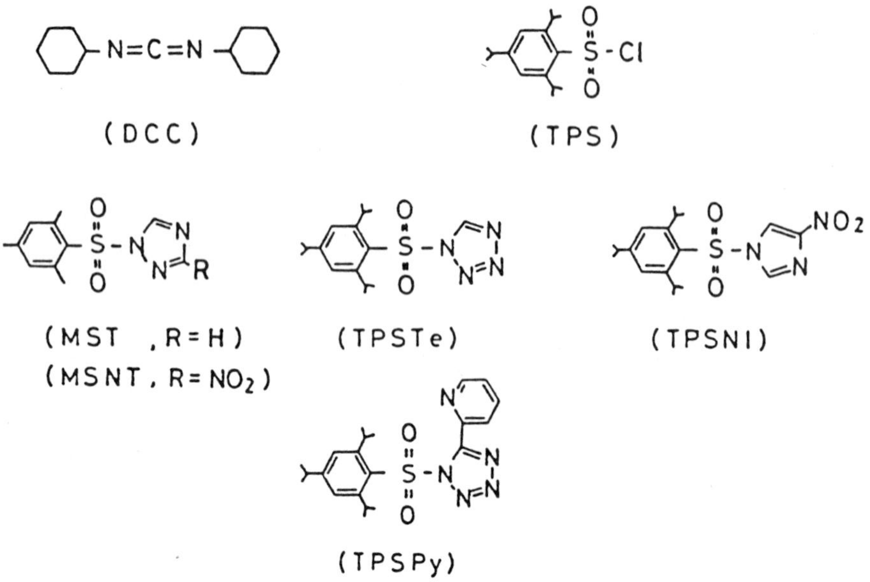

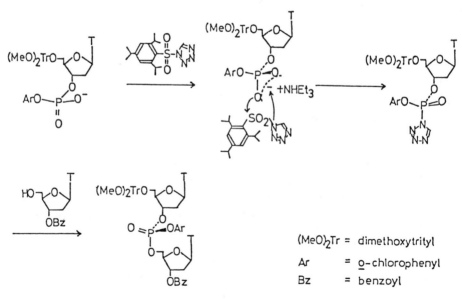

Figure 15. Condensing reagents.

The stereospecific reaction presumably occurred at the mixed anhydride stage of formation, because TPS chloride in the presence of 5(pyridin-2-yl)-tetrazole gave two isomers. According to the previous assumption, a threefold excess of tetrazole should assist in the activation of phosphodiesters by the formation of a tetrazolide intermediate (e.g., Fig. 16).

Figure 16. Activation of phosphodiesters with TPSTe.

3. Synthesis of Oligonucleotides by the Phosphotriester Method

3.1. Synthesis of Oligo(deoxyribonucleotides) in Solution

The first chemical synthesis of a 3'–5'-linked dinucleoside monophosphate was accomplished by Michelson and Todd *via* a triester intermediate, benzyl ester of d(TpT), as illustrated in Fig. 17.[9] As described in Section 2, the benzyl ester was unstable in pyridine and the phosphotriester method was not investigated further until Letsinger and Ogilvie used the 2-cyanoethyl group as a protecting group for the internucleotidic phosphates.[75] However, phosphotriester intermediates containing the 2-cyanoethyl ester were also found to be unstable and unsatisfactory for multistep condensations. Reese and co-workers investigated the alkaline stability of phosphotriesters of phenyl derivatives and found that the *p*-chlorophenyl group had suitable stability as a protecting group for the phosphotriester synthesis.[63] *o*-Chlorophenyl phosphate was found to be slightly more labile, which resulted in a complete deblocking at the final step.[76] Oligothymidylates up to a chain length of 32 were synthesized by Reese and co-workers to investigate the different factors involved in a complete deblocking.[77] They also studied the selective removal of phenyl derivatives from the internucleotidic phosphate and found that an oximate-mediated reaction afforded higher selectivity.[78] Reactions for the removal of the aryl ester from an internucleotidic phosphate are shown in Fig. 18.

Narang and co-workers synthesized oligothymidylates by using a completely substituted triester intermediate[80] as shown in Fig. 19. A 26 mer duplex containing the sequence of the operator for *E. coli lac* has been synthesized[81] by this approach using TPS–tetrazolide as the condensing reagent.[37]

Another approach, which was used to prepare fully substituted phosphotriester intermediates, involved a monofunctional phosphorylating

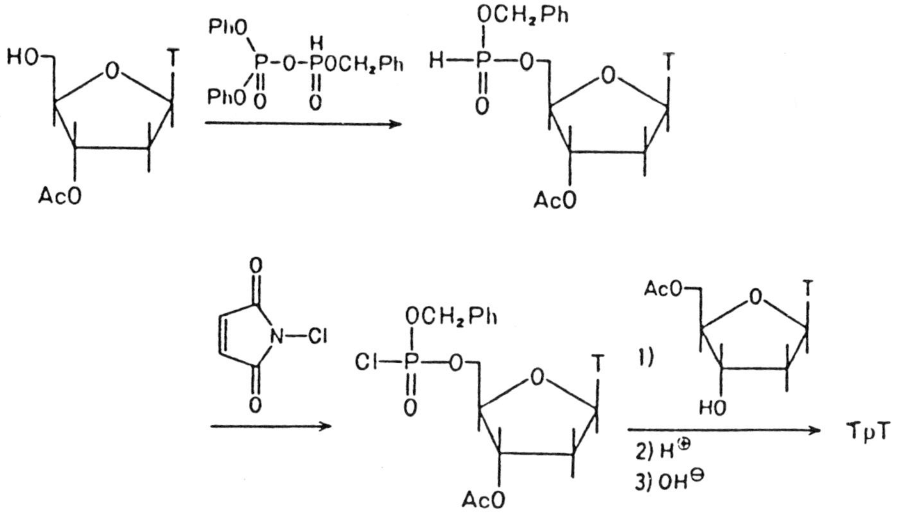

Figure 17. Formation of an internucleotidic linkage.

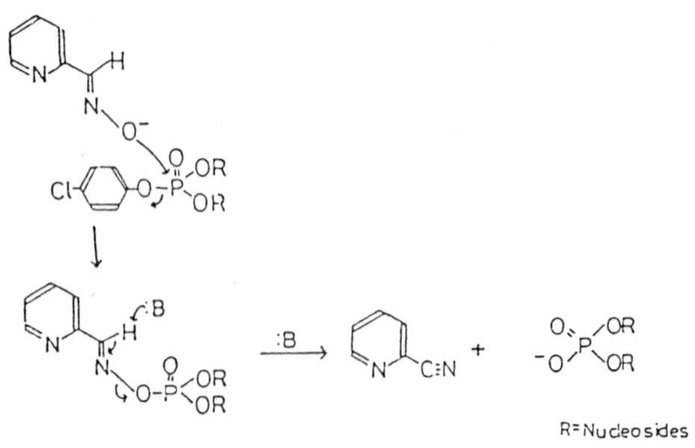

Figure 18. Formation of aryl groups with oximate ions.

reagent[6a,82] (Fig. 20). The fully protected mononucleotide can be converted to the 5'-hydroxyl and 3'-phosphodiester components. These components are then condensed to yield the fully protected dinucleotide. Dinucleotides of this type can be elongated either in the 3' or the 5' direction. Deoxyoligonucleotides corresponding to the gene fragments for somatostatin and insulin were synthesized by using these oligonucleotide building blocks.[6a,c] However, researchers later modified the preparation of the fully protected mononucleotides by using a bifunctional phosphorylating reagent, *o*-chlorophenyl phosphoroditriazolide.[83] Reaction schemes for the preparation of dinucleotides are shown in Fig. 21. The putative active intermediate (monotriazolide) can be used in a condensation with nucleosides in the presence of 1-methylimidazole. Protected dinucleotides

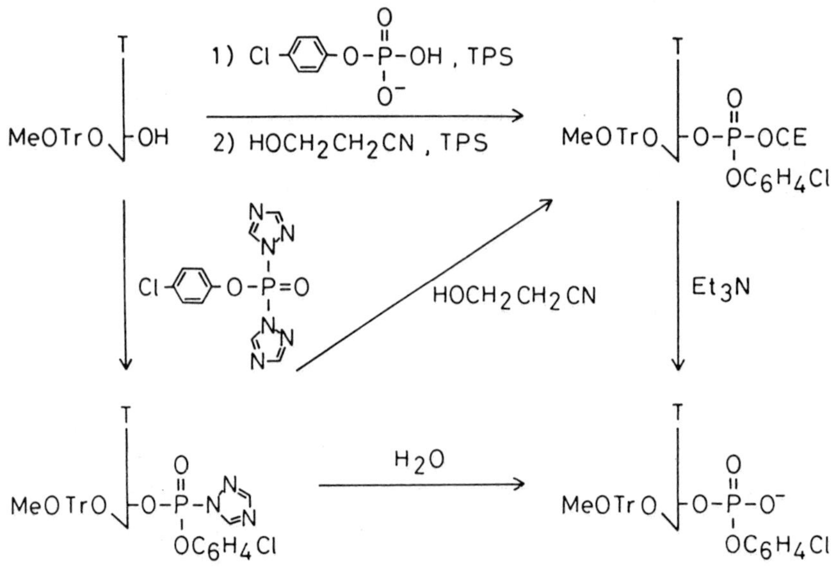

Figure 19. Preparation of protected thymidine 3'-phosphates.

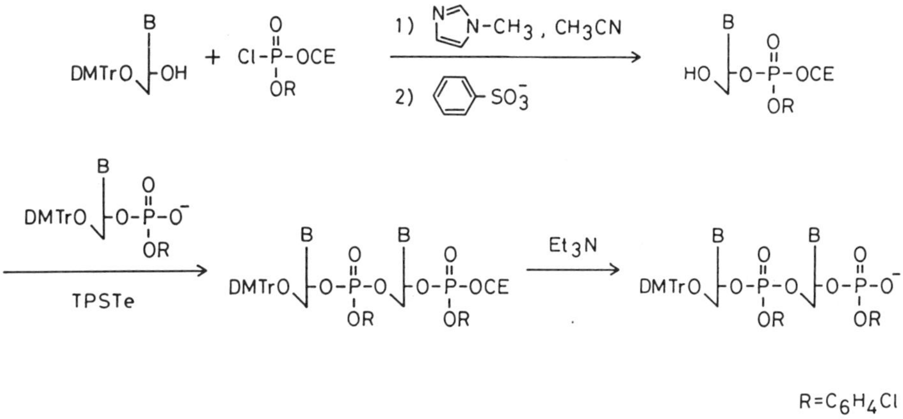

Figure 20. Phosphorylation with a monofunctional phosphorylating reagent.

containing the 3'-phosphoro-*p*-anisidate have also been synthesized by reacting the monotriazolide with a *N*-protected deoxynucleoside 3'-(*o*-chlorophenyl)phosphoro-*p*-anisido chloridate as shown in Fig. 22.[84] The terminal phosphoraromatic amidate of a protected oligonucleotide is more stable than the 2-cyanoethyl group and can be retained for several synthetic steps. In multistep condensations, the protecting groups of oligonucleotide blocks must be retained during chromatography on silica gel or alkylated silica gel. This chromatography is necessary for the purification of intermediates. Several complementary deoxyoligonucleotides to messenger RNAs, including a dodecanucleotide complementary to bacteriorhodopsin mRNA, have been synthesized by the anisidate

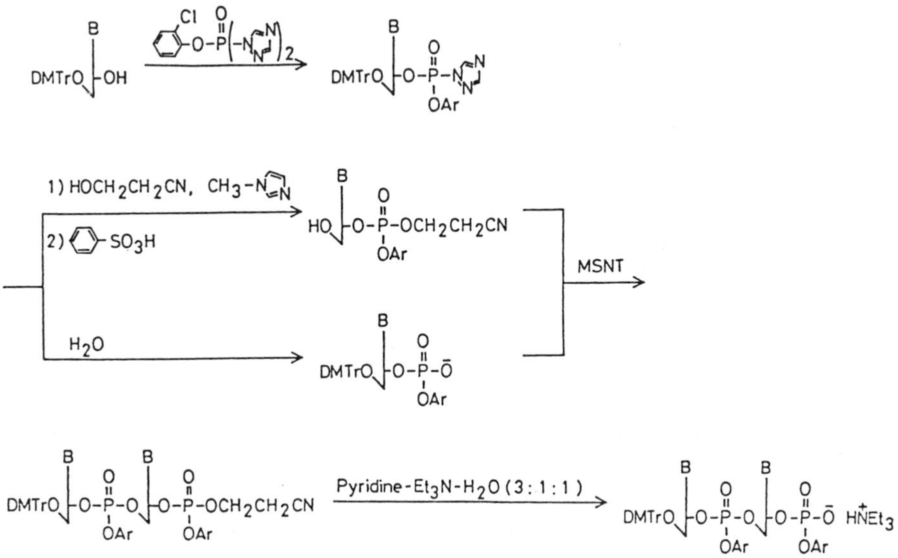

Figure 21. Preparation of dinucleotides.

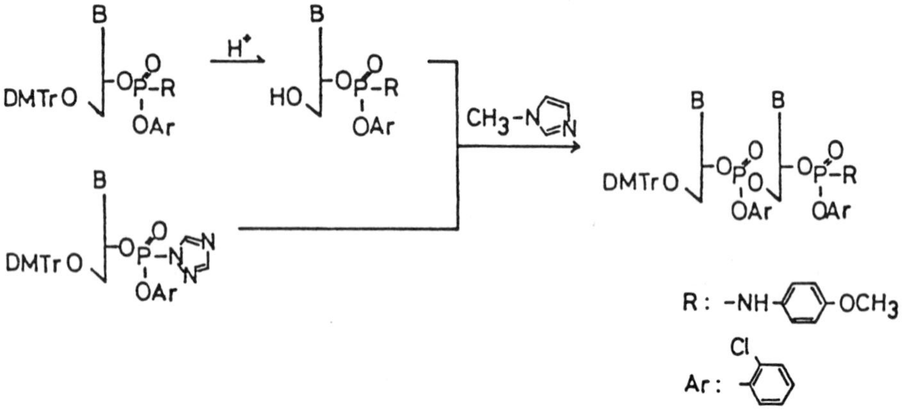

Figure 22. Activation of phosphorotriazolide with methylimidazole.

protection method.[85] Large-scale syntheses of short DNA duplexes, which have the sequence for binding sites of regulatory protein such as λ cro and catabolite gene activator protein (CAP), have also been prepared by the same procedure[14] as described for the synthesis of a pentamer duplex containing the Pribnow sequence[47] (Fig. 23).

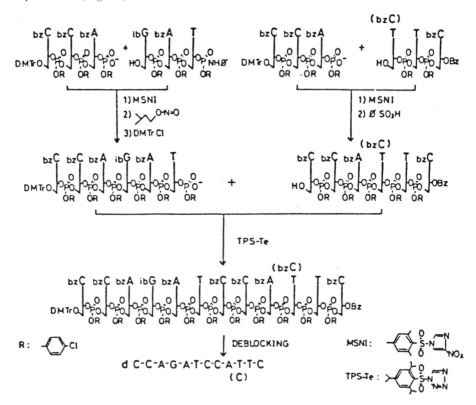

Figure 23. Synthesis by the phosphoroamidate method.

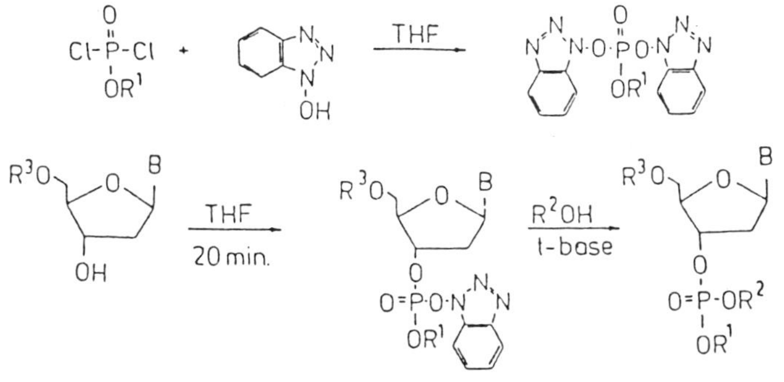

Figure 24. Activation of phosphodiesters with 2-hydroxybenzotriazole.

A unique phosphorylating reagent, *o*-chlorophenyl di-*N*-hydroxybenzotri-azolide, has been used for the preparation of protected dinucleotides as shown in Fig. 24.[86]

Rapid methods for the synthesis of poly(deoxyribonucleotide) in solution have been devised by employing rapid separation methods.[87]

3.2. Synthesis of Oligo(deoxyribonucleotides) on Polymer Supports

As discussed in a previous review, early attempts at polymer support syntheses of polynucleotides on polystyrene gave only partially satisfactory results, primarily because of the incomplete recovery of oligonucleotides from the polymer supports.[1,2] A reexamination of the polymer support synthesis on polyacrylmorpholide[88] and polyaryldimethylamides,[89] using the phosphodiester method, showed encouraging results in terms of recovering products. These polymer supports were then used for the phosphotriester synthesis of poly (deoxyribonucleotides)[90] and Fig. 25 shows the preparation of the nucleoside resin using 3'-succinyl deoxynucleosides as the starting linkage unit.[48] Properly protected mononucleotides, 5'-dimethoxytrityl-*N*-protected deoxynucleoside 3'-arylphosphates, are used as the condensing units. Various oligonucleotides with chain lengths of approximately 20 have been synthesized.[91] Oligonucleotide blocks, prepared as described in Section 3.1 (e.g., Fig. 22), were also found to be useful as condensing units, since overall yields could be improved by using dimer or trimer blocks providing that the oligomers were activated efficiently.[92] A 31mer was synthesized by the condensation of trinucleotide blocks on a polyacrylmorpholide support.[93] A variety of oligo(deoxyribonucleotides), which are complementary to mRNAs, have been synthesized using dinucleotide blocks[94] on a polyacryldimethylamide support. About 60 oligonucleotide fragments, corresponding to the gene for α_1-interferon, have been synthesized by the same procedure. These oligomers with a chain length of ~ 15 were used to compose the DNA duplex of 514 base pairs. This indicated that small quantities of a large number of deoxyoligonucleotides could be synthesized rapidly on polymer supports. Developments in techniques for purification, especially by high-performance liquid chromatography (HPLC)—which is indispensable in these syntheses—have made it possible to obtain pure products. In the solid-phase

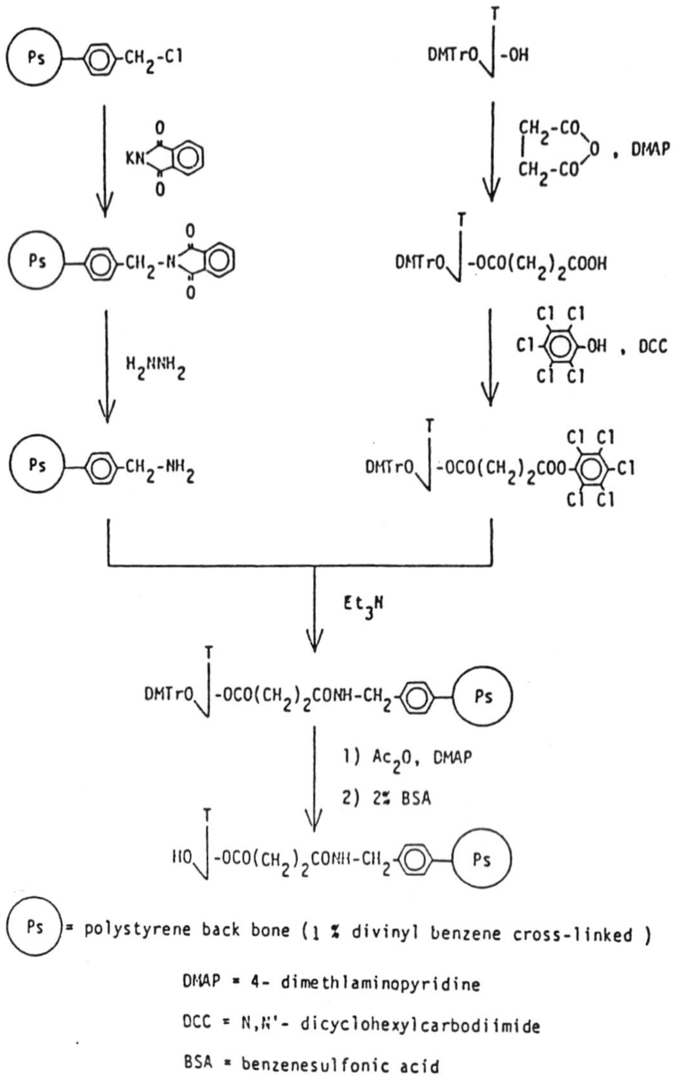

Figure 25. Preparation of nucleoside resin.

method, unchanged starting materials and other accumulated side products are recovered together with the desired products. It is essential to complete each condensation reaction in order to decrease these side products. Capping of unreacted 5′-hydroxyl groups of a growing chain by acetylation or carbamoylation was employed in most cases. The use of reversed-phase chromatography on alkylated silica gel,[95] to isolate highly lipophilic products containing the 5′-dimethoxytrityl group, was found to be very useful for isolating the desired products.[93] Therefore, a small amount of polynucleotides with a chain length of ∼30 can be synthesized by this technique and in most cases the product can be purified by analytical HPLC.

Polystyrene supports were reexamined and compared with polyacryl-morpholide resin in the synthesis of oligo(deoxyribonucleotides) by the phosphotriester method.[96] Cross-linked polystyrene with 2% divinylbenzene was found to be mechanically stronger than the polyacrylamides, but a recovery of oligonucleotides from the 2% cross-linked resin by cleavage of the 3'-succinyl ester was not quantitative. However, it was subsequently found that 1% cross-linked polystyrene gave satisfactory results.[97] By using protected dinucleotides as condensing units, 78 oligonucleotides corresponding to a 564 base pair gene for human growth hormone have been synthesized as described in Fig. 26 (a synthesis of a 15 mer).[6j] A typical cycle for an elongation reaction is shown in Table IV. A study of the condensation reaction established that the time could be shortened by raising the temperature. At 40°C the reaction was completed within 20 min using a threefold excess of the condensing reagent. For the preparation of a relatively large oligonucleotide (26 mer), tetranucleotide blocks have been used as the inter-mediates and the isolation of the product was found to be simpler than that for oligonucleotides synthesized by a condensation of dinucleotide units. A 46 base pair duplex corresponding to a promoter region of *E. coli gal* has been synthesized

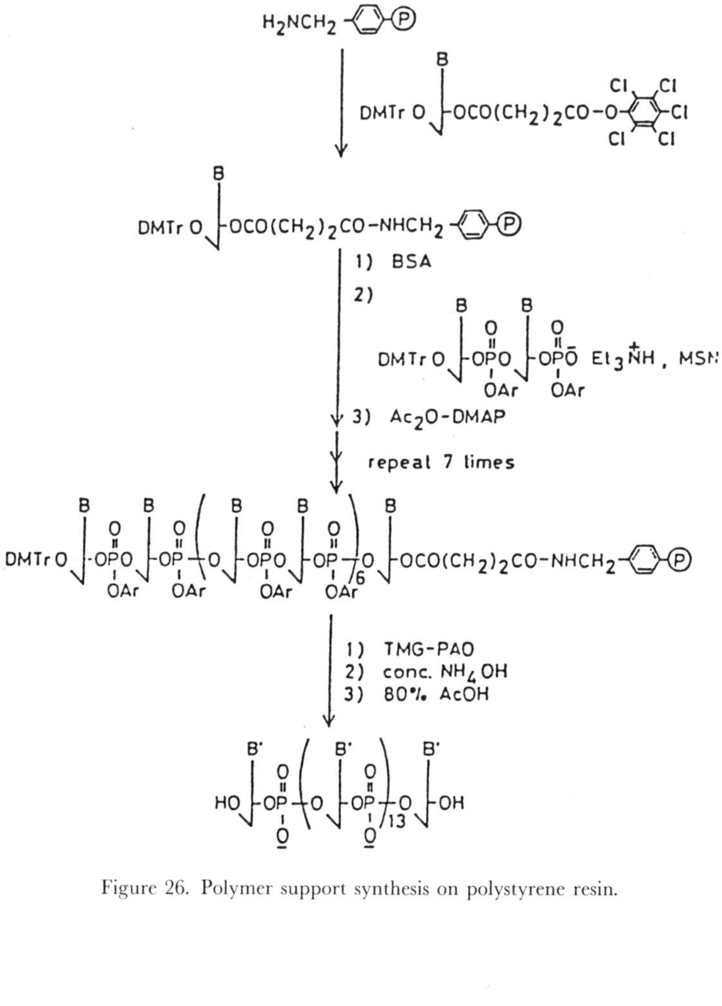

Figure 26. Polymer support synthesis on polystyrene resin.

Table IV. Polymer Support Synthesis of Deoxyoligonucleotides

Step	Solvent or reagent	Amount	Reaction time	Number of operations
1	CH_2Cl_2/MeOH (7 : 3)	2 ml	0.1 min	3
2	2% BSA in CH_2Cl_2/MeOH	2 ml	1 min	1
3	CH_2Cl_2/MeOH (7 : 3)	2 ml	0.1 min	1
4	2% BSA in CH_2Cl_2/MeOH (7 : 3)	2 ml 2 ml	1 min	1
5	CH_2Cl_2/MeOH (7 : 3)	2 ml	0.1 min	2
6	Pyridine	2 ml	0.1 min	3
7	Pyridine	0.3 ml	Coevaporation	1
8	Dimer block in pyridine	200 mg/0.3 ml	Coevaporation	1
9	MSNT in pyridine	200 mg/0.3 ml	20 min (40°C)	1
10	Pyridine	2 ml	0.1 min	2
11	0.1 M DMAP in pyridine Ac_2O	1.8 ml 0.2 ml	3 min	1
12	Pyridine	2 ml	0.1 min	3

chemically by using tetramer blocks on the same support.[98] Although the preparation of tetranucleotides from dinucleotides in solution takes time for chromatography on silica gel, the larger condensing unit affords pure products, which require less time for isolation. Use of trimer blocks for the synthesis of gene fragments (up to 60 nucleotides) on a 1% polystyrene support has been reported.[99] Hydroxybenzotriazole-activated phosphotriester intermediates have been applied to the solid-phase synthesis.[100]

Other polymer supports have been also tested for their properties.[101] Silica gel has been used in the phosphite solid-phase method and can be derivatized with the so-called coupling reagent, (3-aminopropyl)triethoxysilane[102] (Fig. 27). *N*-Acyl-5′-*O*-dimethoxytrityl-3′-*O*-succinyl deoxynucleosides are linked to the silica gel. The yield for the coupling in the first step was found to be low in the case of silica gel supports.[103] Controlled pore size glass (CPG) has been introduced in the phosphotriester synthesis[104] as well as in the phosphite approach.[12] Linking of derivatized CPG with the 5′-hydroxyl group of 2′(3′)-*O*-benzoyluridine was used as an all-purpose support.[105] Silica-gel-coated polyacrylamide was prepared as a suitable support for a mechanized synthesis.[106]

A new approach has been reported[107] for the synthesis of a large number of oligonucleotides simultaneously by using cellulose paper disks. The main idea is to react a variety of chains in the same vessel when the same building block is used. The synthesis of two deoxyoctanucleotides on cellulose filter disks using protected deoxynucleoside phosphodiesters has been described. Over 200 species of larger fragments (8–22mer) have been synthesized by the same procedure.[108] The method is suitable for the preparation of 50–150 nmol of oligonucleotides with chain lengths less than 20. Products of this amount are usually isolated by electrophoresis on polyacrylamide. Homogeneity in reversed-phase chromatography may be required in the synthesis of structural genes for peptides, since any minor modification on the bases may cause mutageneses in the amino acid sequences.

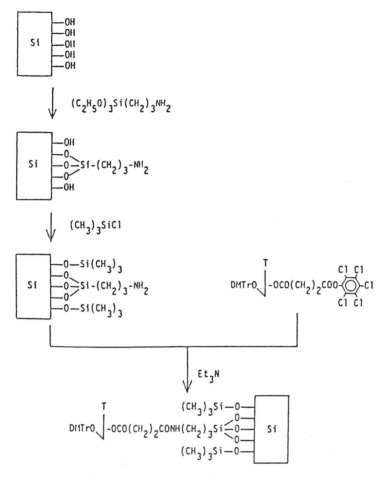

Figure 27. Preparation of silica gel nucleoside.

3.3. Synthesis of Oligo(ribonucleotides) in Solution

For the triester synthesis of poly(ribonucleotides), a combination of protecting groups for the internucleotidic phosphates, the 2'- and 5'-hydroxyl groups, is the major problem.

Protected mononucleotides are prepared by a phosphorylation with phosphorylating reagents having the proper protecting groups. For the preparation of oligonucleotide blocks, the 3'-terminal phosphate requires a temporary protection.

A fully protected mononucleotide is prepared by reacting *p*-chlorophenyl anilidophosphorochloridate[109] with ribonucleosides having an acid and alkaline stable protecting group, *o*-nitrobezyl[59] (Fig. 28). This nucleotide can be deanilidated by treatment with isoamyl nitrite, which yields the 3'-diesterified nucleotide. This nucleotide can also be obtained by phosphorylation with

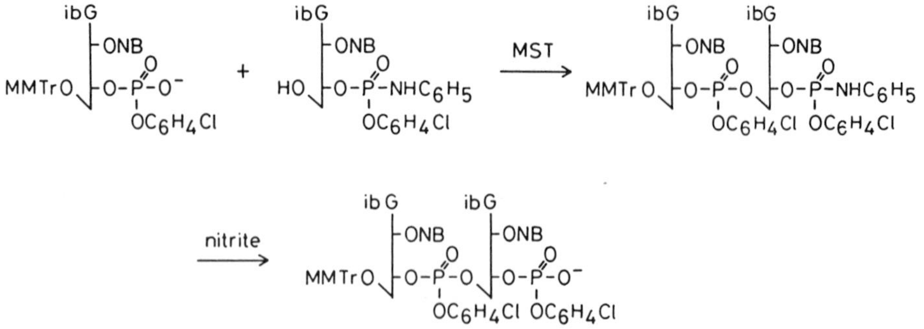

Figure 28. Preparation of protected ribodinucleotides.

p-chlorophenyl phosphate in the presence of DCC. Nucleotides of this type can be de-monomethoxytritylated with acid under more drastic conditions than in the case of deoxynucleosides. Condensation of these monomer units yields fully protected dinucleotides, which can then be elongated either in the 3′ or 5′-direction. This method was applied to the synthesis of tRNA fragments with chain lengths of 10–20 monomer units.[58]

Acid labile protecting groups for the 2′-hydroxyl group are most often used in the synthesis of ribo-oligonucleotides, with 3′–5′-unprotected nucleosides being used as an incoming unit. The *o*-dibromomethylbenzoyl group has been used for 5′ protection since it can be removed selectively by treatment with silver perchlorate.[31] Another approach for a removal of the 5′ protecting group of oligonucleotides is shown in Fig. 29. The levulinyl group on the 5′-hydroxyl function and the trichloroethyl on the 3′-phosphate are removed with dilute hydrazine and zinc, respectively.[45,110] The 2′-(*t*-butyldimethylsilyl)nucleosides, which have been developed in the phosphite approach,[53] are also employed

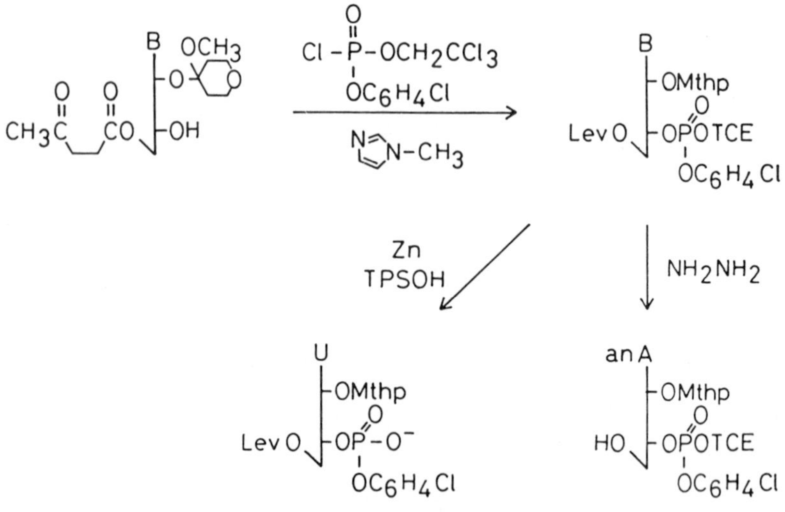

Figure 29. Selective removal of protecting groups.

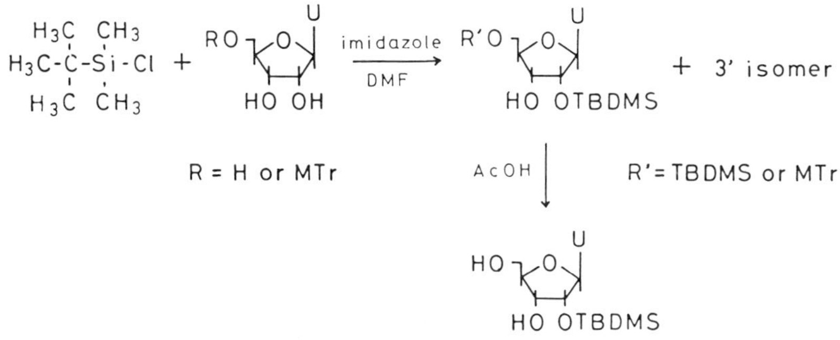

Figure 30. Preparation of a 2'-(*t*-butyldimethylsilyl)nucleoside.

for the synthesis of phosphotriester oligonucleotides[111] as shown in Fig. 30. 2'-O-Tetrahydrofuranyl nucleosides[112] have also been used[20] in combination with the 5'-dimethoxytrityl group, which can be removed by zinc bromide under the conditions described for the solid-phase synthesis of deoxyoligonucleotides.[113] The 5'-dimethoxytrityl group can be removed preferentially from 2'-O-tetrahydropyranyl nucleosides with 2% *p*-toluenesulfonic acid[114] (Fig. 31).

These protected oligonucleotide intermediates are isolated by chromatography on silica gel or alkylated silica gel. The latter gel can resolve the phosphodiester intermediate by reversed-phase chromatography by using increasing amounts of acetonitrile. Larger oligonucleotide fragments can also be purified by these techniques. The activation of larger oligonucleotide fragments with arenesulfonyl azolides (e.g., MSNT, MSTe; see Section 2.4) has enabled investigators to build RNA fragments with a chain length of 33, which has a sequence of the 5' half of the tRNA$_1^{Gly}$ of *E. coli*. Figure 32 shows abbreviated structures of oligonucleotide blocks used in the synthesis of this protected 33 mer. Protecting groups and condensation procedures are the same as illustrated in Fig. 30, except that all di- and trinucleotides with the 3'-phosphodiester group are prepared by a condensation of the 3', 5'-unprotected nucleosides followed by phosphorylation with *o*-chlorophenyl phosphoroditriazolides. These diester components are condensed with 3'-anisidated 5'-hydroxyl components and fully protected oligonucleotides are elongated in the 3' direction or 5'-direction by treatment either with isoamyl nitrite or zinc bromide.

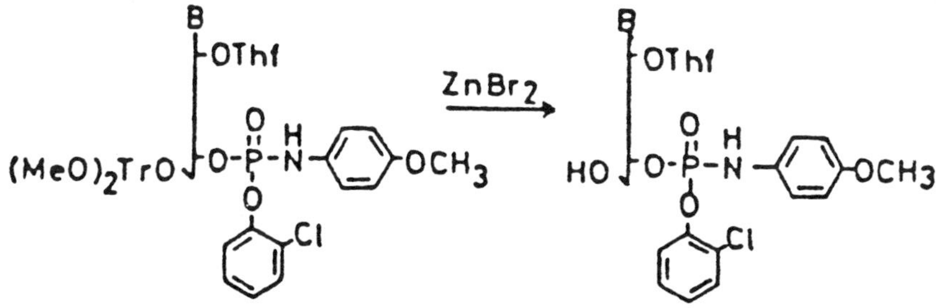

Figure 31. Selective demonomethoxytritylation of 2'-ketal nucleosides.

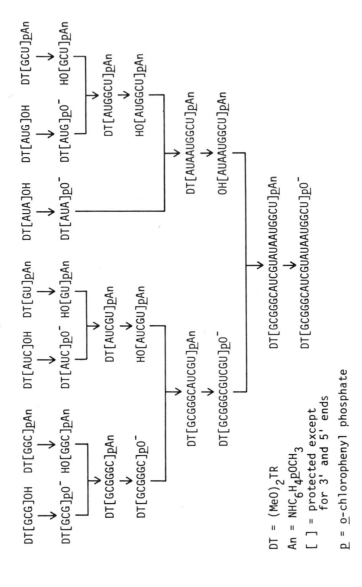

Figure 32. Condensation of protected ribo-oligonucleotides.

A new phosphorylating reagent, 2-chlorophenyl-*O*-bis(1-hydroxybenzo-triazolyl)phosphate, has been used to prepare dimers by a reaction with a 2′, 5′-protected ribonucleoside and then with a 2′-protected nucleoside. A pentamer UACGC was synthesized by this method.[115]

3.4. Synthesis of Oligo(ribonucleosides) on Polymer Supports

In contrast to the triester synthesis of deoxyoligonucleotides on polymer supports, only a few papers on the solid-phase synthesis of ribo-oligonucleotides have been reported.[116,117] The preparation of protected ribo-oligonucleotide blocks requires a large number of steps of reactions because of the added presence of the 2′-hydroxyl group. Since a three- to fourfold excess of incoming nucleotides is used in the solid-phase synthesis, larger amounts of nucleotide blocks are required when compared to the synthesis in solution. The solid-phase synthesis of ribo-oligonucleotides may not be advantageous until the synthesis of protected ribo-oligonucleotides is simplified. In principle, the ribo-oligonucleotide blocks discussed in Section 3.3 can be used as the condensing unit. Figure 33 illustrates a synthesis of a riboheptamer on polyacrylmorpholide resin linked through the 3′-succinyl ester by a condensation of protected mononucleotides.[117] A tridecamer with a repeating sequence has been synthesized by using trinucleotides as shown in Fig. 34.

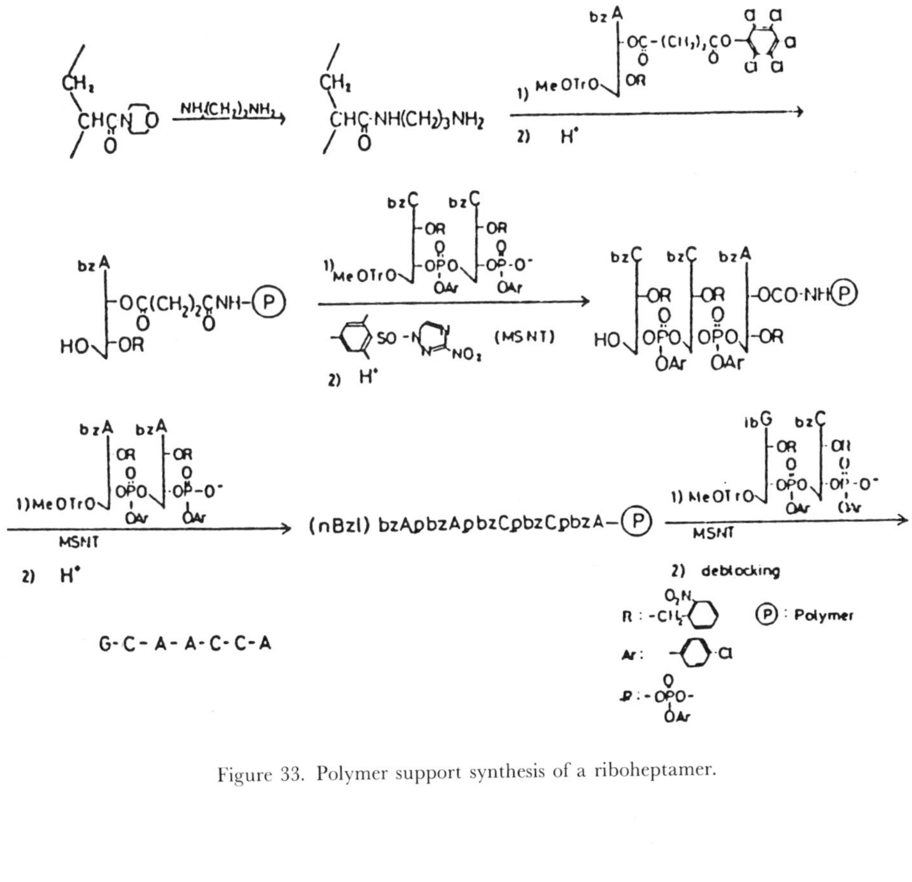

Figure 33. Polymer support synthesis of a riboheptamer.

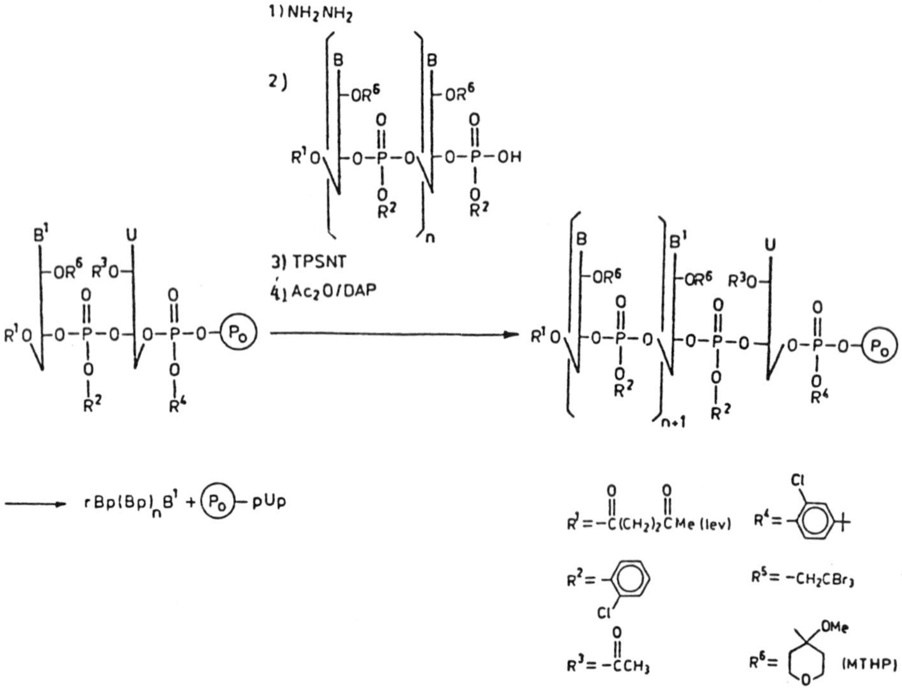

Figure 34. Polymer support synthesis of a repeating tridecamer.

4. Phosphite Method

4.1. Synthesis of Oligo(deoxyribonucleotides) in Solution

The synthesis of oligonucleotides using a phospite method was introduced by Letsinger and co-workers.[10] The major advantage of this method is that a phosphorylation of the hydroxyl groups of the sugar moiety proceeded very rapidly (a few minutes at $-78°C$) in direct contrast to the phosphotriester method (more than 20 min at room temperature). 2,2,2-Trichloroethyl-phosphorodichloridite was the first reagent successfully used in solution (Fig. 35). The 3′-hydroxyl group of the 5′-O-protected thymidine was phosphitylated at $-78°C$ and then condensed with a 3′-hydroxyl-protected thymidine to afford a thymidine dimer with a 3′–5′ linkage (82% yield). The phosphite linkage thus prepared was converted to the phosphate linkage by an oxidation using iodine and water. Other phosphitylating reagents, such as 2,2,2-tribromoethyl-, benzyl-, methyl-, 2-cyanoethyl-, p-chlorophenyl-, phenylethyl-, and p-nitrophenylethyl-phosphorodichloridite, were synthesized and investigated.[118] Daub and van Tamelen[119] used the methyl analogue, where the methyl group was easily removed by treatment with thiophenoxide and showed that the methyl-phosphorodichloridite was suitable for the synthesis of di- and trinucleotides. However, one of the drawbacks in using this reagent is the lack of reaction

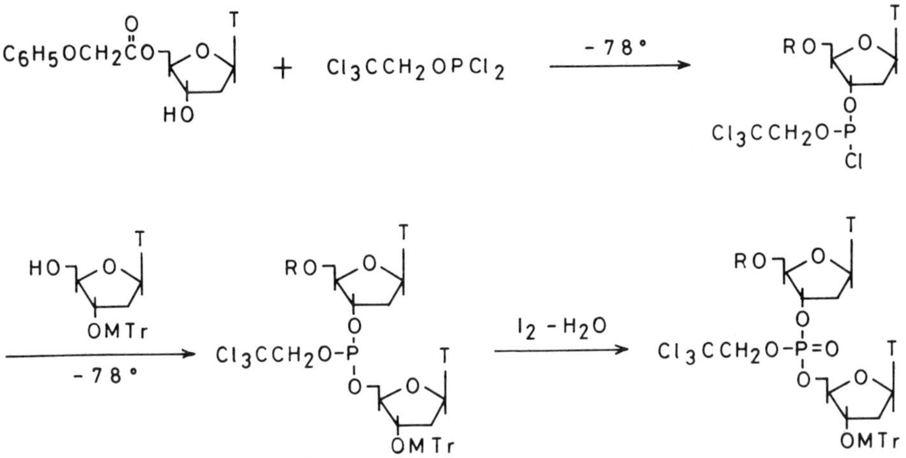

Figure 35. Phosphite-triester synthesis of oligonucleotides.

specificity. Fourrey and Shire[120] prepared the dinucleotide (TpT) using 5'-dimethoxytritylthymidine methylphosphorodichloridite, and 3'-acetylthymidine in an equal ratio and showed that the yield of the desired dinucleotide with the 3'–5' linkage was low (38%) and that an undesired dinucleotide with a 3'–3' linkage was also formed (yield of 15%) (Fig. 36). To control the reaction specificity, they substituted 1,2,4-triazole or 1,2,3,4-tetrazole for chlorine on the phosphitylating reagent. Using methylphosphoroditriazole or tetrazole, the desired dinucleotide was obtained in yields of 70 and 75%, respectively. Another approach that was employed to control the reaction specificity involved the use of

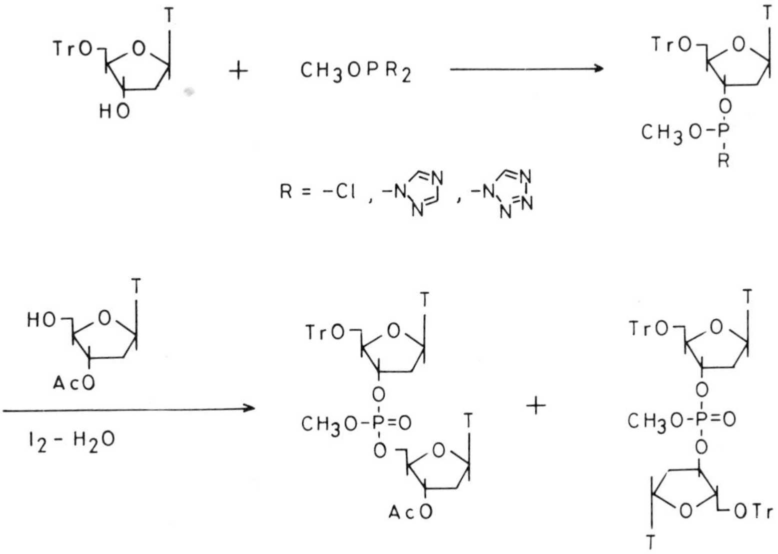

Figure 36. Reactivity of phosphite derivatives.

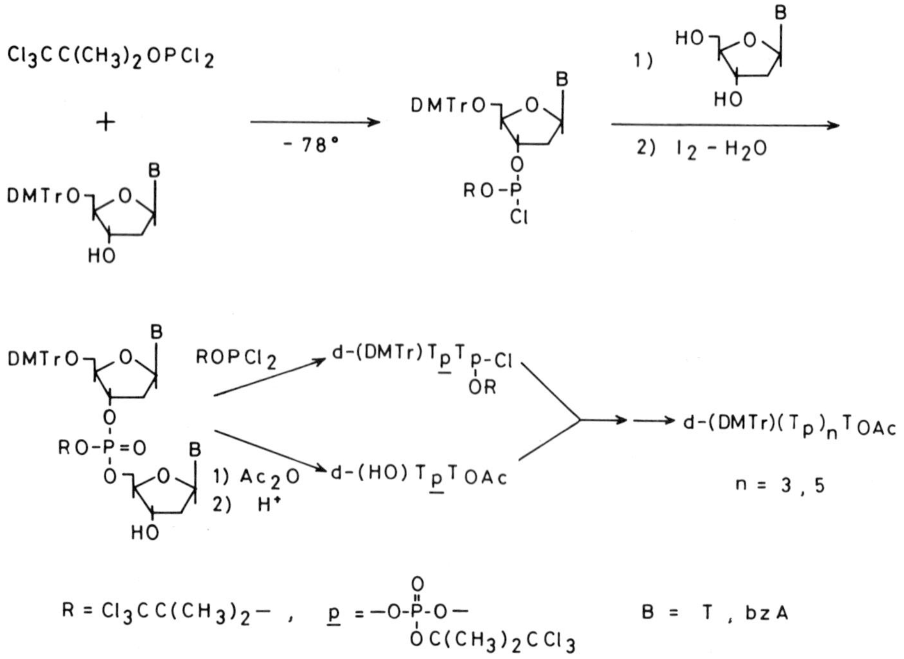

$Cl_3CC(CH_3)_2OPCl_2$

+

DMTrO—[sugar, B]—HO

$\xrightarrow{-78°}$

DMTrO—[sugar, B]—O—$\overset{O}{\underset{Cl}{RO-P}}$

1) HO—[sugar, B]—HO

2) $I_2 - H_2O$ $\longrightarrow$

DMTrO—[sugar, B]—O—$\overset{O}{\underset{HO}{RO-P=O}}$—[sugar, B]—HO

$\xrightarrow{ROPCl_2}$ d-(DMTr) $T_{\underline{p}}T$ $\overset{}{\underset{OR}{P-Cl}}$

1) Ac_2O
2) H^+ $\longrightarrow$ d-(HO) $T_{\underline{p}}T$ OAc

$\longrightarrow$ d-(DMTr)($T_{\underline{p}})_n T$ OAc

n = 3 , 5

$R = Cl_3CC(CH_3)_2-$, $\quad \underline{p} = -O-\overset{O}{\overset{\|}{P}}-O-$ $\quad$ B = T , bzA
$\underset{OC(CH_3)_2CCl_3}{}$

Figure 37. Block condensation by the phosphite method.

a bulky alkyl group such as 2,2,2-trichlorodimethylethyl,[121,122] which was removed by zinc metal or tributylphosphine, to protect the phosphate group (Fig. 37). 2,2,2-Trichlorodimethyl(TCDM)ethylphosphorodichloride reacts easily with the 3'-hydroxyl group of 5'-dimethoxytritylthymidine even at −78°C. In this reaction, the formation of a 3'–3' dimer was avoided because of the steric hindrance of the bulky alkyl group. The phosphitylated 5'-dimethoxytrityl-thymidine was reacted selectively with the 5'-hydroxyl group of another nucleoside. In this way, the desired dinucleotides, such as d-(DMTr)T(TCDM)T and d-(DMTr)bzA(TCDM)bzA, were synthesized in yields of 70 and 62%, respectively, after purification on a silica gel column. These dinucleotides were used for the condensation after phosphitylation with 2,2,2-trichloromethylethyl-phosphorodichloridite and protected dT_4, dT_6, d(AATT), d(TTAA) were synthesized in yields of 93, 81, 71, and 70%, respectively.

    Another problem with the phosphite method was the instability of the nucleoside phosphoromonochloridite toward moisture. However, this problem was solved by Beacage and Caruthers[11] by synthesizing a new phosphitylating reagent, chloro-N,N,-dimethylaminomethoxyphosphine—$CH_3OP(Cl)N(CH_3)_2$. This reagent was prepared from a reaction of methylphosphorodichloridite with dimethylamine and reacts with the 3'-hydroxyl group of 5',N-protected nucleosides to form a nucleoside phosphoramidite. This nucleoside phosphoramidite shows a high stability toward moisture and can be stored for at least a month at room temperature under an inert gas. A ^{31}P-NMR study showed that by the addition of 1,2,3,4-tetrazole the amidite derivatives are converted to a

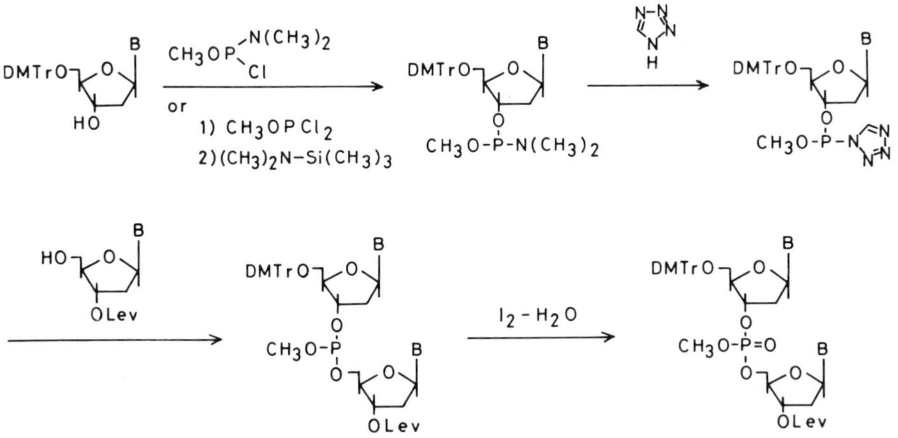

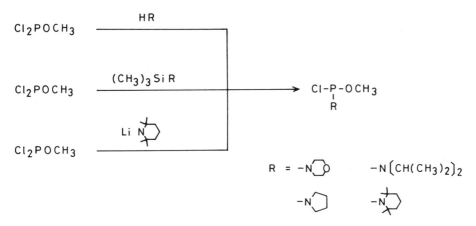

Figure 38. Reactivity of nucleoside phosphoramidites.

tetrazole derivative, which can then react with the 5'-hydroxyl group of 3'-levulinylthymidine (Fig. 38). The nucleoside phosphoramidite can also be prepared from *N,N*-dimethyltrimethylsilylamine and a 5',*N*-protected nucleoside 3'-methylphosphoromonochloridite.

To find better phosphitylating reagents, various chloro-*N,N*-dialkylamino-methoxyphosphines were prepared and investigated[123,124] (Fig. 39). The *N,N*-diisopropylamino and *N*-morpholino derivatives, which are prepared from methylphosphorodichloridite and the corresponding amine or their silyl derivatives, are most promising from the standpoint of high stability and reactivity. Nucleoside phosphoramidites are prepared by the reaction of a 5',*N*-protected nucleoside with chloro-*N,N*-dialkylaminomethoxyphosphine or bisdialkylaminophosphines in the presence of a salt.[125,126] These compounds have also been converted to the tetrazole derivatives by a reaction with 1,2,3,4-tetrazole or 5-*p*-nitrophenyltetrazole[127] (Fig. 40).

Figure 39. Preparation of phosphitylating reagents.

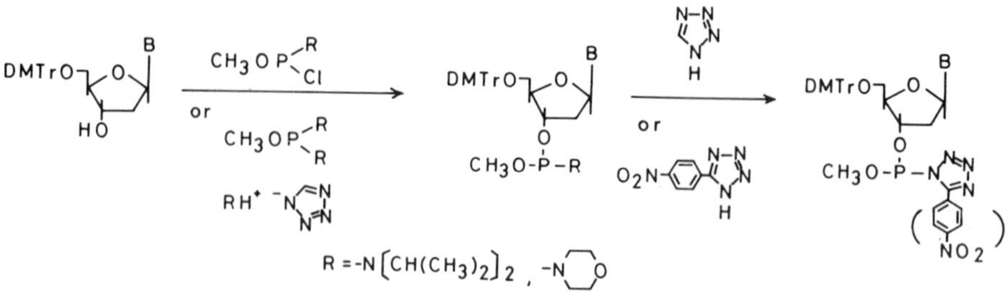

Figure 40. Activation of the phosphoramidite.

Other protecting groups, such as 2-cyanoethyl,[128,129] 2-methylsulfonyl-ethyl,[130] 4-nitrophenylethyl,[131] and 2-chlorophenyl,[132] were introduced and their N,N-diisopropylamino or N-morpholino phosphoramidite derivatives were synthesized.

4.2. Synthesis of Oligo(deoxyribonucleotides) on Polymer Supports

The phosphite triester method has been applied successfully to nucleotide synthesis especially on polymer supports. The reactions consist of a condensation, oxidation, capping reaction and deprotection of the 5'-hydroxyl group. Nulcleoside methylphosphoromonochloridite was first used for a condensation with the 5'-hydroxyl group of a nucleoside bound to the derivatized silica gel through the 3'-hydroxyl group.[132] In general, the condensation reaction is continued for 1 hr, although subsequent investigators found that several minutes were enough to afford completion if a 10–20-fold excess of pure activated nucleotides were used. The resultant dinucleoside phosphite was oxidized to the phosphate derivative by iodine and water. The unreacted 5'-hydroxyl group was blocked with phenylisocyanate or acetic anhydride containing dimethylaminopyridine. The 5'-hydroxyl protecting group can be removed by treatment with various acids. The average time required for one cycle is about 20 min and this cycle is continued until the desired oligonucleotide is obtained (Fig. 41). By this method, Matteucci and co-workers obtained a yield of more than 95% in each coupling during the synthesis of T_7, T_9, d(TCTCTCTTT). In this method, the preparation of a pure activated nucleotide is very important. Various procedures have been used successfully for the preparation of oligodeoxyribonucleotides and mixed oligodeoxyribonucleotides.[37,102,133–135]

On the other hand, the preparation of nucleoside phosphoramidites makes it much easier to obtain purified activated nucleotides. The widely used N-morpholino and N,N-diisopropylamino derivatives can be purified by silica gel column chromatography and give satisfactory results in oligonucleotide synthesis. A 21mer d(CCTTATTTTGGATTGAAGCCA) has been synthesized[136] using the N-morpholino derivatives to provide a total yield of 20%. A 13mer d(AATTCTAGCTGCA)[137] and a 9mer d(AACCAGCAC)[138] have been synthesized on controlled pore glass (CPG) as a polymer support in an overall condensation yield of 67%. On the other hand, usine N-diisopropylamino

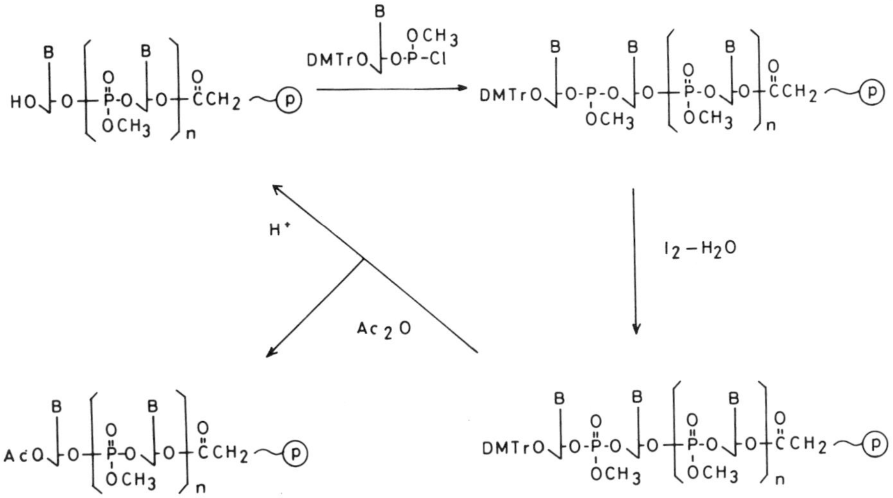

Figure 41. Reaction cycle for oligonucleotide syntheses on a polymer support.

derivatives, Adams and co-workers[12,139] synthesized two 51mer DNA on the long-chain alkylamino-controlled pore glass as a support.

Although a methyl group is the most widely used for the protection of a phosphate group, the use of other protecting groups has also been investigated. When the 2,2,2-trichlorodimethylethyl group is used,[121,122] the condensation reaction takes a much longer time (~ 30 min). However, 1-methylimidazole can accelerate the condensation or reduce the reaction time to 15 min. Using this protecting group, dC_4T, dA_4T, dG_4T, dT_6, dT_{16}, and d(GCAAATATCATTTT) were synthesized on a silica gel support in an average coupling yield of 94%. Moreover, the use of a bulky trichlorodimethylethyl group for the protection of phosphorus enables one to prepare dimer blocks in solution and to convert them to active phosphite derivatives for the oligonucleotide synthesis on polymer support. Using dT bound to a silica support, dT_5 was synthesized in a good yield after two cycles of reaction using the dimer.

β-Cyanoethylphosphorodichloridite was also converted to various amidite derivatives by the reaction of β-cyanoethylphosphorodichloridite and N-trimethylsilylated secondary amines or free secondary amines followed by distillation at a reduced pressure.[128,129] The advantage of using this protecting group is that it can be removed easily from the internucleotidic phosphate moiety with concentrated aqueous ammonia during the final deprotection of oligomer from polymer and de-N-acylation. By using this phosphitylating reagent, the nuleoside phosphoramidites were prepared in more than 90% purity. Using 5'-O,N-protected nucleoside-3'-β-cyanoethyl-N,N-diisopropylphosphoramidite and 1,2,3,4-tetrazole, the octamer d(CGGTACCG),[128] 13 mer d(TCAGTTGCAGTAG), and the two mixed 14 mer d(GGA/GTGA/GATA/GTAC/GAC) and d(GGATGA/TATATAG/A/TAC)[129] were synthesized in overall yields of 55, 45, 40, and 38% on controlled pore glass beads as support.

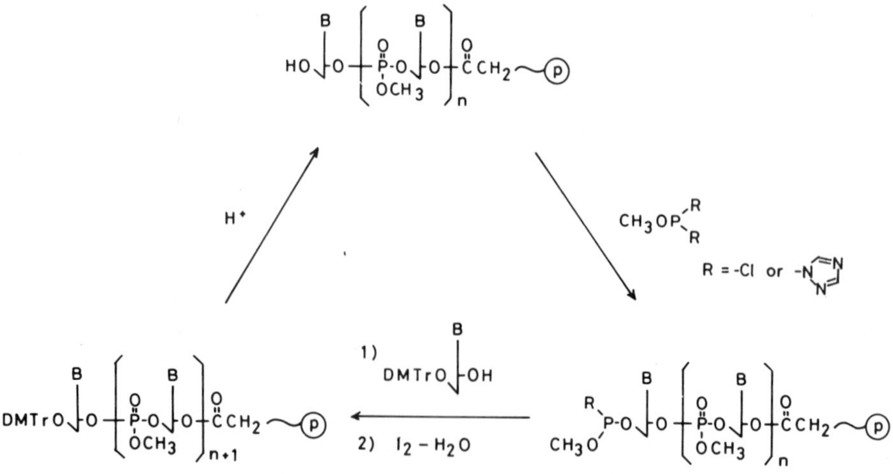

Figure 42. Reaction cycle for oligonucleotide synthesis on a polymer support.

The 2-methylsulfonylethyl group[130] can also be used as a phosphorus protecting group. It has been shown that 2-methylsulfonylethyl N-morpholino phosphomonochloridite has the same phosphitylating properties as the methyl analogue. Phosphitylations of the 5'-O-,N-protected nucleosides proceed quantitatively and the nucleoside phosphoramidites were obtained in 88–97% yields after purification by column chromatography. This compound was condensed with 3'-O-levulinyl nucleosides in the presence of 1-hydroxybenzotriazole followed by oxidation with iodine–water. After workup and purification, the pure dimer was obtained in a 65% yield. This nucleoside amidite was applied to the synthesis of a oligonucleotide on a solid support to afford a hexadecamer d(GGTCGACGTTTTTATT) with each coupling yield being higher than 95%.

An alternative way to synthesize DNA is by an activation of the 5'-hydroxyl group of the nucleoside bound to the resin by methylphosphorodichloridite or its ditetrazolide (Fig. 42). The activated resin was condensed with 5'-O,N-protected 3'-O-free nucleoside followed by an oxidation. After a cycle, it was shown that the condensation of the monomer unit proceeds at a greater than 98% yield from the amount of dimethoxytritanol released by acid. In this way, d(CAAGCTAG) (isolation yield of 37%)[140] and d(GGCGATATAAGGAT) (total coupling yield of 94%)[141] were synthesized.

4.3. Synthesis of Oligo(ribonucleotides) in Solution

Extensive studies in the ribonucleotide series have not been accomplished using the phosphite procedure. The utility of the phosphite procedure[118] has been researched and used for the synthesis of oligoribonucleotides[55] (Fig. 43). N-Protected 5'-O-MMTr-2'-O-t-butyldimethylsilyl nucleosides were phosphitylated with 2,2,2-trichloroethylphosphorodichloridite and used in the condensation procedure. By a stepwise synthesis, a heptaribonucleotide r(GCAACCA) was synthesized in coupling yields at each step of 50–87%.

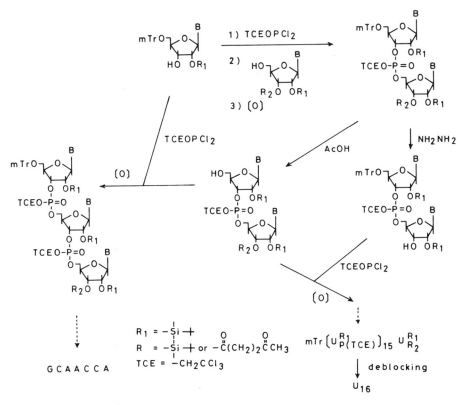

Figure 43. Synthesis of oligoribonucleotides in solution.

For the block condensation synthesis,[142] the 3'-hydroxyl group of protected uridine was phosphitylated and condensed with 2'-*O*-*t*-butyldimethylsilyl-3'-*O*-levulinyluridine followed by oxidation. The resultant fully protected dinucleotide was used as the key intermediate in the synthesis of hexadecauridylic acid.

4.4. Synthesis of Oligo(ribonucleotides) on Polymer Support

For the polymer support synthesis, 5'-*O*-dimethoxytrityl-*N*-protected-2'-*O*-*t*-butyldimethylsilyl-3'-*O*-methylphosphorochloridite was prepared and condensed with the 5'-hydroxyl group of a protected ribonucleoside bound to a silica polymer support *via* a 2'- or 3'-hydroxyl group. The reaction cycle is basically the same as that used in the deoxy series. Using phosphitylated nucleotide, a tetramer r(UGCA)[143] was synthesized by this stepwise procedure. Application of this procedure to the synthesis of longer ribo-oligonucleotides is illustrated in the syntheses of a hexamer r(AGCUCG)[144] and a nonadecamer r(GGAGCAGCCUGGUAGCUCG)[145] corresponding to oligonucleotides from the 21st to 26th and from the 9th to 27th of *E. coli* tRNA$_f^{Met}$, respectively (Fig. 44).

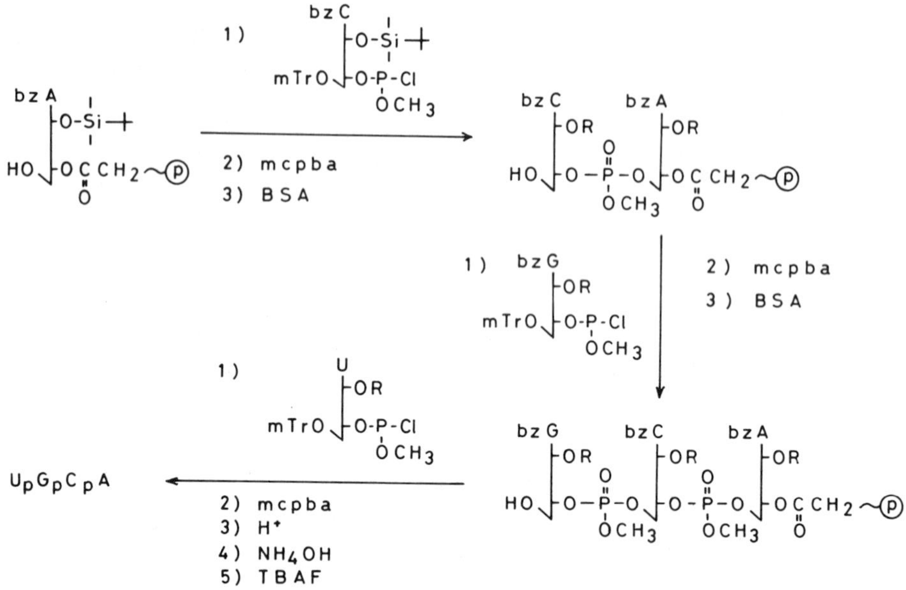

Figure 44. Synthesis of oligoribonucleotide on a polymer support.

4.5. Synthesis of Modified Nucleotides

Applications of phosphite chemistry to the synthesis of modified nucleotides, which have P−C, P−N, P=N, P=S, P=Se, P: bonds at an internucleotidic linkage, have been reported.[145–149] The synthesis of DNA with methyl phosphonate, is accomplished by using methylchloro-*N,N*-dimethylamino-

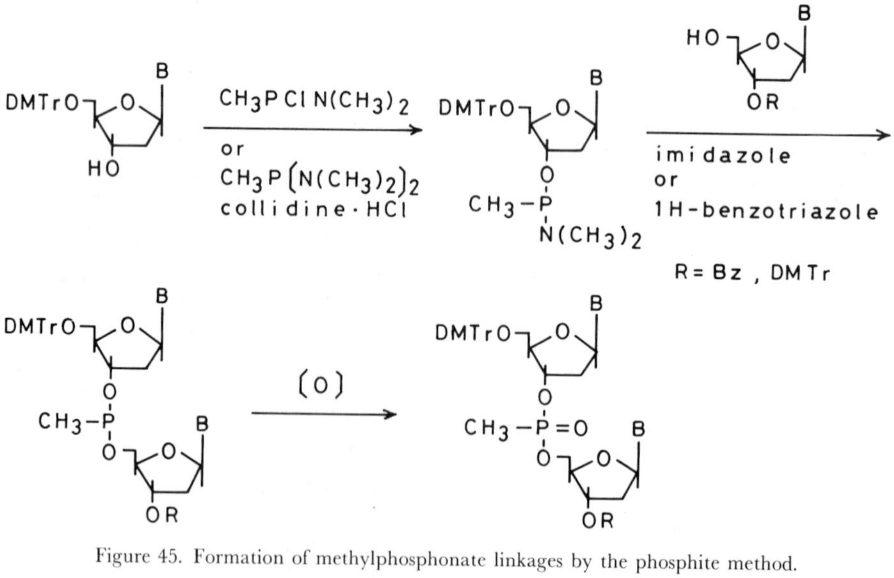

Figure 45. Formation of methylphosphonate linkages by the phosphite method.

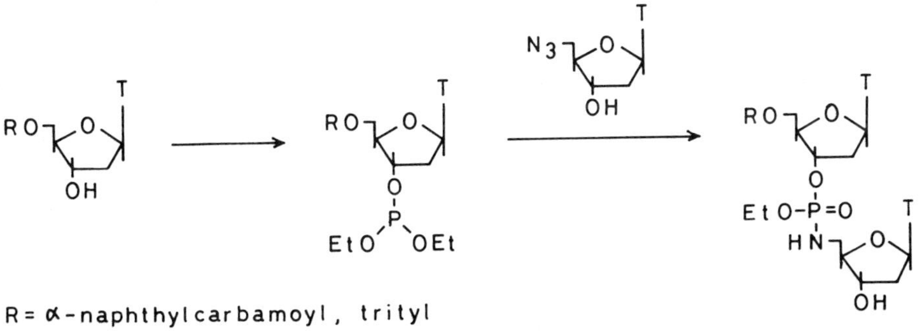

R = α-naphthylcarbamoyl, trityl

Figure 46. Formation of phosphoramidate linkages by the phosphite method.

phosphine[150] or methyl-bis(*N*,*N*-dimethylaminophosphine)[141] as the phosphitylating reagents. Procedures for the synthesis of DNA with methyl phosphonates are the same as those described for the synthesis of natural DNA. However, when nucleoside amidites are used, imidazole or 1*H*-benzotriazole is required for an activation of the amidite, because 1,2,3,4-tetrazole causes side reactions, such as an acid-catalyzed ligand exchange reaction.[151] Using the above reagents, several dinucleotides with methyl phosphonate were synthesized in more than 80% yields and four 21 base pair *lac* operators, which contained one methyl phosphonate, were also synthesized[150] (Fig. 45).

For a synthesis of the phosphoroamidate DNA, the phosphitylating reagent having a P—N linkage, such as tris(dimethylamino)phosphane[152,153] or diethyl-aminophosphorodichloridite,[149] was used. However, the phosphoroamidate linkage was formed more easily by the coupling reaction between phosphites and azides,[154] for example, a reaction of a 5′-protected thymidine with diethyl-phosphorochloridite. The corresponding phosphite was then coupled with 5′-azido-5′-deoxythymidine in the presence of lithium chloride, which furnished the phosphoroamidate dinucleotide in yields of 61–66% (Fig. 46). This method has been applied to the synthesis of phenanthridinium dinucleotide as a model of the intercalation of ethidium into DNA[155] (Fig. 47).

Several automated synthesizers, using the phosphite triester method on polymer supports, have become commercially available and they can be used to obtain oligonucleotides of limited chain lengths. However, the purification and characterization of oligomers prepared by this method require much more time and effort than for oligomers prepared by the synthetic procedure.

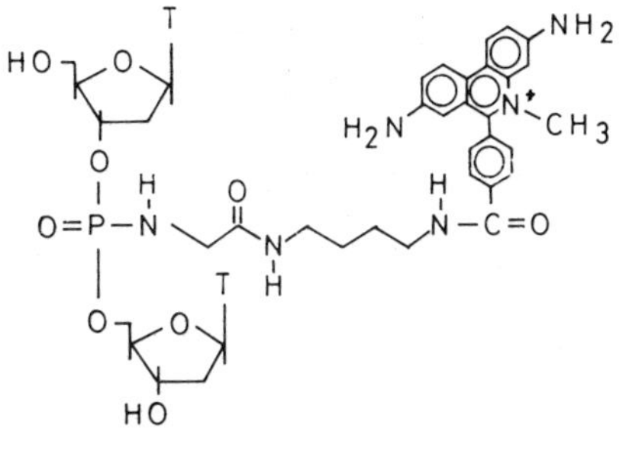

Figure 47. The structure of phenanthridinium dithymidylic acids.

5. Purification and Characterization of Oligonucleotides

5.1. Isolation of Oligonucleotides

5.1.1. Liquid Chromatography

Isolation is one of the most important aspects in the synthesis of protected oligonucleotides. In the phosphodiester synthesis, anion-exchange chromatography was required for essentially each step in order to isolate the protected intermediates as described, for example, in the synthesis of eicosamers having sequences of the tRNA gene.[24] The procedure takes a long period of time, but several improvements, including extraction methods, have been made for the isolation of diester intermediates.[156] In the phosphotriester approach, intermediates are usually isolated by partition chromatography on silica gel. Reversed-phase chromatography can be used for the separation of both phosphodiesters and phosphotriesters. Partially protected oligonucleotides, for example, 5'-*O*-dimethoxytrityldeoxyoligonucleotides and 5'-*O*-dimethoxytrityl-2'-*O*-tetrahydrofuranylribo-oligonucleotides are isolated from other less lipophilic materials by reversed-phase chromatography on alkylated silica gel (most frequently on coarse C-18 silica gel).[95] High-performance liquid chromatography (HPLC) on C-18 silica gel can also be used for the same purpose. In fact, completely deblocked oligonucleotides are often analyzed by this system and fractionated if products are contaminated with impurities. These techniques are

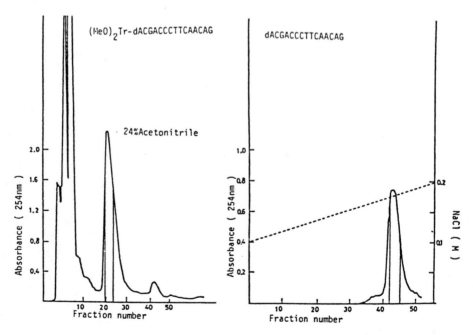

Figure 48. Chromatography of a deoxypentadecanucleotide on columns of reversed-phase (C-18 silica gel) and DEAE–Toyopearl.

most efficiently applied to products obtained by the polymer support synthesis. Oligonucleotides isolated by reversed-phase chromatography must be analyzed by ion-exchange chromatography or gel electrophoresis for their homogeneity in charges or chain lengths. Elution profiles are shown in Fig. 48 for the isolation of a deoxyribo-oligonucleotide obtained in the solid-phase synthesis by the chromatography systems mentioned above.

Gel filtration can resolve oligonucleotides of different size and can always be used to remove impurities of smaller sizes or to desalt. Low-pressure gel filtration (e.g., on Sephadex G-50) has been used for these purposes; however, it has sometimes been shown to resolve oligonucleotides not only by sizes but also by the shape of the molecules. High-pressure techniques on rigid gel provides a more precise resolution of polynucleotides.[157]

5.1.2. Gel Electrophoresis

Gel electrophoresis on polyacrylamide has been developed for the separation of oligonucleotides by charges. Relatively large oligonucleotides (e.g., 20 nucleotides) can be resolved by one charge difference in this system.[158] However, large quantities of material are not held on gel easily and recovery from the gel is not quantitative. The technique was originally applied to labeled oligonucleotides obtained by polymerase reactions in sequencing nucleotides. Up to 200 nucleotide long chains can be resolved by the electrophoresis. Theoretically, different oligonucleotides can be isolated from the gel; however, contamination between oligonucleotides may not always be avoided. The extent of contamination may be negligible for detection of bands for sequence analysis or for the isolation of products in certain instances. Large numbers of deoxyoligonucleotides synthesized on paper disks have been isolated by gel electrophoresis.[108] Polyacrylamide with high cross linkages is called "sequence gel" and is used to resolve radioactively labeled nucleotides in sequencing DNA[159] and RNA.[160] Usually, the denaturing reagents, 7 M urea or formamide, are present in the gel. For the isolation of tRNAs or rRNAs, two-dimensional gel electrophoresis has been developed.[161] Double-stranded DNA from a virus was separated by the denaturing conditions in gel electrophoresis.[162]

5.2. Identification of Oligonucleotides

5.2.1. Enzymes Used for Structural Analysis

E. coli alkaline phosphatase, which has a pH optimum of 8.3, has been used extensively to identify phosphomonoesters.[163] In contrast to 5′-nucleotidase from venom or 3′-nucleotidases that coexist in some nucleases (e.g., nuclease P1, potato nuclease), alkaline phosphatase hydrolyzes essentially any phosphomonoester. In paper chromatography or electrophoresis, oligonucleotides with phosphomono-esters can be distinguished by treatment with the phosphatase. Bacterial alkaline phosphatase is a useful enzyme for structural analyses of nucleic acids and for gene cloning. Intestine phosphatase can also be used for the same purposes and it may be deactivated more easily by heating at 60°C, at which temperature bacterial alkaline phosphatase works faster than at 37°C. Bacterial alkaline phosphatase is inactivated by denaturation only after a removal of zinc ions. This treatment is applicable in postlabeling of polynucleotides.[164] Conversion of a labeled 5'-phosphate of oligonucleotides to phosphodiesters, with joining enzymes, is detected by phosphatase insensitivity.[165]

Polynucleotide kinase is isolated from *E. coli* infected with phage T4 and catalyzes phosphorylation of the 5'-end of polynucleotides or 3'-phosphorylated nucleosides using adenosine 5'-triphosphate (ATP) as a cofactor.[166] $[\gamma^{-32}P]ATP$ can be used to label the 5'-end of polynucleotides with a kinase reaction. Phosphodiesterase from venom (VPDase) cleaves oligonucleotides from the 3'-hydroxyl end exonucleolytically to yield 5'-phosphorylated nucleosides. If the 3'-end is blocked, the hydrolysis occurs endonucleolytically but is more sluggish. The enzyme also catalyzes the hydrolysis of phosphodiesters between 5'–5', 2'–5', and phosphoramidate linkages.[167] Phosphodiesterase from spleen (SPDase) hydrolyzes oligonucleases from the 5'-end to give 3'-phosphorylated nucleosides and recognizes only 3'–5' linkages.[168] These exonucleases are useful reagents for structural analysis of short oligonucleotides. For nearest-neighbor analyses of DNA and RNA, SPDase is used in combination with an endonuclease, micrococcal nuclease, which does not hydrolyze polynucleotides to mono-nucleotides.

Nuclease P1 and similar endonucleases from a variety of sources have been found to be very useful in obtaining 5'-nucleotides from nucleic acids.[169] Ribo- and deoxyribonucleosides can be prepared from nucleic acids by using these nucleases in the presence of phosphatases. Nuclease S1 is a unique enzyme that recognizes the secondary structure of nucleic acids and cleaves single-stranded regions.[170]

RNase A from pancreas has long been known as a pyrimidine nucleotidyl transferase.[171] RNase T1 was isolated from *Aspergillus oryzae* and found to recognize only guanosine 3'-phosphate.[172] These RNases have been used extensively in structure determination of RNAs. The sequence of alanine tRNA from yeast was determined by Holley and co-workers in the first structural study using complete or partial hydrolysis by RNase A and T1.[173] RNases can be used to determine linkage integrity of synthetic RNA. Base-nonspecific RNases (RNase T2,[173] RNase M[174]), which are obtained mostly from fungi, are useful for complete digestion of RNA to provide 3'-phosphorylated nucleosides. Other RNases with base preferences can be used for partial digestion of RNA in structure determination.[175]

RNase H is unique because it hydrolyzes RNA in heteroduplex form,[176] and it has been used to cleave mRNA by hybridizing synthetic deoxyoligo-nucleotides.[177]

In contrast to the presence of base-specific RNases, only a few base preferences of DNase were known.[178] However, the discovery of restriction endonucleases that recognize specific base sequences of DNA opened the way to analyze DNA extensively.[179] These enzymes have become indispensable in the manipulation of genes and have made it possible to sequence DNA in combination with DNA sequence methods.[158,180]

5.2.2. Sequence Analyses

Chemically synthesized oligonucleotides can be characterized by the mobility shift analysis[181] for their sequence and purity. This method was developed by Sanger for the analysis of deoxyribo-oligonucleotides with chain lengths of 10–20, which could not be sequenced at that time because of the absence of base-specific DNases. The principle of the mobility shift analysis consists of a combination of electrophoresis at pH 3.5 and homochromatography (thin layer chromatography on DEAE–cellulose with a developing solution of RNA fragments) of a labeled oligonucleotide that is 5'-labeled and partially degraded by venom phosphodiesterase or nuclease P1 to give a series of mixtures.[164] Detailed procedures for the mobility shift analysis have been described.[182]

The 5'-nucleotide of oligonucleotides can be identified by comigration in paper electrophoresis (pH 3.5) with authentic 5'-nucleotides and 5'-labeled mononucleotide samples, which have been obtained by complete digestion of oligonucleotides with nuclease P1 after 5'-phosphorylation with polynucleotide kinase and $[\gamma\text{-}^{32}\text{P}]\text{ATP}$. Electrophoresis of four major mononucleotides at pH 3.5[183] is illustrated in Fig. 49.

For deoxyribo-oligonucleotides with more than 20 nucleotides, the Maxam–Gilbert method[159] can be applied to confirm the base sequence. Examples of autoradiographs obtained in the analysis of a 46 mer DNA duplex synthesized on a polymer are shown in Fig. 50. Principles and procedures for this sequencing are described in detail.[159b] In combination with restriction endonucleases, the method has been used to determine total sequences of various genes or whole genomic DNA.

Chain termination using dideoxynucleoside triphosphates, which is called the "dideoxy method," has been used extensively for sequence determination of DNA[180] cloned to phage M13 derivatives.[184] The method was found to be

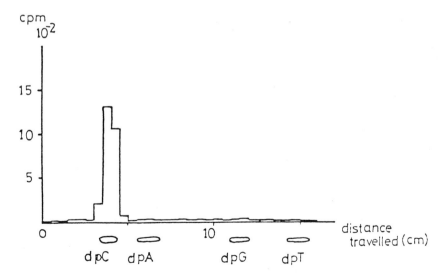

Figure 49. Electrophoresis of mononucleotides at pH 3.5.

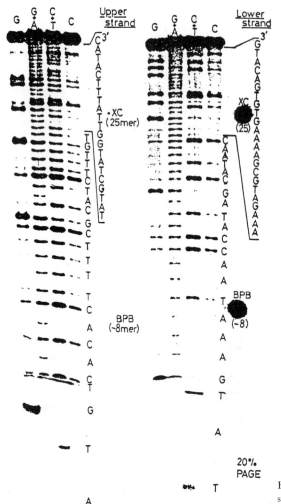

Figure 50. Autoradiograph of the sequence analysis of a 46 mer duplex.

applicable to the analysis of double-stranded DNA, including genes obtained by joining chemically synthesized fragments, after denaturation of the double-stranded DNA followed by annealing with a primer oligonucleotide.

6. Joining Reactions and Other Usage of Synthetic Oligonucleotides

6.1. Synthesis of Genes by Joining of Deoxyoligonucleotides with DNA Ligase

The discovery that DNA ligase[165] catalyzes single-strand breakages in double-stranded DNA enabled the synthesis of genes by chemically joining synthesized deoxyoligonucleotides aligned as a hydrogen-bonded duplex with single-

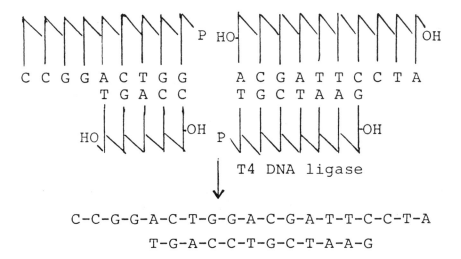

CCGGACTGG   ACGATTCCTA
   TGACC   TGCTAAG

T4 DNA ligase

C-C-G-G-A-C-T-G-G-A-C-G-A-T-T-C-C-T-A

T-G-A-C-C-T-G-C-T-A-A-G

Figure 51. Flash ligation with T4-induced *E. coli* DNA ligase.

strand breakages.[4,5] DNA ligase, isolated from T4-infected *E. coli*, requires ATP[165a] as a cofactor while the DNA ligase from *E. coli* uses NAD[165b] for activation of the 5′-phosphate of deoxyribonucleotides by adenylation.[185] The adenylated terminal phosphate is linked to the 3′-hydroxyl group of the adjacent segment on the complementary "splint" oligonucleotide. T4 DNA ligase can also join two-strand breakages as shown in Fig. 51. DNA ligase from *E. coli*, which lacks this activity, is used for selective ligation at cohesive ends.

6.1.1. Synthesis of Structural Genes

The yeast alanine tRNA gene was synthesized by this method using oligonucleotides with chain lengths of 6–20. These oligonucleotides were prepared by the phosphodiester approach.[4] Prior to this synthesis, properties of DNA ligases in reactions with synthetic short deoxyoligonucleotides had been studied to find the minimum number of hydrogen bonds as cohesive ends, optimum temperature, and base preferences.[186] It was also found that ribo-oligonucleotides were joined with T4 DNA ligase with much less efficiency.[187] Some serious side reactions, with self-complementary strands, were encountered in these early syntheses.[188] The gene for tyrosine suppressor tRNA precursor, from *E. coli*, was designed by considering joining points and optimum chain lengths of oligonucleotides.[189] The gene was synthesized and the suppressor activity of the expressed product in *E. coli* was confirmed by using phage λ with an amber mutation.[5]

The assembly method was then applied to the joining of gene fragments, which had been prepared by the phosphotriester method, for peptides. The first peptide gene synthesized by this approach was the gene for somatostatin.[6a] Amino acid codons, for the 14 amino acids, were selected by considering the contents of tRNA in *E. coli*. Since the codon usage of *E. coli* phage MS2 was known at that time, the synthetic gene was designed according to these data and expressed as a fused protein with a part of *E. coli* β-galactosidase using plasmids containing *E. coli lac* genes. Somatostatin was released by cleavage of the extra methionine placed in front of the amino terminus of the peptide by treatment with cyanogen bromide. Formation of the product was confirmed by radioimmunoassay and the theory proposed in modern molecular biology was proved to be correct. The synthesis of an insulin gene was performed by a similar procedure.[6b] Several peptide genes have been synthesized by the joining of synthetic oligonucleotides, which had been prepared either by the liquid-phase or the solid-phase synthesis.[6]

A structural gene for α₁-interferon, which had 564 base pairs, was synthesized by ligation of oligonucleotides ranging in size from 14 to 15 and prepared by the triester solid-phase method.[6e] This indicated that relatively large proteins could be synthesized in bacteria by a chemical–enzymatic synthesis of the corresponding genes, providing chemically synthesized fragments could be obtained rapidly. A gene coding for human growth hormone, consisting of 192 amino acids, was synthesized by ligations of 78 deoxyribo-oligonucleotides, which

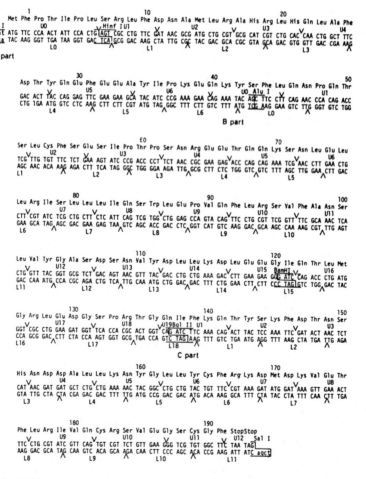

Figure 52. Nucleotide and amino acid sequences for a synthetic gene for human growth hormone.

Table V. Synthetic Structural Genes

Gene product	Synthetic method	Promoter used
1 Yeast alanine tRNA	Phosphodiester[a,(4)]	
2 E. coli suppressor tRNA precursor	Phosphodiester[a,(5)]	Tyrosin tRNA[(190b)]
3 Somatostatin	Phosphotriester[a,(6a)]	E. coli lac UV5[(190c)]
4 Insulin	Phosphotriester[a,(6b)]	E. coli lac UV5[(190c)]
5 Angiotensin	Phosphodiester[a,(6b)]	
6 Thymosin	Phosphotriester[a,(6d)]	E. coli lac UV5[(190c)]
7 Secretin	Phosphotriester[b,(6g)]	pBR322 β-lactamase[(190c)]
8 Poly(L-aspartyl-L-phenylalanine)	Phosphotriester[a,(191a)]	E. coli trp
9 α₁-Interferon	Phosphotriesteer[b,(6e)]	E. coli trp, synthetic[(195)]
10 γ-Interferon	Phosphotriester[b,(6f)]	E. coli lac[(190c)]
11 γ-Interferon	Phosphite[b,(197)]	T5, synthetic[(196)]
12 Human β-urogastron (EGF)	Phosphotriester[a,(6i)]	E. coli trp[(190f)]
13 α₂-Interferon	Phosphotriester[b,(191b)]	E. coli trp, synthetic[(195)]
14 Proinsulin	Phosphotriester[a,(191c)]	Yeast GAL[(190g)]
15 Human EGF	Phosphite[b,(6h)]	Yeast glyceraldehyde-3-phosphate dehydrogenase[(6h)]
16 Human growth hormone	Phosphotriester[b(6j)]	E. coli trp, modified[(6j)]
17 Eglin C	Phosphotriester[b,(191d)]	E. coli trp[(190e)]
18 RNase S	Phosphite[b,(191e)]	
19 RNase T₁	Phosphotriester[b,(191f)]	E. coli trp, modified[(6j)]

[a] In solution.
[b] Solid phase.

had been prepared on polymer supports.[(6j)] The gene was designed to construct with frequently occurring amino acid codons of *E. coli* and contained several restriction sites. Figure 52 shows the synthetic HGH gene. The ligated synthetic duplex was joined to an *E. coli* plasmid carrying the *E. coli* tryptophan promoter[(190a)] and expressed in *E. coli* efficiently upon induction with indole acrylic acid. Other peptides with more than 100 amino acid residues have been produced in *E. coli* or yeast using enzymatically joined synthetic oligonucleotides that were linked into plasmids containing proper promoters.[(191)] Table V summarizes synthetic methods for some of these constructed structural genes. Promoters used for expression of these genes are also listed.[(190)]

By the use of chemically synthesized genes, proteins of around 200 amino acids may be designed and produced in microorganisms by the recombinant DNA technique. Site-directed mutations of amino acids or domains can be performed specifically by replacing nucleotides in the corresponding region of the gene fragments. Mutations can be introduced by using mismatched synthetic oligonucleotides.

6.1.2. Synthesis of Promoters and Operators

Nucleic acids are recognized by proteins to process expression of genetic information. Proteins involving replication or transcription of DNA must recognize fine structures of DNA derived from their base sequence. The base sequence of *E. coli* lactose operator, which is the binding site for the lactose repressor, was sequenced in the early 1970s.[192] The amino acid sequence of the repressor has also been established. Several studies on the interactions of the operator–repressor were performed by synthesizing a deoxyoligonucleotide duplex that had sequences for the operator.[81,193a,b]

The base sequences of promoters that are recognized by RNA polymerase have been determined in various genes. Two consensus sequences located upstream (-10 and -35) of the initiation site for transcription have been proposed by Pribnow and other groups.[193b] To investigate the correlation of these base sequences for the efficiency of gene expressions, synthetic duplex pentadecamers, which contain ideal sequences for the -10 region, have been prepared[47] and ligated to a plasmid pBR322 at its promoter site for a tetracycline-resistant gene.[194] The distance between the consensus sequences was altered and a maximum transcription was observed at the spacer length of 17 base pairs between the -10 and -35 regions.

For the construction of expression vectors for foreign genes in *E. coli*, duplexes containing base sequences of strongly inducible promoters in *E. coli* have been constructed by ligation of chemically synthesized deoxyoligonucleotides. A duplex with 81 base pairs containing *E. coli trp* promoter was synthesized and used for the expression of the synthetic α_1-interferon gene.[195] The secondary structure formed in this region, and the structural gene seemed to affect the efficiency of expression. A bacteriophage T5 (T5P25) early promoter, with 46 base pairs, was synthesized by joining synthetic fragments obtained by the phosphite method. This early promoter was then ligated to the tetracycline-resistant gene of pBR322 at restriction sites for *Eco*RI and *Hind*III.[196] The vector was used to express a synthetic γ-interferon gene with high efficiency.[197]

6.2. Use of Oligonucleotides as Probes or Primers for Sequencing DNA

Synthetic deoxyoligonucleotides are used as primers for reverse transcription of messenger RNAs to form complementary DNA (cDNA), which can then be sequenced for the coding region of DNA.[198] These primer DNAs complementary to mRNAs are also useful as probes in hybridization experiments to identify homologous DNA regions or to detect DNA fragments containing homologous sequences.[199] It was found that a chain length of at least 11 nucleotides was necessary to serve as a primer.[198] As a probe for hybridization, at least 14 nucleotide long oligonucleotides are required to distinguish it from nonspecific bindings. In most cases, 17 mer or longer probes are used for blot hybridization. When only a sequence for amino acids is known, the structural gene can have various base sequences owing to the degeneracy of amino acid codons. Methionine and tryptophan can provide unique base sequences. An example for a primer DNA complementary to a possible messenger RNA of Glu-Trp-Ile-Trp corresponding to the sequence (9–12) of bacteriorhodopsin is shown in Fig. 53. Oligonucleotides, with mixed base sequences d(CCAGATCCACTC), were used for the preparation of cDNA in the gene sequencing.[85] Although these primer

Glu-Trp-Ile-Trp

mRNA 5' GA^A_G-UGG-AU^U_C-UGG 3'
                             A

Figure 53. Possible cDNA sequence for Glu-Trp-Ile-Trp.

primer DNA 3' CT^T_C-ACC-TA^A_G-ACC 5'
                                T

oligonucleotides were previously synthesized in a liquid phase, the solid-phase triester methods were employed more often, with developments in the small scale, for rapid synthesis on polymer supports. Machines can be used either in the phosphodiester or in the phosphoramidite method for the synthesis of deoxyoligonucleotides with mixed base sequences. Premixed intermediates for these syntheses are now commercially available. These probe DNAs are essential compounds for biological studies on DNA and availability of oligonucletoides and should lead to rapid findings in gene structures.

6.3. Use of Deoxyoligonucleotides on Site-Directed Mutagenesis

Base sequences in DNA can be mutated by a replication with DNA polymerase using mismatched synthetic deoxyoligonucleotides. Mutation in structural genes should result in alterations of the amino acid sequence of proteins.

Smith and co-workers have demonstrated site-directed mutagenesis in the phage $\phi \times 174$ gene for E protein using oligonucleotides d(GTATCCCACAAA) and d(GTATCCTACAAA).[200a] Similarly, the DNA of an amber mutant of phage $\phi \times 174$ DNA, whose total base sequence had been determined, was used as a target of site-directed mutagenesis.[200b] These chemically synthesized oligonucleotides were hybridized to the minus strand of the amber mutant DNA. A 17 mer having the complementary sequence to the wild-type minus strand affords revertant in a yield of 30%. Shorter mismatched primers may be unable to form a stable duplex. Correlation between abilities as primers for replication or transcriptions and strength of hydrogen bonds is discussed in a review article.[201]

Synthetic oligonucleotides can also be used as probes in colony hybridization for the selection of a mutant that contains the incorporated synthetic fragments. An intervening sequence of tyrosine suppressor tRNA gene in yeast has been removed by using a synthetic 21 mer DNA complementary to the corresponding region except the intervening region. The experiment was performed by incorporation of the gene into an *E. coli* plasmid pBR322, followed by mispairing to the single-stranded plasmid obtained by *Eco*RI digestion in the presence of ethidium bromide.[199a] The plasmid containing the deleted gene was selected by colony hybridization. The active tRNA was then produced in *E. coli* harboring these plasmids, indicating that the intervening sequence in the tRNA gene was not essential for expression of the gene.

Another example of a site-directed mutagenesis in a tRNA gene revealed that a structural gene served as a promoter.[202] *E. coli* suppressor tRNA gene, cloned to a single-stranded phage M13mp3, was mutated at the common GTAC region to GATC using a synthetic d(TCCATCCTTC), and the transcription of the gene was found to be repressed by the mutation.

Mutations in structural genes for proteins should provide an extremely useful method for studies on the structure–function relationship of proteins. Several mutated proteins have been produced by changing DNA sequences in their structural genes.[15,203]

6.4. Joining of Ribo-oligonucleotides with RNA Ligase

RNA ligase was discovered in phage T4-infected *E. coli* as an activity to circularize polyadenylates.[204] ATP was found to adenylate the enzyme[205] and adenylated 5′-phosphorylated oligonucleotides were identified as active intermediates.[206] A comprehensive review on this enzyme has been published with some earlier references.[16] The discovery of RNA ligase provided a method for joining single-stranded ribo-oligonucleotides to yield larger molecules such as tRNA. Although preferences in chain length and base sequences have been found,[207] this enzyme has served as a useful tool for the elongation of chemically synthesized ribo-oligonucleotides. The nascent formyl methionine tRNA of *E. coli* has been synthesized by joining synthetic ribo-oligonucleotide fragments with RNA ligase.[17] Figure 54 shows the structure of the tRNA and joined

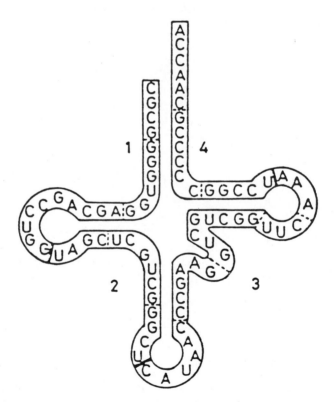

Figure 54. Nascent strand of tRNA$_f^{Met}$ constructed by joining of synthetic fragments.

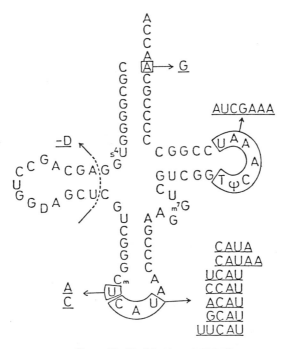

Figure 55. Modification of tRNA$_f^{Met}$.

oligonucleotides prepared by the phosphodi- or triester approach. Yeast alanine tRNA has been synthesized by the joining of small oligonucleotides, including minor nucleosides, which were obtained by either chemical or enzymatic methods.[18]

The joining technique is believed to be useful for the construction of entirely changed tRNA molecules. Examples of several modifications on a tRNA are shown in Fig. 55. *E. coli* formylmethionine tRNA can be cleaved at the anticodon by limited digestion with RNase A. The anticodon loop was modified by replacement with trimers complementary to the nonsense codons UAG and UAA.[208] Loop size of the anticodon and the constant uridine located at the 5'-side of the anticodon have been changed by a similar procedure.[209] Yeast phenylalanine tRNA has also been cleaved at the anticodon loop by partial digestion with RNase A and used for modification.[210,211] Interaction of these analogues with proteins such as aminoacyl tRNA synthetases or ribosome subunits has been investigated.[212] RNA ligase has also been used to construct model substrates for ribosomal RNA maturation endonucleases.[213]

7. Structure and Properties of Oligonucleotides

7.1. Crystal Structure of Oligonucleotides

Nucleic acid structures determined by x-ray crystallography have provided the basis for understanding physical and biological properties of nucleic acids since the proposal of a DNA double-helix model by Watson and Crick.[214] Early crystallographic works were done solely on fibers of nucleic acids and, therefore, the "deduced" structures have some ambiguity (even in the handedness of DNA helices[215]) because of the poor resolution. Nevertheless, it was established from those works that DNA mainly takes two families of double-helical conformations, B form and A form, whereas RNA only takes the A-form structure.[216]

The first successful x-ray diffraction analyses on single crystals of double-helical nucleic acids were achieved with ribodinucleoside monophosphates, AU[217,218] and GC[219,220], and phenylalanine tRNA[221–226] in 1973. No other detailed crystal structure of ribo-oligonucleotide double helix has been published since then. In 1979, a new DNA structure, left-handed Z form, was found in crystals of d(CGCGCG) by Wang *et al.*[227] This discovery created a new field in nucleic acid research ranging from physics to biology. In the following years, many crystal structures of deoxyoligonucleotide duplexes including Z, A, and B forms were published. They are listed in Table VI together with other

Table VI. Crystal Structure of Oligonucleotides

Oligonucleotide		Reference
Single-stranded or unusual duplex		
1. (2′–5′)AU		228
2. UA		229, 230
3. UA		231
4. d(pTT)		232, 233
5. AAA		234, 235
6. d(TA)		236
7. d(ATAT)		237, 238
8. AA		239
9. A^s-A^s		240
10. I^s-A^s		241
11. (2′–5′)AC		242
12. d(CG)		243
Double-stranded		
13. AU		217, 218
14. GC		219, 220
15. d(CGCGCG)	(Z form)	227, 244
16. d(CGCG)	(Z form)	245
17. d(CGCG)	(Z form)	246
18. d(CGCGAATTCGCG)	(B form)	247–251
19. Z-DNA		252
20. d(GGCCGGCC)	(A form)	253
21. r(GCG)–d(TATACGC)	(A form)	254
22. A-, B-, and Z-DNA	(Review)	255
23. d(CGCGAATTbr⁵CGCG)	(B form)	256
24. d(m⁵CGm⁵CGm⁵CG)	(Z form)	257
25. d(io⁵CCGG)	(A form)	258, 259
26. d(m⁵CGTAm⁵CG)	(Z form)	260
27. d(GGTATACC)	(A form)	261
28. Cm⁸GCm⁸G, Cbr⁸GCbr⁸G	(Z-RNA)	262

Table VII. Averaged Torsion Angles and Sugar Conformation

	χ	α	β	γ	δ	ε	ζ	Sugar conformation
B-DNA[a]	−102	−41	136	38	139	−133	−157	C2'-*endo*
d(CGCGAATTCGCG)[b]	−117	−63	171	54	123	−169	−108	C1'-*exo*, C2'-*endo*
A-DNA[a]	−154	−90	−149	47	83	175	−45	C3'-*endo*
d(GGTATACC)[c]	−160	−62	173	52	88	−152	−78	C3'-*endo*
r(GCG)d(TATACGC)[d]	−162	−69	175	55	82	−151	−75	C3'-*endo*
d(CGCGCG)[e] dC	−159	−137	−139	56	138	−94	80	C2'-*endo*
dG	68	47	179	−169	99	−104	−69	C3'-*endo*
RNA-11[f]	−164	−62	180	48	83	−151	−74	C3'-*endo*
tRNA[g]	−162	−74	171	61	84	−156	−73	C3'-*endo*

[a] Fiber data taken from ref. 263.
[b] Taken from ref. 249.
[c] Taken from ref. 261.
[d] Taken from ref. 254.
[e] Taken from ref. 244.
[f] Taken from ref. 264.
[g] Crystal data for amino acid acceptor stem of tRNA[phe]; taken from ref. 265.

oligonucleotides that have structures solved by x-ray analysis. The averaged torsion angles of the monomer unit for selected nucleic acid double helices are presented in Table VII. For definitions of the conformational parameters, see Ref. 266: Recommendations 1982 by IUPAC–IUB Joint Commission on Biochemical Nomenclature (Fig. 56).

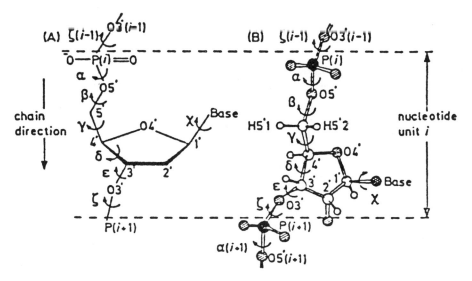

Figure 56. Section of a polynucleotide backbone showing the atom numbering and the notation for torsion angles.[53] (A) Conventional representation; (B) absolute stereochemistry. χ is defined by dihedral angles formed by O4'-C1'-N9-C4 for purine and by O4'-C1'-N1-C2 for pyrimidine derivatives.

7.1.1. B-DNA

So far a B-DNA structure is only found in self-complementary deoxydo-decanucleotide crystals with a base sequence of CGCGAATTCGCG.[247-251,256] From fiber diffraction studies of DNA, it was assumed that DNA makes up a right-handed double helix and the monomer unit in B-DNA takes a C2'-*endo*-furanose puckering conformation ($\delta \approx 140°$) and *anti*-glycosidic conformation ($\chi \approx -100°$) (Table II). The largest discrepancy from the fiber data is seen in ζ, a torsion angle about the $O(3')$—P bond. The crystal data show that the combination of torsion angles about the $O(3')$—P and P—$O(5')$ bonds, ζ and α, is *gauche⁻–gauche⁻*, which is the same as that of the A form, rather than *trans–gauche⁻*. The furanose ring conformations of the dodecamer are mainly C1'-*exo* (11/24) and C2'-*endo* (9/24) forms that belong to a S-type conformation.[267] The crystal data also show that the local helical parameters—twist, tilt, and roll—are influenced by the base sequence[249] and the sugar conformations are heterogeneous dependening on the type of base and on base pairing.[250] Another characteristic of the B-DNA is that the minor groove is hydrated in an extensive and regular manner with a zigzag spine.[248] A B-DNA helix contains 10 base pairs per turn.

7.1.2. A-DNA

So far, A-DNA structures are found in crystals of three deoxyoligonucleotide duplexes, d(io⁵CCGG)[258,259], d(GGCCGGCC),[253] and d(GGTATACC).[261] The averaged torsion angles of a monomer unit for d(GGTATACC) are presented in Table II. From fiber diffraction studies it is assumed that the monomer unit in A-DNA takes a C3'-*endo*-furanose puckering conformation ($\delta \approx 80°$) and *anti*-glycosidic conformation ($\chi = -155°$), which is quite different from that in B-DNA ($\chi \approx 100°$). These predictions are in good agreement with the crystal data ($\chi = -160°$, $\delta = 88°$). The mean values for d(io⁵CCGG)($\chi = -157°$, $\delta = 89°$) and for d(GGCCGGCC)($\chi = -149°$, $\delta = 98°$) are also similar to those data. When compared with the crystal data of B-DNA, the torsion angles other than χ and δ are almost identical except for the ζ's, which are nevertheless in the same *gauche⁻–gauche⁻* domain. Therefore, we can now say that differences in the backbone torsion angles χ, δ, and ζ mainly make up the difference between overall structures of A-DNA and B-DNA helices. The most definitive difference between the two forms is the positioning of base pairs relative to the helix axis.[259] In B-DNA, the base pairs sit on the central helix axis, whereas in A-DNA the base pairs are pushed toward the surface of the helix and the helix axis passes by on the major groove side of each base pair (Fig. 57 and 58). An A-DNA helix contains 11 base pairs per turn.

7.1.3. Z-DNA

Z-DNA was first found in crystals of d(CGCGCG)[227,244] and was later found in crystals of d(CGCG),[245,246] d(m⁵CGm⁵CGm⁵CG),[257] and d(m⁵CGTAm⁵CG).[260] They all contain an alternating pyrimidine–purine

B-DNA   A-DNA   Z-DNA

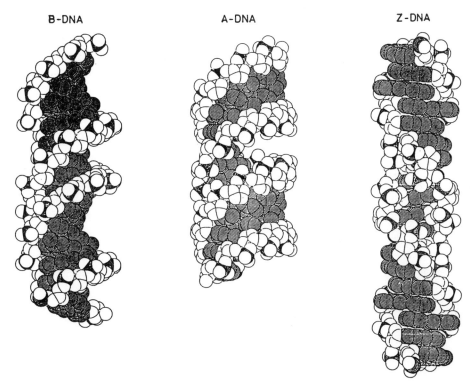

Figure 57. Side view of B-, A-, and Z-DNA helices in space-filling representations. The drawings are kindly supplied by Dr. S. Fujii.

sequence. Z-DNA is a left-handed double helix in contrast to the right-handed helices of B- and A-DNAs. The monomer units take alternating conformations that are quite different from those of A- and B-DNAs. The dC residue takes the C2′-*endo*-furanose puckering conformation ($\delta = 138°$) and *anti*-glycosidic conformation ($\chi = -159°$), while the dG residue takes C3′-*endo* ($\delta = 99°$) and *syn* ($\chi = 69°$) conformations (Table VII; see also Fig. 59). The conformations about

B-DNA   A-DNA   Z-DNA

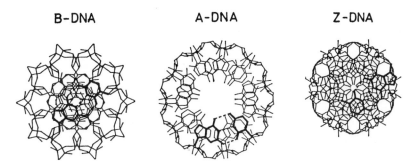

Figure 58. Top view of B-, A-, and Z-DNA helices. Guanine–cytosine base pairs are shown in heavy lines. The drawings were kindly supplied by Dr. S. Fujii.

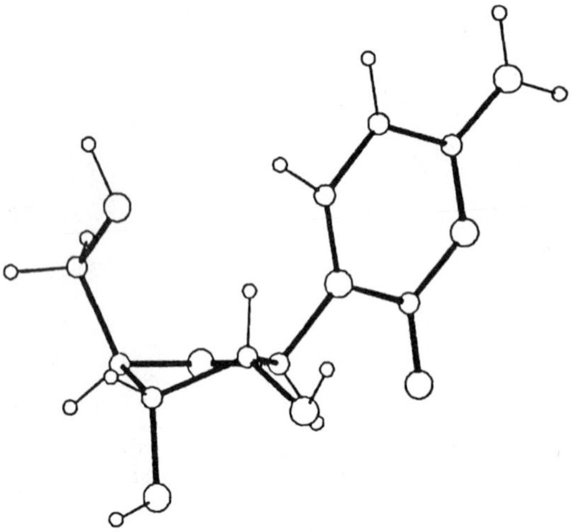

RIBO CYTIDINE WITH C2'-ENDO AND ANTI

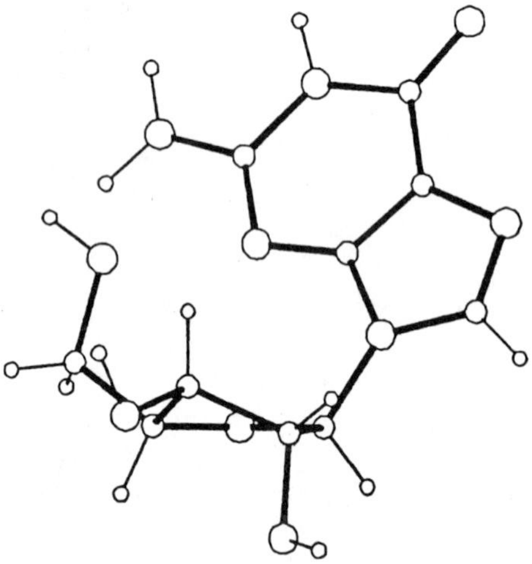

RIBO GUANOSINE WITH C3'-ENDO AND SYN

Figure 59. Conformations of cytidine and guanosine residues in Z-RNA. The drawings were kindly supplied by Dr. S. Fujii.

P–O bonds (*gauche*⁺–*gauche*⁺ for dC–dG and *gauche*⁻–*trans* for dG–dC) are quite different from those of B- and A-DNAs.

The G : C pairs are located on the outer wall of the helix and the helix axis passes by on the minor groove side of each base pair (Fig. 58). The deoxyribose-phosphate backbone follows a zigzag course, from which the name of "Z"-DNA originated. A Z-DNA helix contains 12 base pairs per turn.

7.1.4. RNA–DNA Hybrid and RNA

The crystal structure of r(GCG)–d(TATACGC) has been analyzed by Wang *et al.*[254] This RNA–DNA hybrid molecule forms an A-form duplex with 11 base

Z-RNA HEXAMER (SIDE VIEW) 120 DEG.

Z-RNA HEXAMER (SIDE VIEW) 210 DEG.

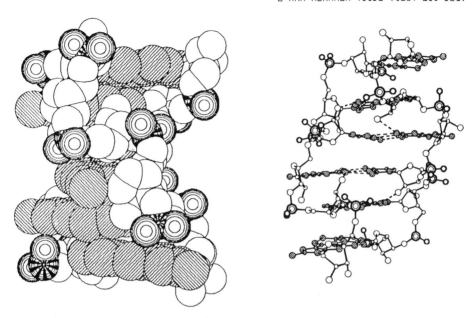

Figure 60. Crystal structure of $Cm^8GCm^8GCm^8G$ duplex (Z-RNA)[56] in space-filling and ball-and-stick representation. The hatched atoms belong to base residues including 8-methyl groups. The broken lines show hydrogen bondings between bases and 2'-OH (acceptor) of cytidine residues and 2-NH_2 (donor) of 8-methylguanosine residues.

pairs per turn. The torsion angles are almost identical with those of A-DNA crystals of d(GGTATACC) (Table VII).

No detailed crystal structures of synthetic ribo-oligonucleotide duplexes have been published except for AU and GC. The double-helical stem regions of tRNA form an A-form helix with 11 base pairs per turn. The torsion angles are surprisingly similar to those of A-DNA (Table VII).

Preliminary x-ray diffraction studies on crystals of GGCUp have been reported.[268] Two GGCUp strands form a short antiparallel double helix with G : U base pairs at the ends of it.

Quite recently, Z-RNA structure has been found in crystals of Cbr^8GCbr^8G[262] and $Cm^8GCm^8GCm^8G$[269] (Figs. 59 and 60).

7.2. Conformation of Oligonucleotides in Solution

Conformation of oligonucleotides can be studied by various spectroscopic techniques such as UV, CD, NMR, and IR. Among these techniques, NMR, especially ^{1}H-NMR, can provide the most detailed structural information. A number of papers on NMR studies of deoxyoligonucleotide duplexes were published during 1983 and 1984. The explosive increase of activities in this field may be due to the recent progress in NMR spectroscopy, including hardware and

Table VIII. NMR Studies on Oligonucleotides

Oligonucleotide	Reference
Single-stranded, deoxyribo-	
1 d(AAA)	274
2 d(ATGT), d(ACATGT)	275
3 d(ATGT)	276
4 d(CCAAG), d(CTTGG), d(CCA), d(CCAA), d(TTG), d(TTGG)	277
5 d(CCA), d(TGG), d(CCAA), d(TTGG), d(CCAAG), d(CTGG), d(CCAAGA), d(CCAAT)	278
Double-stranded, deoxyribo-	
6 d(AACAA) : d(TTGTT)	279
7 d(ATGCAT), pd(GC)$_4$, pd(TA)$_4$, pd(AG)$_4$: pd(CT)$_4$	280
8 d(ATGCAT)	281,282
9 d(AAAGCTTT)	283
10 d(CGCG), d(CGCGCG)	284
11 d(CCGG), d(GGCC)	285
12 d(C$_{15}$A$_{15}$) : d(T$_{15}$G$_{15}$)	286
13 d(GGAATTCC)	287
14 d(GGAATTGTGAGCGG) duplex d(GGAATTG) duplex d(TGAGCGG) duplex (parts of *lac* operator)	288
15 d(CGCG), d(CGCGCG)	289
16 d(CGCGAATTCGCG)	290–292
17 d(CGCAGAATTCGCG)	291–293
18 d(CGTGAATTCGCG)	291, 292, 294
19 d(GAATTCGCG)	295, 291
20 d(CCAAGCTTGG)	296
21 d(TGAGCGG) : d(CCGCTCA)	297
22 d(CGATTATAATCG) duplex (promoter)	298
23 d(CA$_3$CA$_3$G) : d(CT$_6$G)	299
24 d(GGTATACC)	300
25 d(G-C)$_3$	301
26 d(G-C)$_5$	302
27 d(CTAG)	303
28 d(TATCACCGCAAGGGATA) duplex (O$_R$3 operator)	304
29 d(TA)$_5$	305
30 d(GGm^{5}Cm^{5}CGGCC)	306
31 d(CACGTG), d(GTGCAC)	307
32 d(CCAAGATTGG)	308
33 d(TATCACCGCAAGGGATAp) duplex (O$_R$3 operator)	309
34 d(TCACAT) : d(ATGTGA)	310
35 d(ATATCGATAT), d(ATATGCATAT)	311
36 d(ATATCGATAT)	312
37 d(CGTACG), d(ACGCGCGT)	313
38 d(CGCGAATTCGCG)	314
39 O$_R$3 17-mer	315
40 d(CGCG) (P-^{17}O derivative)	316
41 d(TGAGCGG) : d(CCGCTCA)	317
42 d(m^{5}CGCm^{5}CG)	318

Table continued

Table VIII. (continued)

	Oligonucleotide	Reference
43	d(m^{5}CGm^{5}CGm^{5}CG)	319
44	d(CGCGTATACGCG)	320
	d(CGCGAATTCGCG)	
45	d(TGAGCGG) : d(CCGCTCA)	321
46	d(AAAGTGTGACGCCGT) duplex (CAP binding site)	322
47	d(TACCACTGGCGGTGATA) duplex	323
	(OL^1 operator)	
48	d(CGTTATAATGCG) duplex (consensus Pribnow promoter)	324, 325
49	d(GGAATTCC)	326
50	d(CGm^{5}CG), d(CGm^{5}CGCG)	327
51	d(AAGTGTGACAT) duplex (CAP binding site)	328
52	d(CGCGAATTCACG)	329
53	d(CGAGAATTCGCG)	330
54	d(CGCG), d(CGCGCG)	331
55	d(TAAT)	332
56	d(GGCGGAATTGTGAGCGC) duplex	333
	d(GGATAACAATTTGTGGC) duplex	
	d(GGTGAGCGGATAACAGC) duplex	
	(parts of *lac* operator)	
57	d(TATTAATATCAAGTTG) duplex	334
	(gene A protein binding site)	
	Single-stranded, ribo-	
58	A$_n$ ($n = 2 - 5$)	335
59	AAA	336
60	CCA, ACC	337
61	CAX, AGX, (X = A,G,C,U)	338
62	CXG, AXG, CAXUG, AGXC, AGXCU (X = A,G,U,C)	339
63	AUCCA	340
64	CAUAUG, AUGCUA, UCAUGA, AAGCUU	341
65	m$_2^6$AUm$_2^6$A	342
66	ACGU, CAUG, AUGUA	343
67	AACC	344
68	A$_n$G ($n = 2$–4), GAAp, AGAAA, AAAAAAp, AGA,	345
	A$_7$Gp, A$_{11}$Gp	
69	AA, AAC, UAA, UAG, ACG, AUG, GUG,	346
	ACC, AUU, UAU, CUG, UUG, UUC	
70	UUC	347
71	m$_2^6$AUm$_2^6$AUm$_2^6$A	348
72	AACC	349
	Double-stranded, ribo-	
73	CCGG	350
74	AAGCUU	351, 352
75	GAGC : GCUC	353
76	CAUG	354
77	GCA	355
78	AGCU, ACGU	356
79	GGCUp, G*GCUp (G* : ^{15}N-enriched)	357, 358
80	GCX(X = m^{6}A, m$_2^6$A, m^{1}G)	359

Table continued

Table VIII. (continued)

Oligonucleotide	Reference
81 CCGGAp, ACCGGp, ACCGGUp, CCGGp	360, 361
82 AAGCUU	362
83 CGCG, CGCGCG	363
84 AGCU	364
85 GGCC, CCGG, CGGC, GCGC, CGCG	365
86 CGCGCG	366
87 Cbr^{8}GCbr^{8}G, Cm^{8}GCm^{8}G	262
Double-stranded, DNA–RNA hybrid	
88 r(C$_{11}$)–d(C$_{16}$) : d(G$_n$)	367
89 d(CT$_5$G) : r(CA$_5$G)	368
r(CU$_5$G) : d(CA$_5$G)	
90 d(CT$_5$G) : r(CA$_5$G)	369
91 d(CG)–r(CG)–d(CG)	370
92 r(GCG)–r(TATACGC)	371

software, and in deoxyoligonucleotide synthesis. Works (since 1974) on oligo-nucleotides (larger than dimer) by NMR spectroscopy are listed in Table VIII.

The pioneering ^{1}H-NMR studies on all possible diribonucleoside mono-phosphates, except GpG, were reported by Ts'o *et al.*[270] Complete assignment of the proton signals for those ribodimers was reported by Lee *et al.*[271] and Ezra *et al.*[272] with the aid of high-field NMR instruments. A complete assignment of proton signals for 16 possible dideoxyribonucleoside monophosphates has been reported.[273]

7.2.1. Conformation of DNA

Many papers on DNA duplexes have been published.[279–334] Most of the duplexes are those of self-complementary deoxyoligonucleotides. Some of them contain imperfect or nonordinary self-complementary sequences. G : T and G : A pairs are found in d(CGTGAATTCGCG),[291,292,294] d(CCAAGATTGG),[95] and d(CGAGAATTCGCG).[330] Extra adenine in d(CGCAGAATTCGCG) stacks into the duplex[291–293] as if it were an intercalator. Extra thymine in d(CGTCG) is also assumed to stay inside the double helix.[372] On the other hand, the unpaired cytosine in d(CA$_3$CA$_3$G) : d(CT$_6$G) is not stacked in the helix.[299]

The duplexes formed by two different strands contain the recognition sequen-ces for proteins such as repressor, catabolite activator protein (CAP), and RNA polymerase. Some of the duplexes containing alternating pyrimidine–purine sequences adopt a Z-form structure in high salt solution.[289,318,319,327] d(CGCG) and d(CGCGCG) form a Z-DNA duplex in 4 M NaCl[289] while they form a B-DNA in 0.1 M NaCl[289,331,373] (Fig. 61). It is noteworthy that Z-DNA crystals can be obtained from a low-salt solution of oligo(dC-dG)s.[227,245] Therefore, it should be kept in mind that the crystal structure is not always the same as the solution structure. A similar discrepancy is also observed between the crystal

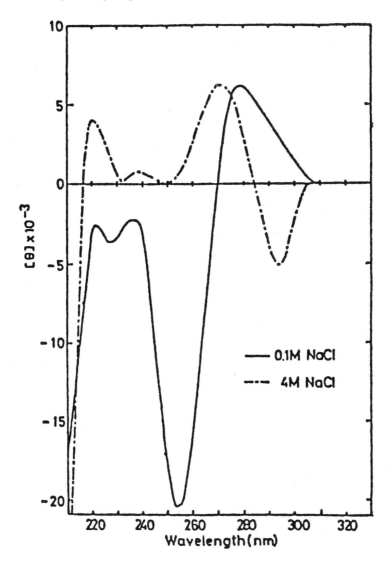

Figure 61. CD spectra of d(CGCGCG)(1 A_{260} unit/ml): —, 0.1 M NaCl, pH 7.5, 5°C; - - -, 4 M NaCl, pH 7.5, 5°C.

(A-DNA)[261] and solution (B-DNA) structure of d(GGTATACC). All other DNA duplexes adopt a B-form structure in solution.

7.2.2. Conformation of RNA

Single-stranded ribo-oligonucleotides take a helical structure similar to that of the A-RNA form by base stacking. The stability of stacking is influenced by substituents on the base.[342,348] It is assumed that very stable stacking between m^6A residues bulges out the interior uridine residues in m^6AUm^6A and m⁶AUm⁶AUm⁶A. From conformational studies on dinucleoside monophosphates containing various 2′-substituents, the stability of stacking in the A form is determined by electronegativity, size, and hydrophobicity of the substituent.[374-377] A 2′-fluoro substituent, with high electronegativity, greatly stabilizes the stacking conformation.

    RNA duplexes almost exclusively take the A-form structure. Poly- and oligo-r(C-G)s show an unusually large, negative CD band in the long-wavelength region (280–310 nm),[378,363] which is very similar to that of Z-DNA[379,289] (Fig. 62). The CD spectrum is also similar to that of the C-br⁸G duplex, in which

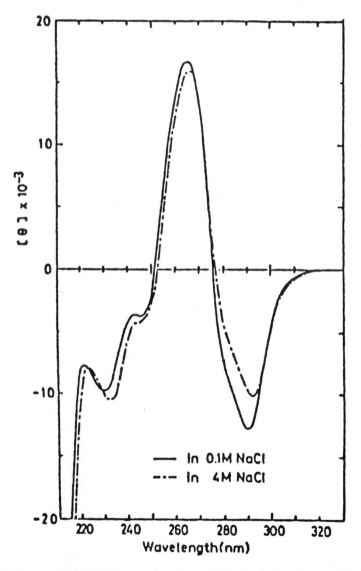

Figure 62. CD spectra of CGCGCG (1 A_{260} unit/ml); —, 0.1 M NaCl, pH 7.5, 5°C; - · -, 4 M NaCl, pH 7.5, 5°C.

the 8-bromoguanosine residue is assumed to take a *syn* conformation.[380] Therefore, it can be assumed that poly- and oligo-r(C-G)s could take a Z-form-like structure. However, ^{1}H- and ^{31}P-NMR studies on CGCG and CGCGCG showed that they take an ordinary A-form structure even in 4 M NaCl despite their unusual CD spectra.

As mentioned above, C-br^{8}G forms a stable duplex, which is assumed to take the same conformation as the CpG part of the Z-form RNA.[380] For the characterization of Z-RNA duplexes, C-G-C-G analogues containing 8-bromoguanosine or 8-methylguanosine residues were synthesized and examined by UV, CD, and NMR spectroscopy.[262] Cbr^{8}GCbr^{8}G and Cm^{8}GCm^{8}G indeed form a Z-RNA duplex in 0.1 M NaCl that shows a positive CD band in the long-wavelength region (Fig. 63). These results suggest that oligo(C-G) sequences in a natural RNA duplex could form left-handed duplexes with the aid of some factor such as protein, salt, or supercoiling as observed in the case of DNA. In fact, it was shown quite recently that poly(G-C)[381] and oligo(G-C)[382] take a Z-RNA form in 6 M NaClO$_{4}$.

Alkema *et al.*[355] and D'Andrea *et al.*[359] showed that a GC duplex can be stabilized by 3'-dangling extra purine nucleoside residues. The dangling bases are assumed to increase base stacking and thus improve overall duplex stability. The

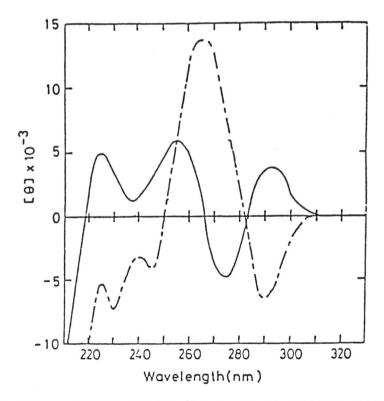

Figure 63. CD spectra of CGCG and Cbr^{8}GCbr^{8}G (1 A$_{260}$ unit/ml) in 0.1 M NaCl, pH 7.5 at 1°C; ---, CGCG; —, Cbr^{8}GCbr^{8}G.

stabilization effects of 3′-dangling nucleotide residues on GGCC and CCGG duplexes are also reported.[383,384]

Imino proton signals of the G : U wobble base pair have been unambiguously assigned by ^{1}H- and ^{15}N-NMR analysis of GGCUp and G*GCUp (G* : ^{15}N-labeled guanosine).[357,358]

Some papers on RNA–DNA hybrid duplexes that contain RNA in one strand or in both strands have been published.[367,371] However, little conformational details have been elucidated.

7.3. Interaction of Oligonucleotides with Drugs

7.3.1. Crystal Structures of Drug–Oligonucleotide Complexes

Crystal structures of many dinucleoside monophosphate–drug commplexes have been solved by x-ray crystallography (Table IX).[385–398,400] The dinucleoside monophosphates are mostly ribodimers that contain pyrimidine–purine base sequences. The drugs are mostly tri- or tetracyclic aromatic compounds that intercalate between base pairs of DNA and/or RNA (Fig. 64). Most of the dimer–drug complexes have common conformational features, alternating sugar puckering, C3′-*endo* (3′–5′)C2′-*endo* puckering conformation, and unwinding of the helical twist.[389,393] Proflavine–ribodimer complexes show exceptional features, C3′-*endo*(3′–5′)C3′-*endo* puckering conformation and no unwinding of the helical twist.

An unusual duplex structure, where protonated C : C and neutral A : A base pairs are involved, is found in a proflavine–CA intercalation complex.[397,398] In 9-aminoacridine–AU[387] and actinomycin D–d(GC)[400] complexes, the dimer molecules are in an extended form and intercalation occurs between the hydrogen-bonded, intermolecularly stacked bases.

Table IX. Crystal Structure of Drug–Oligonucleotide Complexes

	Complex	Reference
1	Ethidium–io^{5}UA	385, 386
2	9-Aminoacridine–AU	387
3	Proflavine–CG	388, 389
4	Ethidium–io^{5}CG	390
5	Terpyridine platinum compound–d(CG)	391
6	9-Aminoacridine–io^{5}CG	392
7	Proflavine–io^{5}CG	393
8	Acridine orange–io^{5}CG	393
9	Acridine orange–CG	394
10	Ellipticine–io^{5}CG	395
11	3,5,6,8-tetramethyl-*N*-methyl phenanthrolinium–io^{5}CG	395
12	Proflavine–d(CG)	396
13	Proflavine–CA	397, 398
14	Daunomycin–d(CGTACG)	399
15	Actinomycin D–d(GpC)	400
16	Cisplatin–d(CGCGAATTCGCG)	401
17	Actinomycin D–d(ATGCAT)	402

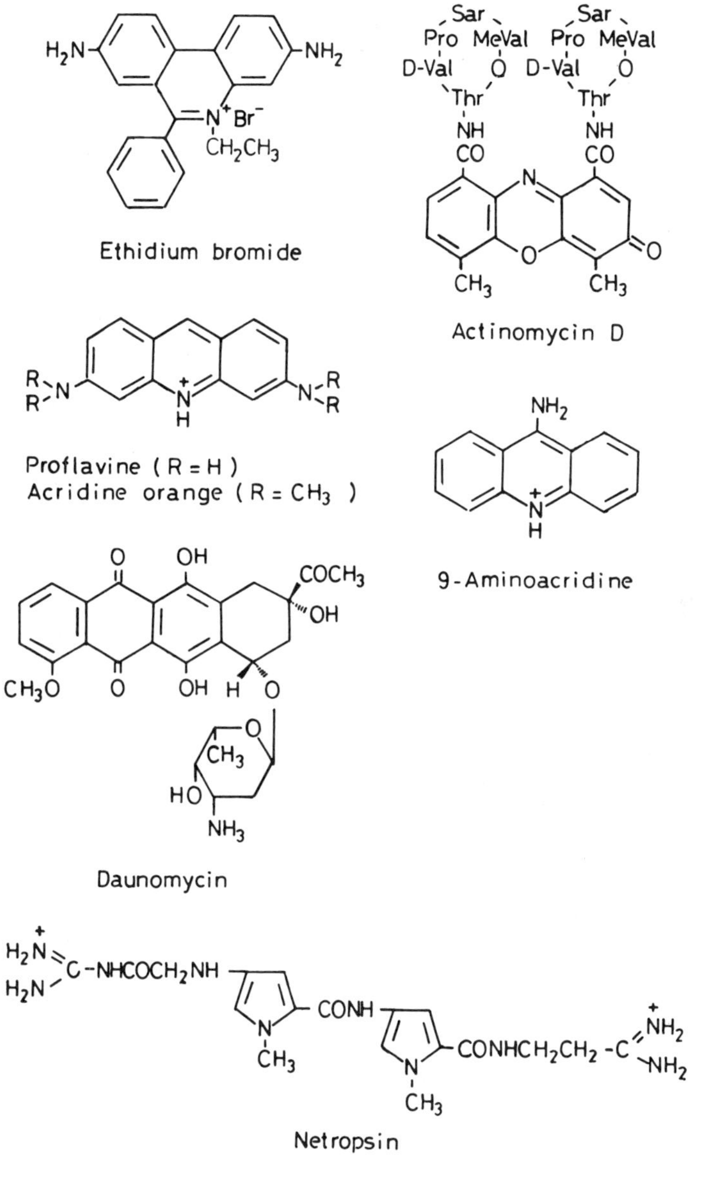

Ethidium bromide

Actinomycin D

Proflavine (R = H)
Acridine orange (R = CH₃)

9-Aminoacridine

Daunomycin

Netropsin

Figure 64. Structures of drugs that bind to nucleic acids.

Only a few x-ray crystallographic studies on a complex of oligonucleotides longer than a dimer have been published (Table IV).[399,401,402] Detailed crystal structure is reported for a daunomycin–d(CGTACG) complex.[399] A unique feature of the complex is that tetracyclic daunomycin molecules intercalate in the d(CG) sequences at right angles to the long dimension of the DNA base pairs. In all other complexes, the long axes of the polycyclic drugs are approximately parallel to those of the base pairs.

7.3.2. Interaction of Oligonucleotides with Drugs in Solution

Papers on oligonucleotide–drug interactions, in solution, by various spectroscopic techniques are listed in Table X. The drugs are mainly those containing polycyclic aromatic groups that can intercalate into double helices of nucleic acids. The ethidium and actinomycin D complexes are the most extensively

Table X. Studies on Drug–Oligonucleotide Interactions

	Complex	Reference
1	Actinomycin D–d(ATGCAT)	403
2	Ethidium–d(pCG), d(pGC), d(pTA), d(pTG) : d(pCA)	404
3	Ethidium–CG, UA, AA : UU, UG : CA, GU : CA	405
4	Actinomycin D–d(CG)$_n$ ($n = 3.4$)	406
5	Ethidium–d(CGCG)	407
6	Actinomycin–d(pTG), d(pGT), d(pGA), d(pAG), d(pGG), d(CG)	408
7	Actinomycin–d(pGG) : d(pCC), d(pGT) : d(pAC), d(pTG) : d(pCA)	409
8	Actinomycin D–d(AC), d(pAC), d(pGC), d(GC) (Mn salt, ESR)	410
9	Ethidium–d(pCG), CG (^{31}P-NMR)	411
10	9-Aminoacridine–CG (^{31}P-NMR)	411
11	Proflavine–d(CCGG), d(GGCC)	412
12	Ethidium–d(pCG), d(CG), d(pTA), d(TG) : d(CA)	413
13	Ethidium–d(pCCGG), d(pCGCG), d(pGCGC), d(pGGCC)	414
14	Ethidium–CUG, GUG : CC	415
15	Actinomycin D–d(GC), d(pGCGC), d(CGCG)	416
16	Netropsin–d(GGAATTCC)	417
17	Daunomycin–d(GCGC)	418
18	Ethidium–poly(dA) : d(pT)$_n$ ($n = 6$–10)	419
19	Netropsin–poly(dA) : d(pT)n ($n = 6$–10)	419
20	4-Nitroquinoline 1-oxide–d(pCG), d(pTG) : d(pCA), d(pGC), d(pTA), d(pAT)	420
21	Daunomycin–d(pAT), d(pTA), d(pGC), d(pCG), d(pCC)	421
22	Ethidium–A^{s}pI^s, A^{s}pUo, (pCo)$_4$: (pI^s)$_4$	422
23	9-Aminoacridine–d(AGCT)	423
24	9-Aminoacridine–GG : CC, AG : CT, CG, GC	424
25	Netropsin–d(CGCGAATTCGCG)	425
26	Actinomycin D–d(CGCGAATTCGCG)	425
27	Phenanthridium–d(TT) derivative–poly(A)	426
28	Carminomycin, marcellomycin, aclacinomycin–CG (MCD)	427
29	Netropsin and/or actinomycin D–d(CGCGAATTCGCG), its analogues	428, 429
30	Ethidium–CUG and GUG : CC, CG, UA, CA : UG, d(TA), AA : UU, AG : CU	430
31	Actinomycin D–d(pGC) (^{15}N-NMR)	431
32	Netropsin–d(CGCGAATTCGCG)	432
33	Ethidium analogues–CG,GC	433
34	*cis*-Platinum–d(GCG)	434
35	Netropsin–(pAT)$_3$, (pAT)$_6$	435
36	Actinomycin D–d(AGCT)	436
37	Ethidium–oligo(U) : poly(A) (^{31}P-NMR)	437
38	Diamino-2-phenylindole (DAPI)–d(AT)$_3$	438
39	Actinomycin D–d(ATGCAT) (^{2}D-NMR)	439
40	Ethidium–CA$_5$G : CU$_5$G, d(CA$_5$G) : d(CT$_5$G)	440
41	Platinated d(GATCCGGC) + d(GCCGGATCGC)	441
42	d(Tp)$_4$(CH$_2$)$_5$–Acridine + poly(A)	442

studied. Ethidium is known to bind tightly to DNA and RNA forming a fluorescent, intercalation complex, and therefore it is widely used for the detection of nucleic acids. Krugh and co-workers showed that self-complementary or complementary pairs of dinucleotides or dinucleoside monophosphates can be used as miniature models of double-helical DNA or RNA.[404,405] They found that ethidium exhibits a definite preference for binding to dinucleoside monophosphates that have a pyrimidine (3'–5') purine sequence when compared to their isomeric purine (3'–5') pyrimidine sequence dimers. The same preference is also observed at the level of tetranucleotides.[407,414] This preference is reversed in the case of a left-handed duplex, formed by a self-complementary, cyclonucleotide dimer with a high *anti*-glycosidic conformation.[422] Among dinucleoside monophosphates containing 8, 2'-S-cycloadenosine (A^s) and 6,2'-O-cyclouridine (U^o) residues, A^{s}pU^o and U^{o}pA^s,[239] only A^{s}pU^o shows strong binding to ethidium. The induced CD spectrum of A^{s}pU^o–ethidium complex in the 300–400 nm region (Fig. 65) is quite different from that of UpA–ethidium complex, though the absorption spectra in the 400–600 nm region (Fig. 66) of both complexes are very similar. The former CD bands in the 300–360 nm region are much larger than that of the latter and the sign of the bands is reversed. The reversal of the sign of the CD bands could be due to the reversed handedness of the miniature double helix. It has been shown that oligonucleotides containing

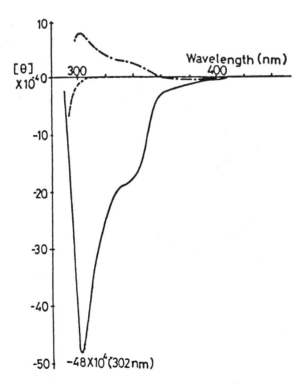

Figure 65. CD spectra of A^{s}pU^o (---), A^{s}pU^o–ethidium bromide (—), and UpA–ethidium bromide (-··-).

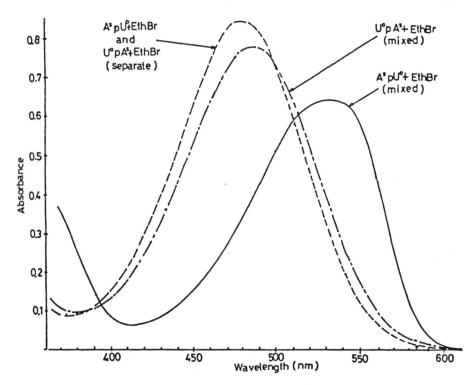

Figure 66. Absorption spectra of ethidium bromide + dimers.

cyclonucleosides with a high *anti*-glycosidic conformation adopt a left-handed helical structure[443–451] and can form a left-handed double helix.[452,453]

Actinomycin D is an antibiotic that contains a tricyclic phenoxazine chromophore and two cyclic pentapeptide residues. It was found that intercalation of actinomycin D preferentially occurs at the dG–dC sites in d(ATGCAT),[403,439] d(CGCG),[406] d(AGCT),[436] and d(CGCGAATTCGCG).[425] The pentapeptide residues fill the minor groove of the DNA double helix.[436,439] Actinomycin D can form a 1 : 1 stacked complex with non-self-complementary dimers.[409] Daunomycin[421] and 9-aminoacridine[424] also can form a 1 : 1 stacked complex.

Netropsin is a peptide antibiotic and is assumed to bind in the minor groove at the dA : dT rich region of DNA duplex by hydrogen bonding. It was found that netropsin indeed binds to the d(AATT) portion of d(GGAATTCC)[417] and d(CGCGAATTCGCG).[432] Netropsin and actinomycin D can bind simultaneously to the latter duplex.[425] Sequence specificity of actinomycin D and netropsin binding to a 142 base pair DNA fragment has been analyzed by protection from DNase I.[454]

7.4. Reactions of Oligonucleotides with Drugs

Studies on reactions of oligonucleotides with drugs are listed in Table XI. In most of these cases, deoxyribo-oligonucleotides are used as the models of double-

Table XI. Reactions of Oligonucleotides with Drugs

Drug–Oligonucleotide	Reference
1. *cis*-Pt(NH$_3$)$_2$(H$_2$O)$_2$(NO$_3$)$_2$–II, GG, AA, GC, AC	455
2. *cis*-Pt(NH$_3$)$_2$Cl$_2$–d(CCGG)	456
3. *cis*-Pt(NH$_3$)$_2$(H$_2$O)$_2$(NO$_3$)$_2$–GG, d(GG),	456
4. *cis*-Pt(NH$_3$)$_2$Cl$_2$–d(GCG)	458, 470
5. *cis*-Pt(NH$_3$)$_2$Cl$_2$–d(CG)	459
6. *cis*-Pt(NH$_3$)$_2$Cl$_2$–d(AGGCCT)	460
7. *cis*-Pt(NH$_3$)$_2$(H$_2$O)$_2$(NO$_3$)$_2$–d(TGGCCA)	461
8. Phenanthroline–d(CG)	462
9. Phenanthroline–d(ATACGCGTAT), d(CG)	463
10. Bleomycin–d(ATACGCGTAT), d(CGCGCG)	463
11. Bleomycin–d(CGCGCG)	464
12. Anthramycin–d(ATGCAT)	465
13. *cis*-Pt(NH$_3$)$_2$Cl$_2$–d(ATGG)	466
14. *cis*-Pt(NH$_3$)$_2$Cl$_2$–d(TCTCGGTCTC)	467
15. *cis*-Pt(NH$_3$)$_2$Cl$_2$–d(ATGG), d(CCATGG)	468
16. *cis*-Pt(NH$_3$)$_2$Cl$_2$–d(GATCCGGC)	469

stranded DNA or single-stranded DNA. The drugs are those forming adducts with DNA (cis-Pt and anthramycin) and those cleaving a DNA chain in the presence of metal ions (bleomycin and 1,10-phenanthroline) (Fig. 67).

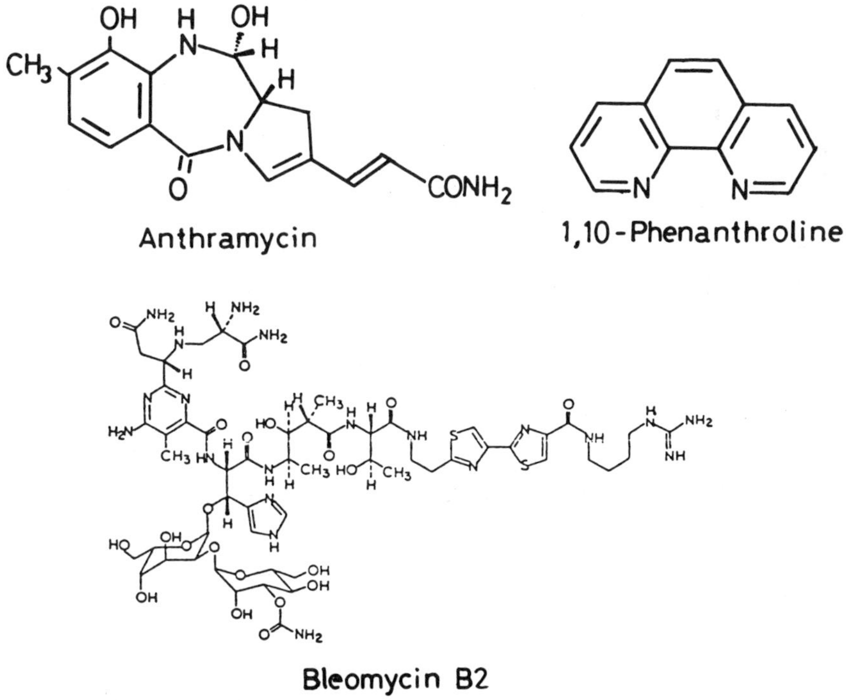

Anthramycin

1,10-Phenanthroline

Bleomycin B2

Figure 67. Structures of drugs that react with nucleic acids.

7.4.1. Oligonucleotide–Drug Adduct Formation

The primary target responsible for the cytotoxicity of the antitumor drug, *cis*-Pt(NH$_3$)$_2$Cl$_2$ (*cis*-Pt), is believed to be DNA. The kinetically most favored binding sites are the N7 atoms of the guanine bases.[471] It is assumed that bifunctional binding of *cis*-Pt between the guanines of the same strand is the predominant lesion. Many single-stranded or double-stranded oligonucleotides containing a G-G sequence are shown to form a 1 : 1 adduct.[455–461,466–470] The *cis*-Pt(NH$_3$)$_2$ moiety binds to the two N7 atoms of the adjacent guanines. It is also shown that an intrastrand cross-linking can occur between two guanines separated by one extra base as in d(GCG).[458,470] The cross-linking disturbs normal base stacking interactions and destabilizes double-stranded structures. The adducts of d(AGGCCT)[460] and d(TGGCCA)[461] are single stranded in conditions where the free hexanucleotides form a duplex. In the case of an adduct of d(TCTCGGTCTC)[467] and d(GATCCGGC),[469] they do form duplexes with the complementary oligonucleotides, d(GAGACCGAGA) and d(GCCGGTACGC), respectively. It was proved by ^{1}H-NMR studies in H$_2$O that even the modified guanine residues are involved in hydrogen bonding. However the Tms of the duplexes are much lower than those of unmodified duplexes. In the case of d(GATCCGGC),[469] the Tm decreases from 50 to 30°C (at 2×10^{-5} M oligomer strand concentration).

It is interesting to note that poly(dG-dC), modified with chloro(diethylenetriamine) platinum(II) chloride (dien-Pt) at N7 of the guanine residues, adopts a Z-form structure under physiological conditions, while the unmodified poly(dG-dC) takes a Z-form structure at much higher salt concentration.[472,473] In this case, the platinum reagent is monofunctional and the product is similar to a N7-methylated derivative.[474]

Anthramycin is an antitumor antibiotic that forms an adduct with DNA at guanine sites and inhibits nucleic acid synthesis. Graves and co-workers showed that the covalent bond is formed between C11 of anthramycin and the N^2 of a guanine residue. This was accomplished by a ^{13}C-NMR study of a ^{13}C-enriched anthramycin–DNA adduct and a ^{1}H-NMR study of an anthramycin–d(ATGCAT) adduct.[465] It was also shown that the d(ATGCAT) duplex is stabilized upon adduct formation where the anthramycin residue lies entirely within the narrow groove of the B-DNA double helix.

7.4.2. Strand Scission Reactions of Oligonucleotides by Drugs

DNA strand scission reactions by a free radical mechanism have been observed for many drugs such as bleomycin,[475] neocarzinostatin,[476] daunomycin,[477] and mitomycin[478] and also in radiolysis.[479] The 1,10-phenathroline–Cu(I) complex, which is an inhibitor of DNA polymerase, has also been reported to cleave DNA in an oxygen-dependent reaction.[480] Bleomycin also requires a metal, Fe(II), and oxygen, which can be replaced by Fe(III) and hydrogen peroxide.[481,482] However, it is only recently that some degradation products of bleomycin[483] and neocarzinostatin[484] reactions have been identified. It is advantageous to use a deoxyoligonucleotide duplex with a defined sequence as a

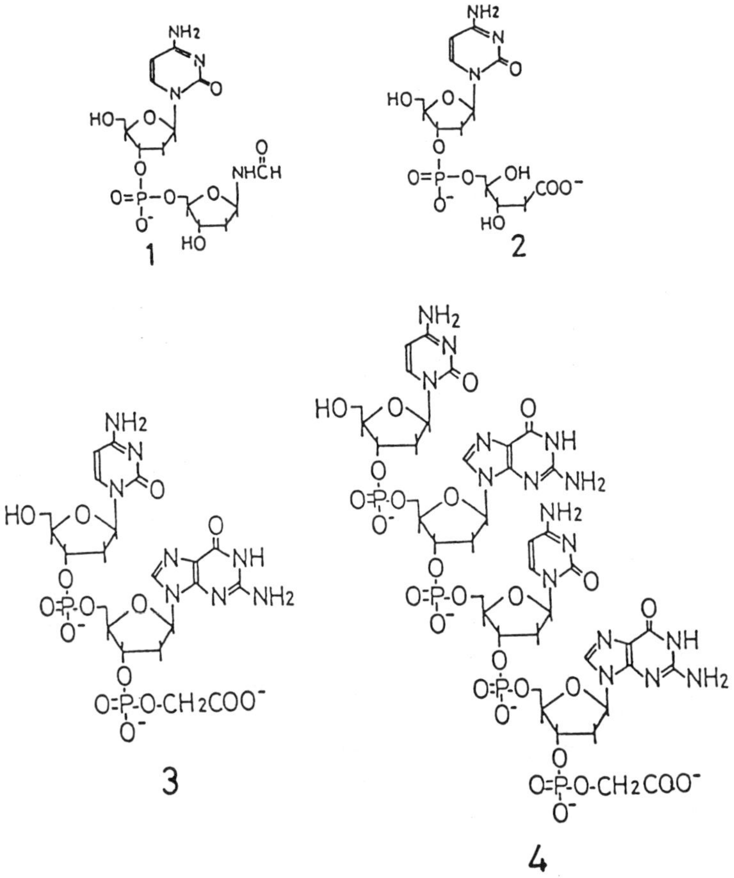

Figure 68. Products of DNA cleavage reactions.

model of DNA, especially for the identfication of the reaction products. Uesugi and co-workers employed this strategy for the identification of the products of DNA cleavage by 1,10-phenanthroline and bleomycin–metal complexes.[462,464]

In the case of phenanthroline, d(C-G) was treated with phenanthroline–Cu(II) in the presence of 3-mercaptopropionic acid at 0°C with aeration and (1) and (2) (Fig. 68) were identified mainly by ^{1}H- and ^{13}C-NMR spectroscopy in addition to cytosine, guanine, and pdG.[462] From these results, it is assumed that strand scission is initiated mainly by an oxidation at Cl′ of the deoxyribose residues as shown in Fig. 69. The initial steps may involve intercalation of phenanthroline–Cu(I) complex between base pairs of DNA and the activation of molecular oxygen at that site. The proposed mechanism predicts the formation of fragments with a 3′-terminal phosphate group and with a 5′-terminal phosphate group. This prediction was confirmed by the results of Pope *et al.*[485] with the identification of phosphomonoester groups at both ends.

Bleomycin has two separated functional sites: (1) a site for metal chelation and (2) a binding site to DNA (probably by intercalation of the bithiazole ring

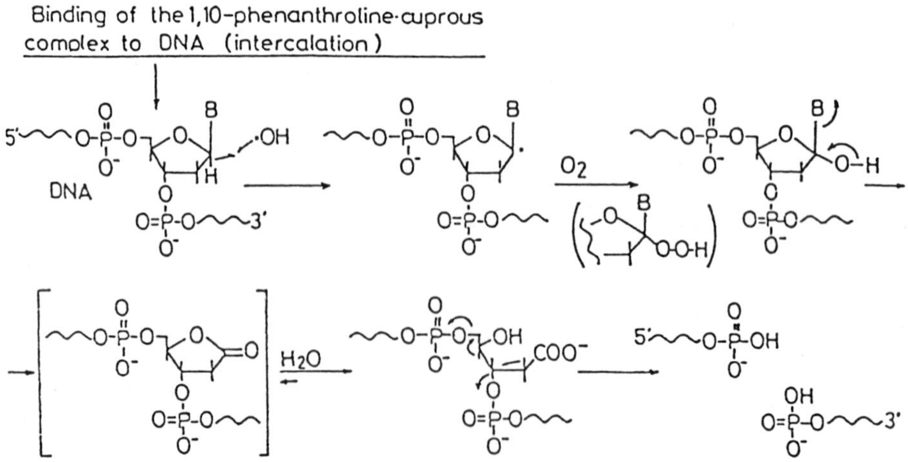

Binding of the 1,10-phenanthroline-cuprous
complex to DNA (intercalation)

Figure 69. Proposed reaction mechanism for DNA cleavage by phenanthroline.

system). It should be noted that phenanthroline also has both sites on a single-ring system.

In the case of bleomycin, d(CGCGCG) was treated with belomycin–Fe(II) complex in the presence of hydrogen peroxide and (3), (4) (Fig. 68), pdG, and cytosine were identified by UV and NMR spectroscopy.[464] These results suggest that the strand scission is caused by a cleavage of the C3'—C4' bond of the deoxyribose moiety of dC residues as shown in Fig. 70. The initial reaction on DNA is oxidation at C4' in this case. Fragments with a 3'-terminal carboxymethyl phosphoryl group and with a 5'-terminal phosphomonoester group are formed in addition to a base propenal.[483] It should be noted that bleomycin cleaves d(CGCGCG), destroying the dC residues, while phenanthroline seems to destroy both dC and dG residues of d(C-G). It is known that bleomycin preferentially cleaves G-T and G-C sequences of DNA.

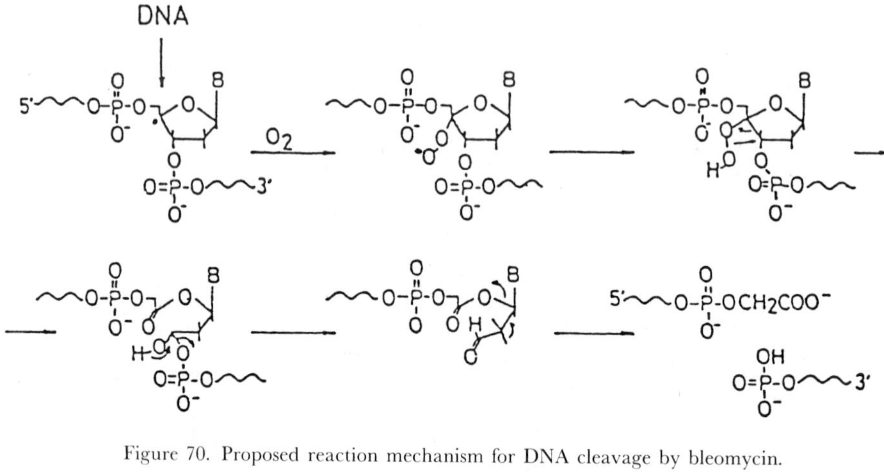

DNA

Figure 70. Proposed reaction mechanism for DNA cleavage by bleomycin.

d(ATACGCGTAT) was cleaved by both drugs much more readily than d(C-G) or d(CGCGCG).[473] The product distribution patterns of phenanthroline and bleomycin reactions are quite different, suggesting a difference in the 3′-terminal groups and in sequence specificities. These results also suggest that phenanthroline does not have a sequence specificity.

Recently, it has been reported that γ-ray-induced DNA strand cleavage produces 3′-termini with carboxymethyl phosphoryl groups and with simple phosphoryl groups.[486] Production of the latter 3′-phosphoryl termini can be explained by an oxidation at C1′ as obtained for the phenanthroline system. Examination of molecular models of a DNA duplex shows that H1′ and H4′ of the deoxyribose residue are exposed to the minor groove of the double helix and relatively close to the outer surface of the helix. Moreover, H1′ and H4′ are more acidic than the other protons of the deoxyribose residue. This characteristic property of H1′ and H4′ may make them vulnerable to hydrogen abstraction by free radicals.

Note added in proof

After this review was written, another promising method, the hydrogenphosphonate method, for the oligonucleotide synthesis was reported.[487] This method was applied only for the polymer support synthesis. By this method, both oligodeoxyribo-,[488] and oligoribonucleotides[489,490] were synthesized.

8. References

1. M. Ikehara, E. Ohtsuka, and A. F. Markham, *Adv. Carbohydr. Chem. Biochem.* **36**, 135–213 (1979).
2. E. Ohtsuka, M. Ikehara, and D. Söll, *Nucleic Acids Res.* **10**, 6553–6570 (1982).
3. H. G. Khorana, *Pure Appl. Chem.* **17**, 349–381 (1968).
4. H. G. Khorana, K. L. Agarwal, H. Büchi, M. H. Caruthers, N. K. Gupta, K. Kleppe, A. Kumar, E. Ohtsuka, U. L. RajBhandary, J. H. van de Sande, V. Sgaramella, T. Terao, H. Weber, and T. Yamada, *J. Mol. Biol.* **72**, 209–217 (1972).
5. H. G. Khorana, *Science* **203**, 614–625 (1979).
6. (a) K. Itakura, T. Hirose, R. Crea, A. Rigs, H. L. Hynecker, F. Boliver, and H. W. Boyer, *Science* **198**, 1056–1063 (1977); (b) H. Köster, H. Blocker, R. Frank, S. Geussenhainer, and W. Kaiser, *Ann. Chemie*, 839–853 (1978); (c) R. Crea, T. Hirose, A. Kraszewski, and K. Itakura, *Proc. Natl. Acad. Sci. U.S.A.* **75**, 5765–5769 (1978); (d) R. Wetzel, H. L. Heynecker, D. V. Goeddel, P. Jhurani, J. Shapiro, R. Crea, T. L. K. Low, J. E. McClure, C. B. Thurman, and A. L. Goldstein, *Biochemistry* **19**, 6096–6104 (1980); (e) M. D. Edge, A. R. Greene, G. R. Heathcliffe, P. A. Meacock, W. Schuch, D. B. Scanlon, T. C. Atkinson, C. R. Newton, and A. F. Markham, *Nature* **292**, 756–762 (1981); (f) S. Tanaka, T. Oshima, K. Ohsuye, T. Ono, A. Mizuno, A. Ueno, H. Nakazato, M. Tsujomoto, N. Higashi, and T. Noguchi, *Nucleic Acids Res.* **11**, 1707–1723 (1983); (g) M. Suzuki, S. Sumi, A. Hasegawa, T. Nishizawa, K. Miyoshi, S. Wakisaka, T. Miyake, and F. Misoka, *Proc. Natl. Acad. Sci. U.S.A.* **79**, 2475–2479 (1982); (h) M. S. Urdea, J. P. Merryweather, G. T. Mullenbach, D. Coit, U. Heberlin, P. Valenzuela, and P. J. Barr, *Proc. Natl. Acad. Sci. U.S.A.* **80**, 7461–7465 (1983); (i) J. Smith, E. Cook, I. Fotheringham, S. Pheby, R. D. Shire, M. A. W. Eaton, M. Doel, D. M. J. Lilley, J. F. Pardon, T. Patel, H. Lewis, and L. Bell, *Nucleic Acids Res.* **10**, 4467–4482 (1982); (j) M. Ikehara, E. Ohtsuka, T. Tokunaga, T. Taniyama, S. Iwai, K. Kitano, S. Miyamoto, T. Ohgi, Y. Sakuragawa, K. Fujiyama, T. Ikari, M. Kobayashi, T. Miyake, S. Shibahara, A. Ono, T. Ueda, T. Tanaka, H. Baba, T. Miki, A. Sakurai, T. Oishi, O. Chisaka, and K. Matsubara, *Proc. Natl. Acid. Sci. U.S.A.* **81**, 5956–5960 (1984).

7. (a) H. G. Khorana, in *The Nucleic Acids* (E. Chargraff and J. N. Davidson, eds.), Vol. 3, pp. 105–146, Academic Press, New York, 1960; H. G. Khorana, in *The Nucleic Acids* (E. Chargraff and J. N. Davidson, eds.), Wiley, New York, 1961; (b) A. M. Michelson, *The Chemistry of Nucleosides and Nucleotides*, Academic Press, New York, 1963; (c) W. W. Zorbach and R. S. Tipson, eds., *Synthetic Procedures in Nucleic Acids Chemistry*, Wiley-Interscience, New York, 1968, Vol. 1, 1973, Vol. 2; *Nucleic Acid Chemistry* (L. B. Townsend and R. S. Tipson, eds.), Wiley-Interscience, New York, 1978; (d) N. K. Kochetkov and J. Budovskii, eds., *Organic Chemistry of Nucleic Acids*, Plenum Press, London, 1971.

8. (a) H. Kössel and H. Seliger, *Fortschr. Chem. Org. Naturst.* **32**, 297–508 (1975); (b) E. Ohtsuka, in *Methodicum Chimicum* (F. Korte and M. Goto, eds.), Vol. 11, pp. 10–22, George Thieme Verlag, Stuttgart, 1976; (c) V. Amarnath and A. D. Broom, *Chem. Rev.* **77**, 183–217 (1977); (d) C. B. Reese, *Tetrahedron* **34**, 3143–3179 (1978); (e) M. Ikehara, E. Ohtsuka, and A. F. Markham, *Adv. Carbohydr. Chem. Biochem.* **36**, 135–213 (1979); (e) E. L. Brown, R. Belagaje, M. J. Ryan, and H. G. Khorana, *Methods Enzymol.* **68**, 109–151 (1979); (f) S. A. Narang, H. M. Hsing, and R. Brousseau, *Methods Enzymol.* **68**, 90–98 (1979); (g) M. H. Caruthers, *Acc. Chem. Res.* **13**, 155–160 (1980); (h) S. A. Narang, *Tetrahedron* **39**, 3–22 (1983); (i) Recent aspects of the chemistry of nucleoside, nucleotides and nucleic acids, Tetrahedron Symposia-in-print, edited by C. B. Reese, *Tetrahedron* **40**, 1–153 (1984).

9. A. M. Michelson and A. R. Todd, *J. Chem. Soc.*, 2632–2638 (1955).

10. (a) R. L. Letsinger, J. L. Finnan, G. A. Heavner, and W. B. Lunsford, *J. Am. Chem. Soc.* **97**, 3278–3279 (1975); (b) R. L. Letsinger and W. B. Lunsford, *J. Am. Chem. Soc.* **98**, 3655–3661 (1976).

11. S. L. Beaucage and M. H. Caruthers, *Tetrahedron Lett.* **22**, 1859–1862 (1981).

12. S. P. Adams, K. S. Kauka, E. J. Wykes, S. B. Holder, and G. K. Galluppi, *J. Am. Chem. Soc.* **105**, 661–663 (1983).

13. H. Buchi and H. G. Khorana, *J. Mol. Biol.* **72**, 251–288 (1972).

14. E. Ohtsuka, M. Shin, Z. Tozuka, A. Ohta, K. Kitano, Y. Taniyama, and M. Ikehara, *Nucleic Acids Res. Symp. Ser.* **11**, 193–196 (1982).

15. (a) D. Shortle, D. Dimaio, and D. Nathans, *Annu. Rev. Genet.* **151**, 265–294 (1981); (b) M. Smith and S. Gillam, in *Genetic Engineering* (J. K. Setlow and Hollaender, eds.), Vol. 3, pp. 1–32, Plenum Press, New York, 1981.

16. R. I. Gumport and O. C. Uhlenbeck, in *Gene Amplification and Analysis* (J. G. Chirikjian and T. S. Papas, eds.) Vol. 2, pp. 313–345, Elsevier/North-Holland, Amsterdam, 1981.

17. E. Ohtsuka, S. Tanaka, T. Tanaka, T. Miyaka, A. F. Markham, E. Nakagawa, T. Wakabayashi, Y. Taniyama, S. Nishikawa, R. Fukumoto, H. Uemura, T. Doi, T. Tokunaga, and M. Ikehara, *Proc. Natl. Acad. Sci. U.S.A.* **78**, 5493–5497 (1981).

18. P. T. Wang, *Kexue Tongbao* **27**, 216–219 (1982).

19. E. A. Miele, D. R. Miele, D. R. Mills, and F. R. Kramer, *J. Mol. Biol.* **171**, 281–295 (1983).

20. E. Ohtsuka, A. Yamane, T. Doi, and M. Ikehara, *Tetrahedron* **40**, 47–57 (1984).

21. P. T. Gilham and H. G. Khorana, *J. Am. Chem. Soc.* **80**, 6212–6222 (1958).

22. R. K. Ralph and H. G. Khorana, *J. Am. Chem. Soc.* **83**, 2926–2934 (1961).

23. R. K. Ralph, W. J. Conners, H. Schaller, and H. G. Khorana, *J. Am. Chem. Soc.* **85**, 1983–1988 (1963).

24. H. Weber and H. G. Khorana, *J. Mol. Biol.* **72**, 219–249 (1972).

25. (a) H. Köster, K. Kulikowski, T. Liese, W. Heikens, and V. Kohli, *Tetrahedron* **37**, 363–369 (1981); (b) S. S. Jones, C. B. Reese, S. Sibanda, and A. Ubasawa, *Tetrahedron Lett.* **23**, 2257–2260 (1982); (c) K. K. Ogilvie, M. J. Nemer, G. H. Hakimelahi, Z. A. Proba, and M. Lucas, *Tetrahedron Lett.* **23**, 2615–2618 (1982); (d) B. E. Watkins, J. S. Kiely, and H. Rapoport, *J. Am. Chem. Soc.* **104**, 5702–5708 (1982).

26. C. S. Ti, B. L. Gaffney, and R. A. Jones, *J. Am. Chem. Soc.* **104**, 1316–1319 (1982); D. P. C. McGee, J. C. Martin, and A. S. Webb, *Synthesis*, 540–541 (1983).

27. (a) R. H. Hall, *Biochemistry* **3**, 769–773 (1964); (b) E. Ohtsuka, E. Nakagawa, T. Tanaka, A. F. Markham, and M. Ikehara, *Chem. Pharm. Bull. (Tokyo)* **26**, 2998–3006 (1978); (c) K. A. Watanabe and J. J. Fox, *Angew. Chem. Int. Ed.* **5**, 579– (1966); (d) T. Sasaki and Y. Mizuno, *Chem. Pharm. Bull. (Tokyo)* **15**, 894–896 (1967).

28. M. Smith, D. H. Rammler, I. H. Goldberg, and H. G. Khorana, *J. Am. Chem. Soc.* **84**, 430–440 (1962).

29. T. Shimidzu and R. L. Letsinger, *J. Org. Chem.* **33**, 708–711 (1968).

30. J. Zemlicka and A. Holý, *Collect. Czech. Chem. Commun.* **32**, 3159–3168 (1967).

31. S. S. Jones, B. Rayner, C. B. Reese, A. Ubasawa, and M. Ubasawa, *Tetrahedron* **36**, 3075–3085 (1980).

32. C. B. Reese and A. Ubasawa, *Tetrahedron Lett.* **21**, 2265–2268 (1980); C. B. Reese and A. Ubasawa, *Nucleic Acids Res. Symp. Ser.* **7**, 5–21 (1980).

33. (a) S. S. Jones, C. B. Reese, S. Shibanda, and A. Ubasawa, *Tetrahedron Lett.* **22**, 4755–4758 (1981); (b) M. Sekine, J. Matsuzaki, M. Satoh, and T. Hata, *J. Org. Chem.* **47**, 571–573 (1982); (c) M. Sekine, J. Matsuzaki, and T. Hata, *Tetrahedron Lett.* **23**, 5287–5290 (1982); (d) F. Himmelsbach, B. S. Schulz, T. Trichtinger, R. Charubala, and W. Pfleiderer, *Tetrahedron* **40**, 59–72 (1984); (e) H. Takaku, S. Ueda, and T. Ito, *Tetrahedron Lett.* **24**, 5363–5366 (1983).

34. (a) C. T. Welch and J. Chattopadhyaya, *Acta Chem. Scand. B* **37**, 147–150 (1983); (b) J. Matsuzaki, H. Hotoda, M. Sakine, and T. Hata, *Tetrahedron Lett.* **25**, 4019–4022 (1984).

35. (a) A. Kume, M. Sekine, and T. Hata, *Tetrahedron Lett.* **23**, 4365–4368 (1982); (b) L. J. McBride and M. H. Caruthers, *Tetrahedron Lett.* **24**, 2953–2956 (1983).

36. M. W. Moon, S. Nishimura, and H. G. Khorana, *Biochemistry* **3**, 937–945 (1964).

37. J. Stawinski, T. Hozumi, S. A. Narang, C. B. Bahl, and R. Wu, *Nucleic Acids Res.* **4**, 353–371 (1977).

38. T. Tanaka and R. L. Letsinger, *Nucleic Acids Res.* **10**, 3249–3260 (1982).

39. E. F. Fische and M. H. Caruthers, *Nucleic Acids Res.* **11**, 1589–15999 (1983).

40. J. B. Chattopadhyaya and C. B. Reese, *J. Chem. Soc. Chem. Commun.*, 639–640 (1978).

41. (a) H. P. M. Fromageot, B. E. Griffin, C. B. Reese, and J. E. Sulston, *Tetrahedron* **23**, 2315–2331 (1967); (b) G. Weimann and H. G. Khorana, *J. Am. Chem. Soc.* **84**, 4329–4341 (1962).

42. C. B. Reese, J. C. M. Stewart, J. H. van Boom, H. P. M. de Leeuw, J. Nagel, and J. F. M. de Rooy, *J. Chem. Soc. Perkin Trans. I*, 934–936 (1975).

43. E. S. Werstiuk and T. Neilson, *Can. J. Chem.* **50**, 1283–1291 (1972).

44. T. Neilson, K. V. Deugan, T. E. England, and E. S. Werstiuk, *Can. J. Chem.* **53**, 1093–1098 (1975).

45. J. H. van Boom and P. M. J. Burgers, *Tetrahedron Lett.*, 4875–4878 (1975); J. H. van Boom and P. M. J. Bergers, *Recl. Trav. Chim.* **97**, 73–80 (1978).

46. J. B. Chattopadhyaya, C. B. Reese, and A. H. Todd, *J. Chem. Soc. Chem. Commun.*, 987–988 (1979).

47. E. Ohtsuka, Y. Taniyama, R. Marumoto, H. Sato, H. Hirosaki, and M. Ikehara, *Nucleic Acids Res.* **10**, 2597–2608 (1982).

48. K. Miyoshi, T. Miyake, T. Hozumi, and K. Itakura, *Nucleic Acids Res.* **8**, 5474–5489 (1980).

49. D. P. L. Green, T. Ravindranathan, C. B. Reese, and R. Saffhill, *Tetrahedron* **26**, 1031–1041 (1970).

50. D. G. Norman, C. B. Reese, and H. T. Serafinowska, *Tetrahedron Lett.* **25**, 3015–3018 (1984).

51. (a) E. Ohtsuka, A. Yamane, and M. Ikehara, *Chem. Pharm. Bull. (Tokyo)* **31**, 1535–1543 (1983); (b) E. Ohtsuka, A. Yamane, and M. Ikehara, *Nucleic Acids Res.* **11**, 1325–1335 (1983).

52. W. T. Markiewicz, *J. Chem. Res.* (s), 24 (1979); W. T. Markiewicz, E. Biala, R. W. Adamiak, R. Kierzek, A. Kyszewski, J. Stawinski, and M. Wiewirowski, *Nucleic Acids Res. Symp. Ser.* **7**, 115–127 (1980).

53. K. K. Ogilvie, in *Nucleosides, Nucleotides and Their Biological Applications*, Academic Press, New York, 1983.

54. S. S. Jones and C. B. Reese, *J. Chem. Soc. Perkin Trans. I*, 2762–2769 (1979).

55. K. K. Ogilvie and N. Y. Theriault, *Can. J. Chem.* **57**, 3140–3144 (1979).

56. E. Ohtsuka, S. Tanaka, and M. Ikehara, *Nucleic Acids Res.* **1**, 1351–1357 (1974); E. Ohtsuka, S. Tanaka, and M. Ikehara, *Chem. Pharm. Bull. (Tokyo)* **25**, 949–959 (1977); E. Ohtsuka, S. Tanaka, and M. Ikehara, *Synthesis*, 453–454 (1977).

57. D. G. Bartholomew and A. D. Broom, *J. Chem. Soc. Chem. Commun.*, 38 (1975); E. Ohtsuka, T. Wakabayashi, S. Tanaka, T. Tanaka, K. Oshie, A. Hasegawa, and M. Ikehara, *Chem. Pharm. Bull. (Tokyo)* **29**, 318–324 (1981).

58. (a) E. Ohtsuka, T. Tanaka, and M. Ikehara, *J. Am. Chem. Soc.* **101**, 6409–6414 (1979); (b) E. Ohtsuka, K. Fujiyama, T. Tanaka, and M. Ikehara, *Chem. Pharm. Bull. (Tokyo)* **29**, 2799–2806 (1981); (c) E. Ohtsuka, K. Fujiyama, and M. Ikehara, *Nucleic Acids Res.* **9**, 3505–3522 (1981).

59. E. Ohtsuka, K. Oshie, and M. Ikehara, *Tetrahedron Lett.*, 3677–3680 (1979).

60. K. K. Ogilvie and N. Y. Theriault, *Tetrahedron Lett.*, 2111–2114 (1979); R. Charubala and

W. Pfleiderer, *Tetrahedron Lett.*, 1933–1936 (1980); S. S. Jones and C. B. Reese, *J. Am. Chem. Soc.* **101**, 7399–7401 (1979); J. A. J. den Hartog, I. Doornbos, R. Crea, and J. H. van Boom, *Recl. Trav. Chim. Pays-Bas* **98**, 469–470 (1979).

61. H. Takaku and K. Kamaike, *Chem. Lett.*, 189–192 (1982).
62. G. M. Tener, *J. Am. Chem. Soc.* **83**, 159–168 (1961).
63. J. H. van Boom, P. M. J. Burgers, P. H. van Daursen, R. Arenzen, and C. B. Reese, *Tetrahedron Lett.*, 3785–3788 (1974).
64. F. Eckstein, in *Protecting Groups in Organic Chemistry* (J. F. W. McOmic, ed.), p. 217, Plenum Press, London, 1973; J. H. van Boom and P. M. J. Burgers, *Tetrahedron Lett.*, 4875–4878 (1977).
65. E. Ohtsuka, T. Tanaka, S. Tanaka, and M. Ikehara, *J. Am. Chem. Soc.* **100**, 4580–4584 (1978); E. Ohtsuka, S. Shibahara, and M. Ikehara, *Chem. Pharm. Bull. (Tokyo)* **29**, 3440–3448 (1981).
66. C. Christodoulou and C. B. Reese, *Tetrahedron Lett.* **24**, 951–954 (1983).
67. E. Uhlmann and W. Pfleiderer, *Tetrahedron Lett.*, 1181–1184 (1980); W. Pfleiderer, E. Uhlmann, R. Charubala, D. Flockerzi, G. Siber, and R. S. Verma, *Nucleic Acids Res. Symp. Ser.* **7**, 61–71 (1980).
68. (a) S. Honda, K. Urakami, K. Koura, K. Terada, Y. Sato, K. Kohno, M. Sekine, and T. Hata, *Tetrahedron* **40**, 153–163 (1984); (b) H. Takaku, K. Kamaike, and K. Kasuga, *J. Org. Chem.* **47**, 4937–4940 (1982).
69. D. T. Elmore and A. R. Todd, *J. Chem. Soc.*, 3681–3683 (1952); M. Ikehara and T. Tezuka, *Nucleic Acids Res.* **2**, 1345–1364 (1975); Y. Hayakawa, M. Uchiyama, and R. Noyori, *Tetrahedron Lett.* **25**, 4003–4006 (1984).
70. T. M. Jacob and H. G. Khorana, *J. Am. Chem. Soc.* **86**, 1630–1635 (1964); R. Lohrmann and H. G. Khorana, *J. Am. Chem. Soc.* **88**, 829–833 (1966).
71. V. M. Clark, *Angew Chem. Int. Ed.* **3**, 678–685 (1964).
72. Yu. A. Berlin, O. G. Chakhmakhcheva, V. A. Efimov, M. N. Kolosov, and G. Korobko, *Tetrahedron Lett.*, 1353–1356 (1973).
73. (a) A. K. Seth and E. Jay, *Nucleic Acids Res.* **8**, 5445–5459 (1980); (b) V. A. Efimov, S. A. Reverdatto, and O. G. Chakhmakhcheva, *Tetrahedron Lett.* **23**, 961–964 (1982); (c) T. Wakabayashi and S. Tachibana, *Chem. Pharm. Bull. (Tokyo)* **30**, 3951–3958 (1982); (d) E. M. Ivanova, L. M. Khalinskaya, V. P. Romanenko, and V. F. Zarytova, *Tetrahedron Lett.* **23**, 5447–5450 (1983); (e) V. F. Zarytova and D. G. Knorre, *Nucleic Acids Res.* **12**, 2091–2110 (1984).
74. E. Ohtsuka, Z. Tozuka, and M. Ikehara, *Tetrahedron Lett.* **22**, 4483–4486 (1981); E. Ohtsuka, Z. Tozuka, S. Iwai, and M. Ikehara, *Nucleic Acids Res.* **10**, 6235–6241 (1982).
75. R. L. Letsinger and K. K. Ogilvie, *J. Am. Chem. Soc.* **89**, 4801–4803 (1967).
76. R. W. Adamiak, R. Arenzen, and C. B. Reese, *Tetrahedron Lett.*, 1431–1434 (1977).
77. R. Arenzen and C. B. Reese, *J. Chem. Soc. Perkin Trans. I*, 445–460 (1977).
78. C. B. Reese, R. C. Titmas, and L. Yau, *Tetrahedron Lett.*, 2727–2730 (1978).
79. A. K. Sood and S. A. Narang, *Nucleic Acids Res.* **4**, 2757–2766 (1977).
80. K. Itakura, N. Katagiri, C. P. Bahl, R. H. Wightman, and S. A. Narang, *J. Am. Chem. Soc.* **97**, 7327–7331 (1975).
81. K. Itakura, N. Katagiri, S. A. Narang, C. P. Bahl, K. J. Marians, and R. Wu, *J. Biol. Chem.* **250**, 4592–4600 (1975).
82. R. H. Scheller, R. E. Dickerson, H. W. Boyer, A. D. Riggs, and K. Itakura, *Science* **196**, 177–180 (1977).
83. C. Broka, T. Hozumi, R. Arenzen, and K. Itakura, *Nucleic Acids Res.* **8**, 5461–5471 (1980).
84. E. Ohtsuka, Y. Taniyama, S. Iwai, T. Yoshida, and M. Ikehara, *Chem. Pharm. Bull. (Tokyo)* **32**, 85–93 (1984).
85. S. H. Chang, A. Majumdar, R. Dunn, O. Makabe, U. T. RajBhandary, H. G. Khorana, E. Ohtsuka, T. Tanaka, Y. Taniyama, and M. Ikehara, *Proc. Natl. Acad. Sci. U.S.A.* **78**, 3398–3402 (1981).
86. G. van der Marel, C. A. A. van Boeckel, G. Wille, and J. H. van Boom, *Tetrahedron Lett.* **22**, 3887–3890 (1980).
87. G. R. Gough, M. J. Brunden, J. G. Nadean, and P. T. Gilham, *Tetrahedron Lett.* **23**, 3439–3442 (1982); B. Chaudhuri, C. B. Reese, and K. Weclawek, *Tetrahedron Lett.* **25**, 4037–4040 (1984).
88. C. K. Narang, K. Brunfeld, and K. E. Norris, *Tetrahedron Lett.*, 1819–1822 (1977).
89. (a) M. J. Gait and R. C. Sheppard, *Nucleic Acids Res.* **4**, 1135–1138 (197); (b) M. J. Gait and R. C. Sheppard, *Nucleic Acids Res.* **4**, 4391–4410 (1977).

90. K. Miyoshi and K. Itakura, *Tetrahedron Lett.*, 3635–3638 (1979).
91. K. Miyoshi, T. Huang, and K. Itakura, *Nucleic Acids Res.* **8**, 5491–5505 (1980).
92. S. Ikuta, R. Chattopadhyaya, and R. E. Dickerson, *Nucleic Acids Res.* **12**, 6511–6522 (1984).
93. P. Dembek, K. Miyoshi, and K. Itakura, *J. Am. Chem. Soc.* **103**, 706–708 (1981).
94. A. F. Markham, M. D. Edger, T. C. Atkinson, A. R. Greene, G. R. Heathcliffe, C. R. Newton, and O. Scanlon, *Nucleic Acids Res.* **8**, 5193–5205 (1980); T. P. Patel, T. A. Millican, C. C. Bose, R. C. Titmus, G. A. Mock, and M. A. W. Eaton, *Nucleic Acids Res.* **10**, 5605–5620 (1982).
95. H.-J. Fritz, R. Belagaje, E. Brown, R. H. Fritz, R. A. Jones, R. G. Lees, and H. G. Khorana, *Biochemistry* **17**, 1257–1267 (1978).
96. K. Miyoshi, R. Arenzen, T. Huang, and K. Itakura, *Nucleic Acids Res.* **8**, 5507–5517 (1980).
97. H. Ito, Y. Ike, S. Ikuta, and K. Itakura, *Nucleic Acids Res.* **10**, 1755–1769 (1982).
98. E. Ohtsuka, T. Ohgi, T. Fukui, and M. Ikehara, *Chem. Pharm. Bull. (Tokyo)* **33**, 1849–1855 (1985).
99. H. Rink, M. Liersch, P. Sieber, and F. Meyer, *Nucleic Acids Res.* **12**, 6369–6387 (1984).
100. J. E. Marugg, L. W. McLaughlin, N. Piel, M. Tromp, G. A. van der Marel, and J. H. van Boom, *Tetrahedron Lett.* **24**, 3989–3992 (1983); J. E. Marugg, M. Tromp, P. Thurani, C. F. Hoyng, G. A. van der Mare, and J. H. van Boom, *Tetrahedron* **40**, 73–78 (1984).
101. A. Rosenthal, D. Cech, V. P. Veiko, T. S. Orenzkaja, E. A. Kuprijianova, and Z. A. Shabarova, *Tetrahedron Lett.* **24**, 1691–1694 (1983).
102. M. D. Matteucci and M. H. Caruthers, *J. Am. Chem. Soc.* **103**, 3185–3191 (1981).
103. E. Ohtsuka, H. Takashima, and M. Ikehara, *Tetrahedron Lett.* **23**, 3081–3084 (1982).
104. H. Köster, J. Biernar, J. McMaus, A. Wolter, A. Stumpe, Ch. K. Narang, and N. D. Sinha, *Tetrahedron* **40**, 103–112 (1984).
105. G. R. Gough, M. J. Brunder, and P. T. Gilham, *Tetrahedron Lett.* **24**, 5321–5324 (1983).
106. M. J. Gait, H. W. D. Mattes, M. Singh, and R. C. Titmus, *J. Chem. Soc. Chem. Commun.*, 37–40 (1982).
107. R. Frank, W. Heikens, G. Heisterberg-Moutsis, and H. Blöcker, *Nucleic Acids Res.* **11**, 4365–4377 (1983).
108. H. W. D. Mattes, W. M. Zenke, T. Grundström, A. Stanb, M. Wintzerith, and P. Chambon, *EMBO J.* **3**, 801–805 (1984).
109. E. Ohtsuka, T. Tanaka, T. Wakabayashi, Y. Taniyama, and M. Ikehara, *J. Chem. Soc. Chem. Commun.*, 824–825 (1978).
110. (a) J. A. J. den Hartog and J. H. van Boom, *Recl. Trav. Chim. Pays-Bas* **100**, 275–284 (1981); (b) J. A. J. den Hartog, G. Wille, and J. H. van Boom, *Recl. Trav. Chim. Pays-Bas* **100**, 320–330 (1981).
111. W. L. Sung and S. A. Narang, *Can. J. Chem.* **60**, 111–120 (1982).
112. E. Ohtsuka, A. Yamane, M. Ohkubo, and M. Ikehara, *Chem. Pharm. Bull. (Tokyo)* **31**, 1910–1916 (1983).
113. R. Kierzek, H. Ito, R. Bhatt, and K. Itakura, *Tetrahedron Lett.* **22**, 3761–3764 (1981).
115. C. T. J. Wreesmann, A. Fidder, G. A. van der Marel, and J. H. van Boom, *Nucleic Acids Res.* **11**, 8389–8405 (1983).
116. E. Ohtsuka, H. Takashima, and M. Ikehara, *Tetrahedron Lett.* **22**, 765–768 (1981).
117. G. A. van der Marel, G. Wille, and J. H. van Boom, *Recl. Trav. Chim. Pays-Bas* **101**, 241–246 (1982).
118. K. K. Ogilvie, N. Y. Theriault, J.-M. Seifert, R. T. Pon, and M. J. Nemer, *Can J. Chem.* **58**, 2686–2693 (1980).
119. G. W. Daub and E. E. van Tamelen, *J. Am. Chem. Soc.* **99**, 3526–3528 (1977).
120. J. L. Fourrey and D. J. Shire, *Tetrahedron Lett.* **22**, 729–732 (1981).
121. R. L. Letsinger, E. P. Groody, N. Lander, and T. Tanaka, *Tetrahedron* **40**, 137–143 (1984).
122. R. L. Letsinger, E. P. Groody, and T. Tanaka, *J. Am. Chem. Soc.* **104**, 6805–6806 (1982).
123. J.-L. Fourrey and J. Varenne, *Tetrahedron Lett.* **24**, 1963–1966 (1983).
124. L. J. McBride and M. H. Caruthers, *Tetrahedron Lett.* **24**, 245–248 (1983).
125. A. D. Barone, J.-Y. Tang, and M. H. Caruthers, *Nucleic Acids Res.* **12**, 4051–4061 (1984).
126. H. J. Lee and S. H. Moon, *Chem. Lett.*, 1229–1232 (1984).
127. B. C. Fraehler and M. D. Matteucci, *Tetrahedron Lett.* **24**, 3171–3174 (1983).
128. N. D. Shinha, J. Biernat, and H. Köster, *Tetrahedron Lett.* **24**, 5843–5846 (1983).
129. N. D. Shinha, J. Biernat, J. McManus, and H. Köster, *Nucleic Acids Res.* **12**, 4539–4557 (1984).

130. C. Claesen, G. I. Tesser, C. E. Dreef, J. E. Marugg, G. A. van der Marel, and J. H. van Boom, *Tetrahedron Lett.*, 1307–1310 (1980).

131. A. H. Beiter and W. Pfleiderer, *Tetrahedron Lett.* **25**, 1975–1978 (1984).

132. M. D. Matteucci and M. H. Caruthers, *Tetrahedron Lett.* **21**, 719–722 (1980).

133. F. Chow, T. Kempwe, and G. Palm, *Nucleic Acids Res.* **9**, 2807–2817 (1981).

134. G. A. Urbina, G. M. Sathe, W.-C. Liu, M. F. Gillem, P. D. Duck, R. Bender, and K. K. Ogilvie, *Science* **204**, 270–274 (1981).

135. A. Elmblad, S. Josephson, and G. Palm, *Nucleic Acids Res.* **10**, 3291–3301 (1982).

136. T. Dörper and E.-L. Winnacker, *Nucleic Acids Res.* **11**, 2575–2584 (1983).

137. H. Seliger, C. Scalfi, and F. Eisenbeiss, *Tetrahedron Lett.* **24**, 4963–4966 (1983).

138. H. Köster, J. Biernat, J. AcManus, A. Wolter, A. Stupme, C. K. Narang, and N. D. Sinha, *Tetrahedron* **40**, 103–112 (1984).

139. S. P. Adams and G. R. Galluppi, *Science* **216**, 398–   (1982).

140. K. Jayaraman and H. McClaugherty, *Tetrahedron Lett.* **23**, 5377–5380 (1982).

141. T. M. Cao, S. E. Bingham, and M. T. Sung, *Tetrahedron Lett.* **24**, 1019–1020 (1983).

142. K. K. Ogilvie and M. J. Nemer, *Can. J. Chem.* **58**, 1389–1397 (1980).

143. K. K. Ogilvie and M. J. Nemer, *Tetrahedron Lett.* **22**, 2531–2532 (1981).

144. K. K. Ogilvie and M. J. Nemer, *Tetrahedron Lett.* **21**, 4159–4162 (1980).

145. K. K. Ogilvie, M. J. Nemer, and M. F. Gillen, *Tetrahedron Lett.* **25**, 1669–1672 (1984).

146. B. P. Melnick, J. L. Finnan, and R. L. Letsinger, *J. Org. Chem.* **45**, 2715–2716 (1980).

147. K. K. Ogilvie and M. J. Nemer, *Tetrahedron Lett.* **21**, 4145–4148 (1980).

148. M. J. Nemer and K. K. Ogilvie, *Tetrahedron Lett.* **21**, 4149–4152 (1980).

149. M. J. Nemer and K. K. Ogilvie, *Tetrahedron Lett.* **21**, 4153–4154 (1980).

150. M. A. Dorman, S. A. Noble, L. J. McBride, and M. H. Caruthers, *Tetrahedron* **40**, 95–102 (1984).

151. A. Jäger and J. Engels, *Tetrahedron Lett.* **25**, 1437–1440 (1984).

152. G. Baschang and V. Kvita, *Angew. Chem. Int. Ed. Engl.* **12**, 71–72 (1973).

153. G. S. Bajwa and W. G. Bentrude, *Tetrahedron Lett.*, 421–424 (1978).

154. R. L. Letsinger and G. A. Heavner, *Tetrahedron Lett.*, 147–150 (1975).

155. R. L. Letsinger and M. E. Schott, *J. Am. Chem. Soc.* **103**, 7394–7396 (1981).

156. K. L. Agarwal, A. Yamazaki, P. J. Cashion, and H. G. Khorana, *Angew. Chem. Int. Ed.* **11**, 451–459 (1972).

157. Y. Kato, M. Sasaki, T. Hashimoto, T. Murotsu, S. Fukushige, and K. Matsubara, *J. Chromatogr.* **266**, 341–349 (1983).

158. F. Sanger and A. R. Coulson, *J. Mol. Biol.* **94**, 441–448 (1975).

159. (a) A. M. Maxam and W. Gilbert, *Proc. Natl. Acad. Sci. U.S.A.* **74**, 560–564 (1977); (b) A. M. Maxam and W. Gilbert, *Methods Enzymol.* **65**, 499–560 (1980); (c) W. Gilbert, *Science* **214**, 1305–1312 (1981).

160. D. A. Peattie, *Proc. Natl. Acad. Sci. U.S.A.* **76**, 1760–1764 (1979).

161. T. Ikemura and J. Dahlberg, *J. Biol. Chem.* **248**, 5024–5032 (1978); T. Ikemura and H. Ozeki, *J. Mol. Biol.* **117**, 419–446 (1977).

162. G. S. Hayward, *Virology* **49**, 342–344 (1972).

163. H. G. Khorana and J. P. Vizzsolyi, *J. Am. Chem. Soc.* **83**, 675–686 (1961).

164. R. T. Walker and U. L. RajBhandary, *Nucleic Acids Res.* **5**, 57–70 (1978).

165. B. S. Zimmerman, J. W. Little, C. K. Oshinsky, and M. Gellert, *Proc. Natl. Acad. Sci. U.S.A.* **57**, 1841–1848 (1967); (b) B. M. Olivera and I. R. Lehman, *Proc. Natl. Acad. Sci. U.S.A.* **57**, 1700–1704 (1967).

166. C. C. Richardson, *Proc. Natl. Acad. Sci. U.S.A.* **54**, 158–162 (1965).

167. W. E. Razzell and H. G. Khorana, *J. Biol. Chem.* **234**, 2105–2113 (1959).

168. H. G. Khorana, *The Enzymes* (P. D. Boyer, H. Lardy, and K. Myrbäck eds.), Vol. 5, pp. 79–94, Academic Press, New York, 1961.

169. A. Kuninaka, M. Fujimoto, and H. Yoshida, *Agric. Biol. Chem.* **39**, 597–602 (1975).

170. R. C. Wiegand, G. N. Godson, and C. M. Radding, *J. Biol. Chem.* **250**, 8848–8855 (1975).

171. C. B. Anfinsen and F. H. White, Jr., in *The Enzymes* (P. D. Boyer, H. Lardy, and K. Myrbäck, eds.), Vol. 5, pp. 95–122, Academic Press, New York, 1961.

172. F. Egami, K. Takashima, and T. Uchida, in *Progress in Nucleic Acids Research and Molecular Biology* (J. N. Davidson and W. E. Cohn, eds.), Vol. 3, pp. 59–101, Academic Press, New York, 1964.

173. R. W. Holley, J. Apgar, G. A. Everett, J. T. Madison, M. Marquisse, S. H. Merrill, J. R. Penswick, and A. Zamir, *Science* **147**, 1462–1465 (1965).

174. M. Imazawa, M. Irie, and T. Ukita, *Biochem. J.* **64**, 595–602 (1968).

175. T. Uchida and F. Egami, *The Enzymes* (P. D. Boyer, ed.), Vol. 4, pp. 205–250, Academic Press, New York, 1971.

176. H. Berkower, J. Leis, and J. Hurwitz, *J. Biol. Chem.* **248**, 5914–5926 (1973); D. Baltimore and D. F. Smoler, *Proc. Natl. Acad. Sci. U.S.A.* **72**, 3369–3372 (1975).

177. H. Donis-Keller, *Nucleic Acids Res.* **1**, 179–192 (1979).

178. P. D. Sadowski and I. Bakyta, *J. Biol. Chem.* **247**, 405–412 (1972).

179. H. D. Smith and W. Wilcox, *J. Mol. Biol.* **51**, 379–391 (1970); A. H. A. Bingham and T. Atkinson, *Biochem. Rev.* **6**, 315–324 (1978); W. Arber, *Angew. Chem. Int. Ed.* **17**, 73–79 (1978); D. Nathans, *Science* **206**, 903–909 (1979).

180. F. Sanger, *Science* **214**, 1205–1210 (1981).

181. G. G. Brownlee and F. Sanger, *Eur. J. Biochem.* **11**, 395–399 (1969); F. Sanger, J. E. Donelson, A. R. Coulson, H. Kössel, and D. Fischer, *J. Mol. Biol.* **90**, 315–333 (1974); E. Jay, R. Bambara, P. Padmanabhan, and R. Wu, *Nucleic Acids Res.* **1**, 331–353 (1974).

182. C. D. Tu and R. Wu, *Methods Enzymol.* **65**, 620–638 (1980).

183. E. Ohtsuka, T. Ono, and M. Ikehara, *Chem. Pharm. Bull. (Tokyo)* **29**, 3274–3280 (1981).

184. J. Messing and J. Vieira, *Gene* **19**, 269–276 (1982); J. Messing, *Methods Enzymol.* **101**, 20–78 (1983).

185. N. P. Higgins and N. R. Cozzarelli, *Methods Enzymol.* **68**, 50–71 (1979).

186. N. K. Gupta, E. Ohtsuka, H. Weber, S. H. Chang, and H. G. Khorana, *Proc. Natl. Acad. Sci. U.S.A.* **60**, 285–289 (1968); N. K. Gupta, E. Ohtsuka, V. Sgaramella, H. Büchi, A. Kumar, H. Weber, and H. G. Khorana, *Proc. Natl. Acad. Sci. U.S.A.* **60**, 1334–1342 (1968).

187. K. Kleppe, J. H. van de Sande, and H. G. Khorana, *Proc. Natl. Acad. Sci. U.S.A.* **67**, 68–71 (1970); G. C. Fareed, E. M. Wilt, and C. C. Richardson, *J. Biol. Chem.* **246**, 925–932 (1971).

188. V. Sgaramella and H. G. Khorana, *J. Mol. Biol.* **72**, 427–444 (1972); C. M. Tsiapalis and S. A. Narang, *Biochem. Biophys. Res. Commun.* **39**, 631–636 (1970).

189. T. Sekiya, P. Besmer, T. Takeya, and H. G. Khorana, *J. Biol. Chem.* **251**, 634–641 (1976); A. Panet, R. Kleppe, K. Kleppe, and H. G. Khorana, *J. Biol. Chem.* **251**, 651–657 (1976).

190. (a) C. Yanofsky, T. Platt, I. P. Crawford, B. P. Nichols, G. E. Christie, H. Horowitz, M. Van-Cleemput, and A. M. Wu, *Nucleic Acids Res.* **9**, 6647–6668 (1981); (b) T. Sekiya, E. L. Brown, R. Belagaje, H.-J. Fritz, M. J. Gait, R. G. Lees, M. J. Ryan, and H. G. Khorana, *J. Biol. Chem.* **254**, 5781–5786 (1979); (c) K. Backman and M. Ptashne, *Cell* **13**, 65–71 (1978); (d) B. Polisky, R. J. Bishopand, and D. H. Gelfand, *Proc. Natl. Acad. Sci. U.S.A.* **73**, 3900–3904 (1976); (e) F. Bolivar, R. L. Rodriguez, P. J. Greene, M. C. Betlach, H. L. Heneker, and H. W. Boyer, *Gene* **2**, 95–113 (1977); (f) W. Tacon, N. Carey, and S. Emtage, *Mol. Gen. Genet.* **177**, 427–438 (1980); (g) P. P. Stepien, R. Brousseau, R. Wu, S. Narang, and D. Y. Thomas, *Gene* **24**, 289–297 (1983); (h) D. V. Goeddel, H. M. Shepard, E. Yelverton, D. Leung, and R. Crea, *Nucleic Acids Res.* **8**, 4057–4074 (1980).

191. (a) M. T. Doel, M. Eaton, E. A. Cook, H. Lewis, T. Patel, and N. H. Carey, *Nucleic Acids Res.* **8**, 4578–4574 (1980); (b) M. D. Edge, A. R. Greene, G. R. Heathcliffe, V. E. Moore, N. J. Faulkner, R. Camble, N. N. Pelter, P. Truemen, W. Schuch, J. Hennam, T. C. Atkinson, C. R. Newton, and A. F. Markham, *Nucleic Acids Res.* **11**, 6419–6435 (1983); (c) R. Brousseau, R. Searpulla, W. Sung, H. M. Hsiung, S. A. Narang, and R. Wu, *Gene* **17**, 279–289 (1982); (d) H. Rink, M. Liersch, P. Sieber, and F. Meyer, *Nucleic Acids Res.* **12**, 6369–6387 (1984); (e) K. P. Nambiar, J. Stackhoure, D. M. Stanffer, W. P. Kennedy, J. K. Eldredge, and S. A. Benner, *Science* **223**, 1299–1301 (1984); (f) M. Ikehara, E. Ohtsuka, T. Tokunaga, S. Nishikawa, S. Uesugi, T. Tanaka, Y. Aoyama, S. Kikyodani, K. Fujimoto, K. Yanase, K. Fuchimura, and H. Morioka, *Proc. Natl. Acad. Sci. U.S.A.* **83**, 4695–4699 (1986).

192. W. Gilbert, N. Maizels, and A. Maxam, *Cold Spring Harbor Symp. Quant. Biol.* **38**, 845–857 (1973).

193. (a) M. H. Caruthers, *Acc. Chem. Res.* **13**, 155–160 (1980); (b) C. P. Bahl, R. Wu, and S. A. Narang, *Methods Enzymol.* **65**, 877–885 (1980); (c) U. Siebenlist, R. B. Simpson, and W. Gilbert, *Cell* **20**, 269–281 (1980).

194. T. Aoyama, M. Takanami, E. Ohtsuka, Y. Taniyama, R. Marumoto, H. Sato, and M. Ikehara, *Nucleic Acids Res.* **11**, 5855–5864 (1983).

195. J. D. Windness, C. P. Newton, J. De Maeyer-Guignard, V. E. More, A. F. Markham, and M. D. Edge, *Nucleic Acids Res.* **10**, 6639–6657 (1982).

196. J. Rommens, D. Macknight, L. Pomeroy-Cloney, and E. Jay, *Nucleic Acids Res.* **11**, 5921–5940 (1983).

197. E. Jay, J. Rommens, L. Pomeroy-Cloney, D. Macknight, C. Lutze-Wallace, P. Wishart, D. Harrison, W.-Y. Lui, V. Asundi, M. Dawood, and F. Jay, *Proc. Natl. Acad. Sci. U.S.A.* **81**, 2290–2294 (1984).

198. S. J. Chan, B. E. Noyes, K. L. Agarwal, and D. F. Steiner, *Proc. Natl. Acad. Sci. U.S.A.* **76**, 5036–5040 (1979); K. L. Agarwal, J. Brunstedt, and B. E. Noyes, *J. Biol. Chem.* **256**, 1023–1029 (1981).

199. (a) B. Wallace, P. F. Johnson, S. Tanaka, S. Schöld, K. Itakura, and J. Abelson, *Science* **209**, 1396–1400 (1980); (b) G. Dalbadie-McFarland, L. W. Cohen, A. D. Riggs, C. Morin, K. Itakura, and J. H. Richards, *Proc. Natl. Acad. Sci. U.S.A.* **79**, 6409–6413 (1982).

200. (a) C. A. Hachison III, S. Phillips, M. H. Edgall, S. Gillam, P. Jahnke, and M. Smith, *J. Biol. Chem.* **253**, 6551–6560 (1978); (b) A. Razin, T. Hirose, K. Itakura, and A. D. Riggs, *Proc. Natl. Acad. Sci. U.S.A.* **75**, 4268–4270 (1978).

201. M. Smith, in *Methods of RNA and DNA Sequencing* (S. M. Weissmann, ed.), Praeger Scientific, New York, 1964.

202. I. Kudo, M. Leineweber, and U. L. RajBhandary, *Proc. Natl. Acad. Sci. U.S.A.* **78**, 4753–4757 (1981).

203. J. E. Villafrance, E. E. Howell, D. H. Voet, M. S. Strobel, R. C. Ogden, J. N. Abelson, and J. Kraut, *Science* **222**, 782–788 (1984).

204. V. G. Malathi and J. Hurwitz, *Proc. Natl. Acad. Sci. U.S.A.* **69**, 3009–3013 (1972).

205. J. W. Cranston, R. Silver, V. G. Malathi, and J. Hurwitz, *J. Biol. Chem.* **249**, 7447–7456 (1974).

206. M. Ikehara, *Nucleic Acids Res.* **3**, 1613–1623 (1976).

207. (a) G. Kaufmann, T. Klein, and U. Z. Littauer, *FEBS Lett.* **46**, 271–275 (1974); (b) G. C. Walker, O. C. Uhlenbeck, E. Bedows, and R. I. Gumport, *FEBS Lett.* **72**, 122–126 (1975); (c) E. Ohtsuka, T. Doi, H. Uemura, Y. Taniyama, and M. Ikehara, *Nucleic Acids Res.* **8**, 3909–3916 (1980); (d) E. Romaniak, L. W. McLaughlin, T. Neilson, and P. J. Romaniuk, *Eur. J. Biochem.* **125**, 639–643 (1982).

208. E. Ohtsuka, T. Doi, R. Fukumoto, J. Matsugi, and M. Ikehara, *Nucleic Acids Res.* **11**, 3863–3872 (1983).

209. T. Doi, A. Yamane, E. Ohtsuka, and M. Ikehara, *Nucleic Acids Res.* **13**, 3685–3697 (1985).

210. A. G. Bruce and O. C. Uhlenbeck, *Biochemistry* **21**, 855–861 (1982).

211. L. Bare, A. G. Bruce, R. Gesteland, and O. C. Uhlenbeck, *Nature* **305**, 554–556 (1983).

212. S. J. Rose III, P. T. Lowary, and O. C. Uhlenbeck, *J. Mol. Biol.* **167**, 103–117 (1983).

213. N. R. Pace, *Proc. Natl. Acad. Sci. U.S.A.* **75**, 3045–3049 (1978).

214. J. D. Watson and F. H. C. Crick, *Nature* **171**, 737–738 (1953).

215. G. Gupta, M. Bansal, and V. Sasisekharan, *Proc. Natl. Acad. Sci. U.S.A.* **77**, 6488–6490 (1980).

216. S. A. Arnott, *Prog. Biophys. Mol. Biol.* **21**, 267–319 (1970).

217. J. M. Resenberg, N. C. Seeman, J. J. P. Kim, F. L. Suddath, H. B. Nicholes, and A. Rich, *Nature* **243**, 150–154 (1973).

218. N. C. Seeman, J. M. Resenberg, F. L. Suddath, J. J. P. Kim, and A. Rich, *J. Mol. Biol.* **104**, 109–114 (1976).

219. R. O. Day, N. C. Seeman, J. M. Resenberg, and A. Rich, *Proc. Natl. Acad. Sci. U.S.A.* **70**, 849–853 (1973).

220. J. M. Resenberg, N. C. Seeman, R. O. Day, and A. Rich, *J. Mol. Biol.* **104**, 145–167 (1976).

221. S. H. Kim, J. Weinzierl, and A. Rich, *Science* **179**, 285–288 (1973).

222. J. D. Robertus, J. E. Ladner, J. T. Finch, D. Rhodes, R. S. Brown, B. F. C. Clark, and A. Klug, *Nature* **250**, 546–551 (1974).

223. J. E. Ladner, A. Jack, J. D. Robertus, R. S. Brown, D. Rhodes, B. F. C. Clark, and A. Klug, *Nucleic Acids Rres.* **2**, 1629–1638 (1975).

224. G. J. Quigley, N. C. Seeman, A. H.-J. Wang, F. L. Suddath, and A. Rich, *Nucleic Acids Res.* **2**, 2329–2341 (1975).

225. J. L. Sussman and S. H. Kim, *Biochem. Biophys. Res. Commun.* **68**, 89–96 (1976).

226. C. D. Stout, H. Mizuno, J. Rubin, T. Brennan, S. T. Rao, and M. Sundaralingam, *Nucleic Acids Res.* **3**, 1111–1123 (1976).

227. A. H.-J. Wang, G. J. Quigley, F. J. Kolpak, J. L. Crawford, J. H. van Boom, G. van der Marel, and A. Rich, *Nature* **282**, 680–686 (1979).

228. E. Shefter, M. Barlow, R. A. Sparks, and K. N. Trueblood, *Acta Crystallogr. B* **25**, 895–909 (1969).

229. N. C. Seeman, J. L. Sussman, H. M. Berman, and S. H. Kim, *Nature New Biology* **233**, 90–92 (1971).

230. J. L. Sussman, N. C. Seeman, S.-H. Kim, and H. M. Berman, *J. Mol. Biol.* **66**, 403–421 (1972).

231. J. Rubin, T. Brennan, and M. Sundaralingam, *Biochemistry* **11**, 3112–3128 (1972).

232. N. Camerman, J. K. Famcett, and A. Camerman, *Science* **182**, 1142–1143 (1973).

233. N. Cammerman, J. K. Fawcett, and A. Camerman, *J. Mol. Biol.* **107**, 601–621 (1976).

234. D. Suck, P. C. Manor, G. Germain, C. H. Schwalbe, G. Weimann, and W. Saenger, *Nature New Biol.* **246**, 161–165 (1973).

235. D. Suck, P. C. Manor, and W. Saenger, *Acta Crystallogr. B* **32**, 1727–1737 (1976).

236. H. R. Wilson and J. Al-Mukhtar, *Nature* **263**, 171–172 (1976).

237. M. A. Viswamitra, O. Kennard, P. G. Jones, G. M. Sheldrick, S. Salisbury, L. Falvello, and Z. Shakked, *Nature* **273**, 687–688 (1978).

238. M. A. Viswamitra, Z. Shakked, P. G. Jones, G. M. Sheldrick, S. A. Salisbury, and O. Kennard, *Biopolymers* **21**, 513–533 (1982).

239. H. Einspahr, W. J. Cook, and C. E. Bugg, *Biochemistry* **20**, 5788–5794 (1981).

240. S. Fujii, K. Hamada, R. Miura, S. Uesugi, M. Ikehara, and K. Tomita, *Acta Crystallogr. B* **38**, 564–570 (1982).

241. K. Hamada, Y. Matuo, A. Miyamae, S. Fujii, and K. Tomita, *Acta Crystallogr. B* **38**, 2528–2531 (1982).

242. R. Parthasarathy, M. Malik, and S. M. Fridey, *Proc. Natl. Acad. Sci. U.S.A.* **79**, 7292–7296 (1982).

243. W. B. T. Cruse, E. Egert, O. Kennard, G. B. Sala, S. A. Salisburg, and M. A. Viswamitra, *Biochemistry* **22**, 1833–1939 (1983).

244. A. H.-J. Wang, G. J. Quigley, and F. J. Kolpak, G. van der Marel, J. H. van Boom, and A. Rich, *Science* **211**, 171–176 (1981).

245. H. Drew, T. Takano, S. Tanaka, K. Itakura, and R. E. Dickerson, *Nature* **286**, 567–573 (1980).

246. J. L. Crawford, F. J. Kolpak, A. H.-J. Wang, G. J. Quigley, J. H. van Boom, G. van der Marel, and A. Rich, *Proc. Natl. Acad. Sci. U.S.A.* **77**, 4016–4020 (1980).

247. R. Wing, H. Drew, T. Takano, C. Broka, S. Tanaka, K. Itakura, and R. E. Dickerson, *Nature* **287**, 755–758 (1980).

248. H. R. Drew and R. E. Dickerson, *J. Mol. Biol.* **151**, 535–556 (1981).

249. R. E. Dickerson and H. R. Drew, *J. Mol. Biol.* **149**, 761–786 (1981).

250. H. R. Drew, R. M. Wing, T. Takano, C. Broka, S. Tanaka, K. Itakura, and R. E. Dickerson, *Proc. Natl. Acad. Sci. U.S.A.* **78**, 2179–2183 (1981).

251. M. L. Kopka, A. V. Fratini, H. R. Drew, and R. E. Dickerson, *J. Mol. Biol.* **163**, 129–146 (1983).

252. H. R. Drew and R. E. Dickerson, *J. Mol. Biol.* **152**, 723–736 (1981).

253. A. H.-J. Wang, S. Fujii, J. H. van Boom, and A. Rich, *Proc. Natl. Acad. Sci. U.S.A.* **79**, 3968–3972 (1982).

254. A. H.-J. Wang, S. Fujii, J. H. van Boom, G. A. van der Marel, S. A. A. van Boeckel, and A. Rich, *Nature* **299**, 601–604 (1982).

255. R. E. Dickerson, H. R. Drew, B. N. Conner, R. M. Wing, A. V. Fratini, and M. L. Kopka, *Science* **216**, 475–485 (1982).

256. A. V. Fratini, M. L. Kopla, H. R. Drew, and R. E. Dickerson, *J. Biol. Chem.* **257**, 14686–14707 (1982).

257. S. Fujii, A. H.-J. Wang, G. van der Marel, J. H. van Boom, and A. Rich, *Nucleic Acids Res.* **10**, 7879–7892 (1982).

258. B. N. Conner, T. Takano, S. Tanaka, K. Itakura, and R. E. Dickerson, *Nature* **295**, 294–299 (1982).

259. B. N. Conner, C. Yoon, J. L. Dickerson, and R. E. Dickerson, *J. Mol. Biol.* **174**, 663–695 (1984).

260. T. Hakoshima, A. H.-J. Wang, J. H. van Boom, and A. Rich, *Nucleic Acids Symp. Ser.* **12**, 213–216 (1983).

261. Z. Shakked, D. Rabinovich, O. Kennard, W. B. T. Cruse, S. A. Salisbury, and M. A. Viswamitra, *J. Mol. Biol.* **166**, 183–201 (1983).

262. S. Uesugi, M. Ohkubo, H. Urata, M. Ikehara, Y. Kobayashi, and Y. Kyogoku, *J. Am. Chem. Soc.* **106**, 3675–3676 (1984).

263. S. Arnott, R. Chandrasekaran, D. L. Birdsall, A. G. W. Leslie, and R. L. Ratliff, *Nature* **283**, 743–745 (1980).

264. S. Arnott, D. W. L. Hukins, S. D. Dover, W. Fuller, and A. R. Hodgson, *J. Mol. Biol.* **81**, 107–122 (1973).

265. A. Jack, J. E. Ladner, and A. K. Lug, *J. Mol. Biol.* **108**, 618–649 (1976).

266. IUPAC-IUB Joint Commission on Biochemical Nomenclature, Abbreviations and Symbols for the Description of Conformations of Polynucleotide Chains Recommendations 1982, *Eur. J. Biochem.* **131**, 9–15 (1983).

267. C. Altona and M. Sundaralingam, *J. Am. Chem. Soc.* **94**, 8205–8212 (1972).

268. H. Mizuno, K. Tomita, E. Nakagawa, E. Ohtsuka, and M. Ikehara, *J. Mol. Biol.* **148**, 103–106 (1981).

269. S. Fujii and K. Tomita, unpublished work.

270. P. O. P. Ts'o, N. S. Kondo, M. P. Schweizer, and D. P. Hollis, *Biochemistry* **8**, 997–1029 (1969).

271. C.-H. Lee, F. S. Ezra, N. S. Kondo, R. H. Sarma, and S. S. Danyluk, *Biochemistry* **15**, 3627–3639 (1976).

272. F.-S. Ezra, C.-H. Lee, N. S. Kondo, S. S. Danyluk, and R. H. Sarma, *Biochemistry* **16**, 1977–1987 (1977).

273. D. M. Cheng and R. H. Sarma, *J. Am. Chem. Soc.* **99**, 7333–7348 (1977).

274. C. S. M. Olsthoorn, L. J. Bostelaar, J. H. van Boom, and C. Altona, *Eur. J. Biochem.* **112**, 95–110 (1980).

275. S. Tran-Dinh, J. M. Neumann, T. Huynh-Dinh, J. Igolen, and S. K. Kan, *Org. Magn. Reson.* **18**, 148–152 (1982).

276. J.-M. Neumann, T. Huynh-Dinh, S. K. Kan, B. Genissel, J. Igolen, and S. Tran-Dinh, *Eur. J. Biochem.* **121**, 317–323 (1982).

277. D. M. Cheng, L.-S. Kan, E. E. Leutzinger, K. Jayaraman, P. S. Miller, and P. O. P. Ts'o, *Biochemistry* **21**, 621–630 (1982).

278. D. M. Cheng, L.-S. Kan, V. L. Iuorno, and P. O. P. Ts'o, *Biopolymers* **23**, 575–592 (1984).

279. D. M. Crothers, C. W. Hilbers, and R. G. Shulman, *Proc. Natl. Acad. Sci. U.S.A.* **70**, 2899–2901 (1973).

280. D. J. Patel and A. E. Tonelli, *Biopolymers* **13**, 1943–1964 (1974).

281. D. J. Patel and C. W. Hilbers, *Biochemistry* **14**, 2651–2655 (1975).

282. C. W. Hilbers and D. J. Patel, *Biochemistry* **14**, 2656–2660 (1975).

283. N. R. Kallenbach, W. E. Daniel, Jr., and M. A. Kaminker, *Biochemistry* **15**, 1218–1224 (1976).

284. D. J. Patel, *Biopolymers* **15**, 533–558 (1976).

285. D. J. Patel, *Biopolymers* **16**, 1635–1656 (1977).

286. T. A. Early, D. R. Kearns, J. F. Burd, J. E. Larson, and R. D. Wells, *Biochemistry* **16**, 541–551 (1977).

287. D. J. Patel and L. L. Canuel, *Eur. J. Biochem.* **96**, 267–276 (1979).

288. E. R. P. Zuiderweg, R. M. Scheek, G. Veeneman, J. H. van Boom, R. Kaptein, H. Ruterjans, and K. Beyreuther, *Nucleic Acids Res.* **9**, 6553–6569 (1981).

289. S. Uesugi, T. Shida, and M. Ikehara, *Chem. Pharm. Bull. (Tokyo)* **29**, 3573–3585 (1981).

290. D. J. Patel, S. A. Kozlowski, L. A. Marky, C. Broka, J. A. Rice, K. Itakura, and K. J. Breslauer, *Biochemistry* **21**, 428–436 (1982).

291. D. J. Patel, A. Pardi, and K. Itakura, *Science* **216**, 581–590 (1982).

292. A. Pardi, K. M. Morden, D. J. Patel, and I. Tinoco, Jr., *Biochemistry* **21**, 6567–6574 (1982).

293. D. J. Patel, S. A. Kozlowski, L. A. Marky, J. A. Rice, C. Broka, K. Itakura, and K. J. Breslauer, *Biochemistry* **21**, 445–451 (1982).

294. D. J. Patel, S. A. Kozlowski, L. A. Marky, J. A. Rice, C. Broka, J. Dallas, K. Itakura, and K. J. Breslauer, *Biochemistry* **21**, 437–444 (1982).

295. D. J. Patel, S. A. Kozlowski, L. A. Marky, J. A. Rice, C. Broka, K. Itakura, and K. J. Breslauer, *Biochemistry* **21**, 451–455 (1982).

296. L. S. Kan, D. M. Cheng, K. Jayaraman, E. E. Leutzinger, P. S. Miller, and P. O. P. Ts'o, *Biochemistry* **21**, 6723–6732 (1982).

297. R. M. Scheek, N. Russo, R. Boelens, and R. Kaptein, *J. Am. Chem. Soc.* **105**, 2914–2916 (1983).

298. D. J. Patel, S. A. Kozlowski, and R. Bhatt, *Proc. Natl. Acad. Sci. U.S.A.* **80**, 3908–3912 (1983).

299. K. M. Morden, Y. G. Chu, F. H. Martin, and I. Tinoca, Jr., *Biochemistry* **22**, 5557–5563 (1983).

300. D. G. Reid, S. A. Salisbury, S. Bellard, Z. Shakked, and D. H. Williams, *Biochemistry* **22**, 2019–2025 (1983).

301. S. M. Freier, D. D. Albergo, and D. H. Turner, *Biopolymers* **22**, 1107–1131 (1983).

302. D. G. Reid, S. A. Salisbury, and D. H. Williams, *Nucleic Acids Res.* **11**, 3779–3793 (1983).

303. A. Pardi, R. Walker, H. Rapoport, G. Wider, and K. Wuethrich, *J. Am. Chem. Soc.* **105**, 1652–1653 (1983).

304. S. H. Chou, D. R. Hare, D. E. Wemmer, and B. R. Reid, *Biochemistry* **22**, 3037–3041 (1983).

305. J. R. Mellema, P. N. van Kampen, C. N. Carlson, H. E. Bosshard, and C. Altona, *Nucleic Acids Res.* **11**, 2893–2905 (1983).

306. M. R. Sanderson, J. R. Mellema, G. A. van der Marel, G. Wille, J. H. van Boom, and C. Altona, *Nucleic Acids Res.* **11**, 333–3346 (1983).

307. S. Tran-Dinh, J. M. Neumann, J. Taboury, T. Huynh-Dinh, S. Renous, B. Genissel, and J. Igolen, *Eur. J. Biochem.* **133**, 579–589 (1983).

308. L. S. Kan, S. Chandrasekharan, S. M. Pulford, and P. S. Miller, *Proc. Natl. Acad. Sci. U.S.A.* **80**, 4263–4265 (1983).

309. E. L. Ulrich, E. M. M. John, G. R. Gough, M. J. Brunden, P. T. Gilham, W. M. Westler, and J. L. Markley, *Biochemistry* **22**, 4362–4365 (1983).

310. D. G. Reid, S. A. Salisburg, T. Brown, D. H. Williams, J. J. Vasseur, B. Rayner, and J. L. Imbach, *Eur. J. Biochem.* **135**, 307–314 (1983).

311. J. Feigon, W. A. Williams, W. Leupin, and D. R. Kearns, *Biochemistry* **22**, 5930–5942 (1983).

312. J. Feigon, W. Leupin, W. A. Denny, and D. R. Kearns, *Biochemistry* **22**, 5943–5951 (1983).

313. G. M. Clore and A. M. Gronenborn, *EMBO J.* **2**, 2109–2115 (1983).

314. D. R. Hare, D. E. Wemmer, S. H. Chou, G. Drobny, and B. R. Reid, *J. Mol. Biol.* **17**, 319–336 (1983).

315. S. J. Lee, H. Akutsu, Y. Kyogoku, K. Kitani, Z. Tozuka, A. Ohta, E. Ohtsuka, and M. Ikehara, *Nucleic Acids Symp. Ser.* **12**, 197–200 (1983).

316. M. Petersheim, M. S. Matthew, and J. A. Gerlt, *J. Am. Chem. Soc.* **106**, 439–440 (1984).

317. R. M. Scheek, R. Boelens, N. Russo, J. H. van Boom, and R. Kaptein, *Biochemistry* **23**, 1371–1376 (1984).

318. S. Tran-Dinh, J. Taboury, J. M. Newmann, T. Huynh-Dinh, B. Genissel, B. L. d'Estaintot, and J. Igolen, *Biochemistry* **23**, 1362–1371 (1984).

319. J. Feigon, A. H. J. Wang, G. A. van der Marel, J. H. van Boom, and A. Rich, *Nucleic Acids Res.* **12**, 1243–1263 (1984).

320. D. J. Patel, S. Ikuta, S. Kozlowski, and K. Itakura, *Proc. Natl. Acad. Sci. U.S.A.* **80**, 2184–2188 (1983).

321. R. M. Scheek, N. Russo, R. Boelens, R. Kaptein, and J. H. van Boom, *J. Am. Chem. Soc.* **105**, 2914–2916 (1983).

322. A. M. Gronenborn, G. M. Clore, M. B. Jones, and J. Jiricny, *FEBS Lett.* **165**, 216–222 (1984).

323. M. A. Weiss, D. J. Patel, R. T. Sauer, and M. Karplus, *Proc. Natl. Acad. Sci. U.S.A.* **81**, 130–134 (1984).

324. S. H. Chou, D. R. Hare, and B. R. Reid, *Biochemistry* **23**, 2262–2268 (1984).

325. S.-H. Chou, D. E. Wemmer, D. R. Hare, and B. R. Reid, *Biochemistry* **23**, 2257–2262 (1984).

326. M. S. Broido, G. Zon, and T. L. James, *Biochem. Biophys. Res. Commun.* **119**, 663–670 (1984).

327. C. Giessner-Prettre, B. Pullman, S. Tran-Dinn, J. M. Neumann, T. Huynh-Dinh, and J. Igolen, *Nucleic Acids Res.* **12**, 3271–3281 (1984).

328. G. M. Clore and A. M. Gronenborn, *Eur. J. Biochem.* **141**, 119–129 (1984).

329. D. J. Patel, S. A. Kozlowski, S. Ikuta, and K. Itakura, *Biochemistry* **23**, 3218–3226 (1984).

330. D. J. Patel, S. A. Kozlowski, S. Ikuta, and K. Itakura, *Biochemistry* **23**, 3207–3217 (1984).

331. D. M. Cheng, L. S. Kan, D. Frechet, P. O. P. Ts'o, S. Uesugi, T. Shida, and M. Ikehara, *Biopolymers* **23**, 775–795 (1984).

332. J.-R. Mellema, A. K. Jellema, C. A. G. Haasnoot, J. H. van Boom, and C. Altona, *Eur. J. Biochem.* **141**, 165–175 (1984).

333. S. Cheung, K. Arnst, and P. Lu, *Proc. Natl. Acad. Sci. U.S.A.* **81**, 3665–3669 (1984).

334. N. Tibanyenda, S. H. de Bruin, C. A. G. Haasnoot, G. A. van der Marel, J. H. van Boom, and C. W. Hilbers, *Eur. J. Biochem.* **139**, 19–27 (1984).

335. P. A. Kroon, G. P. Kreishman, J. H. Nelson, and S. I. Chan, *Biopolymers* **13**, 2571–2592 (1974).

336. F. E. Evans and R. H. Sarma, *Nature* **263**, 567–572 (1976).

337. D. M. Chang, S. S. Danyluk, M. M. Dhingra, F. S. Ezra, M. MacCoss, C. K. Mitra, and R. H. Sarma, *Biochemistry* **19**, 2491–2497 (1980).

338. J. R. Everett, D. W. Hughes, R. A. Bell, D. Alkema, T. Neilson, and P. J. Romaniuk, *Biopolymers* **19**, 557–573 (1980).

339. R. A. Bell, J. R. Everett, D. W. Hughes, D. Alkema, P. A. Hader, T. Neilson, and P. J. Romaniuk, *Biopolymers* **20**, 1383–1398 (1981).

340. M. P. Stone, D. L. Johnson, and P. N. Borer, *Biochemistry* **20**, 3604–3610 (1981).

341. P. A. Hader, T. Neilson, D. Alkema, E. C. Kofoid, and M. C. Ganoza, *FEBS Lett.* **136**, 65–69 (1981).

342. P. P. Lankhorst, C. M. Groeneveld, G. Wille, J. H. van Boom, and C. Altona, *Recl. Trav. Chim. Pays-Bas* **101**, 253–263 (1982).

343. P. A. Hader, D. Alkema, R. A. Bell, and T. Neilson, *J. Chem. Soc. Chem. Commun.*, 10–12 (1982).

344. J. Doornbos, C. T. J. Wreesmann, J. H. van Boom, and C. Altona, *Eur. J. Biochem.* **131**, 571–579 (1983).

345. B. W.-K. Shum and D. M. Crothers, *Biopolymers* **22**, 919–933 (1983).

346. C.-H. Lee, *Eur. J. Biochem.* **137**, 347–356 (1983).

347. A. M. Gronenborn, B. J. Kimber, G. M. Clore, and L. W. McLaughlin, *Nucleic Acids Res.* **11**, 5691–5699 (1983).

348. P. P. Lankhorst, G. Wille, J. H. van Boom, and A. Altona, *Nucleic Acids Res.* **11**, 2839–2856 (1983).

349. J. Doornbos, C. T. J. Wreesmann, J. H. van Boom, and C. Altona, *Eur. J. Biochem.* **131**, 571–579 (1983).

350. D. B. Arter, G. C. Walker, O. C. Uhlenbeck, and P. G. Schmidt, *Biochem. Biophys. Res. Commun.* **61**, 1089–1094 (1974).

351. L. S. Kan, P. N. Borer, and P. O. P. Ts'o, *Biochemistry* **14**, 4864–4869 (1975).

352. P. N. Borer, L. S. Kan, and P. O. P. Ts'o, *Biochemistry* **14**, 4847–4863 (1975).

353. D. W. Hughes, R. A. Bell, T. E. England, and T. Neilson, *Can. J. Chem.* **56**, 2244–2248 (1978).

354. P. J. Romaniuk, T. Neilson, D. W. Hughes, and R. A. Bell, *Can. J. Chem.* **56**, 2250–2252 (1978).

355. D. Alkema, R. A. Bell, P. A. Hader, and T. Neilson, *J. Am. Chem. Soc.* **103**, 2866–2868 (1981).

356. E. Bubienko, M. A. Uniak, and P. N. Borer, *Biochemistry* **20**, 6987–6994 (1981).

357. S. Uesugi, E. Nakagawa, E. Ohtsuka, M. Ikehara, M. Watanabe, Y. Kobayashi, Y. Kyogoku, and M. Kainosho, *J. Am. Chem. Soc.* **104**, 7340–7341 (1982).

358. Y. Kyogoku, M. Watanabe, Y. Kobayashi, M. Kainosho, S. Uesugi, E. Nakagawa, E. Ohtsuka, and M. Ikehara, *Nucleic Acids Res. Symp. Ser.* **11**, 273–276 (1982).

359. P. L. D'Andrea, D. Alkema, R. A. Bell, J. M. Coddington, P. A. Hader, D. W. Hughes, and T. Neilson, *J. Am. Chem. Soc.* **105**, 636–638 (1983).

360. M. Petersheim and D. H. Turner, *Biochemistry* **22**, 264–268 (1983).

361. M. Petersheim and D. H. Turner, *Biochemistry* **22**, 269–277 (1983).

362. H. Fritzsche, L.-S. Kan, and P. O. P. Ts'o, *Biochemistry* **22**, 277–280 (1983).

363. S. Uesugi, M. Ohkubo, E. Ohtsuka, M. Ikehara, Y. Kobayashi, Y. Kyogoku, H. P. Westerink, G. A. van der Marel, J. H. van Boom, and C. A. G. Haasnoot, *J. Biol. Chem.* **259**, 1390–1393 (1984).

364. E. F. McCord, K. M. Morden, A. Pardi, I. Tinoco, Jr., and S. G. Boxer, *Biochemistry* **23**, 1927–1934 (1984).

365. A. Sinclair, D. Alkema, R. A. Bell, J. M. Coddington, D. W. Hughes, T. Neilson, and P. J. Romaniuk, *Biochemistry* **23**, 2656–2662 (1984).

366. H. P. Westerink, G. A. van der Marel, J. H. van Boom, and C. A. G. Haasnoot, *Nucleic Acids Res.* **12**, 4323–4338 (1984).

367. E. Selsing, R. D. Wells, T. A. Early, and D. R. Kearns, *Nature* **275**, 249–250 (1978).

368. A. Pardi, F. H. Martin, and I. Tinoco, Jr., *Biochemistry* **21**, 3986–3996 (1981).

369. A. Pardi and I. Tinoco, Jr., *Biochemistry* **21**, 4686–4693 (1982).

370. C. A. G. Haasnoot, H. P. Westerink, G. A. van der Marel, and J. H. van Boom, *J. Biomol. Struct. Dyn.* **1**, 131–149 (1983).

371. J. R. Mellema, C. A. G. Haasnoot, G. A. van der Marel, G. Wille, C. A. A. van Boeckel, J. H. van Boom, and C. Altona, *Nucleic Acids Res.* **11**, 5717–5738 (1983).

372. J.-R. Mellema, R. van der Woerd, G. A. van der Marel, J. H. van Boom, and C. Altona, *Nucleic Acids Res.* **12**, 5061–5078 (1984).

373. D. Frechet, D. M. Cheng, L.-S. Kan, and P. O. P. Ts'o, *Biochemistry* **22**, 5194–5200 (1983).

374. S. Uesugi, Y. Takatsuka, M. Ikehara, D. M. Cheng, L. S. Kan, and P. O. P. Ts'o, *Biochemistry* **20**, 3056–3062 (1981).

375. S. Uesugi, T. Kaneyasu, and M. Ikehara, *Biochemistry* **21**, 5870–5877 (1982).

376. S. Uesugi, T. Kaneyasu, J. Imura, M. Ikehara, D. M. Cheng, L.-S. Kan, and P. O. P. Ts'o, *Biopolymers* **22**, 1189–1202 (1983).

377. D. M. Cheng, L. S. Kan, P. O. P. Ts'o, S. Uesugi, Y. Takatsuka, and M. Ikehara, *Biopolymers* **22**, 1427–1444 (1983).

378. D. M. Gray, I. Tinoco, Jr., and M. J. Chamberlin, *Biopolymers* **11**, 1235–1258 (1972).

379. F. M. Pohl and T. M. Jovin, *J. Mol. Biol.* **67**, 375–396 (1972).

380. S. Uesugi, T. Shida, and M. Ikehara, *Biochemistry* **21**, 3400–3408 (1982).

381. K. Hall, P. Cruz, I. Tinoco, Jr., T. M. Jovin, and J. H. van de Sande, *Nature* **311**, 584–586 (1984).

382. H. Urata, unpublished work.

383. M. Petersheim and D. H. Turner, *Biochemistry* **22**, 256–263 (1983).

384. S. M. Freier, B. J. Burger, D. Alkema, T. Neilson, and D. H. Turner, Biochemistry **22**, 6198–6206 (1983).

385. C.-C. Tsai, S. C. Jain, and H. M. Sobell, *Proc. Natl. Acad. Sci. U.S.A.* **72**, 628–632 (1975).

386. C. Tsai, S. C. Jain, and H. M. Sobell, *J. Mol. Biol.* **114**, 301–315 (1977).

387. N. C. Seeman, R. O. Day, and A. Rich, *Nature* **253**, 324–326 (1975).

388. S. Neidle, A. Achari, G. L. Taylor, H. M. Berman, H. L. Carrell, J. P. Glusker, and W. C. Stallings, *Nature* **269**, 304–307 (197).

389. H. M. Berman, W. Stallings, H. L. Carrell, J. P. Glusker, S. Neidle, G. Taylor, and A. Achari, *Biopolymers* **18**, 2405–2429 (1979).

390. S. C. Jain, C. Tsai, and H. M. Sobell, *J. Mol. Biol.* **114**, 317–331 (1977).

391. A. H. J. Wang, J. Nathans, G. van der Marel, J. H. van Boom, and A. Rich, *Nature* **276**, 471–474 (1978).

392. T. D. Sakore, B. S. Reddy, and H. M. Sobell, *J. Mol. Biol.* **135**, 763–785 (1979).

393. B. S. Reddy, T. P. Seshadri, T. D. Sakore, and H. M. Sobell, *J. Mol. Biol.* **135**, 787–812 (1979).

394. A. H.-J. Wang, G. J. Quigley, and A. Rich, *Nucleic Acids Res.* **6**, 3879–3890 (1979).

395. S. C. Jain, K. K. Bhandary, and H. M. Sobell, *J. Mol. Biol.* **135**, 813–840 (1979).

396. H. S. Shieh, H. M. Berman, M. Dabrow, and S. Neidle, *Nucleic Acids Res.* **8**, 85–97 (1980).

397. E. Westhof and M. Sundaralingam, *Proc. Natl. Acad. Sci. U.S.A.* **77**, 1852–1856 (1980).

398. E. Westhof, S. T. Rao, and M. Sundaralingam, *J. Mol. Biol.* **142**, 331–361 (1980).

399. G. J. Quigley, A. H.-J. Wang, G. Ughetto, G. van der Marel, J. H. van Boom, and A. Rich, *Proc. Natl. Acad. Sci. U.S.A.* **77**, 7204–7208 (1980).

400. F. Takusagawa, M. Dabrow, S. Neidle, and H. M. Berman, *Nature* **296**, 466–469 (1982).

401. R. M. Wing, P. Pjura, H. R. Drew, and R. E. Dickerson, *EMBO J.* **3**, 1201–1206 (1984).

402. F. Takusagawa, B. M. Goldstein, S. Youngster, R. A. Jones, and H. M. Berman, *J. Biol. Chem.* **259**, 4714–4715 (1984).

403. D. J. Patel, *Biochemistry* **13**, 2396–2402 (1974).

404. T. R. Krugh, F. N. Wittlin, and S. P. Cramer, *Biopolymers* **14**, 197–210 (1975).

405. T. R. Krugh and C. G. Reinhardt, *J. Mol. Biol.* **97**, 133–162 (1975).

406. D. J. Patel, *Biopolymers* **15**, 533–558 (1976).

407. D. J. Patel and L. L. Canuel, *Proc. Natl. Acad. Sci. U.S.A.* **73**, 3343–3347 (1976).

408. P. Davanloo and D. M. Crothers, *Biochemistry* **15**, 5299–5305 (1976).

409. T. R. Krugh, E. S. Mooberry, and Y.-C.C. Chiao, *Biochemistry* **16**, 740–747 (1977).

410. Y.-C. C. Chiao and T. R. Krugh, *Biochemistry* **16**, 747–755 (1977).

411. C. G. Reinhardt and T. R. Krugh, *Biochemistry* **16**, 2890–2895 (1977).

412. D. J. Patel and L. L. Canuel, *Proc. Natl. Acad. Sci. U.S.A.* **74**, 2624–2628 (1977).

413. C. G. Reinhardt and T. R. Krugh, *Biochemistry* **17**, 4845–4854 (1978).

414. R. V. Kastrup, M. A. Young, and T. R. Krugh, *Biochemistry* **17**, 4855–4865 (1978).

415. C.-H. Lee and I. Tinoco, Jr., *Nature* **274**, 609–610 (1978).

416. Y.-C. C. Chia, K. G. Rao, J. W. Hook III, T. R. Krugh, and S. K. Sengupta, *Biopolymers* **18**, 1749–1762 (1979).

417. D. J. Patel, *Eur. J. Biochem.* **99**, 369–378 (1979).

418. D. J. Patel, *Biopolymers* **18**, 553–569 (1979).

419. B. W. Kalisch and J. H. van de Sande, *Nucleic Acids Res.* **6**, 1881–1894 (1979).

420. S. A. Winkle and I. Tinoco, Jr., *Biochemistry* **18**, 3833–3839 (1979).

421. M. E. Nuss, T. L. James, M. A. Apple, and P. A. Kollman, *Biochim. Biophys. Acta* **609**, 136–147 (1980).

422. S. Uesugi, T. Shida, A. Miyamae, and M. Ikehara, *Nucleic Acids Res.* **8**, s147–s150 (1980).

423. P. R. Young and N. R. Kallenbach, *Proc. Natl. Acad. Sci. U.S.A.* **77**, 6453–6457 (1980).

424. R. P. Young and N. R. Kallenbach, *J. Mol. Biol.* **145**, 785–813 (1981).

425. D. J. Patel, S. A. Kozlowski, J. A. Rice, C. Broka, and M. Itakura, *Proc. Natl. Acad. Sci. U.S.A.* **78**, 7281–7284 (1981).

426. R. L. Letsinger and M. E. Schott, *J. Am. Chem. Soc.* **103**, 7394–7396 (1981).

427. M. J. Bell, G. W. Buchanan, B. R. Hollebone, and E. D. Jones, *Can. J. Chem.* **60**, 291–298 (1982).

428. D. J. Patel, A. Pardi, and K. Italura, *Science* **216**, 581–590 (1982).

429. A. Pardi, K. M. Morden, D. J. Patel, and I. Tinoco, Jr., *Biochemistry* **22**, 1107–1113 (1983).

430. K. S. Dahl, A. Pardi, and I. Tinoco, Jr., *Biochemistry* **21**, 2730–2737 (1982).

431. S. C. Brown, R. H. Shafer, and P. A. Mirau, *J. Am. Chem. Soc.* **104**, 5504–5506 (1982).

432. D. J. Patel, *Proc. Natl. Acad. Sci. U.S.A.* **79**, 6424–6428 (1982).

433. P. Laugaa, A. Delbarre, J.-B. LePecq, and B. P. Roques, *Eur. J. Biochem.* **143**, 163–173 (1983).

434. J. H. J. Den Hartog, C. Altona, J. H. van Boom, A. T. M. Marcelis, G. A. van der Marel, and L. Rinkel, *Eur. J. Biochem.* **134**, 485–495 (1983).

435. H. Fritzsche and D. M. Crothers, *Stud. Biophys.* **97**, 43–48 (1983).

436. D. G. Reid, S. A. Salisbury, and D. H. Williams, *Biochemistry* **22**, 1377–1385 (1983).

437. E. M. Goldfield, B. A. Luxon, V. Bowie, and D. G. Gorentein, *Biochemistry* **22**, 3336–3344 (1983).

438. G. Manzini, M. L. Barcellona, M. Avittabile, and F. Quadrifoglio, *Nucleic Acids Res.* **11**, 8861–8876 (1983).

439. S. C. Brown, K. Mullis, C. Levenson, and R. H. Shafer, *Biochemistry* **23**, 403–408 (1984).

440. J. W. Nelson and I. Tinoco, Jr., *Biopolymers* **23**, 213–233 (1984).

441. B. van Hemelryck, E. Guittet, G. Chottard, J.-P. Girault, T. Huynh-Dinh, J.-Y. Lallemand, J. Igolen, and J.-C. Chottard, *J. Am. Chem. Soc.* **106**, 3037–3039 (1984).

442. 442. U. Asseline, M. Dalarue, G. Lancelot, F. Toulme, N. T. Thuong, T. Montenay-Garestier, and C. Helene, *Proc. Natl. Acad. Sci. U.S.A.* **81**, 3297–3301 (1984).

443. M. Ikehara, S. Uesugi, and T. Shida, *Chem. Pharm. Bull. (Tokyo)* **28**, 189–197 (1980).

444. M. Ikehara, S. Uesugi, and M. Yasumoto, *J. Am. Chem. Soc.* **92**, 4735–4736 (1970).

445. S. Uesugi, M. Yasumoto, M. Ikehara, K. N. Fang, and P. O. P. Ts'o, *J. Am. Chem. Soc.* **94**, 5480–5486 (1972).

446. M. Ikehara and S. Uesugi, *J. Am. Chem. Soc.* **94**, 9189–9193 (1972).

447. M. Ikehara, S. Uesugi, and J. Yano, *Nature New Biol.* **240**, 16–17 (1972).

448. M. Ikehara, S. Uesugi, and J. Yano, *J. Am. Chem. Soc.* **96**, 4966–4972 (1974).

449. S. Uesugi, J. Yano, E. Yano, and M. Ikehara, *J. Am. Chem. Soc.* **99**, 2313–2323 (1977).

450. M. M. Dhingra, R. H. Sarma, S. Uesugi, and M. Ikehara, *J. Am. Chem. Soc.* **100**, 4669–4673 (1978).

451. S. Uesugi, T. Shida, and M. Ikehara, *Chem. Pharm. Bull. (Tokyo)* **28**, 3621–3631 (1980).

452. S. Uesugi, T. Tezuka, and M. Ikehara, *J. Am. Chem. Soc.* **98**, 969–973 (1976).

453. M. M. Dhingra, R. H. Sarma, S. Uesugi, T. Shida, and M. Ikehara, *Biochemistry* **20**, 5002–5011 (1981).

454. M. J. Lane, J. C. Dabrowiak, and J. N. Vournakis, *Proc. Natl. Acad. Sci. U.S.A.* **80**, 3260–3264 (1983).

455. J. C. Chottard, J. P. Girault, G. Chottard, J. Y. Lallemand, and D. Mansuy, *J. Am. Chem. Soc.* **102**, 5565–5572 (1980).

456. A. T. M. Marcelis, G. W. Canters, and J. Reedijk, *Recl. Trav. Chim. Pays-Bas* **100**, 391–392 (1981).

457. J.-P. Girault, G. Chotlard, J.-Y. Lallemand, and J.-C. Chottard, *Biochemistry* **21**, 1352–1356 (1982).

458. A. T. M. Marcelis, J. H. J. den Hartog, and J. Reedijk, *J. Am. Chem. Soc.* **104**, 2664–2665 (1982).

459. J. H. J. den Hartog, C. Altona, J.-C. Chothard, J.-P. Girault, J.-Y. Lallemand, F. A. A. M. de Leeum, A. T. M. Marcelis, and J. Reedijk, *Nucleic Acids Res.* **10**, 4715–4730 (1982).

460. J. P. Caradonna, S. J. Lippard, M. J. Gait, and M. Singh, *J. Am. Chem. Soc.* **104**, 5793–5795 (1982).

461. J.-P. Girault, J.-C. Chottard, E. R. Giuttet, J.-Y. Lallemand, T. Huynh-Dinh, and J. Igolen, *Biochem. Biophys. Res. Commun.* **109**, 1157–1163 (1982).

462. S. Uesugi, T. Shida, M. Ikehara, Y. Kobayashi, and Y. Kyogoku, *J. Am. Chem. Soc.* **104**, 5494–5495 (1982).

463. S. Uesugi, T. Shida, H. Miyashiro, K. Tomita, M. Ikehara, Y. Kobayashi, and Y. Kyogoku, *Nucleic Acids Symp. Ser.* **11**, 237–240 (1982).

464. S. Uesugi, T. Shida, M. Ikehara, Y. Kobayashi, and Y. Kyogoku, *Nucleic Acids Res.* **12**, 1581–1592 (1984).

465. D. E. Graves, C. Pattaroni, B. S. Krishnan, J. M. Ostrander, L. H. Hurley, and T. R. Krugh, *J. Biol. Chem.* **259**, 8202–8209 (1984).

466. J. M. Neumann, S. Tran-Dinh, J.-P. Girault, J.-C. Chottard, T. Huynh-Dinh, and J. Igolen, *Eur. J. Biochem.* **141**, 465–472 (1984).

467. J. H. J. den Hartog, C. Altona, J. H. van Boom, G. A. van der Marel, C. A. G. Haasnoot, and J. Reedijk, *J. Am. Chem. Soc.* **106**, 1528–1530 (1984).

468. J. P. Girault, J. C. Chottard, J. M. Neumann, T. D. Son, T. Huynh-Dinh, and J. Igolen, *Nouv. J. Chim.* **8**, 7–9 (1984).

469. B. van Hemelryck, E. Guittet, G. Chottard, J.-P. Girault, T. Huynh-Dinh, J.-Y. Lallemand, Y. Igolen, and J.-C. Chottard, *J. Am. Chem. Soc.* **106**, 3027–3029 (1984).

470. J. H. J. den Hartog, C. Altona, J. H. van Boom, A. T. M. Marcelis, G. A. van der Marel, L. J. Rinkel, G. Wille-Hazeleger, and J. Reedijk, *Eur. J. Biochem.* **134**, 485–495 (1983).

471. S. Mansy, G. Y. H. Chu, R. E. Duncan, and R. S. Tobias, *J. Am. Chem. Soc.* **100**, 607–616 (1978).

472. B. Malfoy and M. Leng, *FEBS Lett.* **132**, 45–48 (1981).

473. B. Malfoy, N. Rousseau, and M. Leng, *Biochemistry* **21**, 5463–5467 (1982).

474. A. Möller, A. Nordheim, S. R. Nichols, and A. Rich, *Proc. Natl. Acad. Sci. U.S.A.* **78**, 4777–4781 (1981).

475. J. W. Lown and S.-K. Sim, *Biochem. Biophys. Res. Commun.* **77**, 1150–1157 (1977).

476. S.-K. Sim and J. W. Lown, *Biochem. Biophys. Res. Commun.* **81**, 99–105 (1978).

477. J. W. Lown, S.-K. Sim, K. C. Majumdar, and R.-Y. Chang, *Biochem. Biophys. Res. Commun.* **76**, 705–710 (1977).

478. J. W. Lown, A. Begleiter, D. Johnson, and A. R. Morgan, *Can. J. Biochem.* **54**, 110–119 (1976).

479. D. S. Kapp and K. C. Smith, *Radiat. Res.* **42**, 34–49 (1970).

480. D. S. Sigman, D. R. Graham, V. D'Aurora, and A. M. Stern, *J. Biol. Chem.* **254**, 12269–12272 (1979).

481. H. Kuramochi, K. Takahashi, T. Takita, and H. Umezawa, *J. Antibiot.* **34**, 576–582 (1981).

482. R. M. Burger, J. Peisach, and S. B. Horwitz, *J. Biol. Chem.* **256**, 11636–11644 (1981).

483. L. Giloni, M. Takeshita, F. Johnson, C. Iden, and A. P. Grollman, *J. Biol. Chem.* **256**, 8608–8615 (1981).

484. L. S. Kappen, I. H. Goldberg, and J. M. Leisch, *Proc. Natl. Acad. Sci. U.S.A.* **79**, 744–748 (1982).

485. L. M. Pope, K. A. Reich, D. R. Graham, and D. S. Sigman, *J. Biol. Chem.* **257**, 12121–12128 (1982).

486. W. D. Henner, L. O. Rodriguez, S. M. Hecht, and W. A. Haseltine, *J. Biol. Chem.* **258**, 711–713 (1983).

487. B. C. Froehler and M. D. Matteucci, *Tetrahedron Lett.* **27**, 469–472 (1986).

488. B. C. Froehler, P. G. Ng, and M. D. Matteucci, *Nucleic Acids Res.* **14**, 5399–5407 (1986).

489. P. J. Garreg, I. Lindh, T. Regberg, J. Stawinski, and R. Stromberg, *Tetrahedron Lett.* **27**, 4055–4058 (1986).

490. T. Tanaka, S. Tamatsukuri, and M. Ikehara, *Nucleic Acids Res.* **15**, 7235–7248 (1987).

Index